TUNNELS AND UNDERGROUND CITIES: ENGINEERING AND
INNOVATION MEET ARCHAEOLOGY, ARCHITECTURE AND ART

PROCEEDINGS OF THE WTC2019 ITA-AITES WORLD TUNNEL CONGRESS, NAPLES, ITALY, 3-9 MAY, 2019

Tunnels and Underground Cities: Engineering and Innovation meet Archaeology, Architecture and Art

Volume 12: Urban Tunnels - Part 2

Editors

Daniele Peila
Politecnico di Torino, Italy

Giulia Viggiani
University of Cambridge, UK
Università di Roma "Tor Vergata", Italy

Tarcisio Celestino
University of Sao Paulo, Brasil

CRC Press
Taylor & Francis Group
Boca Raton London New York

CRC Press is an imprint of the
Taylor & Francis Group, an **informa** business

A BALKEMA BOOK

First published in paperback 2024

First published 2020
by CRC Press/Balkema
4 Park Square, Milton Park, Abingdon, Oxon, OX14 4RN

and by CRC Press/Balkema
2385 NW Executive Center Drive, Suite 320, Boca Raton FL 33431

CRC Press/Balkema is an imprint of the Taylor & Francis Group, an informa business

Publisher's Note
The publisher has gone to great lengths to ensure the quality of this reprint but points out that some imperfections in the original copies may be apparent.

ISBN: 978-0-367-46900-9 (hbk)
ISBN: 978-1-03-283947-9 (pbk)
ISBN: 978-1-003-03187-1 (ebk)

DOI: 10.1201/9781003031871

Cover illustration:

View of Naples gulf

Typeset by Integra Software Services Pvt. Ltd., Pondicherry, India

**Visit the Taylor & Francis Web site at
http://www.taylorandfrancis.com**

**and the CRC Press Web site at
http://www.crcpress.com**

Tunnels and Underground Cities: Engineering and Innovation meet Archaeology,
Architecture and Art, Volume 12: Urban
Tunnels - Part 2 – Peila, Viggiani & Celestino (Eds)
© 2020 Taylor & Francis Group, London, ISBN 978-0-367-46900-9

Table of contents

Tunnels and Underground Cities: Engineering and Innovation meet Archaeology,
Architecture and Art, Volume 12: Urban
Tunnels - Part 2 – Peila, Viggiani & Celestino (Eds)
© 2020 Taylor & Francis Group, London, ISBN 978-0-367-46900-9

Preface

The World Tunnel Congress 2019 and the 45th General Assembly of the International Tunnelling and Underground Space Association (ITA), will be held in Naples, Italy next May.

The Italian Tunnelling Society is honored and proud to host this outstanding event of the international tunnelling community.

Hopefully hundreds of experts, engineers, architects, geologists, consultants, contractors, designers, clients, suppliers, manufacturers will come and meet together in Naples to share knowledge, experience and business, enjoying the atmosphere of culture, technology and good living of this historic city, full of marvelous natural, artistic and historical treasures together with new innovative and high standard underground infrastructures.

The city of Naples was the inspirational venue of this conference, starting from the title Tunnels and Underground cities: engineering and innovation meet Archaeology, Architecture and Art.

Naples is a cradle of underground works with an extended network of Greek and Roman tunnels and underground cavities dated to the fourth century BC, but also a vibrant and innovative city boasting a modern and efficient underground transit system, whose stations represent one of the most interesting Italian experiments on the permanent insertion of contemporary artwork in the urban context.

All this has inspired and deeply enriched the scientific contributions received from authors coming from over 50 different countries.

We have entrusted the WTC2019 proceedings to an editorial board of 3 professors skilled in the field of tunneling, engineering, geotechnics and geomechanics of soil and rocks, well known at international level. They have relied on a Scientific Committee made up of 11 Topic Coordinators and more than 100 national and international experts: they have reviewed more than 1.000 abstracts and 750 papers, to end up with the publication of about 670 papers, inserted in this WTC2019 proceedings.

According to the Scientific Board statement we believe these proceedings can be a valuable text in the development of the art and science of engineering and construction of underground works even with reference to the subject matters "Archaeology, Architecture and Art" proposed by the innovative title of the congress, which have "contaminated" and enriched many proceedings' papers.

Andrea Pigorini   Renato Casale
SIG President   *Chairman of the Organizing Committee WTC2019*

Acknowledgements

REVIEWERS

The Editors wish to express their gratitude to the eleven Topic Coordinators: Lorenzo Brino, Giovanna Cassani, Alessandra De Cesaris, Pietro Jarre, Donato Ludovici, Vittorio Manassero, Matthias Neuenschwander, Moreno Pescara, Enrico Maria Pizzarotti, Tatiana Rotonda, Alessandra Sciotti and all the Scientific Committee members for their effort and valuable time.

SPONSORS

The WTC2019 Organizing Committee and the Editors wish to express their gratitude to the congress sponsors for their help and support.

Tunnels and Underground Cities: Engineering and Innovation meet Archaeology,
Architecture and Art, Volume 12: Urban
Tunnels - Part 2 – Peila, Viggiani & Celestino (Eds)
© 2020 Taylor & Francis Group, London, ISBN 978-0-367-46900-9

WTC 2019 Congress Organization

HONORARY ADVISORY PANEL

Pietro Lunardi, President WTC2001 Milan
Sebastiano Pelizza, ITA Past President 1996-1998
Bruno Pigorini, President WTC1986 Florence

INTERNATIONAL STEERING COMMITTEE

Giuseppe Lunardi, Italy (Coordinator)
Tarcisio Celestino, Brazil (ITA President)
Soren Eskesen, Denmark (ITA Past President)
Alexandre Gomes, Chile (ITA Vice President)
Ruth Haug, Norway (ITA Vice President)
Eric Leca, France (ITA Vice President)
Jenny Yan, China (ITA Vice President)
Felix Amberg, Switzerland
Lars Barbendererder, Germany
Arnold Dix, Australia
Randall Essex, USA
Pekka Nieminen, Finland
Dr Ooi Teik Aun, Malaysia
Chung-Sik Yoo, Korea
Davorin Kolic, Croatia
Olivier Vion, France
Miguel Fernandez-Bollo, Spain (AETOS)
Yann Leblais, France (AFTES)
Johan Mignon, Belgium (ABTUS)
Xavier Roulet, Switzerland (STS)
Joao Bilé Serra, Portugal (CPT)
Martin Bosshard, Switzerland
Luzi R. Gruber, Switzerland

EXECUTIVE COMMITTEE

Renato Casale (Organizing Committee President)
Andrea Pigorini, (SIG President)
Olivier Vion (ITA Executive Director)
Francesco Bellone
Anna Bortolussi
Massimiliano Bringiotti
Ignazio Carbone
Antonello De Risi
Anna Forciniti
Giuseppe M. Gaspari

Giuseppe Lunardi
Daniele Martinelli
Giuseppe Molisso
Daniele Peila
Enrico Maria Pizzarotti
Marco Ranieri

ORGANIZING COMMITTEE

Enrico Luigi Arini
Joseph Attias
Margherita Bellone
Claude Berenguier
Filippo Bonasso
Massimo Concilia
Matteo d'Aloja
Enrico Dal Negro
Gianluca Dati
Giovanni Giacomin
Aniello A. Giamundo
Mario Giovanni Lampiano
Pompeo Levanto
Mario Lodigiani
Maurizio Marchionni
Davide Mardegan
Paolo Mazzalai
Gian Luca Menchini
Alessandro Micheli
Cesare Salvadori
Stelvio Santarelli
Andrea Sciotti
Alberto Selleri
Patrizio Torta
Daniele Vanni

SCIENTIFIC COMMITTEE

Daniele Peila, Italy (Chair)
Giulia Viggiani, Italy (Chair)
Tarcisio Celestino, Brazil (Chair)
Lorenzo Brino, Italy
Giovanna Cassani, Italy
Alessandra De Cesaris, Italy
Pietro Jarre, Italy
Donato Ludovici, Italy
Vittorio Manassero, Italy
Matthias Neuenschwander, Switzerland
Moreno Pescara, Italy
Enrico Maria Pizzarotti, Italy
Tatiana Rotonda, Italy
Alessandra Sciotti, Italy
Han Admiraal, The Netherlands
Luisa Alfieri, Italy

Georgios Anagnostou, Switzerland
Andre Assis, Brazil
Stefano Aversa, Italy
Jonathan Baber, USA
Monica Barbero, Italy
Carlo Bardani, Italy
Mikhail Belenkiy, Russia
Paolo Berry, Italy
Adam Bezuijen, Belgium
Nhu Bilgin, Turkey
Emilio Bilotta, Italy
Nikolai Bobylev, United Kingdom
Romano Borchiellini, Italy
Martin Bosshard, Switzerland
Francesca Bozzano, Italy
Wout Broere, The Netherlands

Domenico Calcaterra, Italy
Carlo Callari, Italy
Luigi Callisto, Italy
Elena Chiriotti, France
Massimo Coli, Italy
Franco Cucchi, Italy
Paolo Cucino, Italy
Stefano De Caro, Italy
Bart De Pauw, Belgium
Michel Deffayet, France
Nicola Della Valle, Spain
Riccardo Dell'Osso, Italy
Claudio Di Prisco, Italy
Arnold Dix, Australia
Amanda Elioff, USA
Carolina Ercolani, Italy
Adriano Fava, Italy
Sebastiano Foti, Italy
Piergiuseppe Froldi, Italy
Brian Fulcher, USA
Stefano Fuoco, Italy
Robert Galler, Austria
Piergiorgio Grasso, Italy
Alessandro Graziani, Italy
Lamberto Griffini, Italy
Eivind Grov, Norway
Zhu Hehua, China
Georgios Kalamaras, Italy
Jurij Karlovsek, Australia
Donald Lamont, United Kingdom
Albino Lembo Fazio, Italy
Roland Leucker, Germany
Stefano Lo Russo, Italy
Sindre Log, USA
Robert Mair, United Kingdom
Alessandro Mandolini, Italy
Francesco Marchese, Italy
Paul Marinos, Greece
Daniele Martinelli, Italy
Antonello Martino, Italy

Alberto Meda, Italy
Davide Merlini, Switzerland
Alessandro Micheli, Italy
Salvatore Miliziano, Italy
Mike Mooney, USA
Alberto Morino, Italy
Martin Muncke, Austria
Nasri Munfah, USA
Bjørn Nilsen, Norway
Fabio Oliva, Italy
Anna Osello, Italy
Alessandro Pagliaroli, Italy
Mario Patrucco, Italy
Francesco Peduto, Italy
Giorgio Piaggio, Chile
Giovanni Plizzari, Italy
Sebastiano Rampello, Italy
Jan Rohed, Norway
Jamal Rostami, USA
Henry Russell, USA
Giampiero Russo, Italy
Gabriele Scarascia Mugnozza, Italy
Claudio Scavia, Italy
Ken Schotte, Belgium
Gerard Seingre, Switzerland
Alberto Selleri, Italy
Anna Siemińska Lewandowska, Poland
Achille Sorlini, Italy
Ray Sterling, USA
Markus Thewes, Germany
Jean-François Thimus, Belgium
Paolo Tommasi, Italy
Daniele Vanni, Italy
Francesco Venza, Italy
Luca Verrucci, Italy
Mario Virano, Italy
Harald Wagner, Thailand
Bai Yun, China
Jian Zhao, Australia
Raffaele Zurlo, Italy

Strategic use of underground space for resilient cities

Tunnels and Underground Cities: Engineering and Innovation meet Archaeology,
Architecture and Art, Volume 12: Urban
Tunnels - Part 2 – Peila, Viggiani & Celestino (Eds)
© 2019 Taylor & Francis Group, London, ISBN 978-0-367-46900-9

Lessons learned from excavation of the Pinchat tunnel in loose ground in an urban environment

N. Kupferschmied, E. Garin & E. Rigaud
BG Consulting Engineers, Lausanne, Switzerland

ABSTRACT: The Pinchat tunnel was completed in spring 2017. Its construction required to address numerous challenges created by:

– A 100 m^2 cross-section with low overburden in an urban environment and loose ground;
– Access from a vertical shaft;
– Top-down construction of an underground station in an area occupied by numerous traffic routes;
– Excavation beneath a river;

Excavation of the 2 km long tunnel was carried out using conventional methods, with partial face excavation, forepoling and face bolts, and a steel braces and concrete support. The design was value-engineered by applying the observational method, what was made easier by a site supervision carried out by the same engineering team as the project. Value-engineering was also applied to adapt structures, techniques and phasing to the resources committed by the Contractor. The care taken to refine the design, combined with the improvements implemented during construction, allowed to finish the construction on time and on budget.

1 INTRODUCTION

Excavation of large section tunnels in loose ground in an urban environment presents numerous challenges, from design to construction. The care taken during design of support structures, monitoring of soil-structure interaction and the permanent review of the design based on conditions encountered allowed to excavate safely a 100m^2 section tunnel, with due respect given to adjacent structures, quality, cost and time constraints.

2 THE CEVA PROJECT

The new CEVA rail link will allow the commissioning in 2019 of the cross-border "Leman Express" regional rail service which will connect the French and Swiss rail networks. Of a total length of 16 km, the 14 km in Swiss territory is being developed jointly by CFF and Canton Geneva, while the 2 km in French territory is supervised by SNCF Réseau.

The cost of construction in Swiss territory is 1.6 bn CHF, and cost of the French section is 234 m Euro.

The alignment being mainly situated in urban and suburban environments, the 14 km in Swiss territory are mainly underground (cut-and-cover and tunnel sections).

Figure 1. Alignment of the CEVA project.

3 PINCHAT TUNNEL

The CEVA alignment includes two bored tunnels: Champel Tunnel and Pinchat Tunnel. The latter, a single tube double track tunnel, is part of Civil Works package 3.

The Package includes the 2,036 km long bored tunnel, the 235 m long Carouge-Bachet station and the Voirie de Carouge, of length 90 m (cut-and-cover sections).

Design was undertaken by the GE-Pinchat engineering design team, including BG Consulting Engineers as lead consultant, with partners GADZ, SD Ingénierie Genève and Solfor.

Civil works were carried out by the CTP Consortium, comprising Walo (lead contractor), Rothpletz-Lienhard, Implenia, Prader-Losinger and Infra Tunnel.

Construction began at the start of 2012 and finished in Spring 2017.

3.1 *Overall description of the structures*

The objective of the tunnel is to house two tracks allowing passage of passengers and goods trains.

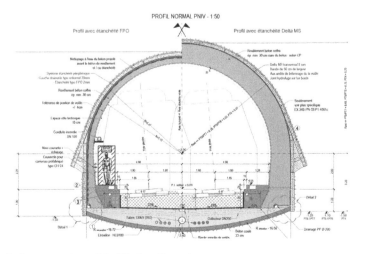

Figure 2. Typical cross-section of the Pinchat Tunnel.

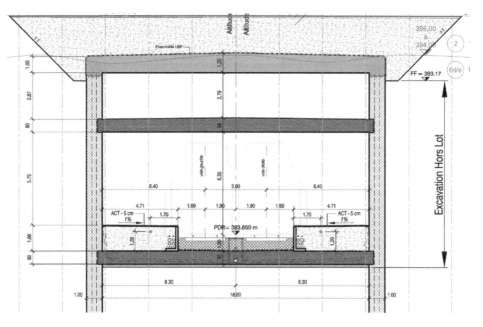

Figure 3.   Carouge Bachet Station.

Design velocity is 100 km/h.

The tunnel is equipped with twin ballastless tracks fitted with mitigation measures to limit ground-borne sounds and vibrations.

The tunnel includes 3 emergency exits and 3 pairs of service alcoves facing each other.

At one end, the Carouge-Bachet station, excavated by cut-and-cover will serve Geneva Stadium and the Bachet transport hub. It includes two platforms with vertical circulations at either end. Connection to the platforms is provided by means of stairs linking them to a footbridge at the level of the adjacent streets. The platforms are 235 m long, with width between walls of 16.6 m. the volume between the intermediate slab and the cover slab is used for technical facilities, including the smoke extraction system.

Access to the tunnel on the Bachet side comprises a 32 m side length square section shaft. There are 3 levels in the shaft: platform, mezzanine and exit level. Propping of the diaphragm walls is provided by the concrete floor slabs for each level.

The Voirie de Carouge cut-and-cover trench passes beneath an existing structure, which was partially dismantled for trench construction. It emerges in another trench constructed beforehand by the adjacent Civil Works Package. A widened section was designed at the West extremity of the cut-and-cover trench to allow connection with the tunnel profile.

Cross-section of this section is rectangular, with a width between walls of 10.20 m, a cover slab and an intermediate slab.

3.2 Ground Conditions

The ground beneath the Geneva Basin comprises 3 main formations: the molasse bedrock, several layers of glacial deposits composed of moraines and retreat deposits linked to the Riss and Würm glaciers, and finally recent formations such as alluvium, colluvial deposits and made ground.

From Carouge-Bachet station, the Pinchat tunnel crosses 400 m of very soft to soft clayey silt and clay, which is compressible, plastic, of poor permeability and saturated, with overburden increasing from 5 m to 15 m.

The tunnel then encounters progressively the Wurm moraine, comprising silty clay with cobbles and blocks, stiff and poorly permeable, and then a layer of very dense gravels, which

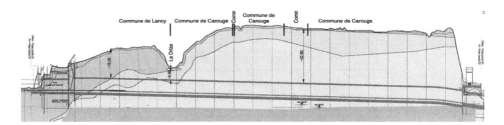

Figure 4.   Geological cross-section.

is overconsolidated and locally cemented (a formation called "Alluvions Anciennes"). This Formation is very permeable and contains the Arve River aquifer.

The distribution of the various formations is given in the following longitudinal profile:

4   PARTICULAR CHALLENGES

Execution of the various structures of Civil Works Package 3 required to find a solution to the numerous challenges posed by their unique nature, such as:

4.1   *Large section tunnels with low overburden in an urban context*

The excavation of a 100m^2 cross-section tunnel in an urban environment with overburdens from 5 to 40 m, with numerous adjacent buildings, required reliable predictions of anticipated settlements in order to design a structure which allows to reduce them to their minimum. During the construction stage, monitoring of settlements allowed to verify the geotechnical model which was adopted and to calibrate the design parameters by back-analysis.

4.2   *Excavation in loose ground*

In a geology composed of heterogeneous fluvio-glacial deposits, comprising highly compressible zones, mixed faces, loose soil and very dense cemented strata, advance support methods and appropriate support methods were required to ensure the stability of the face and excavation profile. The selected methods required to be applicable in the different possible geological configurations without changing plant while guaranteeing adequate advancement rates.

4.3   *Logistical restrictions*

The Bachet face was only accessible by a shaft during most of the works. The other face required sharing the installation zone with another tunnel package, and access to the face passed through the cut-and-cover trench of another package executed previously. The location of the sites on two main accesses to the city generated many constraints in terms of supplying the site due to traffic jams. On the Bachet side, a rail line under operation also had furthermore to be taken into account.

4.4   *Staged construction of an underground station beneath existing structures*

Top-down construction of an underground station at one end of the tunnel, in a space occupied by numerous transport networks (motorway, 6 lane highway, tramway), without service interruption, implied developing a complex construction phasing plan of the surface works.

Underground services (gas, drinking water, electricity, optical fiber, wastewater, road and rail signals) also required to be maintained in operation by means of temporary bypasses and permanent alternatives without interruption to services.

Architectural constraints limited the number of intermediate propping for the retaining walls and thus led to the installation of prestressed Pre-flex type waling beams.

4.5 *Tunnelling beneath a river*

The alignment of the tunnel passed beneath a valley with overburden limited to 5 m. The river running was thus temporarily diverted in a concrete canal and the advanced support was also adapted to limit the risk of cave-ins.

4.6 *Groundwater*

At the Val-d'Arve end of the tunnel and in the Voirie cut-and-cover section, the presence of groundwater in the very permeable gravels required advanced waterproofing, as ingress flowrates would not be manageable. The aquifer being tapped for drinking water purposes, measures were put in place to mitigate the risk of pollution during waterproofing works.

5 STRUCTURAL CONCEPT

5.1 *Cut-and-cover trenches*

The overall poor quality of the rock mass, in particular on the Bachet side, the presence of sensitive adjacent buildings and the need to minimise impact on surface transport links implied constructing stiff boxes and forming a permanent structure.

The structure of the tunnel is thus composed of diaphragm walls, a top level slab and an intermediate slab serving as propping. A reinforced concrete raft blocks deformations as toe level and supports the platforms and tracks. The top level slab is concreted directly against formwork resting on the ground.

, The diaphragm walls are 1 m thick for a depth of 16 to 20 m for Carouge-Bachet station, 0.8 m thick for a depth of 18 to 20 m at the Voirie cut-and-cover trench. The top level slab of Carouge-Bachet station, spanning 18 m and up to 1.5 m thick, must support tramway lines, a highway and a main road. The top level slab at the Voirie is 1 m thick for a span of 11 m and supports the working loads of the building above it

5.2 *Tunnel*

Excavation of the 2 km long tunnel was undertaken using conventional methods, in split section (top heading/bench for most part) and using the German method in the clays (sidewall drifts, top heading, bench). A pre-support comprising forepoling with bored metallic tubes and glass fiber bolts, and a support composed of concrete and steel braces ensured the stability of the excavation.

Excavation of the 240 m in the clays was particularly challenging, with overburden increasing progressively from 5 to 15 m, and sensitive structures directly adjacent to the tunnel (an industrial tunnel situated 3 m above the tunnel, residential buildings build above it, a 2 lane road with wastewater networks superimposed above the tunnel alignment).

These clays and clayey silts are compressible, with low bearing capacity, leading to important settlements during tunnel excavation and difficulty supporting the arch foundation loads.

Advance excavation of sidewall drifts allowed to cast a foundation slab 1 m thick and a wall capable of supporting the loads from the steel braces and diffusing them on a sufficient surface, to keep settlements to an acceptable level.

To ensure a good bond between the support concrete and the steel, the use of dry-mix shotcrete was particularly effective.

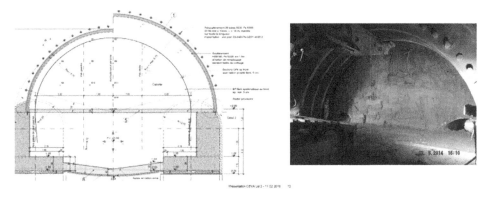

Figure 5. Standard section in the clay formations.

Excavation of the top heading was carried out beneath the protection of a forepoling umbrella. A temporary invert was concreted to stiffen the support structure. The bench was then excavated and the invert slab poured to close the support ring.

At the other extremity of the tunnel, the presence of an aquifer in the inferior half-section of the tunnel lead to adopting split section excavation (top heading/invert). Once the zone crossed, excavation was to be carried out in full section after creation of a ramp. Waterproofing works were then carried out from the top heading section, alternatively on the left and right sides of the tunnel, to maintain access to the tunnel face.

This treatment involved creating secant cells using jet grouting columns, and then injecting the cells with a low-pressure cement-bentonite grout.

The remainder of the tunnel was to be excavated in full-face under protection of the forepoling umbrella and fiber glass face bolts. The advantage of the full-face excavation was to avoid underpinning the top heading, a delicate procedure which is liable to produce large settlements.

Lining is achieved by an unreinforced concrete arch, with reinforced side walls and raft.

In terms of drainage and waterproofing, the concept of the structure takes into account the high level of the aquifer when it is not being drawn down. Groundwater is drained and pumped at the low point in the Val d'Arve. The waterproofing concept is different depending on the tunnel sections. For the clayey zones, the top heading is waterproofed; for zones above the groundwater level, no waterproofing is provided; finally, in the portal zones, full waterproofing was implemented.

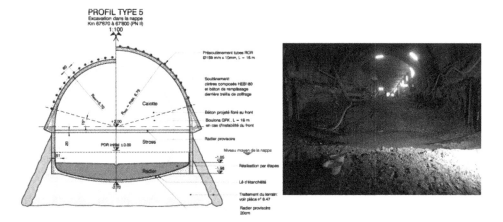

Figure 6. Standard section in the aquifer.

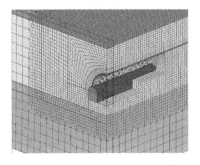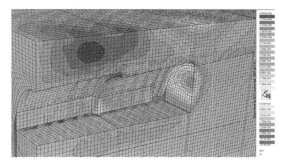

Figure 7.   Finite elements computations.

6   DESIGN

The design underwent a number of successive value engineering phases, by applying the observational method. The objective was to guarantee maximum security for workers and adjacent structures, efficient use of the advance support and support structures, and keeping to the construction schedule.

Based on the geotechnical investigation and testing (in particular triaxial tests), a HSS constitutive model was calibrated to represent the behaviour of the clays. For the granular materials ("alluvions anciennes" formation), a Mohr-Coulomb behaviour was adopted.

Design of the support and lining was undertaken using 2D and 3D finite element soil-structure interaction computations. These models allowed to account for successive excavation stages as well as the face effects. They were also essential to calculate the anticipated settlements at the various stages of advancement and to define the warning and alarm levels in the surveillance plan.

7   CONSTRUCTION

The modification of the construction methods based on encountered conditions was made easier by the fact that the detailed design and site supervision were both carried out by the design team which led design from concept stage.

For construction of the Carouge-Bachet station, diaphragm wall and cover slab execution required work to be carried out in 5 different sectors over a length of 200 m. These 5 sectors required more than 30 different traffic modes over a period of 36 months.

After construction stages at existing ground level, excavation was required to be carried out in full section between the ground level slab and the intermediate slab. After concreting of the latter, excavation was then deepened up to 2 m above the bottom level, and the ground bearing slab was executed in alternating sections of 5 m to ensure sufficient toe depth for the diaphragm walls.

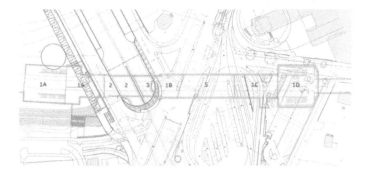

Figure 8.   Staged construction at surface level.

Value engineering was also undertaken to adapt the structures, construction techniques and staging to the plant and equipment provided by the Contractor in charge of the works.

Excavation of the central section of the tunnel, foreseen in full section, was finally undertaken in half-section stages. Indeed, the height of the tunnel (over 10 m) put the top heading above the reach of the jumbo drills which installed the forepoling and face bolts, as well as the excavators used for both advance headings for the tunnel.

Adopting a top-heading/bench approach for both headings allowed to optimize the sequencing of tasks after breakthrough and to secure the construction schedule. Observation of the good stability of the headings also allowed to increase excavation to steps of 2 m with installation of 2 braces at a time.

Face support was initially carried out with Ø32/10 steel self-drilling bolts. The high deviations observed during drilling led to replacing them with glass fiber Ø76/60 bolts.

After breakthrough, excavation of the bench was carried out from Bachet towards Voirie, ensuring optimized logistics for this work station from the portal. The lining work stations (invert, arch foundations, waterproofing, arch and side walls) followed in the same direction, supplied from the Bachet side.

Bench excavation was carried out in 2 m sections, with immediate installation of the support (including invert sets). The duration between excavation and installation of support was less than 6 hours, and no settlements or disturbance were reported.

Thanks to these optimizations, it was possible to avoid having to source a second arch formwork set.

The speeding up of these processes allowed to absorb without any impact to schedule the interruption to excavation due to a 1 in 100 year groundwater flooding event, which rose above the level of the implemented waterproofing.

Figure 9.   Diaphragm wall construction.

Figure 10.   Tunnel support works.

Figure 11.   Benching and concreting work stations.

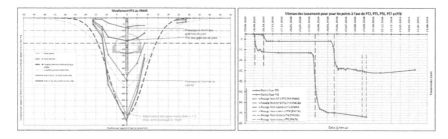

Figure 12.   Comparison between measured and calculated values.

The saw-tooth profile of the excavation due to the forepoling and the friction of the support steel mesh, the lack of a waterproof liner, the large variations in thicknesses and internally induced stresses led to the development of shrinkage cracks in the unreinforced arch lining. The formulation of a concrete with a low concentration of CEM I and the installation of a sliding layer (created using Delta MS sheets) in the unwaterproofed zones allowed to manage this phenomenon, with cracks concentrating in construction joints between segments.

Finally, the concept developed for managing settlements was shown to be effective. Ground level surveys confirmed the finite-element computations, in particular in the clayey zone with high settlement potential, for which a back-analysis was done during excavation, and a calibration of the model and methods was allowed to be carried out.

8   CONCLUSIONS

The care taken during design, coupled with the value engineering undertaken during construction, allowed to deliver the structure on time and on budget, and this despite a number of unforeseen events, as is often the case with projects of this scale.

Figure 13.   Civil works completed.

Tunnels and Underground Cities: Engineering and Innovation meet Archaeology,
Architecture and Art, Volume 12: Urban
Tunnels - Part 2 – Peila, Viggiani & Celestino (Eds)
© 2019 Taylor & Francis Group, London, ISBN 978-0-367-46900-9

The construction of the tunnel Kennedy in Santiago de Chile. A major challenge of an urban tunnel in soils

J. Kuster & R. Núñez
Costanera Norte, Santiago de Chile, Chile

E. Chávez, J.M. Galera & D. Santos
Subterra Ingeniería, Madrid, Spain

ABSTRACT: Costanera Norte is an urban expressway concessionary in Santiago de Chile connecting its Western and Eastern sides. To enhance this connectivity several measures were necessary, including four new tunnels. The most singular one is 'Tunnel Kennedy', that has a length of 1,166 m with two sections (4 and 5 lanes), 17.45m and 20.95m wide, and 11.8m height. The tunnel was excavated in alluvial gravels, partially under the water table. The main challenges were: its section over 250m^2, the ground conditions (requiring SEM), its urban environment and low overburden (10 to 16m). The support consists in a 3cm sealing, HEB steel arches spaced 1.0m and 27cm shotcrete. Systematically buttress and occasional canopy tubes have been used. A concrete lining (20cm to 40 cm, depending on the water column), was added, and a third layer 10cm with polypropylene fibers for fire protection. The excavation took place from October 2014 to February, 2017.

1 TUNNEL DESCRIPTION

The tunnel is located in the NE of Santiago de Chile, close to the financial centre of the city. The whole tunnel alignment is parallel to the current Kennedy Avenue, mostly below a golf course called "Los Leones". Figure 1 shows the location of the tunnel.

The access to both portals is solved by mean of two trench structures composed by piles, crossbeams and slabs. At the beginning portal, the road configuration at the surface has evolved, from a roundabout to a bridge and several connections with high complexity.

The typical cross section of the tunnel provides four lanes, 3.5 m wide each one, and two shoulders of 0.75 m respectively. Below the pavement a maintenance gallery is provided. The inner dimensions with respect to the lining are 17.45 m wide and 11.30 m height.

At the beginning of the tunnel, there is a connection lane so for around 50 m, a special section of five lanes have been constructed. For this section, the inner dimensions are 21.0 m width and 11.8 m height. An additional section was also envisaged, which was called "4.5 lanes", designed in order to soften the transition between the sections of five and four lanes. Figure 2 shows the geometrical definition of the two existing sections of the tunnel.

As mentioned before, three emergency exits have been constructed for the evacuation plan of the tunnel. The first one is located close to the access structures at the portal. The second and third ones were solved by shafts of 8,3 m diameter from surface and pedestrian gallery connections.

2 GEOLOGICAL AND GEOTECHNICAL CHARACTERISTICS

The entire tunnel has been excavated in the well-known alluvial materials composed by gravels and pebbles of the second and first deposition of the Mapocho River.

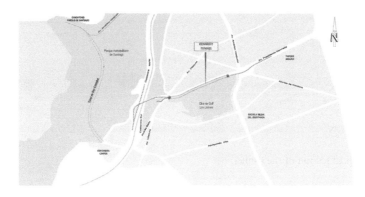

Figure 1.   Location and layout. Kennedy Tunnel.

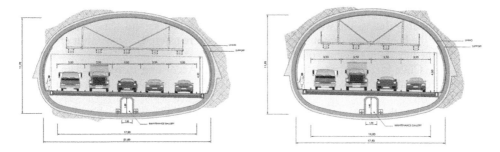

Figure 2.   Functional sections (4-lanes and 5-lanes). Kennedy Tunnel.

In the lower part of the tunnel, the presence of weathered andesites and lutites of Abanico Formation, was expected between Km 1,046 and Km 1,050 in the invert of the tunnel section. However, just some andesite blocks have been excavated

In terms of the soil parameters, Table 1 shows strength and deformational parameters considered for the support calculations.

The water table is located at 623 MASL. Accordingly, it is placed above the tunnel from its beginning until approximately Dm 1350. The water table passes through the tunnel from the level of the top section to the level of the invert in the Dm 1800. From this point, it continues below the level of the invert until to the end of the tunnel.

Table 1.   Strength and deformational parameters of the soil in the Kennedy Tunnel.

| SOIL PARAMETERS | ANTHROPIC FILL | GEOTECHNICAL UNIT | |
		2nd DEPOSITION	1st DEPOSITION
γ (kN/m³)	17,0	22,5	22,5
Cmax (kPa)	10	25	35
φ_m (°)	33	45	48
Ec (kPa)	20.000	42.000xZ0,5(Z<17)	42.000xZ0,55(Z<17)
		(*)	55.000xZ0,53(Z>17)
v	0,30	0,25	0,25
Ψ (°)	0	12	12
K_0	0,43	0,70	0,9-0,0533X(Z-6)(6<Z<18)
			0,26 (Z>18)

Z = Depth with respect to level ground measured in meters
(*) In gravels a minimum value of Ec=100.000 kPa is considered to a depth of 5 m

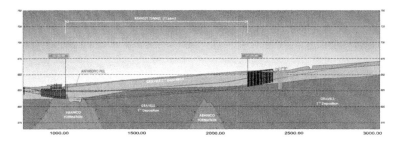

Figure 3. Geological section of the tunnel.

Figure 3 includes a synthetic geological longitudinal section of the tunnel in which the water table has also been highlighted.

During the tunnel construction, the water table was really encountered at 615 MASL, eight meters below the foreseen height. At the top heading excavation, only few infiltrations were registered, probably due to the watering system of the golf course and seasonal rains. At the excavation of the bench a continuous water flow has been registered, but without creating stability problems due to the presence of water.

Santiago de Chile is located at a high seismicity area, According to the Chilean standards the seismic acceleration to be considered are 0.3g and 0.15g respectively as horizontal and vertical acceleration.

On September 16th, 2015, an earthquake of magnitude 8.4 Mw was recorded in the region of Coquimbo. Consequently, an intensity VII in the Mercalli scale was registered in Santiago. Once tunnel inspection and monitoring analysis were performed, it was determined that this event did not affect the structure.

3 SPECIAL DESIGN REQUIREMENTS

Regarding the design of the tunnel, the following technical aspects were crucial during the development of the project:

- To minimize the subsidence induced at surface,
- To reduce traffic impact during construction, and
- To minimize the maintenance costs associated to water pumping during exploitation.

The minimization of the induced settlements in surface was a key aspect of the design. This is a consequence of the alignment of the tunnel, that goes nearly in its entire length, below a private property as the golf course "Los Leones" and tangent to Kennedy Avenue. Any damage over both elements must have been strictly avoided. Therefore, the design of the support and the construction phases of the tunnel was carefully accomplished for the mentioned purposes. Four numerical models were carried out using FLAC 3D code focusing in particular to the stresses acting on the support elements and the induced settlements.

A second major issue has been to minimize the impact on the traffic during the construction. In order to reduce this issue, a temporary underground roadway was excavated. The length of this access to the main tunnel was 190 m, with a cross section of 5.0 m x 6.5 m. This access goes with an extremely low overburden under Vitacura Avenue and Perez Zujovic roundabout but it has allowed to start the tunnel excavation on his 4-lane section without any disturbance to the existing traffic.

This underground access finally connects with the central drift of the 5-lane section, but the excavation of this largest section of the project was postponed until the traffic over the surface in the roundabout was dismantled. Figure 4 illustrates the complexity of the temporary access previously excavated to initiate the construction of the tunnel.

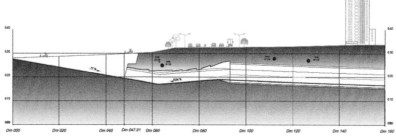

Figure 4. Temporary access tunnel alignment and section. Kennedy Tunnel.

Waterproofing of the tunnel was another main aspect considered in the design phase. For this reason, and in order to minimize the maintenance costs, a pumping rate lower than 500 l/min was stablished in the construction contract. Thus, along all the tunnel and between the concrete of the support and the lining was foreseen to install a waterproofing system composed of a geomembrane and a waterproof sheet.

4 CONSTRUCTION CHALLENGES

There are several aspects of this project that entails construction challenges, among them it can be emphasized the following:

- The dimensions of the tunnel excavation cross section are extraordinarily large as the five lanes section has 22.9 m wide and 13.65 m height which entails an area over 250 m².
- The tunnel was entirely excavated in soils. This fact obliges to take into account several factors such as face stability, limitation of the unsupported span, induced deformations, the stress acting on the support, or safety factors at the support, and systematization of the excavation cycle.
- The tunnel was located in an urban environmental with a quite low overburden (10 to 16 m) and below a private property with no allowed access. Therefore, the limitation of the induced deformations and the monitoring at surface involved several difficulties for the design and also for the control during the construction.

5 DEVELOPED SOLUTIONS

The following solutions were developed for the excavation, support and lining of the tunnel:

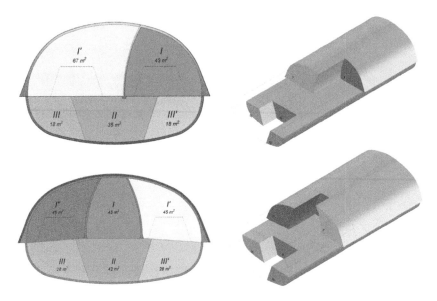

Figure 5. Excavation phases used for the 4 and 5 lanes sections. Kennedy Tunnel.

5.1 *Excavation*

The construction method has been a Sequence Excavation Method, following NATM philosophy, using an elephant foot after finalising the top heading excavation. Depending on the geometry this top heading has been excavated using two or three side drifts, while the benching was always excavated using three phases. The excavation has been done using hydraulic excavator, backhoe and wheel loader.

5.2 *Support*

The support of the typical section consists in a 3 cm shotcrete sealing, HEB-120 steel arches with elephant foot spaced 1.0 m and 27 cm shotcrete Sh35. Systematically buttress at the face and occasional canopy tubes have been used in specific section in which a minimum induced deformation at surface, was allowed.

Figure 6 show the construction sequence for the four-lane section.

The support of the five lanes section consists in a 5 cm of shotcrete sealing, for the vault Norwegian arches spaced 0.75 m were envisaged, additionally it was disposed a ring of HEB-160 arches spaced 0.75 m and 20 cm of shotcrete Sh35. Both arches share an elephant foot foundation. Figure 7 show the construction sequence for the five-lane section, in which the top was constructed in three excavation phases and the bench in three phases similar to the four-lane section.

Finally, a transition was necessary between the four lane section and the five-lane section, as shown in Figure 8.

5.3 *Lining*

Finally, and separated from the support by the waterproofing system a reinforced concrete lining was envisaged. Its thickness varies from 20 cm to 40 cm depending on the water column to be resisted. Three different lining sections have been designed and constructed:

– Lining A, 40 cm thick and heavily steel reinforced, for the design water table located above the tunnel vault,

Figure 6.   Construction sequence for the four-lane section.

Figure 7.   Construction sequence for the five-lane section.

Figure 8.   Transition between the four-lane and the five-lane sections.

– Lining B, 35 cm thick and with a medium reinforcement, for the design water located from the tunnel vault to its invert, and
– Lining C, 20 cm thick with light reinforcement, for the stretch in which the water table is located below the tunnel invert.

A third lining layer with an homogenous thickness of 10 cm with polypropylene fibres for fire protection was also considered.

In the five lanes section the thickness of the lining will be 50 cm.

Figure 9 show the construction sequence used to implement the lining.

Figure 9.   Construction sequence for the lining.

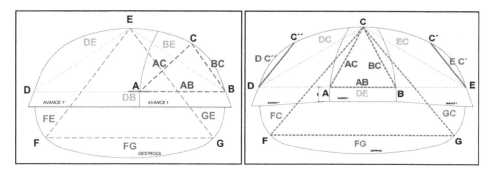

Figure 10. Convergence arrays disposition for 4-lanes and 5 lanes sections.

5.4 *Monitoring*

Due to the extremely low overburden of the tunnel and the number of excavation phases, the control of the induced deformations and subsidence during construction have required a very strict control. This control has been done with the following monitoring elements:

– Inclinometers and extensometers installed from surface prior to the excavation of the tunnels along the tunnel alignment.
– One section of pressure cells (Dm 1374) installed around the emergency exits structures.
– Convergence sections, every 25 m.
– Settlement profiles at surface

Along the typical section of the tunnel, the maximum settlement measured at surface have been 12.4 mm at the Dm 1145, showing its higher slopes after the demolition of the temporary support wall of the left side drift.

In the 5-lanes section of the tunnel the maximum settlement measured at surface was 12.8 mm at the Dm 1037, with the higher slopes related to the excavation of right side drift and the second to the demolition of the temporary support walls of the central drift.

The maximum convergence along the typical section was 14.4 mm at the Dm 1465, with a maximum velocity of 3.66 mm/day. This behaviour was related to the demolition of the temporary support walls. It is significant that after it, a "divergence" deformation was registered at AC and AB arrays.

In relation to the 5-lanes section of the tunnel the maximum convergence registered was 57.52 mm at the Dm 1070 with a convergence velocity of 2.03 mm/day. Also, it was related to the demolition of the temporary support walls.

Figures 10 show the disposition of the convergence arrays for the two existing sections of the tunnel.

6 CONCLUSIONS

The excavation of the tunnel started on October 24th, 2014 and it finished on February 16th, 2017. If we consider the total constructed length of 1,166 m. The average advance rate is 1,4 m/day.

– The main challenges during the construction of the Kennedy Tunnel have been:
– The dimensions of the excavation section, over 250 m^2; 23.2 m width.
– The tunnel is entirely excavated in alluvial gravels and partially under the water table, implying several factors such as face stability, canopy tubes, elephant foot, several excavation phases, to ensure a safety construction.

- The tunnel is located in an urban environmental with an extremely low overburden (10 to 16 m), so the limitation of the induced deformations and the monitoring at surface has been a paramount aspect.
- Because its location, in one of the densest urban areas of Santiago, any affection to the existing surface traffic was not allowed by the Authorities.

Tunnels and Underground Cities: Engineering and Innovation meet Archaeology, Architecture and Art, Volume 12: Urban Tunnels - Part 2 – Peila, Viggiani & Celestino (Eds)
© 2019 Taylor & Francis Group, London, ISBN 978-0-367-46900-9

The construction of the tunnel Lo Saldes in Santiago de Chile. A major challenge, a wide tunnel with an extremely low overburden

J. Kuster & R. Núñez
Costanera Norte, Chile

E. Chávez, J.M. Galera & D. Santos
Subterra Ingeniería, Spain

ABSTRACT: Costanera Norte is an urban expressway concessionary in Santiago de Chile connecting its Western and Eastern sides. To enhance this connectivity several measures were necessary, including four new tunnels. One of these tunnels, is 'Tunnel Lo Saldes', a three-lane one that has a total length of 65 m, and an excavation rectangular section 14 x 7.2m. It has been excavated in anthropic materials, colluvial deposits, and hard tuffs. But the main singularity of the tunnel comes from its extremely low overburden, 2m to the surface, were an existing operative motorway is located. This low overburden has demanded a special construction method, based on two lateral side-drifts and a central one, a pre-grouting sequence and fore-poling, and in some cases, a prior grouting of the ground and/or a face reinforcement with fibre-glass bolts. The support consists in wide flange steel ribs spaced 0.75m and 60 cm of shotcrete Sh35.

1 PROJECT DESCRIPTION

The tunnel is located in the NE of Santiago de Chile, close to the financial centre of the city. The tunnel provides a considerable improvement in the connection between Kennedy and Costanera axes. Figure 1 shows the location of the tunnel.

The access to the South portal is solved with a trench structure composed by piles and crossbeams. The access to the North portal is solved, in the construction phase, by mean of an oval shaft located in a very constrained emplacement which is surrounded by several roads. Finally this shaft was replaced by a trench structure and a cut and cover that crosses Costanera Highway.

The typical cross section of the tunnel provides three lanes, 3.5 m wide each one. The inner dimensions with respect to the lining are 14.0 m wide and 5.7 m height. Figure 3 shows the geometrical definition of the tunnel.

The alignment crosses below Kennedy Street with an extremely low overburden that ranges between 4.7 m at North portal to reach a minimum of 1,60 m at South portal.

Following, the main project characteristics are described:

a) Geology: The tunnel has been excavated in weathered volcanic tuffs of Abanico Formation, alluvial materials composed by gravels, pebbles silt and clay. Covering these deposits, there was antrophic fill materials. In some extension of the tunnel alignment, the crown and walls are excavated in these fills. The water table was located below the level of the invert. Figure 4 includes a synthetic geological longitudinal section of the tunnel and Figure 5 shows the fills at the tunnel face.

b) Geotechnics: Table 1 shows the materials strength and deformational parameters considered in the design. A hardening soil model was considered for the fills and a Mohr-Couolomb one for the Abanico formation.

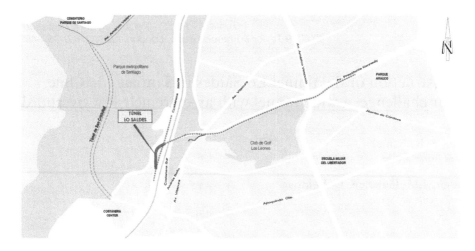

Figure 1.　Location and layout. Lo Saldes Tunnel.

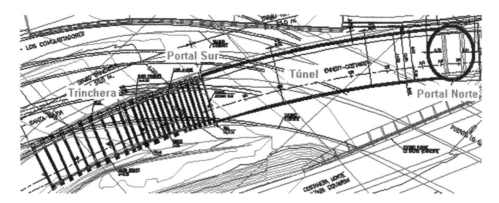

Figure 2.　Layout. Lo Saldes Tunnel.

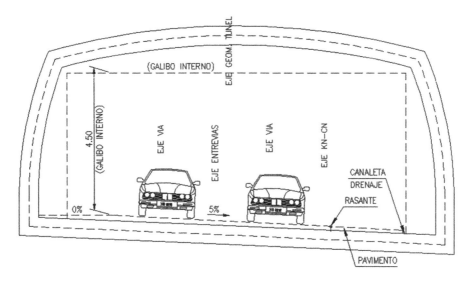

Figure 3.　Functional section. Lo Saldes Tunnel.

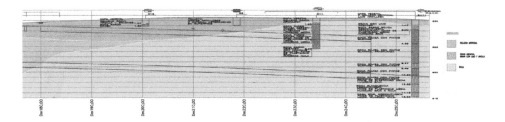

Figure 4. Geological section of the tunnel.

Figure 5. Antrophic fills materials on tunnel face.

Table 1. Strength and deformational parameters of the materials in the Lo Saldes Tunnel.

Suelo	Modelo	γ	c'	φ'	E_{ref}	E_{50}^{ref}	E_{oed}^{ref}	E_{ur}^{ref}		ko
		[kN/m^3]	[kN/m^3]	[°]	[kN/m^2]	[kN/m^2]	[kN/m^2]	[kN/m^2]	γ	(1)
Fill	Hardening Soil	19,0	10	30	-------	17.000	17.000	51.000	0,20	0,53
Rock	Mohr Coulomb	23,5	50	35	500.000	-------	-------	-------	0,20	1,50

c) Seismicity: Santiago the Chile is located at a high seismicity area. According to the Chilean standards a ground shear distortion of $3.8 \cdot 10\text{-}5$ rad was considered.

d) Construction method and support: The construction method has been a sequential excavation method, following NATM philosophy. The complete section has been excavated using three drifts. The advance spam in all of them was 70 cm. Figure 6 shows the three excavation sections.

First side-drift to be excavated was the left one (Section A). A distance of 20 m has been respected between excavation faces of section A and B. Same gap was respected between section B and C.

For the demolition of the temporary walls, the sequence was as follows. Demolition of 5 m of the temporary wall of section B. Installation of the secondary support in section B

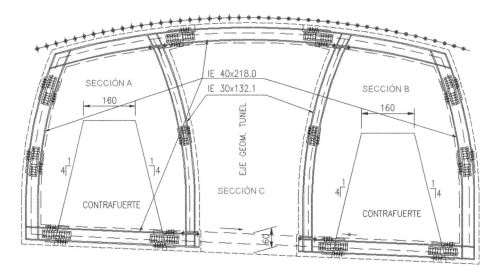

Figure 6. Excavation phases. Lo Saldes Tunnel.

and C. The demolition of the wall continue when the concrete reached 60 % of the UCS. The demolition of the temporary wall in section A started when the process reached 20 m of distance in section B. The process ended with the execution of the whole secondary support in the tunnel section. Figure 7 shows the demolition of the temporary walls and the excavation face at sections B and C.

e) Support and lining: The support is emplacement in two layers. Primary support consists of 30 cm of shotcrete reinforced by mean of double T steel ribs. The arches were closed except for section C were a concrete slab was executed due to the difficulties of the joint between the steel beams. Secondary support consists of an additional layer of 30 cm of shotcrete reinforced by electrowelded mesh.

In order to assure the face stability several actions were envisaged as systematic forepoles (marchiavanti), buttress and fiberglass bolts at the face.

Figure 7. Mixed face and excavation of the tunnel face.

2 KEY CONSTRUCTION DATA

The construction method has been a sequence Excavation Method, following NATM philosophy. The construction of the North portal and the excavation of 40,60 m of the left side drift were carried out by another contract. Then, as part of the SCO2 contract, the excavation and primary support of the tunnel was continued, it started in 20-10-2015 and finished in 11-10-2016. So the tunnel was constructed in 357days. But considering each excavation section of the tunnel, there was an average advance of 0.2 m/day. Table 2 shows the excavation dates by each section of the tunnel.

All the equipment used have been standard trucks and auxiliary machines, while a jumbo has been used for the execution of the systematic forepoles and a robot for the sprayed concrete operations. The excavation has been done using hydraulic hammer and a backhoe loader.

The whole tunnel alignment has a very low overburden with a road crossing at surface, therefore a very strict control of the induced deformations and subsidences has been accomplished. This control has been done by mean of the following monitoring elements:

– 12 Convergence sections by reflex targets. Figures 8 shows the disposition of the convergence arrays in the tunnel section.
– 17 Settlement profiles at surface and structures. Figures 10 shows a settlement evolution of a control point.

The maximum settlement measure has been 10.6 mm at Dm 235 m. The maximum convergence was 11.89 mm at Dm 240 m. The deformation was related to the different phases of excavation and the demolition of the temporary support walls.

The major issues of the construction were the beginning of the excavations at the portals and the dealing with the mixed face, with the presence of fills at the upper part.

The North Portal consists of a shaft with ovoid shape supported by mean of thicknesses of 20 and 30 cm of shotcrete. The shaft had three rings with more thickness (50 cm) in order to provide more stiffness at certain heights. One ring was at the surface, other immediately over

Table 2. Excavation dates by section.

Section	PKi	Date	PKf	Date	Length (m)
B	241,65	20-10-2015	180,75	08-09-2016	60,9
C	241,65	19-11-2015	180,75	11-10-2016	60,9
A	201,05	25-06-2016	180,75	30-09-2016	20,3

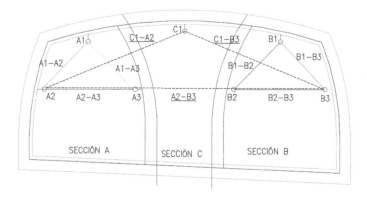

Figure 8. Convergence arrays disposition.

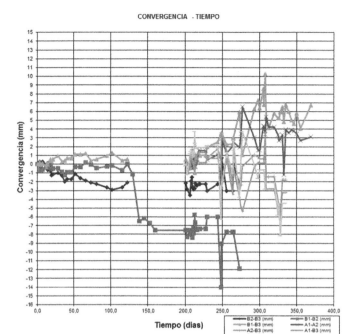

Figure 9.   Convergence evolution.

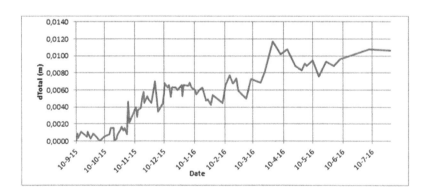

Figure 10.   Settlement evolution.

the tunnel section and the last at the bottom. The support of the shaft has a reduction of its thickness up to 15 cm in the intersection with the tunnel.

At each portal micropile forepoles covering all the ceiling of the tunnel were executed. The tubes geometry was $\emptyset_{ext}$ 142 mm and $\emptyset_{int}$ 122 mm, spaced 0.30 m and 15.0 m of length. A double micropile forepole was executed at the corners of the section. Figure 11 shows a portal view.

From the inner section of the tunnel marchiavanti forepoles were executed ($\emptyset_{ext}$ 76 mm and $\emptyset_{int}$ 58 mm). The length of the marchiavanti were 7 m, spaced 0.30 m and the overlapping was 3.5 m. Furthermore, the tunnel face was reinforced with fiberglass bolts ($\emptyset_{ext}$ 22 mm), in a net of 1.0x1.0 m, with length of 9.0 m and overlapping of 4.5 m. The execution of the marchiavanti and the fiberglass bolts with that level of overlapping in these small space was very complicated.

Figure 11. Portal view. Sections A and B in excavation process.

In some length along the tunnel section, ground improvement injections were envisaged. Most cases they were developed from inside the tunnel (Figures 12 and 13) but in some critical sections the injection had to be executed from the ground surface. In that cases the very different permeability of the fills and the Abanico Formation conditioned the leakage of the whole grout along the contact between these two layers. After several trials, a special procedure for the ground treatment was designed. The solution consisted of two pile mortar walls constructed at both sides of the tunnel in order to prevent the leakage and helping the confinement of the ground injections. That solution was a success. Figure 14 shows the problem before developing the solution, with all the grout injection in the contact between de fills and the Abanico Formation.

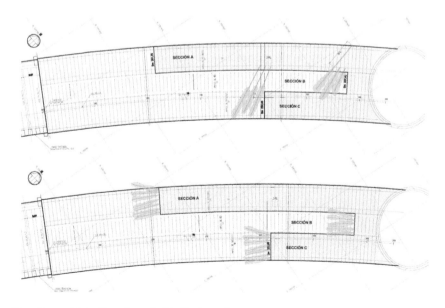

Figure 12. Ground injections from inside the tunnel.

Figure 13. Injection from the tunnel.

Figure 14. Leakage of the injection.

Figure 15. Demolition of the temporary support.

Figure 16. Execution of the secondary support.

Figure 17.   Tunnel ready for its inauguration.

3   CONCLUSIONS

The excavation of the tunnel Lo Saldes started on October 20th, 2015 and finished on October 11th, 2016. The average advance rate for the construction was 0,2 m/day. Nevertheless the existing difficulties, basically due to the extremely low overburden and the mixed face, the excavation has been successfully accomplished without any affection to the traffic of the surrounding and crossing roads.

The construction method used in the construction, as the split of the excavation section, the ground improvement injections or the systematic forepoles, have been proved to be very useful and effective to minimize the subsidence induced and to assure the stability of the tunnel face.

Tunnels and Underground Cities: Engineering and Innovation meet Archaeology,
Architecture and Art, Volume 12: Urban
Tunnels - Part 2 – Peila, Viggiani & Celestino (Eds)
© 2019 Taylor & Francis Group, London, ISBN 978-0-367-46900-9

DSSI-MTS-01 Tunnels in Doha (Qatar) – a showcase of construction techniques for tunnels, shafts, galleries and junctions between them

R. Lamand
Bouygues Travaux Publics – Design Office, Guyancourt, France

C. Penot
Bessac, Doha, Qatar

X. Perrin
Bouygues Travaux Publics, Doha, Qatar

T. Lockhart
Bouygues Travaux Publics – Technical Department, Guyancourt, France

ABSTRACT: The DSSI-MTS1 project, located in Doha (Qatar), consists of 16.4km of sewer tunnels, 11 shafts, 9 mined galleries and connections between these elements, awarded to the Bouygues-UCC JV. The tunnels have been excavated by 2 EBP TBMs, operated jointly by Bouygues Travaux Publics and Bessac, in the specific geological context of the Qatar peninsula, consisting of soft to medium sedimentary rock. The purpose of this article is to describe the Design and Construction of this project, highlight the innovative solutions developed to meet the stringent requirements of durability and hydraulics performance, and illustrate the challenges that have been overcome during the construction.

1   INTRODUCTION

1.1   *The DSSI – MTS-01 project*

As part of the development of the southernmost part of the capital city of Doha, the Ministry of Infrastructures of the state of Qatar has launched the transition to a deep, gravity-based system: Doha South Sewage Infrastructure Project. The Main Trunk Sewer 1 (MTS-01) project comprises 11 shafts and three tunnels with a total length of 16.2km and an internal diameter of 3m.

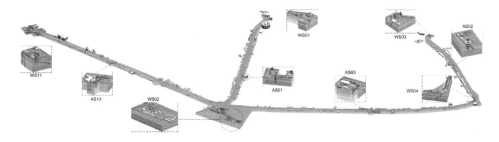

Figure 1.   Overall view of the MTS-01 project: three tunnels and 11 shafts.

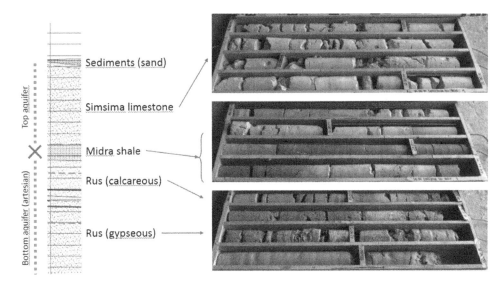

Figure 2. Typical geological configuration in Doha.

1.2 *The hydrogeological context*

The geological configuration of the project is typical of the Qatar peninsula: Upper Dammam Formation (Simsima Limestone), Lower Dammam Formation (Midra Shale), underlain by the Rus Formation, either calcareous or gypseous, all with various degree of alteration. The impervious Midra shale separates two aquifers; the bottom one being slightly artesian.

2 DESIGN

2.1 *General design*

The DSSI-MTS01 is a Design & build contract, and the comprehensive design scope includes the hydraulics studies (physical modelling & CFD simulations), durability, environmental studies, structural design, operation & maintenance, HAZOP, etc.

2.2 *Design of the shafts*

The 11 shafts, scattered in the Southern part of Doha, serve 2 main purposes:

– Collect the flow from the primary (shallow) sewer network
– Provide access points into the sewerage system, for inspection & maintenance.

The shape of the shafts is mainly governed by the hydraulic performance of the system, to minimize the release of corrosive H_2S gas; the structural design showed that the governing load case is the limitation of the crack width to 0.2mm for durability reasons.

2.3 *Design of the tunnel lining*

The universal rings are made of 5+1 segments. The steel-fibres reinforced (50MPa, class 4c) lining is 300mm thick, including 120 mm of sacrificial concrete, with calcareous aggregates. To comply with the 100-year service life of the project, BUJV has developed an innovative technical solution to embed the corrosion protection liner (CPL) directly in the segment moulds (Najder-Olliver & Lockhart, 2017; Najder-Olliver & al., 2019).

2.3.1 *Segments structural design*

Segmental linings are designed for temporary and permanent load cases using 2D elastoplastic soil-structure and an innovative nonlinear 3D model developed in-house taking into account the SFCR tensile constitutive law (Gauguelin & al., 2017). The total lining ovalization is shown to remain below 0.4% (ultimate state) and 0.8% (accidental state).

At detailed design stage, the longitudinal joint distortion and so the bending moment due to the imposed ovalisation has been assessed with the Janssen theory. The Janssen method gives a relationship between the rotation of the longitudinal joint and the bending moment applied on it.

2.3.2 *TBM confinement pressures*

Given the geological configuration, the estimated settlements are negligible and the tunnel face stability is always ensured. However, in such grounds, zones of weathered rock could be encountered, as well as karstic situations with high water inflows.

During the construction of the tunnels, the measured settlements where less than 5 mm, and actually close the practical limit of survey accuracy.

2.4 *Design of the galleries & junctions*

2.4.1 *Excavations temporary supports*

Non-TBM galleries are designed based on the observational NGI Q-system (Lamand & al., 2018). During the excavation of the galleries, a minimum 9cm of shotcrete & systematic crown bolting have been applied as contingency measures for safety.

The design A-A-A convergence thresholds have been set as 25mm, 37mm and 50mm respectively, and have never been reached throughout the project.

2.4.2 *Permanent lining*

The permanent lining of the galleries consists of 500mm-thick precast rings, including 250mm of sacrificial concrete & a HDPE liner at the intrados to provide a dual protection against the H_2S-induced corrosion over the 100 years.

2.4.3 *Junctions*

2.4.3.1 JUNCTIONS BETWEEN SHAFTS AND GALLERIES

Openings are drilled at the bottom of the access shafts (Lamand & al., 2018). Around the opening, the ground is pretreated and drained to release the water pressure.

Figure 3.   CFD modeling of the adits & galleries [left] - Precast permanent lining [right].

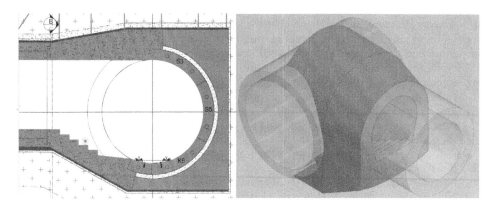

Figure 4.   3d junction between a gallery and the TBM tunnel.

2.4.3.2 JUNCTIONS BETWEEN GALLERIES AND MAIN TUNNELS

The connection of the gallery to the main tunnel has a complicated shape, which required a full 3d modeling to correctly render the geometry and perform the structural design.

2.4.4 *Waterproofing*
Different types of waterproofing are installed depending on the construction solution chosen: epoxy tar coating for precast gallery pipes, flexible PVC membrane for the mined junctions with complex shapes.

3   CONSTRUCTION

3.1   *Shafts*

3.1.1 *Temporary works & Excavation*
The temporary works have been presented in a companion paper, Lamand & al. 2018.

3.1.2 *Permanent structures*

3.1.2.1 BASE SLAB

All the permanent chambers are below the water table and must be waterproofed using a PVC membrane + 50mm protective concrete.

Figure 5.   Shaft chamber waterproofing + Protective concrete.

Figure 6.   Formwork arrangement & gallery connection.

3.1.2.2 WALLS

All permanent chambers have different shapes and dimensions. A specific single-side form-work has been designed to cast the walls, while ensuring the integrity of the membrane.

3.1.2.3 TOP SLAB

The 1250m- to 2250mm-thick top slabs close the bottom chamber like a lid.

3.1.2.4 ACCESS CHIMNEY WITH MANHOLE

Precast pipes are used for the vertical, 2000mm-diameter chimney providing access to the system, up to the ground level where a manhole is provided. The temporary shaft is then back-filled with a cement-stabilized gravel.

3.1.2.5 BENCHING & HYDRAULIC SHAPES

All permanent chambers differ in shape, to channel the flow with minimal turbulence.

Polystyrene formworks have been chosen due to the complexity of the benching with double curves, as well as their light weight making the installation easier.

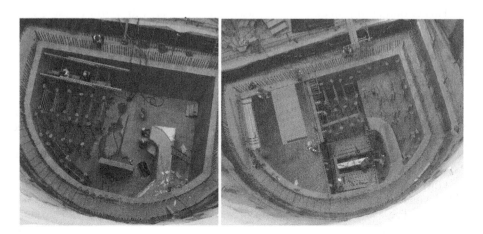

Figure 7.   Top view of the bottom chamber just before the top slab is built.

Figure 8. Top of permanent shaft, backfill & Reinstatement at ground level.

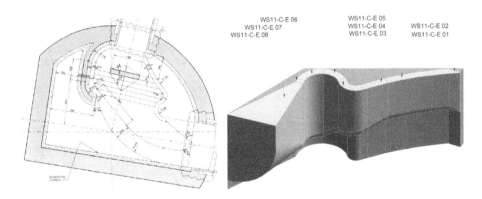

Figure 9. Plan view of a typical benching & 3d BIM used for the polystyrene moulds.

3.2 *Tunnels*

Completed in barely 22 months, the 16.2 km of 3.0mID tunnels brought its share of challenges and difficulties. Heterogeneous geology, ground water management, sustained logistics, demanding guidance requirements, as well as a very tight space to accommodate all the technologies required for tunneling works with high safety and quality standards, were among the main challenges of DSSI – MTS-01 tunnels.

Two Tunnel Boring Machines, 3.895 m excavation diameter, with Earth Pressure Balance mode were used to excavate 190,000m^3 of ground and install 13,513 rings.

The mucking out was performed using rail-bound with electrical locomotives capable of running the long distance tunnels several time before changing the batteries. The trains also provided in-tunnel transportation of men, mortar grout, segments and materials.

3.2.1 *Overview of the Tunnel Boring Machine*

EPB technology was selected as the optimum choice to master the cohesive soft rocks: limestones, siltstones and shales. Soil conditioning with foam and polymers has been required to manage clayey conditions (sticky) or high ground water pressure in weathered areas.

The mixed ground cutter head with a 22% opening ratio is dressed with 21 discs cutters 14" (9 double ring, 8 single ring and 4 inner discs) that penetrate the soil ahead at about 35 mm/min.

The TBM has two articulations, one shield active articulated steering articulation via hydraulic cylinders and tail skin passive articulated articulation at the rear of the shield middle section.The 400 kW hydraulic main drive develops a torque of 1,600kNm, and the total thrust is 16MN.

Fitted with 12 cylinders, the machine has the capacity to push up to 15,966 kN.

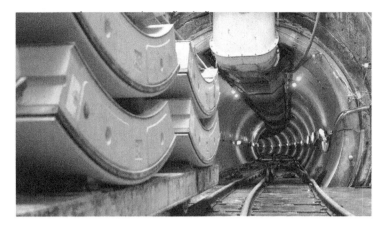

Figure 10.   Segments on the wagon, close to the tunnel entrance.

Figure 11.   Picture of the Mixed Ground Cutter Head [left] – Clogging in Midra shale [right].

3.2.2 *Break-in*

The 3 consecutive drives started from the WS2 shaft, at the junction of the 3 branches. The combination of a 5m-long plug with a launching seal prevents groundwater ingress at break-in.

The TBMs were lifted down the shafts in a single piece, with a spreader beam equipped with load cells to balance the force in each chain.

By lifting the complete shield, BUJV was able to start the excavation in less than one week, commissioning included. The TBM was designed to be launched with a reduced back-up and a bundle of hoses and cables called "umbilical". While the TBM shield progresses alone, all the fluids (water, air, foam, cooling, grease, mortar), electrical power and communication are supplied through this umbilical which is uncoiled gradually according to the advance.

3.2.3 *TBM in operation*

Progress across all the three tunnels showed a peak performance at 33m/day and average advance speed of 170m/week.

These performances were possible partly thanks to:

- A successful design of the cutter head and suitable choice of cutter discs;
- An efficient logistic process: an automatic overhead crane to empty the muck car and then load the segment on the train prepared one production train in less than 50 min, matching with the excavation/ring building cycle;

Figure 12.   Lifting of the TBM [left] – the TBM just in front of the launch seal [right].

Figure 13.   TBM "umbilical" [left] – View from the shaft with thrust frame [right].

- A rigorous maintenance to keep high availability coefficient despite the long drives.

The teams faced the followings major challenges along the drives:

- Water ingress: especially in the interfaces above and below the Midra shale, strong water ingress made the soil conditioning difficult and forced full-EPB mode.
- Clayey/'sticky' conditions: Clogging frequently occurred (**Figure 11**) where Midra shale was encountered at the face, leading to an increase in torque and requiring the use of water and foam, and frequent cleanings of the cutterhead
- Guiding tolerances: the hydraulic profile imposed tight vertical tolerances of ±35mm, which brought the need of careful surveying and the use of regular gyroscopic surveys.

3.2.4 Break-out

The first breakout, for the Western Branch occurred in an empty shaft. In order to prevent any unexpected groundwater ingress, the last 16 m along the tunnel alignment were fissure-grouted.

The bottom half of the TBM pushes against a cement-stabilized backfill, which provides the necessary support for guiding while providing a working platform for the recovery of the machine.

3.3 Galleries & adits

Galleries & adits are mined after preliminary fissure-grouting of the ground. The temporary support (rock bolts, shotcrete) is placed after each excavation step (Lamand & Lockhart, 2018).

After the temporary excavation has been performed, precast pipes are pushed in place on skidding rails to create the permanent lining. The main constraint is to ensure the continuity of the waterproofing; therefore the outer face of the precast pipes are epoxy-tarred and each junction between successive pipes is glued.

The gap between the temporary excavation profile and the extrados of the pipes is then filled with a cement-based mortar.

3.4 Junctions between galleries and TBM tunnel

The junction between the TBM tunnel and the adit is first excavated, then filled with lean mortar; once the TBM has crossed the area, a temporary support is installed inside the tunnel and the segments are cut open using a diamond saw and removed.

Figure 14.   TBM inside the retrieval shaft.

Figure 15.   Installation of precast pipes in the galleries.

Figure 16. 3D model of the junction [left] and opening in the TBM tunnel lining [right].

The reinforced concrete (dark grey on the Figure above) is cast in situ, after the waterproofing complex has been installed.

4 CONCLUSION

The DSSI-MTS1 project has mobilized a wide variety of construction techniques, illustrated in this article, to realize on site a sewage system that meets all performance requirements. From Design to Construction, the JV has

ACKNOWLEDGMENTS

The Authors wish to acknowledge Daniel Clert, Project Director, Olivier Suteau & Georges Pires, Construction Managers of Shafts and Tunnels respectively, for their contribution and proofreading of this article.

The Authors also express their gratitude to the teams of Designer of Record (AECOM), Quality Checking Engineer (COWI) and Project Management Consultant (CH2M – now Jacobs) for their support in the development and implementation of the main innovative solutions.

REFERENCES

Gauguelin, G., Dabet, L., Minec, S. & Taibi, Y. 2017. Study of the effect of the TBM thrust on steel fibre reinforced concrete segmental tunnel lining. *Proceedings of the AFTES congress 2017, Paris, 13–15 November 2017*

Lamand, R. & Lockhart, T. 2018. Three different techniques to build a shaft: the DSSI-MTS01 project, case study in Doha, Qatar. *Proceedings of the WTC2018, Dubai, 23–25 April 2018*

Najder-Olliver A.M. & Lockhart T., *Innovative One-pass Lining Solution for Doha's Deep Tunnel Sewer System*, Proceedings of the World Tunnel Congress 2017 – Surface challenges, underground solutions, Bergen, Norway

Najder-Olliver, A.M., Lockhart, T., Azizi A. & Singh K.G., 2019. Practical experience from the installation of a one-pass lining in Doha's deep tunnel sewer system. *Proceedings of the WTC2019, Naples, 3–9 May 2019* TO BE PUBLISHED

*Tunnels and Underground Cities: Engineering and Innovation meet Archaeology,
Architecture and Art, Volume 12: Urban
Tunnels - Part 2 – Peila, Viggiani & Celestino (Eds)*
© 2019 Taylor & Francis Group, London, ISBN 978-0-367-46900-9

Settlement prediction of an urban shield tunnel using artificial neural networks

L. Li
Shanghai Tunnel Engineering Co., Ltd., Shanghai, China

L.K. Dong & F.F. Wang
Tongji University, Shanghai, China

ABSTRACT: ANNs have become a useful method for predicting tunnel settlement. This paper presents the concept of and procedure for the Artificial Neural Networks (ANNs) method in predicting urban shield tunnel settlement. In the process, apart from soil types, buried depth, ground water table and tunnel diameter, there are many other factors have been taken into consideration, such as grouting pressure, thrust force, volume loss etc. Many existing papers have chosen to use the Finite Element Method (FEM) results as input data sets instead of real monitoring data. In this paper, the application process is based on real monitoring data obtained from a shield tunnel project in Shanghai. The results of the predicted settlement are of high quality which demonstrates its potential to be recommended as a tunnel prediction tool for similar projects in future.

1 INTRODUCTION

For shallow shield tunnels crossing dense urban areas, settlement may be the most challenging problem. This is because, compared with other underground structures, the longitudinal stiffness and integral stiffness of a shield tunnel is relatively small, which will result in large differentiate settlement and ground settlement. Excessive settlement will then cause secondary internal forces and increase the opening width of joints, leading to decreased tunnel bearing capacity, spelling and bursting of segment, water leakage as well as shortening of a tunnel normal service life.

By the end of the first quarter of 2017, Shanghai has completed 13 large-scale cross-river tunnels and 617 kilometers of subway tunnels. With deeper and larger underground space network, the importance of appropriate construction techniques, early warning mechanism and sound judgment has been revealed in many projects. It is essential for engineers to plan and execute effective prevention methods to control shield tunnel settlement.

However, the factors which influence the settlement of tunnels are complex, including geological conditions, construction methods, environmental factors etc. Each parameter could affect the others which makes it impractical using empirical and analytical methods. With the help of computing algorithms, these limitations can be overcome. This is not only an effective way to predict settlement but also an intelligent method to help engineers determine the most influencing parameters and modify construction process if required.

In recent years, the use of artificial neural networks (ANNs) has become a useful method for predicting tunnel settlement. However, most of tunnels studied using intelligent methods are excavated by the NATM method. Few papers have discussed shield tunnel settlement prediction. When it comes to shield tunnels, apart from soil types, buried depth, ground water table and tunnel diameter, there are many other factors that should be taken into consideration while conducting an ANNs analysis, such as grouting pressure, thrust force, and volume loss. These factors have a significant influence on shield tunnel settlement and need more effort to collect and analyze than an analysis based on NATM method tunnels. During the application of this research, many key factors mentioned above have been taken into consideration. Besides, many papers choose to use Finite Element Method (FEM) results as input data sets instead of using real monitoring data. This study goes further using real monitoring data obtained from a shield tunnel project.

2 ENGINEERING BACKGROUND

Many papers intend to derive a general settlement prediction method or conclusion by using the ANNs method with limited parameters and many tunnels' datasets without specific engineering background. In our opinion, however, the best way to utilize the ANNs method is to build different influencing factor models based on different tunnel projects, and to execute prediction analysis for the same tunnel during all construction period. The advantage lies in the increase of accuracy and better understanding of the influencing factor weight.

The South Hongmei Tunnel is the first cross-river road tunnel connecting two administrative districts of Shanghai, Minhang District and Fengxian District (shown in Figure 1). The two-way six-lane tunnel is excavated by two shield tunneling machines with an outer diameter of 14.5m. Pre-fabricated segments have a thickness of 600mm and a width of 2m. The total length of the tunnel shield section is 3390m while the buried depth varies from 9.2m to 43m.

The South Hongmei Road Tunnel crosses a region where engineering environments and geological conditions (building foundations, geological parameters etc.) vary little. The soil layer types that the tunnel crosses under are mainly composed of clay and sandy silt (shown in Figure 2). After eliminating the adverse influence of drastic changes of geological parameters and engineering environmental factors, it effectively enhances the prediction accuracy and reflects the influence of construction factors.

Figure 1. Location of South Hongmei Tunnel.

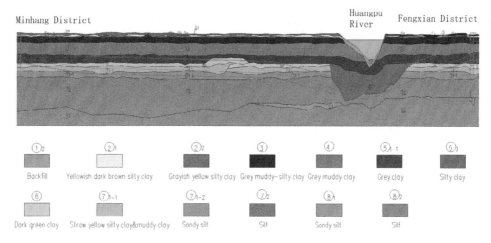

Figure 2.   Engineering geology along the South Hongmei Road Tunnel alignment.

3   DATA SETS AND EFFECTIVE PARAMETERS

The intelligent method is studied on the basis of data obtained from the South Hongmei Road Tunnel on-site monitoring system and early-phase geological survey. The monitoring system records not only the ground settlement caused by tunneling but also certain human factors and machinery working data. The values of Settlement (S) are predicted by using geological classifications G, buried depth Z and construction parameters. Construction parameters

Table 1.   Strength parameter classifications of soil layer which lies within the scope of 0-0.5D above the shield tunnel crown.

Classifications	Soil type	E (Mpa)	C(kPa)	φ
U1	Muddy clay	2.36	12	14
U2	Clay	3.35	15	15
U3	Silty clay	7.71	48	16.5
U4	Silty clay	3.8	15	16.5
U5	Clayey silt imbedded with silty clay	7.82	10	28.5

Table 2.   Strength parameter classifications of soil layer which lies within the scope of 0-D under the shield tunnel invert.

Classifications	Geological condition within the scope of 0-0.5D				Geological condition within the scope of 0.5-D			
	Soil type	E (Mpa)	C (kPa)	φ	Soil type	E (Mpa)	C (kPa)	φ
L1	Silty clay	7.71	48	16.5	Clayey silt imbedded with silty clay	7.82	10	28.5
L2	Clayey silt imbedded with silty clay	7.82	10	28.5	Silty clay	12.06	5	33
L3	Silty clay	12.06	5	33	Silt	14.57	2	35
L4	Silt imbedded with silty clay	14.57	2	35	Silt	14.57	2	35
L5	Silt imbedded with	10.94	2	34	Silt	14.57	2	35
L6	Silt	14.57	2	35	Silt	14.57	2	35

include air cabin pressure P, thrust force F, advance rate V_a, cutter head torque T, excavated volume V_0 and grouting volume V_v. Geological classifications are divided into two categories:

a) The first are the strength parameters of soil layer which lies within the scope of 0-0.5D (0.5 Diameter of the tunnel) above the shield tunnel crown. According to different strength parameters (E, C, φ), geological conditions above the tunnel crown can be classified into 5 types named as U1~U5 (shown in Table 1).

b) The second are the strength parameters of soil layer which lies within the scope of 0-D under the shield tunnel invert. According to different substratum strength parameters (E, C, φ), geological conditions below the tunnel invert can be classified into 6 types named as L1~L6 (shown in Table 2).

Altogether, 86 datasets are utilized for modeling with other 36 others used for testing and evaluation.

4   SIMULATION OF ANNS AND RESULTS

Artificial neural network intimates human brain's working process by using interconnected processing elements (neurons). According to different aims, ANNs can be utilized in different ways, such as data classification, data prediction, data association etc. Generally speaking, an ANNs structure consists of three interconnected layers: input, hidden and output. With pre-processed input dataset and output dataset, Multi-layer networks can be trained and the accuracy can be testified by test dataset.

In ANNs model, the signal at a connection between artificial neurons is a real number, and the output of neuron is computed by non-linear function of the sum of its inputs. The connections between neurons are called 'edges'. Neurons and edges have a weight that adjusts as learning proceeds. The weight increases or decreases the strength of the signal at a connection.

The training process (or called learning process) is fulfilled based on various learning techniques. In this paper, back-propagation approach is adopted. This is a popular and proven method by continuously adjusting the weights of input factors to minimize the error between correct answers and output values.

To achieve ideal ANNs performance, determining the ANNs architecture is equally important which means selecting the number of hidden layers and assigning the number of nodes layers can also have a great influence on the efficiency and accuracy of final results. To some extent, building ANNs structure and selecting influential factors as input data is experience dependent and problem dependent. This is because ANNs method has no prior knowledge of any problem. Its performance may vary from person to person but still, it is a very applicable method specialized in dealing with problems of which the underlying physical relationships are obscure. It is also well-suited in predicting dynamic systems on a real-time basis.

There are a few important steps, which contribute to generalization and accuracy of the ANNs model. These steps are as follows:

4.1   *Feature scaling*

Feature scaling is used to standardize the range of all data features, and it's generally performed during data preprocessing. In this case, raw data is comprised of attributes with varying scales. For example, air cabin pressure P ranges from 2 Bar to 6 Bar while total thrust of the shield F ranges from 20000kN to 140000kN. As total thrust of the shield has a wide range of values, the settlement may be governed by this feature. Therefore the range of all features should be normalized in order to make each feature contributing relatively proportionately to the final settlement. Another benefit of this is that gradient descent converges much faster with feature scaling than without.

The method we used here is standardization. Standardization can make the values of each feature have unit-variance and zero-mean. This method is widely used for normalization in machine learning. Firstly, we calculate the standard deviation and mean for each feature.

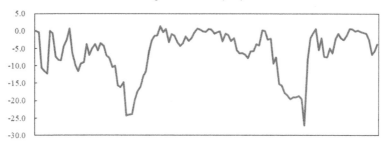

Figure 3a.   Original settlement.

Processed Settlement

Figure 3b.   Processed settlement.

Then we subtract the mean from each feature. Then we divide the values of each feature by its standard deviation. For example:

$$S' = \frac{S - \overline{S}}{\sigma} \tag{1}$$

Where S is the original value of settlement, $\overline{S}$ is the mean of settlement, σ is the standard deviation and S' is the processed settlement which is non-dimensionalized. The following figure 3a, b show the original settlement curve and processed settlement curve.

4.2 *Cross-validation*

K-fold cross-validation (a method of cross-validation) is used in predicting problems, as we did in this paper. We randomly shuffle the dataset into 5 equal sized subsamples x1, x2, x3, x4 and x5. Of the 5 subsamples, x1 is retained as the validation data for testing the model, and then we train on x2, x3, x4 and x5. The cross-validation process is repeated 5 times, for example, by validating on x2 and training on x1, x3, x4 and x5 on the second step, with each of the 5 subsamples used exactly once as the validation data. The 5 results can then be averaged to produce a single prediction.

By using the 5-fold cross-validation, on the one hand, we can solve problems like overfitting. On the other hand, even with only 122 datasets in this project, we can also maximize our ability to evaluate the neural network's performance as all observations are used for both training and validation, and each observation is used for validation exactly once.

4.3 *Dropout*

Dropout is another way to reduce overfitting. The key idea is to randomly drop units (along with their connections) from the neural network during training. At each training stage,

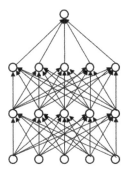

Figure 4a.   Standard neural net.

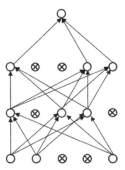

Figure 4b.   After applying dropout.

individual nodes are either kept with probability (1-p) or dropped out with probability p (in the range 0–1), so that a thinned network is left. Only the reduced network is trained on the data during that stage. The removed nodes are then reinserted into the network with their original weights. As we can see, a higher p results in more neurons being dropped during training. In this research, we set the probability to 0.5 in the hidden layers. But for input layer, we set the probability to 0 in order to prevent information being directly lost.

Just like cross-validation mentioned above, on the one hand, dropout can decrease overfitting by avoiding training all nodes on all training data. On the other hand, dropout also improves training speed.

As shown in Figure 4, the circles represent neural network units and arrows represent data transfer paths. The N-level unit performs weighted calculation on the data transmitted by N-1, and transmits it to the unit node of the N+1 layer according to the direction of the arrow. The crossed circles (the unit drawn in Figure 4b.) represent the neural network unit that is randomly dropped during a training session. Once dropped, it means that the unit does not participate in the calculation in this training which can be assumed as non-existence. Experience in computer science has shown that this approach helps to reduce overfitting. This will render the model more powerful and increase its generalization ability.

4.4 *Results*

In this project, we set the number of subsamples k in k-fold cross-validation to 5 and probability p of dropped out to 0.5 as mentioned above, then we set the learning rate to 0.1 while different learning rates do not contribute significantly to differences in training and testing of the network in general. The max number of epochs is 1000. By tuning hyper parameter, we determine the number of hidden layers is 2 and the number of neurons in the 2 layers is 8 and 4 respectively.

The structure of the ANNs is as follow:

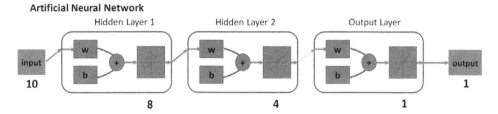

Artificial Neural Network

Figure 5.   Structure of the ANNs.

Figure 6 shows that the simulated curve is very close to the observed curve, and we can also measure of how well the neural network has fit the data by calculating the correlation coefficient R between simulated settlement and observed settlement. The definition of R is shown blow, Cov means covariance and D means variance.

$$R = \frac{Cov(\text{simulated,observed})}{\sqrt{D(\text{simulated})}\sqrt{D(\text{observed})}} \qquad (2)$$

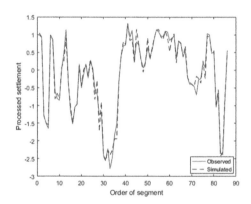

Figure 6.   Comparison between simulated and observed settlement in training dataset.

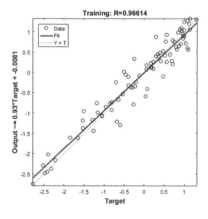

Figure 7.   R of training dataset.

The following regression plot shows the actual network outputs plotted in terms of the associated target values. The network has learned to fit the data well, as the linear fit to this output-target relationship closely intersect the bottom-left and top-right corners of the plot. Here the regression is plotted across training samples (Figures 7–9).

The following figures show the result of test:

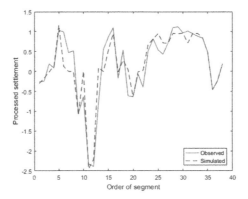

Figure 8.   Comparison between simulated and observed settlement in testing dataset.

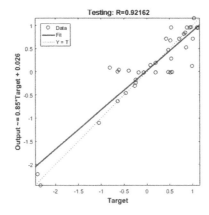

Figure 9.   R of testing dataset.

Figure 8 visually shows that the predicted value (dotted line) is very close to the actual value (solid line). The predicted value of in-sample and out-of-sample are both very close to the actual observation. Figure 7 and Figure 9 show the results of linear regression of the actual value as the abscissa and the predicted value as the ordinate. R is the Pearson correlation coefficient, and the value range is between 0 and 1. The closer it is to 1, the smaller the deviation between the actual value and the predicted value is. In sample, R = 0.966 while R = 0.922 out of the sample. This further confirms that the predicted value and the actual value are close. Besides, the R value out of the sample is not much lower than the R in sample, indicating that the model has good accuracy and generalization capabilities.

5   CONCLUSIONS

In this paper, we use Matlab2016a to simulate the settlement of the South Hongmei Tunnel, creating artificial neural networks based on geological classifications G, buried depth Z and actual construction parameters. In order to reduce overfitting and improve training speed, standardization, K-fold cross-validation and dropout were used. Procedures proposed by this research can facilitate risk-management and expedite tunnel settlement assessment process during construction. The result shows that such methods work well as the simulated curve is close to the observed curve, and the R is 0.92 which is close to 1. And the most suitable number of neurons in the 2 layers are 8 and 4 respectively. The results of the predicted settlement are ideal which demonstrates its potential to be recommended as a tunnel prediction monitoring tool for future similar projects.

REFERENCES

Ahangari, K., Moeinossadat, S.R. & Behnia, D. 2015. Estimation of tunnelling-induced settlement by modern intelligent methods. *Soils & Foundations* 55(4):737–748.
Berrar, D., Granzow, M., Kerr, K. F., Bolstad, B. M., Coombes, K. R., Baggerly, K. A., et al. 2007. Fundamentals of data mining in genomics and proteomics. *Fundamentals of Data Mining in Genomics & Proteomics.*
Bo, L.F., Wang, L. & Jiao, L.C. 2006. Feature Scaling for Kernel Fisher Discriminant Analysis Using Leave-one-out Cross Validation. *Neural Computation (NECO)* vol. 18(4), pp. 961–978
Kolay, P.K. 2008. Settlement Prediction of Tropical Soft Soil by Artificial Neural Network (ANN). *Strahlentherapie* 133(1):9–14.
Srivastava, N., Hinto, G., Krizhevsky, A., Sutskever I. & Salakhutdinov, R. 2014. Dropout: A Simple Way to Prevent Neural Networks from Overfitting. *Journal of Machine Learning Research* 15(1): 1929–1958

Tunnels and Underground Cities: Engineering and Innovation meet Archaeology,
Architecture and Art, Volume 12: Urban
Tunnels - Part 2 – Peila, Viggiani & Celestino (Eds)
© 2019 Taylor & Francis Group, London, ISBN 978-0-367-46900-9

TBM launch under the protection of a closed steel sleeve: A case study

X. Li, D. Yuan & D. Jin
School of Civil Engineering, Beijing Jiaotong University, Beijing, China

Y. Zhou
Shandong Provincial Communications Planning and Design Institute, Jinan, China

ABSTRACT: Because of their many merits such as low disturbance to surroundings, high advance rate and safety, pressurized-face tunnel boring machines (TBMs) are widely used to excavate metro tunnels in urban areas throughout the world. TBM launching is a challenging work especially in crowded area where ground reinforcement isn't allowed or only partially fulfilled. To solve the problem, machines were launched under the protection of a closed steel sleeve, and the launches won success in the cities such as Guangzhou and Shenzhen of China. Based on an earth pressure balance TBM launch in Shangmeilin metro station site in Shenzhen, the launch under the protection of a closed steel sleeve is presented and explained. With the subsurface instrumentation, the earth pressure and pore water pressure was monitored over the whole period of TBM launching. The ground response is investigated with the recorded results. The findings can provide valuable references for similar projects.

1   INTRODUCTION

Metro tunnels are being constructed more frequently than ever to address the issue of traffic congestions in large cities throughout the world. As a versatile tunnelling method in urban areas, pressurized-face tunnel boring machines (TBMs) are widely used to excavate metro tunnels. However, launch of the TBMs increasingly has to contend with very limited surface space, existing surface and subsurface utilities and structures, and poor ground conditions. During the initial launch stage, ground improvement is usually required in critical areas to mitigate tunneling risks due to lack of effective support pressure at the cutting face. TBM launching is a challenging and demanding work in the jobsite where ground reinforcement isn't allowed or only partially fulfilled because of the existing utilities and structures. To deal with the difficult situation, a novel TBM launch scheme was created and used in the large cities in South China such as Guangzhou and Shenzhen. In the scheme, the TBM is launched under the protection of a closed steel sleeve. Taking an earth pressure balance TBM launch in Shenzhen of China for example, the novel TBM launch scheme is presented and explained herein.

As shown in Figure 1, Shangmeilin station on Line 9 of Shenzhen Metro is at the east side of the crossroad between Meilin road and Zhongkang road, and many buildings and underground utilities such as pipelines and tunnels are in the immediate vicinity. This station was constructed by cut and cover method, and the shield launch pit is at the station excavation. The soils adjacent to the wall of the launch pit consists of 2.2 m thick backfill (1-1), 2.6 m thick sandy clay (3-2), 5.7 m thick plastic gravelly clay (6-1), 7.5 m thick hard plastic gravelly clay and 6.5 m thick completely decomposed migmitite (11-1) from the ground surface downward, as shown in Figure 2. The ground water is about 1.2 m underneath the ground surface. Ground improvement was only partially fulfilled by using jet grouting due to the so many

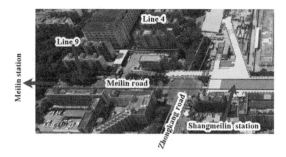

Figure 1. Plane view of Shangmeilin station on Line 9.

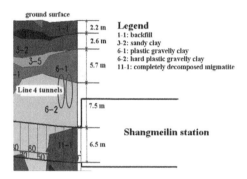

Figure 2. Ground profile adjacent to the launch pit.

buried pipelines and underground structures. Especially, the least horizontal distance is only 16.7 m from the launch pit wall to the Line 4 running tunnel. To guarantee the construction safety and protect the underground utilities and the Line 4 tunnels in operation, the scheme was employed of the shield machine launched under the protection of a closed steel sleeve.

2 THE LAUNCH SCHEME WITH A CLOSED STEEL SLEEVE

2.1 *Working-principle and application of the steel sleeve technology*

The launch scheme with a closed steel sleeve is to launch the shield machine according to the equilibrium principle. In the scheme, the shield machine is assembled in the lower part of the steel sleeve at first, and then the upper part of the steel sleeve is installed. Next, the annular gap between the shield machine and the steel sleeve is filled with solid bulk materials such as sand, and water, resulting in the water and earth pressure in the confined space of the steel sleeve to balance the cutting face pressure. The advance of the shield machine in the steel sleeve is more or less like a driving of the machine in ground. The installed steel sleeve undoubtedly increases the driving length, which augment the space and increase the time to sufficiently backfill the gap between the excavation diameter and the segment outer diameter at the tunnel portal, thus greatly reducing the reliance on enough ground improvement at the tunnel end.

The closed steel sleeve scheme was firstly used to receive a slurry shield machine on Lot 3 of Lines 2 and 8 extension of Guangzhou Metro (Zheng and Ju, 2010). The installed steel sleeve equivalently extended the tunnel length and guaranteed the construction safety. This technology was also employed for the first time to help launch two EPB shield machine in Nanning Rail Transit Line 1 (Cui, 2014).

2.2 *Make-up of the steel sleeve launch system*

As shown in Figure 3, the steel sleeve launch system, consisting of a hollow cylinder that is 11200mm in length and 6500 mm in inner diameter, a pad, a rear end-plate, a reaction frame, a wheel set and etc., is a barrel-shape structure made of A3 steel plate 16 mm in thickness. The cylinder is made up of two parts: the upper part and the lower part, and each part is composed three components, namely Component A, Component B and Component C, as shown in Figure 4. A connecting plate is mounted between the hollow cylinder and the pre-installed steel ring at the tunnel port, as shown in Figure 5, and this plate is welded to the steel ring. As show in Figure 3(a), three feeding gates 600×600 mm are at the upper part of the cylinder; for each component of the cylinder lower part, three grouting pipe with ball valves are designed, as shown in Figure 6, and there are 9 equally-spaced grouting pipes in total. With the grouting pipes, two-component grout can be injected to raise the settled shield machine.

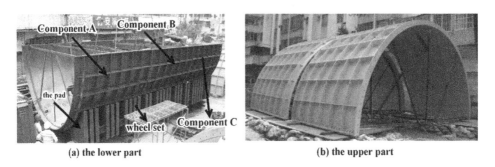

(a) top view

(b) side view

Figure 3. The steel sleeve launch system.

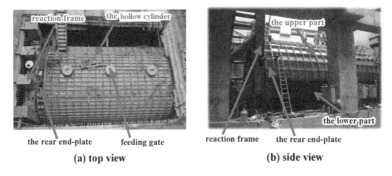

(a) the lower part

(b) the upper part

Figure 4. Components of the hollow cylinder.

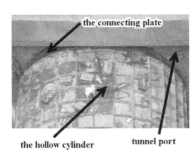

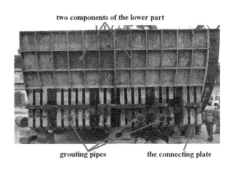

Figure 5. The connecting plate.

Figure 6. The grouting pipes at the bottom.

2.3 Steps of the shield launch with a closed steel sleeve

Construction of the shield launch with a closed steel sleeve can be divided into the following steps:

(1) After inspecting the tunnel portal, the backup system including the gantries are lowered into the launch pit;

(2) A steel plate is installed between the steel ring at the tunnel portal and the steel sleeve;

(3) The lower part of the steel sleeve is installed and the reaction frame is set up;

(4) Rails are laid in the sleeve;

(5) The lower part of the steel sleeve is filled with sand and water for the first time;

(6) The shield machine is assembled above the lower part of the steel sleeve;

(7) The upper part of the steel sleeve is fixed;

(8) The partial segment rings are erected and the cutter head is pushed onto the face;

(9) The steel sleeve is filled with sand and water for the second time;

(10) The gap behind the partial segments is backfilled with grout;

(11) Check airtightness of the launching system until leak test of the steel sleeve under pressure is passed;

(12) Auxiliary measures are taken to protect the buried pipelines if possible and permitted;

(13) Start the shield launch, and the shield machine is pushed into soil.

2.4 Key technologies

As mentioned above, a pressurized environment is created around the shield machine and within the steel sleeve by injecting sand and water in the gap between the shield machine and the steel sleeve. The injection is of paramount importance because of the supporting pressure to the cutting face from the injection. Then how to fill the steel sleeve and create an airtight system remains critical. The filling is usually completed in two stages and grouting behind the partial segments is done to create an airtight system that is examined by hydraulic test.

(1) The first-stage injection

The first-stage injection is undertaken after installing four 30-kg rails on the lower part of the steel sleeve. Sand is filled in-between the four rails and 15 mm higher than the rail surface. As shown in Figure 7, the shield machine is supported by the filled sand, resulting in compaction of the sand and enough friction to prevent the machine from reversing.

(2) The second-stage injection

After the upper part of the steel sleeve is fixed and the cutter head of the machine is pushed onto the excavation face, the second-stage injection is to be performed by means of the three feeding gates at the upper part of the steel sleeve. As shown in Figure 8, the remaining gap between the shield machine and the steel sleeve is filled with sand during the second-stage injection. And meantime water is also injected to make the sand compact. A pipe is laid from

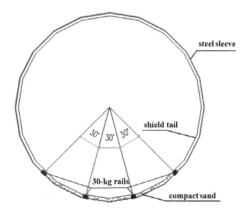

Figure 7.   The first-stage injection.

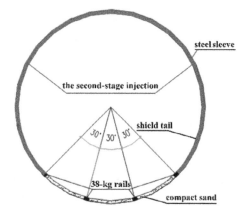

Figure 8.   The second-stage injection.

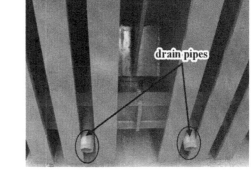

Figure 9.   Injecting sand and water.

Figure 10.   Drain pipes at the bottom.

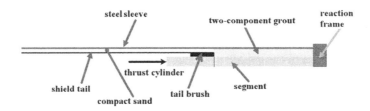

Figure 11.   Grouting behind segment.

the surface to the steel sleeve and a funnel is installed at the top of the pipe. Sand and water are poured into the funnel and conveyed into the steel sleeve along the pipe, as shown in Figure 9. At the bottom of the steel sleeve, drain pipes are designed to discharge water and make sand compact, as shown in Figure 10. To make the filling material uniform and dense, injecting sand and water alternates between the three feeding gates.

(3) Grouting behind the partial segments

To guarantee the sealing between the partial segments and the steel sleeve, grouting behind the segments with a pressure less than 350 kPa is performed through the segment holes of the two ring segments adjacent to the reaction frame, resulting an airtightness ring behind the segments, as shown in Figure 11.

(4) Hydraulic test

Injecting water into the gap between the steel sleeve and the shield machine is continued until the water pressure is reached 3 bar. When maintaining the pressure, leakage detection is carried out for the linking parts, including the connecting plate at the tunnel portal, the horizontal and circumferential linking device of the steel sleeve, the junction of the reaction frame and the steel sleeve, and etc.

The water pressure is stage-controlled, and the duration to detect leakage for each stage pressure varies. Ten minutes or so is necessary for attaining 1.0 bar pressure, and another ten minutes is required to maintain the pressure of 1.0 bar to detect leakages; fifteen minutes or so is necessary from 1.0 bar to 2.0 bar pressure, and the duration to detect leakage for the 2.0 bar pressure is twenty-five minutes; twenty-five minutes or so is necessary to increase the pressure from 2.0 bar to 2.5 bar, and the required duration to detect leakage is forty-five minutes; forty-five minutes is necessary from 2.5 bar to 3.0 bar pressure, and the required duration to detect leakage is one hundred and twenty minutes. Once leakage and sealing off of the soldering seam are found, water pressure must be released in time and counter measures such as tightening bolt and re-welding are taken. Afterwards, the hydraulic test is done again until the water pressure is maintained at 3 bar and no leakage is found and airtightness of the steel sleeve is ensured.

3 MONITORING OF THE GROUND RESPONSE DURING THE SHIELD LAUNCHING

3.1 *Layout of monitoring points and shield launch construction*

In view of importance of the surrounding ground response, subsurface instrumentation was installed to monitor the earth pressure and the pore water pressure, and an information-orientated construction of the shield launch was realized with the monitored results. As shown in Figure 12, four monitoring cross sections were planned before the shield machine that would be launched. In each monitoring cross section, four or three monitoring holes were designed, and a total of 14 monitoring holes was prepared, in which monitoring holes of D1–D7 were for earth pressure and monitoring holes of K1–K7 for pore water pressure. In each monitoring holes, seven monitoring points for earth pressure or pore water pressure were arranged, resulting a total of 98 monitoring points.

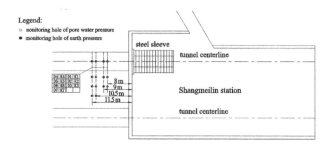

Figure 12.   Planned monitoring cross sections before the shield machine.

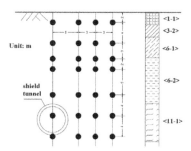

Figure 13.   Layout of monitoring points along the monitoring holes.

Table 1.   The schedule of the construction.

Date	Time	The completed ring number
December 9, 2016		Starting shield launch
December 10, 2016	16:40	-2
	20:35	-1
December 11, 2016	6:19	+1
	9:55	+2
	16:20	+3
December 12, 2016	0:00	+4
	10:47	+7
	15:30	+7
December 13, 2016	23:30	+8
	3:30	+9

3.2 *Shield launch construction*

Shield launch construction commenced on December 9, 2016 and shied machine left the monitoring zone on December 13, 2016. The variations of the earth pressure and the pore water pressure with the shield advance were measured. The schedule of the shield launch construction is listed in Table 1.

4 RESULTS AND DISCUSSIONS

4.1 *The measured earth pressure*

Taking the monitoring holes of *D1* and *D2* for example, the recorded results of the earth pressures over the period of the shield launch at the different monitoring points are presented in Figures 14 and 15. The measured earth pressure at the different depth displayed the varied development. The measured earth pressures at the depth of 11 m, 16 m, 20 m and 24 m gradually decreased at first, reaching the lowest point, and then restored and even exceeded the original level with the shield advance because of action of the backfilling pressure. It is noted the latter parts of the results at the depth of 20 m and 24 m in the monitoring hole of *D1* vanished due to these two monitoring points were excavated by the shield machine. The measured earth pressures at the depth of 2 m, 6 m and 9 m remained almost unchanged at the first half, but experienced a little increase at the second half perhaps on account of the backfilling pressure.

(1) The measured results at the monitoring hole of *D1*

For the two monitoring points at the bottom, the earth pressure variations with the shield advance can divided into two stages by the vertical dashed line of *EA* plotted in Figures 14: Stage I and Stage II. During Stage I, the shield was in the steel sleeve and minor decrease in earth pressure was observed; During Stage II, the shield began excavating the natural soil and a faster decrease in earth pressure resulted. The shield pushed forward, the two monitoring points were excavated, and no results were obtained any longer of the two points.

For the two monitoring points in the middle, the earth pressure variations with the shield advance can divided into five stages by the four vertical dashed lines of *EA, EB, EC* and *ED* plotted in Figures 14: Stage I, Stage II, Stage III, Stage IV and Stage V. During Stage I, the shield drove in the steel sleeve and minor decrease in earth pressure was observed; During Stage II, the shield began excavating the natural soil and a faster decrease in earth pressure with the shield advance resulted due to lack of the enough supporting pressure to the cutting face; During Stage III, the shield crossed from the side of the monitoring points, and the earth pressure fluctuated to rise in general due to the mutual interaction of the soil excavation and the friction between the shield and the surrounding soil; during Stage IV, the shield was leaving the monitoring points and

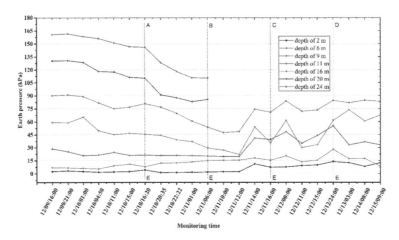

Figure 14.   Recoded earth pressures of the monitoring point at different depth in the hole of D1.

backfilling the annular gap between the excavation diameter and the outer diameter of the segments was performed near the monitoring points, resulting some peak values of the measured earth pressures; during Stage V, the shield machine was a little far from the monitoring points and some decline of the earth pressure existed due to the release of the grouting pressure.

For the three monitoring points at the top, the earth pressure variations with the shield advance can be divided into two stages by the vertical dashed line of *EB* plotted in Figures 15: Stage I and Stage II. During Stage I, the earth pressure remained almost unchanged for the points were far from the shield driving; During Stage II, some increase in earth pressure was observed for the backfilling grouting pressed the surrounding soil.

(2) The measured results at the monitoring hole of *D2*

For the four monitoring points at the bottom, similar development can be found like the variations of the two middle monitoring points at the hole of D1, and the earth pressure development can divided into five stages by the four vertical dashed lines of *EA, EB, EC* and *ED* plotted in Figures 15.

In the same way, the earth pressure variations of the three top monitoring points can divided into two stages by the vertical dashed line of *EB* plotted in Figures 15, as is like the development of the earth pressures at the three top monitoring points of the hole of *D1*.

4.2 The measured pore water pressure

Taking the monitoring holes of *K1* and *K2* for example, the recorded results of the pore water pressures during the shield launch construction are presented in Figures 16 and 17.

The measured pore water pressure at the different depth showed the different development. The measured pore water pressures at the depth of 2 m, 6 m and 9 m remained almost constant at the first half, but experienced a little increase at the second half perhaps because of action of the backfilling pressure. The measured earth pressures at the depth of 11 m, 16 m, 20 m and 24 m gradually decreased at first, reaching the lowest point, and then restored and even exceeded the original level with the shield advance because of action of the grouting pressure. It is noted the latter parts of the results at the depth of 20 m and 24 m of the monitoring hole of K1 vanished due to these two monitoring points were excavated at that time by the shield machine.

(1) The measured results at the monitoring hole of *K1*

For the two monitoring points at the bottom, the pore water pressure variations with the shield advance can divided into two stages by the vertical dashed line of *WA* plotted in Figures 16: Stage I and Stage II. During Stage I, minor influence on pore water pressure was observed because most of the shield was still in the steel sleeve; During Stage II, the observed pore water pressure fluctuated to rise under the mutual interaction of the stress release due to the soil excavation and the friction between the shield and the surrounding soil. With the sustained forward driving of the machine, these two monitoring points were excavated.

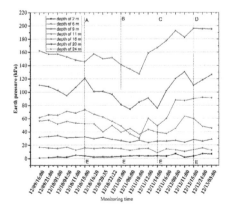

Figure 15.   Recoded earth pressures of the monitoring point at different depth in the hole of D2.

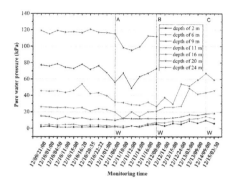

Figure 16. Recoded pore water pressures at different depth of the hole of K1.

For the two monitoring points in the middle, the pore water pressure variations with the shield advance can divided into four stages by the three vertical dashed lines of *WA, WB* and *WC* listed in Figure 16: Stage I, Stage II, Stage III and Stage IV. The pore water pressure development in the first two stages was similar to that of the two bottom monitoring points. During Stage III, the pore water pressure continued to rise under the action of the backfilling grouting pressure; during Stage IV, the backfilling grouting of the shield was leaving the monitoring points and the pore water pressure fell.

For the three monitoring points at the top, the pore water pressure variations with the shield advance can divided into two stages by the vertical dashed line of *WB* plotted in Figures 16: Stage I and Stage II. During Stage I, the observed pore water pressure was almost unaffected by the shield driving; During Stage II, some increase in pore water pressure was due to the backfilling grouting pressing the surrounding soil.

(2) The measured results at the monitoring hole of *K2*

For the four monitoring points at the bottom, similar development can be found like the variations measured at the two middle monitoring points in the hole of *K1*, and the pore water pressure development can divided into four stages by the three vertical dashed lines of *WA, WB* and *WC* listed in Figures 17.

Similarly, the pore water pressure variations of the three top monitoring points can divided into two stages by the vertical dashed line of *WB* listed in Figure 17, as is like the development of the pore water pressures at the three top monitoring points in the monitoring hole of *K1*.

5   SETUP OF THE EARTH CHAMBER PRESSURE

The cutting face pressure is balanced by the earth chamber pressure during soft ground shield tunneling. Therefore, a rapid setup of the earth chamber pressure over the period of shield

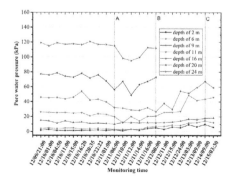

Figure 17.   Recoded pore water pressures at different depths of the hole of K2.

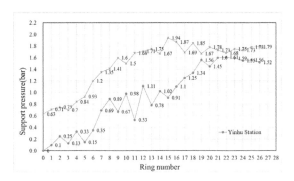

Figure 18.   Recorded earth chamber pressure in Shangmeilin Station as well as in Yinhu Station.

launch is of great importance to guarantee the construction safety. The method of shield launch under the protection of a closed steel sleeve is in favor of setup of the earth chamber pressure, as explained below. Figure 18 presents the earth chamber pressure development during shield launch in this Station as well as in Yinhu Station of Line 9 in Shenzhen. These two stations are similar in ground conditions and tunnel overburden depth. The conventional shield launch method was used in Yinhu Station, and the surrounding ground was reinforced with grouting. As shown in Figure 18, in the conventional method the earth chamber pressure was zero at the beginning, increased slowly with large fluctuations thereafter, and tended to be stable at the shield advance of about 15 segment width. The final stable earth chamber pressure was around 1.5 bar. In this new method, certain initial earth chamber pressure existed, increased rapidly with little fluctuations thereafter, and reached the final stable pressure at the shield advance of about 10 segment width. The final chamber pressure was about 1.7 bar.

6   CONCLUSIONS

The shield launch under the protection of a closed steel sleeve in Shangmeilin metro station site in Shenzhen is presented herein. The used steel sleeve launch system and the launch scheme are elaborated. Based on the construction case, the following conclusions can be drawn:

(1) With the elaborate steel sleeve system, the good preparation work and the well-organized construction, the shield launch can be performed with safety in the complex urban environment where ground improvement isn't allowed or can be only partially fulfilled.

(2) Ground response varies with the sustained shield pushing-forward, as is the result of the mutual interaction in-between the soil excavation induced stress releases, the friction caused by the contact of the shield with the surrounding soil, and the loading caused by the backfilling grouting pressure. The stress release causes the decrease in the observed earth pressures and pore water pressures; the friction and the loading give rise to the increase in the recorded earth pressure and pore water pressure. The observed earth pressure and pore water pressure gradually restore to the original level with the shield away from the monitoring points.

(3) The influence of the shield excavation is limited into a certain range that is about 9 m underneath the ground surface, as concluded from the observed results of the top three monitoring points in the monitoring holes.

REFERENCES

Zheng, S., Ju, S.J. 2014. Technology of steel reception sleeve for slurry shield. Modern Tunnelling Technology 47(6): 5–56.

Cui, Q. 2014. Application of shield equilibrium launch technology – the close launch scheme of steel sleeve. Delta (6): 182-183. Doi: 10.3969/j.issn.1003-9643.2014.06.123.

*Tunnels and Underground Cities: Engineering and Innovation meet Archaeology,
Architecture and Art, Volume 12: Urban
Tunnels - Part 2 – Peila, Viggiani & Celestino (Eds)
© 2019 Taylor & Francis Group, London, ISBN 978-0-367-46900-9*

Red Line tunnels in Israel – engineering highlights

S. Liberman & M. Dvir
Gash Bridge & Buildings Engineering

ABSTRACT: The Light Rail Red Line runs from the city of Bat Yam passing through the cities of Tel Aviv, Ramat Gan, and Bnei Brak reaching finally Petah Tikva to the northeast. In the present paper, some of the engineering challenges are described and analyzed. The main challenges were: (1) The crossing under Ayalon River, cutting heavily reinforced piles serving as retaining wall foundations between the railroad and the river. (2) The TBM crossed through the stations' shafts before they were excavated; a detailed program of excavation and dewatering stages was prepared for this purpose. (3) The alignment crosses under an existing bridge built more than fifty years ago, thereby cutting the foundation piles with the TBM cutter. (4) From the Herzl launching shaft the tunnels are in close proximity and under low overburden, crossing next to more than 100 year old buildings; protection works or other mitigations for buildings in critical or deficient conditions were performed before the TBM crossing.

1   INTRODUCTION

The Red Line has 12 km of at-grade alignment and 12 km of underground tunnels. There are 10 stations along the underground section of the alignment, at a distance of approximately 750 meters from one to the other.

The excavation of the tunnels was performed using 8 Tunnel Boring Machines (TBM) commencing works in 2017.

The Red Line tunnels are the first urban tunnels in soft ground and below the underground water table constructed in Israel and there was no previous experience to rely on. The alignment is located mainly under existing roads, bridges, utilities and close or under existing buildings, in Tel Aviv Metropolitan area.

The Client prepared three TBM launching shafts at strategic points along the alignment that allowed the beginning of the works in 2011 (See Figure 1).

From the hydrological point of view, the Red Line is located mainly in the Mediterranean Coast Aquifer where the underground water level varies from approximately 0.0 m adjacent to the sea to the west, up to approximately +4.0 m at the Em Hamoshavot launching shaft to the east. Most of the tunnels are partially or entirely below the underground water level.

The geological formations consist mainly of sandstone, gravel, sand and lenses of clay. The types of sandstone existing in the Israeli Mediterranean coastal area are known as Kurkar, and they vary from a very soft rock (K1) to a rather hard rock (K4).

The alignment crosses through very populated areas of the Tel Aviv Metropolitan Area, including the cities of Bat Yam, Tel Aviv, Ramat Gan, Bnei Brak and Petah Tikva. The construction bid of the Red line, was divided into two segments, the West Segment starting at the Herzel Launching Shaft and ending at the Ben Gurion Station (See Figure 1) and the East Segment from Ben Gurion Station to the depot. Each segment was constructed by different Contractors.

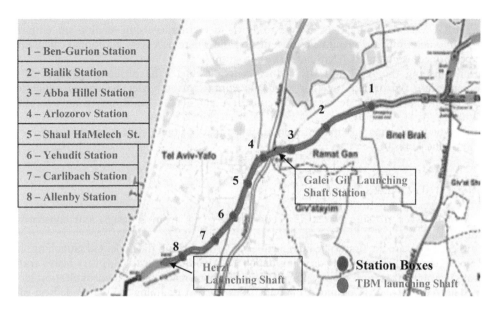

Figure 1. Red line West Segment alignment map.

The present paper will refer to the Western Segment that begins crossing the Tel Aviv Metropolitan area, Ramat Gan and reaching Bnei Brak. The contract was granted to a joint venture between Shikun Binui from Israel, and CRTG from China. Colaborating with the design and as design checker was Eng. Nicola Della Valle from TunnelConsult.

Building Survey Conditions (BSC) and Building Risk Assessment (BRA) reports were prepared in order to advise the Contractor on monitoring programs and risk mitigation measures for each building within the zone of influence (ZOI) of the planned route. More than three hundred buildings and about five thousand apartments were surveyed and documented in the Western Segment.

The BSC was performed in two stages; first external surveys were prepared for each Building including available existing plans, review of the structure and nonstructural elements conditions, pictures, determination of the vulnerability index, etc. Afterwards, and according to Client's request, most of the apartments (more than 80%) were surveyed and documented. The completion of these works, by at least three survey teams, took more than one year.

Figure 2. Ayalon Crossing.

The BRA was determined according to Burland's criteria, using at first an empirical approach to determine the green field movements, and where necessary additional calculations were performed using the PLAXIS numerical program.

After finishing the BSC and BRA documents, visits to the different sections of the alignment were performed together with the contractor in order to decide which risk mitigation measures were necessary to reduce the impact on buildings to an acceptable minimum.

Automatic monitoring instruments were installed in all of the buildings in addition to other mitigation measures.

2 CROSSING UNDER AYALON RIVER

The Ayalon River crossing was a very complicated task from a design and construction standpoints.

TBM 5 and 6 were the first to be launched in the project, from the Galei Gil launching shaft in Ramat Gan, and immediately crossing under the Ayalon Highway with very heavy traffic, railways, sewage pipes, and the Ayalon River in order to reach the Arlozorov Station in Tel Aviv. The Ayalon river is bounded on both sides by two retaining walls founded on piles, which were constructed at different stages, and strengthened later to allow the placing a railroad immediately behind the walls.

The tunnel crown is located about 8 meters under the river bed and about 12 meters under the railway and the highway.

2.1 *Cutting Piles Methodology*

One of the main design and technical issues was to cut the piles under the retaining walls without major impact to the railways, the highway, and the walls. Many possible solutions were considered including grouting, jet grouting, and shaft construction to engage the piles from the river and others.

Ultimately, a very challenging approach was adopted; the piles were cut by the TBM during the crossing. At first, it was considered that a small underwater chamber will be implemented from the TBM chamber using air pressure, bentonite and/or grout to allow reaching the piles and cutting them, but this solution was considered too risky. Afterwards, the contractor suggested the possibility of constructing a special, heavy duty cutterhead capable of demolishing the piles.

The question was how the cutter would deal with the heavily reinforced piles and cross through them without causing major damage to the machine, the structures or the surroundings. To demonstrate the feasibility of the proposed solution two decisions were taken: first to make a simulation test at the TBM factory showing that the cutter would be able to cut the reinforced piles and secondly to improve the ground conditions by injecting jet grout between the piles in order to avoid any possibility that the piles would move during cutting. The grouted block would prevent possible pile movement during cutting and also serve to strengthen the existing foundation after cutting the piles, thus protecting the tunnel. (See Figure 4)

The simulation test took place in China and the proposed cutter was used to cut reinforced concrete as shown in Figure 3. The simulation test allowed the following conclusions to be reached:

- Using either disc cutters or drag bits it is possible to cut reinforced concrete piles.
- Discs are very efficient in cutting but the length of the cut rebars is not uniform. The length of rebars after cutting with drag bits is more uniform

Figure 3. Reinforced concrete, cutting simulation equipment.

It was recommended to operate at an excavation speed of 3 to 5 mm/min and a rotation speed of 1 to1.3 radian/min. The concept is that low excavation speed combined with a high rotation velocity produces fewer disturbances and a smaller length of cut reinforcement.

Another recommendation was to add more quantities of foam and bentonite slurry in order to reduce the cutter head temperature and increase tool's lubrication.

In order to enter the river bed and perform jet grout operations, a small dam was constructed between the site and the sea in order to work in dry conditions.

It was necessary to inject about ten thousand cubic meters of grout in a rather short time. Two sets of grouting equipment were used in parallel in order to meet the deadline of finishing the works before the start of the rainy season and water flow in the river. It was impossible to add more equipment because of the space limitation of the site. (See Figure 4)

There were several issues that arose during the jet grouting process due to the above mentioned reasons and the very limited working site, especially regarding the backflow control of the grout. The backflow entered through an existing drainage layer under the river concrete floor, producing the uplift of the floor which was necessary to repair.

Monitoring instruments on the retaining walls and the railway tracks showed very small movements of no more than three millimeters.

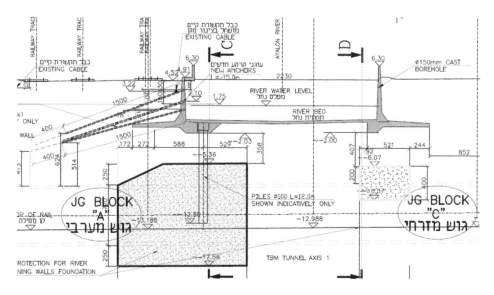

Figure 4. Piles, Ground Anchors and Jet Grouting under the retaining walls.

Drilling to extract samples from the jet grout was performed. The samples UCS (Unconfined Compression Stress) had a mean value of 25 MPa more than required in the specifications.

The Jet Grout work was completed a few days before the site was flooded due to continuous rains.

The TBM crossing and cutting of the piles were performed without any problems. The methodology worked as planned, producing minor settlements of the retaining walls and without affecting their stability.

2.2 *Authorities Authorizations*

This was the first TBM works performed in an urban area in Israel, so there was not any previous experience on this subject. The authorities were very reluctant to provide the required approvals, needed time to study the submitted documents, and required additional explanations and reports, thus producing delays at the beginning of the works.

The Highways Authority (Netivei Ayalon) considered the possibility of stopping vehicular traffic before the crossings, but ultimately the highway remain opened.

The Railroad stopped train transit in this section before the first TBM crossing and only during the second TBM crossing the railroad was opened.

The sewer pipes of the Shafdan were closed during the crossings and a large number of monitoring instruments were installed.

The crossings were performed without any damages or problems of any type. The measured subsidence was only a few millimeters on any point of the alignments.

3 TUNNEL ANTECEDENT

The original design was to construct and excavate the stations' external box walls at a first stage and afterwards to cross with the TBM through the stations' walls. The advantages of this method are that the stations are constructed in parallel with the tunnelling and enable possible checking and maintenance of the TBM at the stations during the crossing. In most of the tunnel projects, this is the preferred and adopted methodology.

In the present project, the stations' construction was very time consuming due to utility relocations and traffic arrangements. If the TBM had to wait until the stations' excavations were completed, more than a year would be added to the timetable.

For this reason, the tunnel antecedent methodology was proposed by the Contractor and approved by the Client. In this case the TBM crosses through the stations after all the external shaft diaphragm walls were cast, or at least partially cast. The excavation of the stations was performed after the TBM crossing and the tunnel segments inside the stations were demolished. (See Figure 5)

3.1 *Stations verification*

When the tunnel antecedent methodology was adopted, parts of the stations were already under construction. In this situation, it was not possible to introduce any modifications, and it was necessary to verify that the existing diaphragm walls had sufficient reinforcement to stand the new construction stages.

This was verified using Plaxis 2D software. The new design bending moments' envelope and shear forces that were calculated considering the different construction stages and loadings were compared with the capacity bending moments and shear forces of the existing reinforcement. Small changes in the positions of struts and the construction stages were necessary to reach an adequate safety factor in all of the stations.

3.2 Uplift forces

In order to ensure the stability of the tunnels during the excavation it was necessary to verify that the groundwater uplift forces were less than the counterweight to avoid buoyancy.

The stages of excavation, strut construction and dewatering were redesigned in order to obtain a sufficient safety factor. The groundwater level was maintained high enough to avoid excessive pumping and low enough to avoid buoyancy at the different stages of construction.

3.3 Segments stability

The inner diameter of the tunnel is 6.5 meters and each ring is composed of six segments. The rings have good stability due to compression forces introduced by the surrounding soil pressure. When the compression forces are large enough, the eccentricity due to moments in the ring is small, and all the connections between the segments are in compression.

During the excavation, the compression forces are reduced due to smaller overburden and the bending moments at the segment connections produce greater eccentricity. As a result, at a certain level of the excavation, the connections behave similarly to hinges reducing the ring rigidity. This problem should be checked and properly calculated to avoid an early collapse of the tunnel.

The following protection measures were taken to reduce the risks:

- After the excavations reach two meters above the crown no entrance to the tunnels underneath was permitted, and no traffic of heavy equipment was allowed.
- When the excavation reached one meter and twenty centimeters above the crown no equipment above the tunnel was allowed.

Figure 5. Demolishing one of the tunnels at Shaul Hamelech Station.

- The excavation shall be performed in parallel on all the sides of the tunnel.
- No asymmetric excavation was allowed.
- The groundwater level inside the stations was maintained under the bottom level of the tunnel.

4 SHEFA TAL BRIDGE CROSSING

The Shefa Tal Bridge is located at the limit between the cities of Tel Aviv and Ramat Gan, and very close to the Galei Gil launching shaft. The road above the bridge has a very large traffic volume serving as a main connection between the two cities. The bridge was constructed in two stages; in 1930 an arch structure was built and in 1960 the bridge was widened by adding two lateral bridges at the southern and northern sides.

The superstructure of these bridges has continuous pre-stressed girders, and the foundation consists of driven piles. According to the design drawings from 1960, the length of the piles was 7 meters, but before the crossing, the length of the piles was measured using sonic tests with the result that the piles are 12 meters in length instead of the designed 7 meters.

This increase in the length has serious consequences for the tunnel crossing under the bridges because if the piles were 7 meters the TBM would cross under them, but because the piles are 12 meters long, the TBM crosses through the piles.

Many utilities were found between the foundations of the bridge including electricity, drainage pipes, communication cables, sewage pipes and others complicating the possible treatment of the foundations.

Only one of the two parallel tunnels, the southern one, crossed under the bridge; the northern one crossed very close to the bridge but not cutting any pile.

The selected construction methodology consisted of the following:

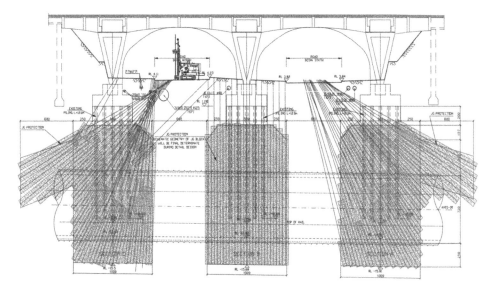

Figure 6.   Shefa Tal Bridge including jet grouting design.

4.1 Jet Grouting

Soil treatment with jet grouting was performed under the bridge having two purposes. The first one was to treat the soil under the pile caps in order to strengthen the foundation, allowing the cutting of the piles without causing major settlements or weakening of the foundation support. The second purpose was to allow the TBM to cut the piles with a minimum impact on the bridge.

A detailed design was prepared for the jet grouting including bridge monitoring, tests, and traffic arrangements.

4.2 Settlements compensation

Settlements calculations showed that ground treatment combined with limited volume loss will reduce the differential settlements in the longitudinal and transversal directions of the bridge to a maximum of 8 mm. The expected additional tensile stresses in the bridge beams, due to these settlements, will be 0.94 MPa.

Extensive continuous monitoring was performed during the crossing.

Superstructure differential settlements of more than 8mm were to be compensated using hydraulic synchronized jacks during Jet Grout works and during tunnelling works.

4.3 Design vs. Reality

According to the design, the first TBM crossing near the bridge was to be the southern one that crosses under the bridge. But due to a delay of the jet grout works the Contractor decided to cross first with the northern one, before completing the ground treatment under the eastern pier.

During the crossing of the northern TBM, due to lack of adequate protection, the eastern pier settled about 6mm at the northern column, close to the tunnel and a negligible settlement occurred at the southern column. The settlement under the northern column was very close to the threshold of 8mm as explained above, and it was expected that additional settlements would happen during the second TBM crossing. The bridge was thoroughly checked and no additional damage was found at this stage.

After finishing all the ground treatments, the southern TBM crossed under the bridge cutting the foundation piles, with only very small settlements of about 3 mm. This proved that the jet grouting solution provided good protection, even better than predicted.

After the TBM crossing, it was required to compensate settlements at the eastern pier and to replace the neoprene bearings that were already in poor condition.

5 TUNNELS IN CLOSE PROXIMITY

The western underground section of the Red Line originates in southern Tel Aviv. This is one of the city's oldest and more crowded areas, with narrow streets and buildings about one hundred years old. Only a small part of them have been refurbished and/or strengthened over the years.

The TBM works started at the Herzl launching shaft, which is very narrow and surrounded by old and partially unstable buildings. During the construction of the shaft by the client at a previous stage, the residents of one of the buildings were temporarily evacuated from their homes.

As noted earlier all the buildings along the alignment were inspected (BCS), a risk assessment report (BRA) was prepared, and in this area, special precautions including monitoring, mitigation works, and emergency programs were implemented.

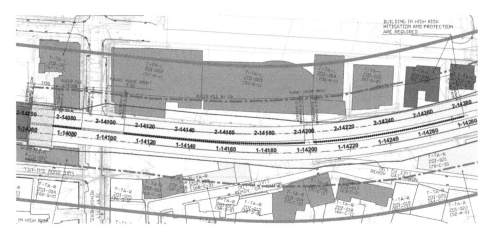

Figure 7. Section with tunnels in close proximity.

5.1 Close proximity

Tunnels in close proximity, means that the distance between two parallel tunnels is very small in relation to their diameter. Normally a minimum distance equal to twice the tunnel diameter center to center is kept and preferable two and a half diameters, depending on the type of soil or rock.

The problem with tunnels being in close proximity is that after the crossing of the first TBM, the second one affects the stability of the first tunnel, since the excavation is executed at a short distance; this problem should be properly addressed or avoided.

At the Herzl launching shaft, the net distance between the tunnels is only 0.8 meters, and the close proximity continues for 220 meters between chainage 14+060 to 14+280. In this alignment section, the minimum overburden is approximately 6.5 meters (at Herzl launching shaft) and the maximum 10.75 meters.

Due to the close proximity of the tunnels, special mitigation measures were implemented including construction of a wall of concrete piles of 0.8 meter diameter and 1.0 m spacing between the tunnels. In some places this solution of pile walls was not adequate due to existing utilities that were not possible to relocate. Instead of pile walls, it was necessary to inject grout from the first tunnel, before the crossing of the second one. The permeation grouting was

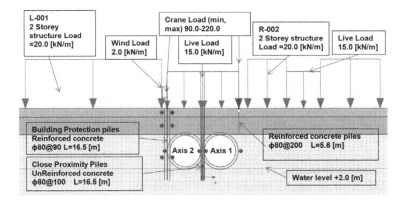

Figure 8. Calculation of a cross section of tunnels in close proximity using PLAXIS 2D software.

performed using small dimension equipment in order to allow the continuation of the tunnel works.

5.2 *Building protection*

As stated above, many of the buildings in this section were in poor condition. As an example, near the Herzl shaft, a two story building was strengthened by the owners with steel struts in order to provide stability, without any connection to the project. The ground floor of the building contains commercial space, and the first floor residences. Considering the critical situation of the building it was decided to protect it by a wall of piles to reduce settlements produced by the TBM. The crossing was scheduled for a weekend when the stores are closed, and the residents were temporarily evacuated.

Risk mitigation measures for buildings in this area included among others: pile walls between the tunnel and the buildings, continuous monitoring with several control parameters, continuous inspection of the buildings by a certified engineer, evacuation of residents, emergency programs, etc.

5.3 *Detailed calculations*

Detailed calculations of the settlements at each section were performed, using PLAXIS 2D software. The model included the different layers of the soil, the protection piles and loads due to building foundations and traffic.

The results of these calculations were used to determine tensile strain induced in each building and the expected damages according to Burland classification.

6 CONCLUSIONS

Tunnelling in urban areas has a variety of challenges and problems that shall be properly treated in all stages of the design and the construction. It is necessary to take special care of existing structures as buildings, bridges, tunnels, utilities, etc. Other problems could be receiving the authorities' approvals and dealing with complicated approach from the surface or even not having such an approach.

In the present paper there is a description of some difficulties and problems during the design and construction of the tunnels for the light-rail Red Line in Tel Aviv. We hope that the way they were treated and finally successfully solved will serve as an example that maybe used in other urban tunnels in the future.

REFERENCES

AASHTO, 2010. *Technical Manual for Design and Construction of Road Tunnels, CIVIL ELEMENTS.*
British Tunnelling Society and the Institution of Civil Engineers, 2004. *Tunnel lining design guide.*
British Tunnelling Society and the Institution of Civil Engineers, 2010. *Specifications for tunnelling.*
British Tunnelling Society, 2016. *Tunnel design, Design of concrete segmental tunnel linings, Code of practice.*
DAUB German Tunnelling Committee (ITA-AITES), 2016. *Recommendations for Face Support Pressure Calculations for Shield Tunnelling in Soft Ground*
DAUB German Tunnelling Committee (ITA-AITES), 2013. *Recommendations for the design, productions and installation of segmental ring.*
FHWA, 2009. *Technical Manual for Design and Construction of Road Tunnels, CIVIL ELEMENTS.*
Guglielmetti, Vittorio et al. 2007. *Mechanized Tunnelling in Urban Areas.*
Gruebl, Fritz, ITA, 2012. *Segmental Ring Design.*
ITA, 2000. *Guidelines for the Design of Shield Tunnel Lining.*
ITAtech, 2015. *ITAtech Guidelines on Monitoring Frequencies in Urban Tunnelling.*

Maidl, Bernhard et al. 2013. *Handbook of Tunnel Engineering, Structures and Methods.*
Netivei Israel (Israel Public Roads Department), 2018. *Specifications for the Design of Road Tunnels.*
SIA 197/2-2004. *Design of Tunnels, Basic Principles.*
SIA 197/2-2004. *Design of Tunnels, Road Tunnels.*

Tunnels and Underground Cities: Engineering and Innovation meet Archaeology,
Architecture and Art, Volume 12: Urban
Tunnels - Part 2 – Peila, Viggiani & Celestino (Eds)
© 2020 Taylor & Francis Group, London, ISBN 978-0-367-46900-9

Embedded barriers as a mitigation measure for tunnelling induced settlements: A field trial for the Line C in Rome

N. Losacco
Università degli Studi di Roma 'Tor Vergata', Rome, Italy

E. Romani
Metro C S.c.p.A., Rome, Italy

G.M.B. Viggiani
University of Cambridge, Cambridge, UK

G. Di Mucci
formerly Metro C S.c.p.A., Rome, Italy

ABSTRACT: Line C of Rome underground railway is currently under construction. Stretch T3 of the line will cross the archaeological area of the historic centre, with significant interferences with existing monuments of utmost historic value; hence, protective measures will be adopted in order to prevent damage on the most sensitive structures. For Line C, in the framework of the European cooperative project NeTTUN, an embedded wall of bored piles was installed in an instrumented area without pre-existing buildings, in order to assess the effectiveness of such a measure in reducing the displacements induced by the excavation. In this study, a class A finite element prediction of the performance of the barrier is discussed and compared with preliminary data gathered during the passage of the TBMs.

1 INTRODUCTION

Line C is the third underground railway line of Rome. Once completed, it will cross the city from Northwest to Southeast, for a total length of 25.6 km and 30 stations, almost doubling the extent of the currently existing underground network. It is also the first fully automated underground line in Rome. The 30 trains provided by Metro C are driverless with automated platform doors, which increase station safety and improve service quality.

The activities started in 2006 with the archaeological surveys and the final design. At the moment there are 21 stations and 18 km of line in operation. The first stretch, between Monte Compatri/Pantano and Parco di Centocelle - 15 stations and 12.5 km of line - was opened to the public in November 2014; the second stretch, from Parco di Centocelle to Lodi - 6 stations and 5.4 km of line - was opened in June 2015. San Giovanni Station, which connects the new Line C to the existing Line A, was opened in May 2018. Two more stations, Amba Aradam and Fori Imperiali, are currently under construction in T3 stretch.

The tunnels of T3 stretch are being excavated by two EPB TBMs with a cutterhead diameter of 6.71 m, at a depth between 25 m to 60 m below the ground surface. The T3 stretch is characterised by the presence of historic buildings and monuments of great value such as the Colosseum, the Aurelian Walls and Basilica of Maxentius. For this reason, it was required that the general contractor, Metro C SCpA, set up a multidisciplinary Steering Technical Committee (STC), in charge of evaluating the effects of the construction of the line and implementing all necessary procedures to safeguard the monumental heritage.

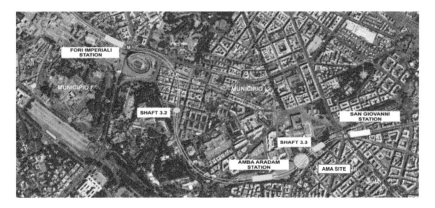

Figure 1. Aerial view of T3 stretch of Line C with indication of the AMA site.

A fully instrumented greenfield control section was established by the STC at the so-called AMA site, in ground conditions representative of those encountered on stretch T3. Figure 1 is an aerial view of the stretch indicating the position of the AMA site, at the very beginning of the contract. In addition to the greenfield control section, in the context of a large cooperative EU funded project (NeTTUN), an embedded wall of bored piles (Harris, 2001; Di Mariano *et al.*, 2007; Bilotta & Russo, 2011) running parallel to the tunnel axis and an instrumented control section were installed at the AMA site to study the effectiveness of barriers as settlement mitigation measures.

This paper presents Class A predictions (Lambe, 1973; Negro, 1998; Boone, 2006) of the settlements induced by the excavation of the tunnels and of the effect of the embedded barrier on those settlements, obtained through finite element analyses. The simulations were carried out using a technique recently proposed by Kavvadas *et al.* (2017), as a part of the NeTTUN project, modelling in detail the most relevant physical processes taking place around the shield. Such detailed simulation approaches have been adopted in the past (*e.g.* Kasper & Meschke, 2004; Litsas *et al.*, 2018) although they have been applied to idealised conditions only, such as *e.g.* uniform soil layers with assumed mechanical behaviour and properties, and, therefore, their ability to reproduce quantitatively rather than qualitatively the actual performance has not been properly assessed. On the contrary, in this paper the predictions are compared with preliminary data acquired in the field, thus allowing a proper validation for the proposed technique.

2 DESCRIPTION OF THE PROBLEM

2.1 *Site*

A longitudinal geological section of the whole T3 contract is shown in Figure 2. Based on the boreholes cored in five campaigns of geotechnical investigation and more recently for the installation of the instruments in the two monitoring sections at the AMA site, a simplified geotechnical model with horizontal layers and ground surface was assumed as follows:

1. R, coarse grained made ground, from ground surface to 17 m depth;
2. LSO, alluvial silty clay and sandy clay, from 17 m to 30 m depth;
3. SG, sands and gravels, from 30 m to 42 m depth;
4. Apl, very stiff clay, from 42 m to indefinite depth.

It is worth noting that, due to their similar mechanical properties, the LSO Holocene layer and the Ar Pleistocene clay layer shown in Figure 2 were considered as a single layer (LSO in the previous list). Figure 3 summarises the main physical properties of the soil layers at the site.

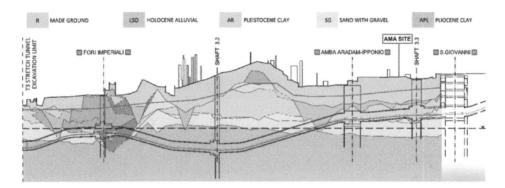

Figure 2. Geological profile of T3 stretch.

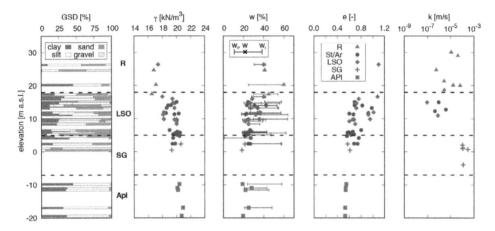

Figure 3. Main physical properties at AMA site.

As shown in Figure 1, the AMA site is located at the beginning of contract T3, about 300 m east of Amba Aradam/Ipponio station. At the site, the tunnel axes run approximately 14.5 m apart at a depth z_0 of about 25 m below ground level. The North Tunnel is excavated first, from multifunctional Shaft 3.3 towards Amba Aradam/Ipponio Station; the excavation of the South Tunnel starts about 60 days after the North Tunnel.

Figure 4a shows a plan view of the site and the position of the instruments installed in the monitoring section, along the transverse centreline of the experimental barrier. The latter consists of 48 equally spaced cast-in-situ bored piles with a diameter $d = 0.6$ m, a spacing $s_p = 1.5\ d$. The total length of the barrier is $L_{barr} = 43.2$ m, including the ends of the capping beam. Figure 4b shows the assumed cross-section along the centreline of the barrier, including the location of the monitoring instrumentation, consisting of 3 Trivecs and one inclinometer embedded in a pile. Surface levelling points are also installed along the monitoring section. The piezometers installed in the area permitted to recognise an existing downwards seepage flow in the LSO layer, due to the difference in hydraulic head between the R and SG layers, as also shown in Figure 4b.

2.2 Numerical model

The analyses were carried out using the finite element software Abaqus 6.14, adopting the technique described in Kavvadas *et al.* (2017) to simulate the tunnelling process. The proposed technique accounts for the main features of the mechanised tunnelling process, namely:

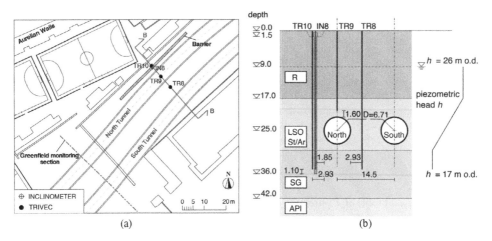

Figure 4. (a) Plan view of the AMA site and (b) monitoring section for the barrier.

application of a face support pressure, geometry of the shield, including overcut at the cutter-head and tapering of the shield, pressurized grouting of the tail void, progressive hardening of the backfill grout. This method differs substantially from the commonly employed approach in which a displacement field is applied at the excavation boundary, calibrated in order to obtain an expected volume loss and a realistic shape of the surface settlement trough in green-field conditions (Rampello *et al.*, 2012; Losacco *et al.*, 2014; Boldini *et al.*, 2018).

In particular, the two EPB shields are 11.8 m long and the diameter of their cutting wheels (*i.e.* the nominal diameter of the excavation) is $D = 6.71$ m; they consist of three cylindrical segments whose diameters decrease towards the shield tail, *i.e.* 6.69 m, 6.68 m and 6.67 m. For sake of simplicity, the shields in the numerical model are slightly shorter ($L_{shield} = 11.2$ m) than the real ones, so that their length is a multiple of the length of one lining ring (*i.e.* $L_{ring} = 1.4$ m); in addition, a linear tapering of the shield was assumed, from maximum diameter 6.69 m behind the tunnel face, to 6.67 m at the tail.

Distinct FE meshes are employed for the shields and for the soil. A contact law is enforced to simulate the interaction between the shield and the soil around the excavation boundary: a bi-linear pressure-overclosure relation is adopted for contact in the normal direction, while frictionless contact is assumed in the tangential direction. The contact constraints are enforced using a penalty method that minimises overclosure and improves the convergence of the FE solver. A concentrated force $W = 3.46$ MN, representing the weight of the shield plus the enclosed machinery and the muck in the excavation chamber, is applied to the shields.

At each excavation stage the shield is advanced a distance $L_{exc} = 2.8$ m, *i.e.* the length of two lining rings, then elements representing two 30 cm thick precast concrete lining rings are activated right behind the tail, over the same length. Pressurised backfill injections are per-formed during the excavation using a two-components grout. In the numerical model, this is simulated through the application of a uniformly distributed pressure p_{grout} over the length L_{exc} immediately behind the shield tail, acting both on the excavation boundary and on the extrados of the freshly activated lining ring, assuming that right after the injection the grout is in a liquid state. Starting from a distance L_{exc} behind the shield tail, hardening of the grout is simulated by removing the radial pressure and activating a layer of solid elements with initial isotropic stress equal to p_{grout} and with Young's modulus E_{grout} increasing with time. A value $p_{grout} = 250$ kPa was used, that is the average between the design value of 400 kPa targeted by the general contractor and the minimum alarm threshold of 100 kPa. Some preliminary ana-lyses, not reported in this paper, were undertaken to assess the effect of a variation of p_{grout} within this range.

Consistently with the indications of the general contractor, a support pressure equal to the *in situ* horizontal total stress is applied at the excavation face at all stages, assuming optimal operation of the EPB shield so that little or no ground loss is induced at the tunnel face. Given the design tunnelling rate of 10 m/day in the examined area, carried out on a 6 days/week basis, a corresponding average advancement rate was assumed such that each L_{exc} long excavation step was carried out in 8 hours.

A distinct mesh was also adopted for the barrier; a no-penetration/frictional contact interaction is enforced between the barrier and the surrounding soil, with friction equal to the angle of shearing resistance of the soil. The employed FE mesh is depicted in Figure 5. The bottom boundary of the mesh corresponds to the roof of the stiff clay deposit, 42 m below the ground surface. The finite element mesh extends 154.5 m in the x direction, transversal to the tunnel axis, and 170.0 m in the longitudinal y direction; the maximum excavation length from the initial boundary is 120.4 m and the monitoring section is located at approximately half way along this length.

The analyses were carried out in terms of effective stress, accounting for the hydro-mechanical coupling of the soil response, and comprise the five following phases: (i) geostatic equilibrium, (ii) barrier activation, (iii) North Tunnel excavation, (iv) time gap between the two excavations, and (v) South Tunnel excavation.

Two different hypoplastic constitutive models, able to reproduce the main relevant features of soil response, were used for the LSO and R layers: a model for clays (Mašín, 2005) and a model for granular materials (von Wolffersdorff, 1996), respectively. The material constants of the former were calibrated on the extensive set of data obtained from *in situ* and laboratory tests on LSO and Ar samples. Model parameters for the coarse-grained R layer were calibrated from the results of *in situ* tests and values inferred from the literature (Herle & Gudehus, 1999), as undisturbed sampling was not possible. Both models use the Inter Granular Strain concept (Niemunis & Herle, 1997) to improve predictions in the small strain range. The SG layer was modelled as linear elastic-perfectly plastic, with a constant elastic shear modulus equal to one third of the mean value of the data from cross-hole tests, to account for the expected shear strain level, while the angle of friction was derived from the SPT tests. A linear elastic model was used for the lining and for the barrier, while an elastic model with time-dependent Young's modulus and Poisson ratio was adopted for the backfill grout. Values of model constants for all the materials involved in the simulations, together with assumed physical properties, are listed in Tables 1–5. A description of the meaning of material constants for the hypoplastic models is beyond the scope of this paper; the reader is referred to the works cited above for clarification.

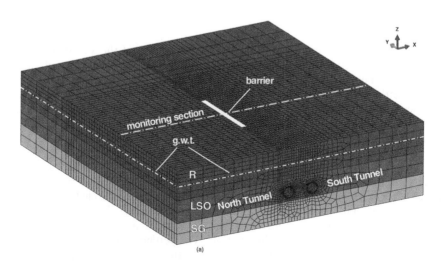

Figure 5.   Finite Element mesh for analyses with the barrier.

Table 1.   R layer, material properties of hypoplastic model for granular materials.

h_s [GPa]	n	e_{i0}	e_{c0}	e_{d0}	φ'_c [°]	a	β	m_R	m_T	R	βr	χ	γ [kN/m³]	k [m/s]	K_0
3.4	0.24	1.1	0.9	0.525	34	0.19	1.5	13.5	2.0	1.E-4	0.5	6.0	17.0	1.E-5	0.441

Table 2.   LSO layer, material properties of hypoplastic model for clays.

φ'_c [°]	N	λ^*	κ^*	ν	A_g	n_g	m_{RAT}	R	βr	χ	γ [kN/m³]	k [m/s]	K_0
33	0.942	0.075	0.012	0.2	18384.0	0.427	0.5	1.E-4	0.4	1.3	19.5	1.E-6	0.525

Table 3.   SG layer, materal properties of linear elastic-perfectly plastic law.

φ'_c [°]	c' [kPa]	E' [MPa]	ν'	γ [kN/m³]	k [m/s]	K_0
45	0.0	316	0.2	20.0	1E-4	0.293

Table 4.   Tunnel lining and barrier, material properties of linear elastic law.

E_{lining} [GPa]	$E_{barrier}$ [GPa]	ν'	γ [kN/m³]
31.0	20.0	0.2	25.0

Table 5.   Tail grout, material properties of elastic law with time-dependent moduli.

E(1 day) [GPa]	E(28 days) [GPa]	ν (t0 < t < 5 hours)	ν (t > 10 hours)	γ [kN/m³]
0.5	1.0	0.5	2.0	21.0

3   RESULTS AND COMMENTS

Class A predictions of the settlements induced by the passage of the TBMs were performed both in greenfield conditions and in the vicinity of the barrier. In this paper, only the excavation of the North Tunnel, closer to the barrier, is considered. The beneficial role of the wall in reducing the settlements is highlighted in Figure 6, comparing the transverse troughs calculated after the excavation at three depths, $z = 1.5$ m (i.e. the head of the piles), 9.0 m and 17.0 m, with those obtained without the barrier. The mitigating effect seems to fade with distance from the barrier while it does not change significantly with depth. Remarkably, also settlements on the other side of the barrier are reduced significantly and an asymmetry of the settlement trough is induced, with the abscissa of maximum settlement slightly shifted away from the piles. The computed response can be ascribed to the mechanism of load transfer exerted by the barrier: due to its roughness and large axial stiffness, the embedded pile wall restrains soil settlements on both sides by transferring shear stresses along the shaft to the tip of the piles, embedded in the stiff base gravelly layer. Figures 7a and 7b show the contours of computed settlements in the monitoring section. The settlements are reduced substantially behind the barrier, while their reduction in front of the barrier is less significant and only occurs above the tunnel crown.

The local efficiency η of the barrier at a given abscissa is commonly expressed as the relative reduction of vertical displacements with respect to those obtained in greenfield conditions w_{gf}, *i.e.*:

$$\eta = \frac{w_{gf} - w_{barr}}{w_{gf}} \tag{1}$$

Figure 8 shows the profile of efficiency with distance behind the barrier, for three depths: 1.5 m, 9.0 m and 17.0 m. The computed efficiency, always smaller than 50%, is maximum immediately behind the barrier and decreases almost linearly with distance up to 40 m from

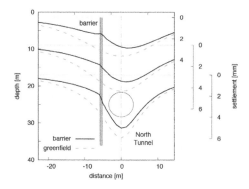

Figure 6.   Computed settlement troughs at various depths with and without barrier.

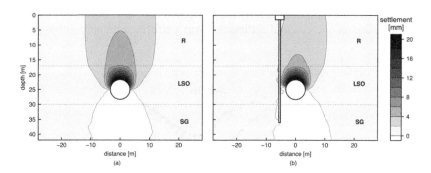

Figure 7.   Contours of settlement in monitoring section: greenfield conditions (a) and with barrier (b).

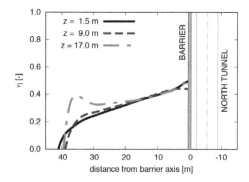

Figure 8.   Profiles of local efficiency η for various depths behind the barrier.

the barrier, where it plummets to zero. No significant variation of efficiency with depth is obtained from the analysis. The results are in fair agreement with those obtained by Fantera *et al.* (2016) for a similar problem.

Preliminary data, gathered after the passage of the North Tunnel TBM through the AMA site, have been processed and are compared with the numerical results to validate our class A predictions, assessing the adequateness of the detailed simulation technique employed in this study. Hence, Figure 9 compares the distributions of settlements at the ground surface obtained numerically with those measured at the site, both in greenfield conditions and with the barrier. Predicted greenfield settlements match fairly the field measurements, especially in the central part of the settlement trough, both in terms of maximum settlement ($w_{max,exp}$ = 3.82 mm, $w_{max,num}$ = 3.47 mm) and volume loss ($V_{L,exp}$ = 0.25%, $V_{L,num}$ = 0.28%). The experimental curve, though, is quite narrower than the numerical one: fitting of the two distributions with a Gaussian function (Peck, 1969) for $w > 0.36\ w_{max}$ yields a trough width parameter $K = 0.37$ for the field, while $K = 0.45$ is obtained from the analysis.

The results of the analysis with the barrier agree qualitatively with the monitoring data. Behind the barrier, the settlements measured in the field are smaller than those predicted, and hence measured efficiencies are larger than predicted ($\eta_{max,exp}$ = 0.65, $\eta_{max,num}$ = 0.50). On the tunnel side of the barrier, instead, the settlements measured in the field are significantly larger than the predictions. Starting from a distance of 0.65 D from the tunnel axis, measured settlements become even larger than those observed in the greenfield monitoring section. The same result was obtained numerically by Di Mariano *et al.* (2007) and Bilotta & Russo (2011) although assuming plane strain conditions and for larger volume loss (V_L = 1–2%). In order to clarify this discrepancy between predictions and measurements, the consistency of the operational parameters of the TBM between the two sections (*e.g.* face support pressure, tail grouting pressure) should be carefully checked.

The profiles of vertical displacements with depth right behind the barrier (TR10) and above the tunnel centreline (TR9) plotted in Figure 10 show a very good agreement between the numerical results and the data obtained by the Trivecs in the field. In this case, for the Trivec TR10, the integration of incremental settlements was performed assuming that the tip of the

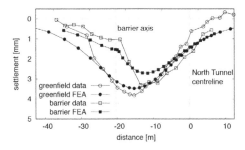

Figure 9. Comparison between predicted and measured settlements, with and without barrier.

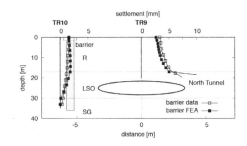

Figure 10. Comparison of computed and measured settlement profiles with depth for two verticals.

instrument was fixed, as it reaches into the stiff SG layer. The sharp increase in settlements in the LSO layer obtained numerically is not confirmed by the monitoring data, suggesting that the stiffness of this layer was possibly underestimated and that an improved calibration of the material properties might be required.

4 CONCLUSIONS

A recently developed technique for the simulation of mechanised tunnelling in finite element analyses was applied to the real case of tunnel excavation with an EPB shield through an experimental site installed at the beginning of stretch T3 of Line C of Rome underground. The technique involves detailed modelling of the most important processes taking place around the shield and has so far been applied only to carry out parametric studies in ideal cases, such as a single tunnel in free field conditions and assuming relatively simple geotechnical settings. On the contrary, in this work, the technique was applied to study the interaction between the excavation and an experimental barrier, purposely installed to study the effectiveness of such a kind of measure in reducing tunnelling induced settlements, in rather complex geotechnical conditions. An advanced constitutive model calibrated on a large set of data from laboratory and in situ tests was adopted for the simulations. Furthermore, the extensive monitoring system installed at the experimental site provided the rare opportunity to perform a thorough comparison between numerical predictions and field measurements.

The results of real class A predictions of the problem provided realistic results, if compared with experimental evidence reported in the literature, both in greenfield conditions and when the barrier is included. When compared to preliminary experimental results from the monitoring system the predictions appear satisfactory. The most notable differences are the smaller efficiency of the barrier predicted by numerical analyses and the underestimation of surface settlement on the tunnel side of the barrier. These differences might be due to a number of factors, including calibration of material parameters for the hypoplastic model for granular soils adopted for the thick made ground layer. For this soil, in fact, average values from the literature were assumed as a first trial, given the inability to collect undisturbed samples for laboratory testing. The arbitrarily reduced Young's modulus assumed for the base sandy and gravelly layer might also play a role, as, thanks to its large axial stiffness, the barrier transfers the shear stresses developed along this shaft down to this layer. An improved calibration of material constants for the silty layer, in which the excavation is actually carried out, should also be performed, in particular focusing on the small strain range, given the small volume loss developed during the excavation. Finally, in the analyses it was assumed that the operational parameters of the TBM were the same for the passage through the greenfield section and when passing close to the barrier, which should be thoroughly checked, to interpret the experimental data better.

REFERENCES

Bilotta, E. & Russo, G. 2011. Use of a Line of Piles to Prevent Damages Induced by Tunnel Excavation. *Journal of Geotechnical and Geoenvironmental Engineering* 137(3): 254–262.

Boldini, D., Losacco, N., Bertolin, S. & Amorosi, A. 2018. Finite Element modelling of tunnelling-induced displacements on framed structures. *Tunnelling and Underground Space Technology* 80:222–231.

Boone, S.J. 2006. Deep Excavations: General Report. In: K.J. Bakker, A. Bezuijen, W. Broere & E.A. Kwast (eds.), *Geotechnical Aspects of Underground Construction in Soft Ground; Proc. intern. conf., Amsterdam, 15–17 June 2005.* London: Taylor & Francis.

Di Mariano, A., Gesto, J., Gens, A. & Schwarz, H. 2007. Ground deformation and mitigating measures associated with the excavation of a new Metro line. In V. Cuéllar & E. Dapena (eds.), *Geotechnical engineering in urban environments: Proc. XIV European Conference on Soil Mechanics and Geotechnical Engineering, Madrid, 24–27 September 2007.* Madrid: Millpress.

Fantera, L., Rampello, S. & Masini, L. 2016. A Mitigation Technique to Reduce Ground Settlements Induced by Tunnelling Using Diaphragm Walls. *Procedia Engineering* 158: 254–259.

Harris, D.I. 2001. Protective measures. In: F. M. Jardine, J. R. Standing, & J. B. Burland (eds.), *Building response to tunnelling*. London: Thomas Telford.

Herle, I. & Gudehus, G. 1999. Determination of parameters of a hypoplastic constitutive model from properties of grain assemblies. *Mechanics of Cohesive-frictional Materials* 4(5): 461–486.

Kasper, T. & Meschke, G. 2004. A 3D finite element simulation model for TBM tunnelling in soft ground. *International Journal for Numerical and Analytical Methods in Geomechanics* 28(14): 1441–1460.

Kavvadas, M., Litsas, D., Vazaios, I. & Fortsakis, P. 2017. Development of a 3D finite element model for shield EPB tunnelling. *Tunnelling and Underground Space Technology* 65: 22–34.

Lambe, T. 1973. Predictions in soil engineering. *Géotechnique* 23(2),151–202.

Litsas, D., Sitarenios, P. & Kavvadas, M. 2018. Parametric investigation of tunnelling-induced ground movement due to geometrical and operational TBM complexities. *Italian Geotechnical Journal* 51(4): 22–34.

Losacco, N., Burghignoli, A. & Callisto, L. 2014. Uncoupled evaluation of the structural damage induced by tunnelling. *Géotechnique* 64(8): 646–656.

Mašín, D. 2005. A hypoplastic constitutive model for clays. *International Journal for Numerical and Analytical Methods in Geomechanics* 29 (4): 311–336.

Negro, A. 1998. General report: Design criteria for tunnels in metropolises. In: A. Negro & A.A. Ferreira (eds.), *Proc., World Tunnel Congress 1998 on Tunnels and Metropolises*, São Paulo, 25–30 April 1998. Rotterdam: Balkema.

Niemunis, A. & Herle, I. 1997. Hypoplastic model for cohesionless soils with elastic strain range. *Mechanics of Cohesive-frictional Materials* 2(4): 279–299.

Peck, R.B. 1969. Deep excavations and tunneling in soft ground. In: *Proceedings of the seventh International Conference on Soil Mechanics and Foundation Engineering, Mexico City, 22–29 August 1969*.

Rampello, S., Callisto, L., Soccodato, F. & Viggiani, G. 2012. Evaluating the effects of tunnelling on historical buildings. In: G.M.B. Viggiani (ed.), *Geotechnical Aspects of Underground Construction in Soft Ground; Proc. intern. conf., Rome, 17–19 May 2011*. London: Taylor & Francis.

von Wolffersdorff, P.A. 1996. A hypoplastic relation for granular materials with a predefined limit state surface. *Mechanics of Cohesive-frictional Materials* 1(3): 251–271.

Tunnels and Underground Cities: Engineering and Innovation meet Archaeology,
Architecture and Art, Volume 12: Urban
Tunnels - Part 2 – Peila, Viggiani & Celestino (Eds)
© 2019 Taylor & Francis Group, London, ISBN 978-0-367-46900-9

Grand Paris Express, Line 15 East – predictive damage analysis combining continuous settlement trough modelling, risk management, automated vulnerability checks and visualization in GIS

S. Mahdi & O. Gastebled
TRACTEBEL ENGIE, Paris, France

H. Ningre & M. Senechal
EGIS, Paris, France

ABSTRACT: Since November 2016, TRACTEBEL and EGIS have been in charge of the design of the metro line 15 East, which is part of the future Grand Paris Express Network. The project that involves a 23km bored tunnel is located in a densely urbanized area and faces various geotechnical risks. In order to manage risks and achieve cost and delay control, necessary geotechnical investigations and building surveys have been carried out by Société du Grand Paris. Detailed modelling of the geotechnical profile was prepared for the continuous assessment of the settlement trough along the tunnel following an advanced methodology, which combines finite element analysis, empirical models and regional feedback from similar projects, to increase the reliability of the predictions. Sources of uncertainty such as face pressure variability and local geotechnical anomalies were considered and combined in several scenarios for risk assessment.

1 PRESENTATION OF THE PROJECT – LINE 15 EST

Line 15 East is a key element of the circular line 15, which will be the backbone of the future Grand Paris Express network. It includes 23 km of 9.8 m diameter bored tunnels, 11 underground stations from Stade de France to Champigny Centre, 20 auxiliary shafts and 2 branching structures. This line will traverse 12 localities of the north-eastern periphery of Paris and will cross two departments, Seine Saint Denis and Val de Marne. Nine out of the eleven stations will have connections with other public transport lines: regional rail network RER, underground metro network, tramway lines and bus stations. The main line is linked to the future rolling stock maintenance centre and depot at Rosny through a 1.2 km underground service tunnel.

The owner, Société du Grand Paris (S.G.P), appointed Koruseo as design engineer and architect of the infrastructure of line 15 East. Koruseo is a consortium led by EGIS and including TRACTEBEL Engineering, INGEROP and six architectural practices: BORDAS-PEIRO, GRIMSHAW, BRENAC & GONZALEZ, SCAPE, VEZZONI and EXPLORATIONS ARCHITECTURE.

The design of the project started in November 2016 with the pre-project design phase completed in December 2017. The awarding of the 2024 Olympic Games to Paris at the start of 2018 induced a reordering of the construction milestones of the Grand Paris Express network, leading to a new target of 2030 for the commissioning of the complete line 15 East. This delay opened the door to reducing the project cost, notably by optimising TBM drives and the design of branching structures.

Figure 1. Line 15 East.

2 GENERAL CONTEXT OF THE PROJECT

2.1 *Geological and geotechnical context*

The geology encountered by line 15 East consists, from the surface down, of quaternary deposits (fills, alluvions) and the succession of Eocene strata:

– Priabonian marls (Masses et Marnes du gypse MMG)
– Bartonian marls and silty sands (Sables verts SV, Calcaire de Saint-Ouen SO and Sables de Beauchamp SB)
– Lutetian calcareous marlstones and limestones (Marnes et Caillasses MC and Calcaire grossier CG)
– Ypresian argillaceous and silty sands (Sables supérieurs SS)

One of the main geological risks in the sector is related to presence of gypsum which has been subject to dissolution phenomena in the Bartonian and Lutetian strata, due to both natural water flows and industrial pumpings.

The main geotechnical characteristics derived from site investigations are summarised below and the geological profile is presented in Figure 2:

– Quaternary deposits and the most superficial stratum (Bartonian or Priabonian marls) are relatively weak with pressure meter moduli under 30 MPa (median value)
– The silty sand stratum, SB, is compact and stiff, with moduli of 62 MPa (median value)
– The layered marlstones, MC, have particularly heterogeneous characteristics due to their nature (limestone, marlstone and argillaceous marl intercalations) and due to the particularly active gypsum dissolution phenomena in this layer
– CG limestones have very high moduli

In terms of hydrogeology, the TBMs bore mostly under a high water table. The water height at the tunnel axis is in the order of 20m, with peaks reaching 27m at the low points of the profile.

Over several site investigation programs, with the first one starting in 2012/2013 and the following ones carried out between 2016 and 2018, geotechnical data has been collected from about 300 boreholes, of which 85 core drillings, 73 pressure meter tests and 134 destructive drilling. On this basis it was possible to establish:

– the geologic profile
– the classification in geotechnical sub-units
– and the geotechnical characteristics for settlement analysis

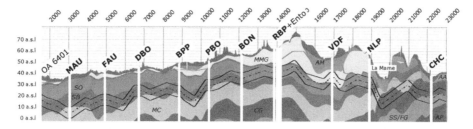

Figure 2. Synoptic of the geotechnical profile of line 15 East.

Each geotechnical sub-unit is associated with a range of values obtained for the pressure meter moduli. This classification identifies more or less compact formations and local anomalies (altered SO, blocky to highly altered MC, sandy CG...).

2.2 *Urban insertion*

Line 15 East is located in a densely urbanised context. The urbanisation encountered along the tunnel alignment can be classified into 3 categories. Areas where urbanisation is dense, such as city centers and densely residential areas representing about 50% of the tunnel's length, areas where urbanisation is less dense, such as suburban areas representing about 25%, and areas of industrial or tertiary activities, factories, warehouses, office buildings, representing the last 25%.

The building survey was initiated in the preliminary studies phase. It was carried out by an assistant to the owner (AMO-B) on more than 4000 buildings located in the geotechnical influence area of the project (GIA). Its purpose is to collect administrative as well as technical data on the type of structure and foundations of buildings, their state of conservation, etc... Based on these data, the AMO-B assigns to each building a level of intrinsic structural sensitivity and a level of global sensitivity.

The AMO-B grouped buildings into 4 categories according to the type of construction. Buildings with less than 4 floors represent about 84%, buildings with 4 to 10 floors represent 6%, and those with more than 10 floors represent only 1% of the total buildings. Technical or industrial buildings represent 9% of buildings. The type of foundation and the depth of building foundations are also important, especially for the 680 buildings (around 17%) identified within the GIA and located directly above the tunnel. By the end of 2017, the building survey had identified 70 buildings (1.7%) with deep foundations and 2244 (56%) with shallow foundations. The foundation mode remains unknown for 1698 buildings (42%).

The intrinsic structural sensitivity rank reflects the structural and functional ability of a building to withstand external mechanical stresses and depends on the type of structure (masonry, reinforced concrete, etc...) and its conservation conditions. Three intrinsic structural sensitivity rankings exist: low-sensitivity (LS), regular-sensitivity (S) and high-sensitivity (HS). The overall sensitivity ranking integrates the concept of social value of the building, defined by its function and its strategic level for the municipality in which it is located. Again, three overall sensitivity rankings exist: low-sensitivity (LS), regular-sensitivity (S) and high-sensitivity (HS). For example, a police station or a hospital will have a higher overall sensitivity ranking than an individual dwelling, regardless of their structural sensitivity ranking.

At the end of the design studies, the available building data remained incomplete since the 2 sensitivity rankings were provided for only 30% of the 4012 buildings located in the GIA. Table 1 presents statistics on known sensitivity rankings.

The design phase schedule of the Grand Paris Express lines is very tight. The design of such a huge urban infrastructure, which is geographically expanded and requires the processing of large amounts of data, poses several challenges for engineering companies.

In addition to these complexities, those related to uncertainties and the lack of data led Koruseo design teams to test various assumptions to ensure that the design remained robust.

Table 1. Statistics on known sensitivity rankings at the end of the current design phase.

Sensitivity rank	Structural sensitivity		Overall sensitivity	
	Total (1194)	Percentage	Total (1473)	Percentage
Low-sensitivity	1 054	88,3 %	1 219	82,7 %
Regular -Sensitivity	132	11 %	230	15,7 %
High-sensitivity	8	0,7 %	24	1,6 %

To cope with these challenges, it has been necessary to adapt both tools and analysis methods. Building data integration in a GIS (Geographic Information System), allowing quick updating and visualisation on an interactive map, proved to be a good answer to these problems. Beyond the data visualisation function, the GIS tool also offers other useful features for the damage assessment analysis, including geographical data crossing. The usefulness of this functionality is described in detail in Section 4.6.

3 SPECIFIC ANALYSIS FOR URBAN TUNNELS DESIGN

Two design checks are necessary to ensure the feasibility, safety and economy of urban tunnel boring: face stability and settlement limitation, both provided by the TBM support pressure.

3.1 *Tunnel face stability checking with analytical methods*

For the line 15 East project Koruseo design team implemented the Anagnostou and Kovari (1994) method, supplemented by Broere (2001), to define the minimum support pressure providing the tunnel face stability. The failure mechanism assumes a sliding wedge in front of the tunnel face that is loaded by a rectangular prism. In order to prevent a blow-up of the overburden, particularly when the tunnel drive has low overburden with poor mechanical parameters, we limited the support pressure at 90% of the total in-situ vertical stress value at the tunnel crown, in accordance with the German recommendation of the DAUB (2016).

The common operational limit for the support pressure in standard operating mode of an urban Earth Pressure Balance tunnel boring machine (EPB) is of the order of 3.0 bars at the tunnel axis, DAUB (2010); it is higher for a Slurry Pressure Balance machine (SPB). Support pressure deviations are of the order of magnitude of +/- 30 kPa for EPB and +/- 10 kPa for SPB, DAUB (2016). In order to not limit the type of machine adapted to the project, the design aimed at keeping the confining pressure below the limit of 3 bars as far as possible, and we considered a +/- 30 kPa support pressure deviation. The lower limit of the operation range was therefore defined as the face stability pressure increased by 30 kPa, and the upper limit as the blow-up pressure decreased by 30 kPa.

The analytical calculation of the operating range was programmed and implemented along the 23 km line 15 East, with a discretisation step of 5 m, Figure 4. This completeness of the analysis made it possible to identify all the singular points regarding support pressure management. In a global way, for this project, tunnel face support pressure is not determinant, support pressure is rather determined either by the water pressure, or by settlement control at sensitive spots, such as railway crossings. In some areas where the tunnel progresses with a low overburden under existing underground structures, the support pressure must not induce any surface heave risk, while providing the tunnel face stability with a 1.5 safety factor, which leads to a shrinkage of the operation range. This is particularly the case for the crossing below the existing underground stations Fort d'Aubervilliers on line 7 and Bobigny-Pablo-Picasso on line 5, but also at the A86 highway cut location.

3.2 *Estimation of settlements*

Two types of methods exist to predict the magnitude of settlements during tunnel boring.

The first one is the analytical and empirical method. The most commonly used is the method proposed by Peck (1969) and Schmidt (1969). Although it has a simple formulation, this method uses two parameters (the trough width parameter k and the volume loss V_{Loss}) which cannot be derived simplistically from the common geotechnical data. These parameters must result from a choice, based on experience from comparable contexts when they exist. Recent projects in the Paris area, such as line 12 extension or the underground section of T6 tramway, make it possible to estimate these parameters for the geotechnical contexts encountered by line 15 East, particularly in the context of the Plaine de France. Nevertheless, these experiences are not sufficient to cover the entire geotechnical context of line 15 East nor its variability. Moreover, it does not allow a fine estimation of the sensitivity of settlements to TBM driving parameters, particularly the support pressure that plays a major role. This method was used in the first stage of the design to estimate the order of magnitude of settlements under all the buildings located within the GIA. Then, during the second stage of the design, to analyse only buildings on shallow foundations.

Finite element modelling is the second group of methods dedicated to settlement trough definition. It makes it possible to model the soil using elastoplastic constitutive law, to model the soil-structure interactions, to take into account loads due to existing buildings and structures and TBM parameters. It also gives access to the deformations of the ground mass. This method is therefore particularly suitable for the analysis of buildings and structures that generate significant loads, when the buildings have deep foundations, or for the analysis of interactions with existing underground structures (metro stations, sewers, gas pipes ...).

4 ESTIMATION OF SETTLEMENTS INDUCED DURING BORING OF LINE 15 EAST TUNNELS

4.1 *Settlement estimation at singular points using 2D finite element modelling*

During the second stage of the design studies, once the accurate geotechnical models were established and the building condition survey data received and integrated into the GIS, settlements and building damage analysis were updated. At the crossings of major infrastructure and specific buildings with deep foundations and/or bringing significant loads, 2D finite element modelling (2D FEM) was systematically conducted. Sixty four sections made it possible to cover all of these singular points of the project and to define horizontal and vertical deformations and displacements at the base of the foundations or at strategic infrastructure spots (rail network, water network, electrical network).

By carrying out these calculations for several values of the support pressure inside the operation range defined in §3.1, it was possible to choose the most suitable support pressure and to quantify the sensitivity of the settlements to the support pressure. This approach made it possible to ensure the robustness of the design with respect to support pressure deviation.

4.2 *Settlement estimation for the common buildings with Peck's calibrated method*

We saw in Section 2.2 that the buildings were predominately superficially-founded, or on around 3m deep foundations. Thus, about 95% of the buildings encountered do not require finite element modeling to estimate the settlements they may undergo. Peck's analysis method is more than enough to provide an estimation of horizontal and vertical displacements and deformations at their foundation level.

Contrary to the assumption made during the first stage of the design regarding the volume loss that suggest a constant value at 0.5%, the volume loss considered in the second stage of the design studies was varied. The volume loss values were defined with respect to the geotechnical context, see Section 4.4, the latter being better known thanks to the results of the geotechnical survey, and according to a support pressure profile, called "setpoint profile", chosen at any point of the tunnel alignment within the operation range defined in Section 3.1.

4.3 Selection of the support pressure

The analysis performed at the 64 cross sections carried out with 2D FEM to estimate the settlements at the singular points was also carried out under greenfield conditions, that is to say without considering any load or structural element. Each section was recalculated for different support pressures within the operation range. The analysis of the results of these new calculations made it possible to make the link between the volume loss and the support pressure for the different geotechnical contexts encountered at the face. Eleven "laws" of evolution of the volume loss with the confinement ratio (ratio between the support pressure and the in-situ stress) were defined, each of them associated with a part of the tunnel alignment located in a homogeneous geotechnical context. As a result of these laws, the volume loss profile associated with the support pressure profile was established over the entire 23 km long tunnel alignment.

The selection of the most suitable support pressure was then the result of an iterative process, in order to define the optimal support pressure to ensure volume loss control, and therefore settlement control. The purpose of this analysis is to ensure that with a standard urban TBM, with a usual support pressure limit of the order of 3 bars at the axis, settlements remain acceptable.

DAUB (2016) recommends a +/-30kPa support pressure deviation for EPB machines. To ensure the robustness of the design against this technological limit, the estimation of the volume loss was made for two scenarios, the support pressure at its nominal value and the support pressure decreased by 30 kPa.

4.4 Estimated volume loss

On the northern half of the line, which is part of the geotechnical context of the Plaine de France, the volume losses estimated by the calculation are in the order of 0.2% to 0.4%, with values very locally higher, around 0.6%, as a result of an alteration of the Sables of Beauchamp. In addition, these volume losses are not very sensitive to changes in support pressure within the operation range: +/- 0.1%.

The southern half of the line is less geologically uniform, the tunnel progresses successively under the Rosny plateau, crosses a depression located around the Val-de-Fontenay station, crosses the Nogent plateau and then the Nogent hillside, crosses the Marne valley then enters under the Champigny plateau before finishing at the Champigny-Center station located in the Marne plain. Due to this high variability of geotechnical context but also to a variable topography, the volume losses variations have a large range. In the plateau areas at Rosny and Nogent, their values are estimated between 0.40% and 0.65%. In the Fontenay-sous-Bois basin, they are estimated to be lower, between 0.25% and 0.40% because of the lower overburden and therefore a higher achievable confining ratio. In these zones, the sensitivity of volume losses to the support pressure is estimated no more than approximately +/- 0.15%. From the Marne crossing when the tunnel enters the Meudon anticline and progress in the Calcaire grossier and the Marnes et Caillasses, the volume losses are estimated quite small, generally between 0,10% and 0,30% and not very sensitive to support pressure, +/- 0.05%; locally when the overburden increases, they can reach 0.50%

4.5 Trough width parameter

The results of the 2D FEM calculations carried out under greenfield conditions also made it possible to determine a calibrated value of the apparent trough width parameter (k_{eq}) at each of the 64 cross sections. Solving the inverse problem with a probabilistic method, Mahdi (2016), the trough width parameter k associated with each one of the geotechnical units was then deduced, knowing the stratigraphy and the position of the tunnel. Unsurprisingly, and as formalised in the ITA/AITES report (2006), k_{eq} values obtained by the FEM calculation are generally higher than those observed in-situ and tend to widen the settlement trough and decrease its amplitude. In order to avoid the underestimation of settlements due to the use of 2D FEM, the k values calibrated on the finite elements were clipped to a 0.5 value to remain coherent with experience feedback, and in particular with those of the extension of line 12 and the underground section of the T6 tramway project.

The k_{eq} values obtained after simple calibration of the results of 2D FEM calculations reached values on the order of 0.55 to 0.70. The clipping of the k parameters greater than 0.5 allowed to return to k_{eq} values between 0.40 and 0.50, more in accordance with those usually measured in the Paris region.

4.6 Settlement mapping and spatial analysis

At each design phase, and based on the assumptions and parameters of Peck and Schmidt's method presented in the preceding paragraphs, a prediction of the surface displacement field (horizontal and vertical) was made at all points within the GIA of the project. Figure 4 shows the maximum settlement at tunnel axis due to volume loss estimated with 2D FEM analysis, see Section 4.4, and k_{eq} values obtained with the methodology presented in previous Section 4.5.

These predictions have been implemented in a GIS tool to allow their mapping and their cross referencing with the building data, see Figure 3.

The GIS tool allowed to automatically select the maximum values of displacements (horizontal and vertical) under the footprint of each one of the 4012 buildings located within the GIA.

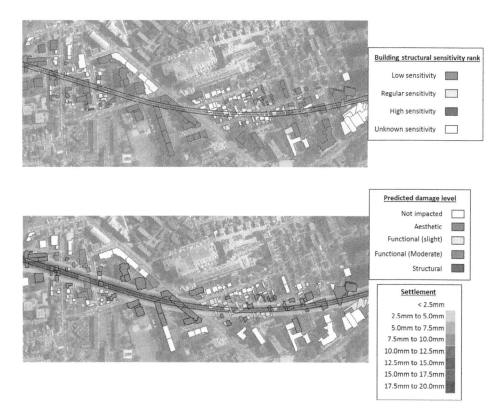

Figure 3.   Example of predicted settlement mapping.

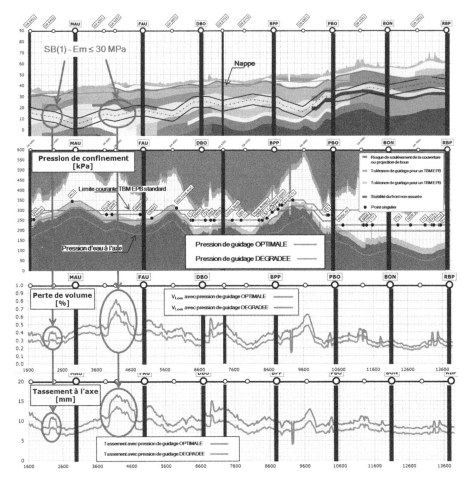

Figure 4. Support pressure operation range/Volume loss and maximum settlements – L15 East – north part.

5 METHODOLOGY FOR BUILDING DAMAGE ANALYSIS

5.1 Overall methodology implemented by the Société du Grand Paris for the definition of building damage and vulnerability

Once the settlement analysis has been completed, the methodology followed by the Société du Grand Paris to define reached damage level and vulnerability level is conducted in two stages. The retained damage criteria are the absolute settlement, the slope and the horizontal deformation.

The first stage, called building damage analysis, consists of determining the predicted damage level for each building confronting the three estimated damage criteria with their threshold values, depending on the structural sensitivity level. Four damage levels exist, from the lightest one inducing only architectural disturbances (microcracks of the non-structural elements) to the severest one reflecting structural disorders endangering the building's stability, with 2 intermediate levels of functional damages, slight (opening of doors and windows ...) and medium (rupture of water pipes ...) related to the proper functioning of the building.

The second stage, called building vulnerability analysis, consists of defining the achieved level of vulnerability by comparing the achieved damage level with the acceptable level defined by the Owner. The maximum accepted damage level is defined by the overall sensitivity level.

Three vulnerability levels exist and they define the preventive procedures to be adopted during design and construction. The lowest vulnerability level does not require any project adaptation or special procedures. The second level requires several procedures, including an extensive building monitoring system. The last vulnerability level is not considered acceptable by the Owner. If reached, it is necessary to implement before the TBM passage preventive measures of structural reinforcement, often heavy and invasive for the occupants, and/or soil and foundation reinforcement/treatment.

When such preventive measures are either not feasible or not economical, it is necessary to adapt the tunnel alignment in plan and/or in profile to avoid the highly vulnerable buildings. It is desirable to identify these hard points as early as possible during the design process, as late revisions to the tunnel alignment during the design are generally synonymous with degradation of the functional performance of the line (reduction of commercial speed).

5.2 Dealing with uncertainties related to building data

For the 70% of buildings that have not yet been classified in terms of sensitivity by the AMO-B during the building condition survey, it was necessary to make assumptions to ensure the robustness of the design. Instead of defining the base cost very conservatively by considering all buildings whose sensitivity levels are unknown to be "Highly Sensitive", several assumptions were made about unknown levels of sensitivity. These assumptions were combined with the two scenarios of TBM support pressure, optimal or decreased by 30 kPa.

Two risk scenarios were derived from this: a medium scenario considering the undefined sensitivity levels as "Sensitive" and the support pressure at its optimal value, and an unfavorable scenario considering the undefined sensitivity levels as "Very sensitive" and the support pressure at its decreased value.

Under the medium risk scenario, the high vulnerability level is not accepted. The cost of extensive building monitoring, preventive measures and light repair work resulting from a medium vulnerability level is taken into account in the base cost of the tunnel.

Provisions for risk (PRI) are estimated on the basis of the unfavorable risk scenario. A medium-functional damage level leads to additional repair work costs to the PRI, and higher damage leads to add structural reinforcement or soil treatment costs.

6 PROJECT OPTIMISATION

6.1 Optimisation of station and shaft depths owing to reduction of the tunnel depth

During the studies, the design of the tunnel has evolved to pursue the optimisation of the depths of the stations and the emergency/ventilation shafts. The shafts' urban integration constraints were also more flexible during the second design phase in accordance with Société du Grand Paris and the local authorities that allowed the construction of buildings, on top of 10 of the 20 shafts for housing the technical equipment instead of locating the equipment inside the shafts belowground. This means that both the diameter and the depth of these structures have been reduced. The depth of 3 of these 10 shafts have been reduced significantly. The changes to the depth of the shafts have resulted in significant raising of the tunnel, while maintaining sufficient overburden. The tunnel being shallower, its boring could induce a potentially increased risk of damage to existing buildings and structures. The analysis method described in Section 4.2 and the use of the GIS tool for the building damage analysis made it possible to quickly validate the raising of the tunnel alignment by ensuring the acceptability of the potential damage for hundreds of buildings.

6.2 Determination of the monitoring cost and the provisions for risk

At the end of the design studies, provisions for potential repair work were added in the PRI for 113 buildings, for which we still did not know the sensitivity levels. These are buildings located in areas where settlements may be significant due to the geotechnical context. However, we have assigned to these PRIs a low probability of occurrence, in the order of 5%, as they are conditioned by the definition of the structural sensitivity level "Very Sensitive". Table 1 shows that the percentage of "Very sensitive" buildings is statistically low, ranging from 1% to 2%.

The cost of extensive building monitoring and repair works of the buildings firstly depends on the floor surface area, the façade perimeter and the number of floors; these data are integrated into the building database. Nevertheless, depending on the extent and position of the building related to the tunnel alignment, a large part of an impacted building may be located outside of the area of potential damage. The calculation of the reduced building footprint that may suffer damage can be done using the GIS tool. The cost evaluation of extensive monitoring and repair works could thus be automated using the GIS tool. The optimisation of the cost of both extensive building monitoring and repairs works resulting from this detailed analysis is 25%.

7 CONCLUSION

This article aims to present the different steps followed during the design studies of the tunnels of line 15 East to deal with the issue of settlements induced by tunnel boring, and the risks that it represents for existing construction. The Koruseo consortium implemented tools and methods adapted to the project's constraints and which led to an optimisation of tunnel costs, but integrating however a certain number of provisions for risks to deal with the remaining uncertainties. Ideally, to continue to reduce the project costs these uncertainties should be reduced, especially those related to building data.

The article focuses mainly on settlement, but the type of foundation of the buildings located directly above the tunnel is also particularly important.

Knowing the sensitivity level of the 113 buildings identified as priorities by Koruseo should allow a decrease in the provision for risks, and the determination of the type of foundation of the 680 buildings located directly above the tunnel should secure the design of the project.

Finally, complementary geotechnical surveys will improve the reliability of the models used to perform the settlement analysis.

REFERENCES

Anagnostou.G. Kovári.K. 1994. The Face Stability of Slurry-shield-driven Tunnels. *Tunnelling and Underground Space Technology*, Vol. 9.

Broere.W. 2001. Tunnel face stability and new CPT Applications. PhD Thesis, Technical University of Delft.

DAUB. 2010. Recommandations for selection of tunnelling machines.

DAUB. 2016. Recommendations for Face Support Pressure Calculations for Shield Tunnelling in Soft Ground.

Mahdi.S. Houmymid.F.Z. Chiriotti.E. 2016. Use of numerical modelling and GIS to analyse and share the risks related to urban tunnelling - Greater Paris - Red Line – South section, ITA-AITES WTC 2016.

Peck. R.B. 1969. Deep excavations and tunnelling in soft ground, 7th International Conference on Soil Mechanics and foundation Engineering, Mexico City, State-of-the-art Volume.

Schmidt.B. 1969. Settlements and Ground Movements associated with Tunnelling in soils, PhD Thesis University of Illinois, Urbana.

Tunnels and Underground Cities: Engineering and Innovation meet Archaeology,
Architecture and Art, Volume 12: Urban
Tunnels - Part 2 – Peila, Viggiani & Celestino (Eds)
© 2020 Taylor & Francis Group, London, ISBN 978-0-367-46900-9

Digital project: Practical application of an integrated process in TBM tunnel design

F. Maltese, A. Ghensi, G. Eccher & A. Konstantinou
SWS Engineering spa, Trento, Italy

ABSTRACT: The behaviour of tunnelling in the urban environment is mainly influenced by TBM geometry, face pressures, geotechnical parameters and hydraulic conditions. The calculation of settlements and face pressures is usually performed in an uncoupled fashion, without a direct correlation between face pressures and volume losses. A further difficulty in the assessment of existing structures arises from the significant amount of data that must be processed, such as building type, foundation type/depth, height and vulnerability to settlements and rotation. The "Digital Project" has been developed by SWS Engineering in order to overcome the above difficulties. The tool creates a database from BIM, GIS and CAD files, avoiding the tedious task of manual data insertion. Coupled analyses are then performed to determine the optimal face pressures, the settlement contours and evaluate the resulting damage class for each structure. These can then be visualised with widely known software (such as open-source QGIS) in an interactive and user-friendly way. Traditional building reports and spreadsheets are also easily generated. This paper illustrates the results obtained in the detailed design of an urban tunnel. The tool will also be used during the construction phase in order to adjust the results based on the encountered stratigraphy and data obtained from monitoring.

1 INTRODUCTION

The concept of digitalization, well-established in mechanical engineering design, is relatively new in infrastructure engineering. BIM and GIS technologies are the key tools allowing the digitalization of conventional civil engineering design, with design data being converted into 3D geo-referenced parametric geometries, and non-geometric information into databases.

In the recent years SWS Engineering has invested in research projects centred on this process, with emphasis on tunnelling. Digital Project is the name of the process that is the tangible outcome of this research. This paper presents the application of the Digital Project to urban tunnelling, including the study of settlements, face pressures and building and utility damage assessment. The background theory as well as the interface of the software are presented. The importance of GIS technologies to create, share and study the available input and output data is highlighted.

2 DIGITAL PROJECT

Digital Project is the name given to the automated process for tunnelling involving digitalisation of design input data (reports and drawings), application of best design practices and production of 3D geo-referenced models (BIM and GIS) as well as a central database with all relevant data.

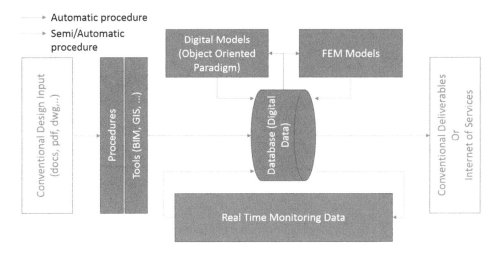

Figure 1. Schematic of Digital Project workflow.

The intent is to progressively abandon manual spreadsheet input to favour object-oriented software code (Python) able to collect data directly from building reports, drawings and mapping data, forming a central database. The method has the following advantages:

- Input data is inserted using GIS and/or BIM technologies by expert modellers and all changes are immediately available to analytical models that can be quickly re-calculated
- data flow automation allows to perform optimizations or parametric analyses to fine tune design parameters and obtain the optimal combination (multivariate-analyses)
- the automation of data flow and analytical analyses allow to perform extensive statistical/ sensitivity analyses

An extremely large number of simulations are performed in a reduced timeframe. The purpose is not to provide a black-box tool promising a fast solution to complex problems, but to provide an efficient tool to support decision making, saving time from tedious manual input of data, allowing the user to make sound assumptions and critically assess results.

3 PROJECT

In this paper the application of the Digital Project on a detail project is presented. The name is confidential as a number of design tasks are pending approval. The project involves the construction of a single double-track tunnel having an overall length of approx. 4km. An 9.0m diameter EPB TBM tunnel is used to cross an area having relatively uniform and predominantly clayey soil strata. The stratigraphy is almost horizontal and constant. The overburden is variable from 1 to a maximum of 2.5 tunnel diameters with an average value of 1.5. The geotechnical survey has highlighted a strong undrained behaviour of the clay deposits. The value

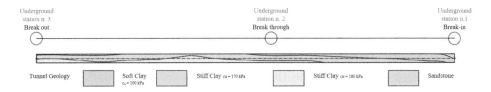

Figure 2. Synoptic of the soil at the face of tunnel along the alignment.

of undrained cohesion at the level of tunnel, Figure 2, is variable from 100 to 180 kPa with a secant Young Modulus variable from 30 to 85 MPa. The characteristics of the shallow soil layers can be assimilated to an equivalent sandy soil with no cohesion, a friction angle equal to 30 degrees and a Young modulus equal to 50 MPa.

The alignment is crossing mainly below an airport area, including runways and taxiways, bridge foundations and several buildings and significant buried services. One of the main design activities of SWS Engineering for this project is the evaluation of settlements, building damage and calculation of face pressures.

4 THEORY

The excavation of a tunnel in soft ground inevitably produces ground movements and consequently surface settlements. Their magnitude and evolution depend on the following two factors, often approached in a disjunct fashion:

- face pressure
- geology, in-situ stress, TBM geometry

As far as face pressure is concerned, several analytical solutions are available for stability and blow-up pressures, widely established in the design process. Among them, there is a varying degree of sophistication, including the consideration of aspects such as excess pore pressure, multiple layers and arch effect.

For the other main factors, three main different approaches can be adopted: empirical, analytical and numerical. The selection of the appropriate method largely depends on the complexity of the problem as well as on the stage of the project. The Digital Project considers analytical methods as they are widely established, they can be easily programmed, and because they have the significant advantage of being able to consider different parameters and understand the different relationships between them. Moreover, they can cover both horizontal and vertical displacements. The main disadvantages, instead, are related to the limited number of available solutions and their application to specific types of ground conditions.

Numerical models (FEM 2D or 3D) were used to validate the results in critical sections. In addition, consolidation effects were evaluated, which are currently not taken into account in the Digital Project. However, in the specific case study it was found that settlements due to consolidation are negligible. The results of these models are not in the scope of this paper.

It is important to underline that in the Digital Project the workflow is based on the work of Loganathan (2011) that has introduced the possibility to relate volume loss, settlement and expected damage with face pressures.

4.1 *Face Pressure*

The operational range of face pressures is evaluated considering the following extremes:

- the lower limit pressure to ensure the minimal support pressure
- the upper limit pressure to avoid a break-up of the overburden or blow-out of the support medium.

The operational range of support pressure defined by the two limits is illustrated in Figure 3. Depending on the excavation mode of the machine, an additional safety margin has been introduced to account for possible oscillations in the pressure value associated with the excavation technology (DAUB, 2016). If the face is stable and no water is present a minimum face pressure of 30 kPa at the crown must be guaranteed, with the assumption that the excavation chamber is completely filled.

Several limit equilibrium models have been developed to determine the minimum stability pressure. Based on the specific conditions of the case study the following methods are used:

- Broms, B.B. & Bennemark, H. 1967 solution with stability number $N = 2$ for the tunnel excavated in stretches with fine soil
- Broere, W. 2001 solution for excavation in stretches with granular soil. This solution takes into account the heterogeneity of ground at the face, soil arching effect for the vertical load and the penetration of the support medium into the tunnel face in terms of excess pore water pressures.

Based on the design recommendations of ITA, AFTES and DAUB, as well as common design assumptions, the following amplification factors are adopted to guarantee a safety margin against failure mechanisms:

- 1.5 for the solid skeleton and 1.1 for water pressure.

The minimum design pressure at the crown is thus:

$$p_{min,\ crown} = max \begin{cases} U_{crown} \\ p'_{min\ stability,\ crown} \end{cases} \quad (1)$$

Where $p'_{min,\ stability,\ crown}$ = minimum stability pressure; U_{crown} = water pressure at the tunnel crown.

The upper support pressure limit involves avoiding the failure/uplift of the overburden (Break-up/Blow-up) or the blow-out of the support medium. For this reason, the total vertical stress at the tunnel crown was corrected with a safety factor of 0.9. The operational limit pressure of the machine must also be respected. Based on these considerations, the maximum pressure at the crown is evaluated as:

$$p_{min,\ crown} = min \begin{cases} 0.9 \cdot \sigma_{v,crown} \\ p'_{max,\ TBM} - \gamma_s \cdot R \end{cases} \quad (2)$$

Where $p_{max,TBM}$ = maximum operational pressure of the machine; γ_s = unit weight of the support medium; R = excavation radius.

Having defined the limits, the methodology for the selection of the operational face pressure is explained in Chapter 5.

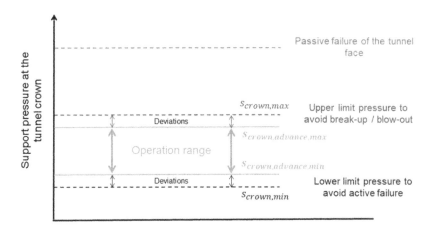

Figure 3.   Allowable operational pressures at the tunnel crown of a shield machine (DUAB 2016).

4.2 Induced Settlement

The analytical approach developed by Loganathan (2011) forms the basis to calculate settlements in the Digital Project. As described by the author, the proposed solution was verified for accuracy using historical cases, centrifugal models and numerical analyses.

In his work, Loganathan links the settlement to the movement of ground around the tunnel, which is not uniform due to the oval vault shape deviation caused by the effect of the force of gravity. The Figure 4 shows the "uniform" and "real" profiles of the gap around a tunnel. In general, the equivalent mean volume loss at the level of tunnel in undrained conditions (V_L), which is also referred to as ε_0 in this study, is defined as follows (Loganathan, 2011).

$$\varepsilon_0 = V_L = \frac{\pi R^2 - \pi \left(R - \frac{g}{2}\right)^2}{\pi R^2} \times 100\% = \frac{g}{R} - \frac{g^2}{4R^2} \times 100\% \qquad (3)$$

Where ε_0 = volume loss at the level of tunnel; R = excavation radius of the tunnel; g = estimated gap at the crown defined according to equation n.4,

Gap g has been defined in a different way, Figure 5, from the author's solution. In detail:

$$g = g_f + \max\ (g_s, g_b + g_v) \qquad (4)$$

Where four different values of gap are evaluated as follows:

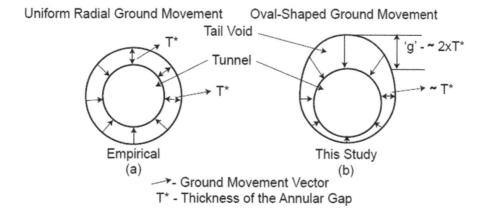

Figure 4.   Circular and oval ground deformation patterns around a tunnel section (Loganathan, 2011).

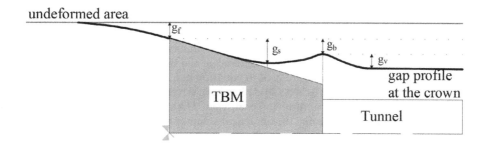

Figure 5.   Evaluated gap profile – not to scale.

1. g_f is the gap due to face loss evaluated according to the original formula given by Loganathan (2011). This value is directly proportional to the face pressure and to the soil characteristics at the face.
2. g_s is the gap due to the shield conicity modified from the author's solution. In this approach we have decided to calculate g_s with the convergence-confinement curve, as the difference between the radial displacement at the front (g_s) and the one along the shield (function of the reduced pressure p_{shield} and of the shield length).
3. g_b is the gap due to the backfill grout injection pressure. This value is a development from the original solution proposed. The same approach of point 2 is applied, with the second term being at the end of the shield as a function of the backfill grout injection pressure p_b (always greater than p_f) and of the length of shield plus two rings.
4. g_v is the gap due to shrinkage of the backfill grout injected in the annular void.

$$g_v = \delta \cdot (\Delta - g_b) \tag{5}$$

Where δ = volumetric shrinkage of backfill grout; Δ = geometric free space, defined as the difference between the head diameter and the lining diameter at the extrados; g_b = gap described at point number 3.

Regarding the convergence-confinement curve, the solution of Panet (1995) has been used instead of the elastic solution proposed by the author. Regarding the gap due to the backfill grout injection pressure, it is an additional development of the original solution, which was introduced to take into account the positive effect of the backfill grout injection pressure that is able to recover and to reduce the volume loss at the level of tunnel (Dias T. & Bezuijen A. 2015). Furthermore, convergence at the level of the tunnel has been evaluated along each point of the perimeter and three different values of pressure are defined instead of a single one used by Loganathan solution.

In general, the scenario where the backfill grout injection pressure p_b leads to a significant recovery of the convergence, schematised in Figure 5, is the best. In this case the final volume loss ε_0 was less than the maximum developed during shield passage. However, we have decided, in a conservative way, to evaluate the final gap according to the most critical situation, as shown equation n. 4.

Surface, subsurface, and lateral ground movements are evaluated with the original formula. The equation of Loganathan (2011) to evaluate vertical displacement is given below:

$$U_z = \varepsilon_0 R^2 \cdot \left(-\frac{z-H}{x^2+(z-H)^2} + (3-4v)\frac{z+H}{x^2+(z+H)^2} - \frac{2z\left[x^2-(z+H)^2\right]}{\left[x^2+(z+H)^2\right]^2} \right)$$
$$\cdot \exp\left\{ -\left[\frac{1.38x^2}{(Hcot\beta + R)^2} + \frac{0.69z^2}{H^2} \right] \right\} \tag{6}$$

Where: R = TBM head radius; H = depth of tunnel axis; z = depth of analysed level from ground surface; v = Poisson's ratio of soil; x = lateral distance from tunnel axis; $\beta = 45 + \varphi/2$; ε_0 = ground loss ratio defined according to equation n.3.

It is clear that while all relevant factors such as face pressure, geometry of TBM, drop pressure and elasto-plastic soil response are considered in the defining of the ground loss ratio, only some soil parameters (φ and v) are used in equation n. 6.

As the solution given in Loganathan (2011) has been developed to evaluate the final settlement in a cross-section a further multiplying coefficient M has been introduced to evaluate the settlement curve longitudinally, using the analytical solution proposed by Attewell and Woodman (1982):

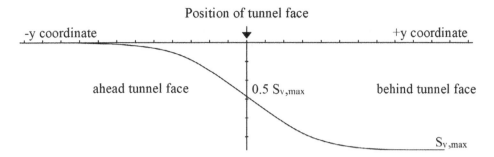

Figure 6. Longitudinal settlement distribution.

$$M = \left\{ G\left[\frac{y - y_i}{i}\right] - G\left[\frac{y - y_f}{i}\right] \right\} \tag{7}$$

$$G(\alpha) = \frac{1}{\sqrt{2\pi}} \int_{-\infty}^{\alpha} e^{-\frac{\alpha^2}{2}} d\alpha \tag{8}$$

$$\alpha = \frac{y - y_i}{i} \tag{9}$$

Where i = defines the width of the settlement trough according to Loganathan and Poulos solution 1998; subscript *(i)* is used to denote 'initial location of tunnel'; subscript *(f)* means 'final location of tunnel'; $G(\alpha)$ is a cumulative normal distribution function.

Therefore, by calculating the term α, the following values of G are obtained:

- $G(0) = 0.5$ at the tunnel front ($y = y_f$)
- $G(1) = 1.0$ for ($y - y_i$) $\rightarrow \infty$

5   ITERATIVE PROCEDURE TO ACHIEVE OPTIMAL FACE PRESSURES

The calculations presented in the previous chapter have been performed for every single calculation point along and across the tunnel alignment (n. 42800 for the case study). This allows to obtain an estimation of settlements for each point according to the calculated face pressures.

The following step is a fundamental element of the Digital Project: starting from the minimum operational value, the face pressure is adjusted in such a way to guarantee that buildings and existing services do not exceed the allowable damage class. This is done with the following iterative procedure:

1. Processing of input data, discretization of the tunnel in portions of 10 m and selection of structures that can be affected by the induced effects of the excavation.
2. Evaluation of the operative pressure range of the TBM
3. Definition of the face pressure of the TBM assuming a first value equal to the minimum pressure estimated at the previous step.
4. For each structure selected in step 1:
   a. In the case of buildings and structures:
      - Discretization of the boundaries of the structures in segments (with a defined length) and definition of the effective surcharge acting on the foundation level
      - Definition of the load acting on the tunnel crown according to the Boussinesq equations
   b. In the case of utilities and railways:

- Discretization in segments (length depending on the characteristics of the neighbouring area). The analysis is then carried out for each segment.

5. Assessment of expected volume loss and estimation of the values of maximum settlement, rotation and horizontal deformation measured under the footprint of the structure (or for each segment of utilities/railways)

6. Comparison of the calculated values with the limits imposed by the specific sensitivity of the structure and definition of the damage class and the vulnerability level:

   a. If the vulnerability level exceeds the maximum value to be respected for the sensitivity class of the structure, the TBM pressure is iteratively increased "step-by-step" by a value of 0.1 bar, ensuring that the adopted value is lower than the blowout pressure and the maximum operating pressure of the TBM, including the safety margin (0.3 bar for an EPB and 0.1 for a Slurry TBM – DAUB 2016).

   b. The damage class and vulnerability level of each structure is updated considering the new values of settlements: the damage class and the general vulnerability level of the structure are the maximum values measured for each chainage intercepting the structure.

7. Definition of the TBM excavation pressure and induced settlements for the section analysed. Where the damage class is still not acceptable with the maximum operational face pressures, additional mitigation measures such ground improvement or underpinning will be necessary.

6 INPUT DATA

GIS (Graphical Interface) and MongoDB (Digital Interface) are used to handle the significant amount of data involved. The input parameters and activities for the creation of a database mainly involve:

- Digitalisation of data related to the location of the project using Autodesk Civil 3D:
  - Horizontal and vertical alignment of the tunnel
  - Excavation and lining geometry
  - Digital Elevation Model (DEM)
  - Stratigraphy and piezometric levels.

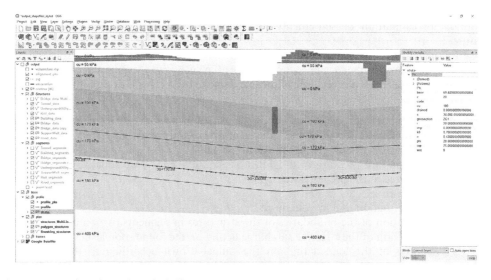

Figure 7.   Stratigraphy and vertical alignment (QGIS).

Figure 8.   Building, rail and calculation points along the alignment (QGIS).

- The geotechnical parameters of the soil
- Digitalisation of existing structures using the Q-GIS tool
  - Building condition survey sheets for buildings and interfering structures.
  - Geometry and positioning of the building structural frames in relation to the tunnel.

For the project in consideration the information digitalised involves the geometry and building condition data of 70 buildings, more than 5000 precast concrete pavement slabs, 9 underground utilities, 8 bridges with the relative piers, which produces a total of 35000 segment of analysis. In addition, the Digital Project input database contains for each 10m-section (in total 420 sections) a complete set of information regarding overburden, geology, water table, presence of buildings/structures, their relative distance from the tunnel axis, foundation depth, sensitivity, and structure type. The vertical alignment and stratigraphy as appearing in QGIS after the input process is illustrated in Figure 7. A significant amount of information can be visualised to study and compare different sections. In Figure 8, some buildings, a railway and the calculation points are shown in plan. These can be visualised using different criteria, such as sensibility class, threshold values and foundation or building types.

7   OUTPUT DATA

At the end of the iterative procedure an extensive output database is created. Information such as the residual building risk class and contour plots of greenfield settlements can be visualised with GIS software, as shown in Figure 9. Non-geometric information was saved and combined with geometric elements. For example, for each building, utility or bridge it is possible to visualise information such as blow-up, stability and excavation pressures, sensitivity data, damage class, volume loss and settlement. Furthermore, reports have been automatically produced to share the building risk assessment data for every structure.

Standard graphs, such as the one shown in Figure 11, were automatically produced to represent the prescribed EPB and backfilling pressure, expected volume loss values, maximum settlements, etc. along the alignment profile. The calculated face pressures, initially calculated in 10m steps, are "smoothened" along longer stretches, responding to a request from the contractor to apply a less variable pressure profile. For this reason, Digital Project has been developed to carry out pressure-imposed analysis. The final design face pressure is shown in

Figure 9. Settlement surface, damage class and output information (QGIS).

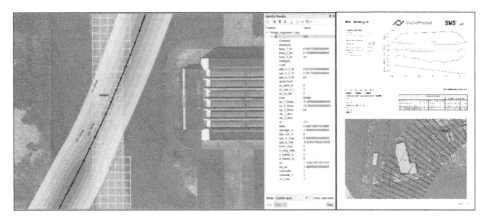

Figure 10. Settlement surface, damage class for concrete slabs and output information (QGIS) – Automatic sheet with BRA information for a specific building (on the right).

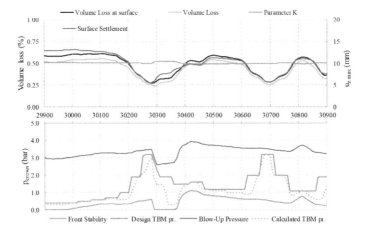

Figure 11. Maximum settlement, volume loss and excavation pressure along the alignment.

Figure 11. As a result, the initial analysis applied in 10m steps was useful to estimate a preliminary face pressure profile that allowed to define a smoother distribution.

8 CONCLUSIONS

This paper has presented SWS Digital Project and its application to an urban tunnel project. The Digital Project approach offers significant advantages, such as automatic elaboration of project input data and quick performance of analyses and representation of results for an extremely large number of simulations, where conventional tools would result highly uneconomical.

Results are fitting well with settlement profiles and face pressures obtained from similar projects. Face, shield and backfill grout injection pressures have been optimised to reduce settlements up to acceptable limits in areas with highly vulnerable buildings or with critical conditions such as poor geology and shallow cover. Furthermore, by studying the face pressure profile it was possible to identify critical sections with detailed monitoring and specific precautions, such as no TBM stoppage, ground improvement and underpinning. Surface settlements are coherent with the calculated volume losses and the geotechnical and in-situ stress conditions.

In the construction phase the Digital Project will be used to follow the advance of the EPB TBM, back-analysing soil parameters, fine tuning face pressures, allowing to quickly define potential mitigation measures.

The Digital Project is an ongoing research, where future developments include considering more recent convergence-confinement curves, consolidation and other long-term effects, additional details of backfill grout behaviour and automatic connection with FEM software.

ACKNOWLEDGEMENT

This paper has been developed in memory of our dear colleague Gabriele Eccher, who passed away last year. His enthusiasm and passion in innovative engineering has been the starting point to conceive and develop all SWS Digital Project Tools. Working with him was a privilege for us, representing an incentive to pursue this research with the same spirit.

REFERENCES

Broere, W. 2001. Tunnel face stability and new CPT Applications. *PhD Thesis, Technical University of Delft.*

Loganathan N. 2011. An innovative method for assessing tunnelling-induced risks to adjacent. *PB 2009 William Barclay Parsons Fellowship Monograph 25; Parsons Brinckerhoff.*

Broms, B.B. & Bennemark, H. 1967. Stability of clay at vertical openings. *ASCE, Journal of Soil Mechanics and Foundation Engineering Division* (SMI 93): 71–94.

DAUB, 2016. Recommendations for face support pressure calculations for shield tunnelling in soft ground.

Attewell P.B. and Woodman, J.P. 1982. Predicting the dynamics of ground settlement and its derivatives caused by tunnelling in soils. *Ground Engineering. 15 (8):* 13–22.

Panet, M. 1995. Le calcul des tunnels par la méthode convergence-confinement. *Pressed de l'ENPC, Paris.*

Tunnels and Underground Cities: Engineering and Innovation meet Archaeology,
Architecture and Art, Volume 12: Urban
Tunnels - Part 2 – Peila, Viggiani & Celestino (Eds)
© 2019 Taylor & Francis Group, London, ISBN 978-0-367-46900-9

Ground response to mini-tunnelling plus ground improvement in the historical city centre of Rome

L. Masini, S. Rampello & S. Carloni
University of Rome La Sapienza, Rome, Italy

E. Romani
Metro C, Rome, Italy

ABSTRACT: The construction of a new underground metro line is being undertaken in the historical city centre of Rome. As part of the project, two tunnels will be excavated for a length of about 114 m following a three-step procedure: excavation of a small diameter tunnel with a mini slurry-shield machine; soil improvement via low-pressure cement grouting, performed radially from the mini-tunnels; and conventional excavation in the improved soil of the two tunnels of the line. The two mini-tunnels reach the *San Giovanni* station passing at a short distance from the ancient *Aurelian Walls*. The paper describes the results of the monitoring activity undertaken during the excavation of the two mini-tunnels and the cement grouting, focusing on observed field of ground displacements. Data are analysed to evaluate the volume loss induced by the mini-tunnelling and the soil heave induced by the low-pressure cement grouting.

1 INTRODUCTION

The traffic congestion in the city of Rome required the construction of a new underground metro line to ease the connection between the south-eastern suburbs and the city centre. Therefore, Municipality of Rome, in October 2002, appointed "*Metro C S.C.p.A.*" the General Contractor, with *Astaldi* as leading company, for the construction of the new Line C of Rome underground. The easternmost part of the line is completed, while the central part of the route is currently under construction in the historical city centre, facing several problems for the presence of archaeological artefacts and historical monuments along the route. Among these, two tunnels are being excavated close the *Aurelian Walls*, between the underground stations of *Amba Aradam* and *San Giovanni*, following a three step procedure: two small-diameter tunnels ($D = 3$m) have been first excavated using a mini slurry-shield machine at a depth of about 25 m; then, soil improvement via low-pressure cement grouting has been carried out using tubes à manchettes installed in boreholes excavated radially to the bored tunnels. Tunnel construction will be finally completed through conventional excavation of the main tunnels, about 7m wide, in the improved soil. Figure 1 shows a plan view of the site; the distance between the axis of the two tunnels varies from 12.5 m to 16.5 m. The mini slurry-shield boring machine was launched and remotely controlled from the "*Multifunctional pit 3.3*", which is located between *Amba Aradam* and *San Giovanni* stations. The "even" (Southern) tunnel was excavated first, and the "odd" (Northern) one after, for a total length of about 114 m.

The new metro line will reach *San Giovanni* station passing at a short distance (25.5 m) from the ancient *Aurelian Walls* (3rd century A.D.); therefore, an extensive monitoring system has been laid out to control ground movements induced by the excavation activities and the pre-treating operations.

Figure 1.   Plan view of the site.

This paper describes the results of the monitoring activity, focusing on the displacement field observed during the excavation of the two mini-tunnels, the drilling of the radial boreholes and the low-pressure cement grouting. The modification of the subsidence profile induced by the excavation of the second tunnel is also discussed. The displacement data are back-analysed using the classical empirical relationships calibrated on green-field settlement measurements, to evaluate the volume loss induced by mini-tunnelling and the operative values of the width parameter K.

2   SITE CONDITIONS

Ground conditions close to the *Aurelian Walls* at *Porta Asinaria* are typical of the historical centre of Rome (Rampello et al., 2012; Fantera et al., 2016). A detailed geotechnical

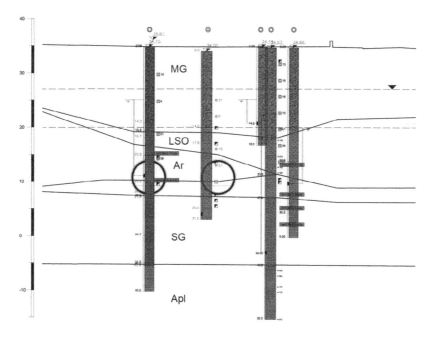

Figure 2.   Geological cross section of the site.

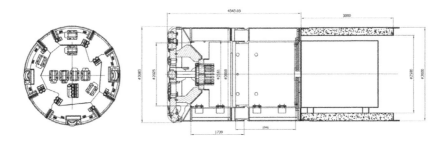

Figure 3.   Cutting head and longitudinal section of the mini TBM.

investigation was carried out, involving site tests (CPT and Cross-Hole tests) and deep boreholes from which undisturbed samples were retrieved and tested in the laboratory. Figure 2 shows a geological cross section of the site: a 15m-thick layer of made ground (MG) is first encountered, from ground surface, at about +35 m a.s.l., mainly consisting of coarse-grained material, sand and gravel; recent alluvial deposits are found underneath, extending down to a depth of 32 m (+3m a.s.l.). The alluvia are variable in grading involving clayey silt and sandy silt (St/Ar) as well as clayey silt (LSO); they overly a layer of sand and gravel of Pleistocene age (SG), with a thickness of about 10m, from +3.0m to -6.6m a.s.l., followed by a thick deposit of stiff and overconsolidated clay of Pliocene age, the "Argille Vaticane" (Apl).

The profile of hydraulic head, evaluated from measurements of pore water pressure, shows a downwards seepage in the silty soil, from the made ground towards the gravel, between +26 m to 17 m a.s.l.

To avoid interaction with the archaeological layer, the tunnels were excavated at depths of about 25m from ground surface, developing mostly through the fine-grained soils (St/Ar).

3   CONSTRUCTION SEQUENCES

3.1   *Mini-tunnelling*

The excavation of small diameter tunnels to perform ground treatment before that of the main tunnels has proven to be capable of reducing the effect of tunnelling in urban areas (e.g. Colombo et al., 1994). For the Metro C of Rome, the two mini-tunnels were excavated using a remote-controlled TBM with an outer diameter of about 3m. Figure 3 depicts the cutting head and a schematic cross section of the slurry shield boring machine in which the excavated soil is collected into the excavation chamber and added with a water-bentonite mixture to form a slurry. The stabilisation of the tunnel face was obtained by keeping the slurry into the chamber to a pressure of 150–200 kPa higher than the pore water pressure at the excavation depth.

The mini-TBM has two shields: an inner shield, which was recovered at the completion of the tunnel together with the electric and hydraulic equipment, and an external shield that will be recovered at the end of the excavation of the large diameter line-tunnel, together with the cutting head.

Figure 4 depicts a conceptual layout of the setup. The machine is advanced by the thrust transmitted by the lining made of precast concrete rings, which are pushed from a jacking station installed into the launching pit. This is composed by four hydraulic jacks mounted on a counter-wall. The thrust of the jacks is applied to the concrete lining rings through a steel ring (Figure 5a). An intermediate jacking station, about 40 m away from the TBM, reduces the thrust acting on the lining (Figure 5b).

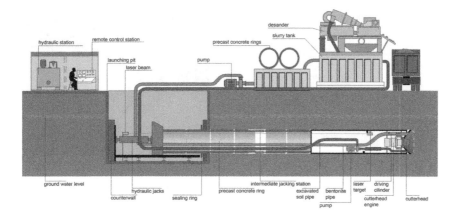

Figure 4. Schematic layout of the excavation machinery.

| (a) | (b) |

Figure 5. Jacking station at the launching pit (a) and (b) the intermediate jacking station.

The concrete rings of the lining have an outer diameter of 3 m, a thickness of 0.25 m and a length of 2.35 m. A total number of 55 and 57 rings were installed for the "*odd*" and the "*even*" tunnel (northern and southern tunnel), respectively. The lining will be completely dismantled with the excavation of the line tunnels.

The gap between the soil and the linings was filled with a water-bentonite mixture to reduce friction during lining jacking. The filling mixture was injected through 3 valves mounted every two lining segments. Special male-female joints between tunnel segments were adopted to guarantee sealing due to the bending radius of the tunnel line as low as 210 m.

After completion of the mini-tunnels the soil-lining gap was filled with cement grout which was injected from the same valves previously used to inject the water-bentonite mixture.

3.2 *Soil Improvement*

At the end of the excavation of the two mini-tunnels, soil improvement was carried out via low-pressure cement grouting using *tubes à manchettes* installed in boreholes excavated radially to the bored tunnels. Twenty boreholes per section were drilled using a longitudinal spacing of 0.6 m; the boreholes had a diameter of 80 mm, a length of 5–7m, and a spacing of 18° in the radial direction (Figure 6).

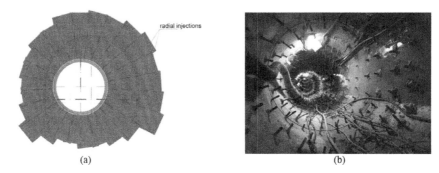

(a)   (b)

Figure 6.   Schematic cross section of the radial injections from within the mini-tunnels (a), view from the inside of the tunnel.

Grout injections were performed through 50 mm-diameter PVC tubes, with 4 manchettes per meter. Soil-tube gap was firstly filled using a water-cement grout with 8÷10% of bentonite content and a water cement-ratio of 2÷3 (32.5 grade cement). Soil improvement was obtained injecting first a *Mistrà*-type cement grout, with 10÷15% of bentonite content and a water-cement ratio of 2.5÷3.5 (52.5 grade cement). Flowability of *Mistrà* cement grout was improved using chemical additives which also reduce the *pressure filtration* that may lead to significant losses of water from the grout during injection, thus reducing grout capability of penetrating into the soil (Masini et al., 2014). In a second stage, a chemical mixture of silica components (*Litosil*-type) was injected to reduce further the permeability of the improved soil.

3.3   Excavation of the line tunnels

The large-diameter line tunnels will be excavated from within the improved soil, enlarging the mini-tunnels via a conventional excavation procedure. Fan-like, overlapping pipe umbrellas, made by 114.7 mm-diameter steel pipes, will be drilled and grouted at the roof of the tunnel, parallel to the direction of advancement of the excavation face. Each roof shield is composed by 41 steel pipes, 12 m-long, covering a maximum excavation span of 8m. The full-face excavation of the tunnel will be carried out along with the installation of the primary support, which consists of IPN 200 steel ribs and 25 cm-thick shotcrete. The final concrete lining has a

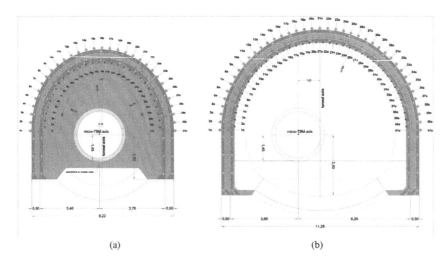

(a)   (b)

Figure 7.   Cross sections of the line tunnels: (a) intermediate position and (b) at the entrance of San Giovanni Station.

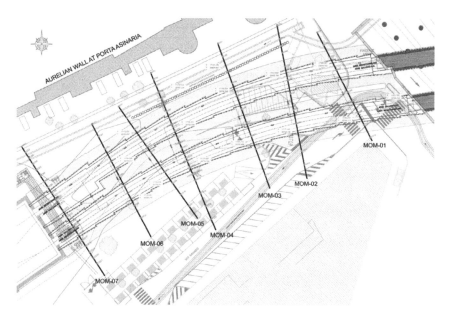

Figure 8. Plan view of the monitoring arrays.

thickness of $0.6 \div 0.1$ m, a width of $6.72 \div 7.72$ m and a height of $7.70 \div 8.20$ m (Figure 7). During the excavation of the main tunnels the linings of the mini-tunnels will be dismantled.

4 INSTRUMENTATION LAYOUT

An extensive monitoring was performed to control ground movement induced by the construction stages in the proximity of the ancient city walls. A plan of the instrumentation layout is shown in Figure 8. The instruments are mainly aligned along 7 arrays transverse to the tunnels axis. Two arrays have been instrumented with vibrating-wire piezometers, settlement markers at the ground surface, inclinometers and *Trivecs*. The last are a combination of a portable strain meter with a biaxial accelerometer measuring the angle of the probe axis to the gravity direction, thus allowing to measure all the three components of the displacement vector. Five arrays have been instrumented only with settlement markers at the ground surface. A total of 96 settlement markers, 4 inclinometers, 10 *Trivecs* and 8 piezometers have been installed in the 7 monitoring sections.

5 MONITORING RESULTS

Table 1 reports the main stages of construction: excavation of the "*even*" (Southern) mini-tunnel started on January 20th, 2017 and went on for about one month; the "*odd*" (Northern) tunnel was excavated from March 13rd, 2017 to March 31st, 2017 (19 days). Ground improvement was carried out in the two mini-tunnels simultaneously; boreholes drilling was first performed from June 27th, 2017 to 22nd of January 2018, injections being carried starting one day after the end of borehole drilling. At the present time, ground improvement is completed but the excavation of the main tunnels has not yet started. In the following, the vertical displacement measured by the settlement markers during mini-tunnelling and ground improvement stages are discussed and interpreted using empirical relationships currently adopted in engineering practice.

Typical trends observed at the instrumented arrays are depicted in Figure 9, which shows the time histories of the vertical displacements measured at section MOM-03, shown in

Table 1. Construction excavation sequences.

Phase description	from	to
"*even*" (Southern) mini-tunnel excavation	2017/01/20	2017/02/16
"*odd*" (Northern) mini-tunnel excavation	2017/03/13	2017/03/31
radial borehole drilling and tube installation	2017/06/27	2018/01/22
injections	2018/01/23	2018/10/15

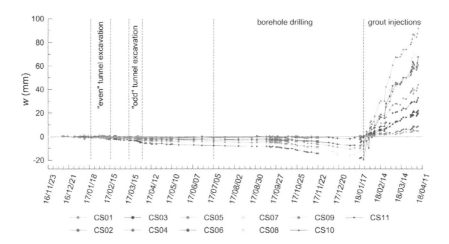

Figure 9. Time histories of the vertical displacement measured at section MOM-03.

Figure 8; negative values indicate settlements. Measurements are referred to the readings taken 2 months before beginning the excavation of the "even" tunnel. Excavation of both tunnels induced a maximum settlement of about 8 mm. During borehole drilling, vertical displacements are seen to increase gradually as drilling of radial boreholes approaches the instrumented section, with a final settlement of about 20 mm, which is more than twice the value observed at the end of mini-tunnels excavation. The subsequent injections caused a massive heave, as high as 92 mm. The maximum heave was measured at the settlement marker located above the axis of the "odd" tunnel, while the heave observed above the axis of the "even" tunnel was equal to about 65 mm. It can be reasonably anticipated that part of the measured heave will be lost as the main tunnels will be excavated.

5.1 Excavation of the "even" (Southern) tunnel excavation

The effect of tunnel excavation is discussed with reference to the settlements observed in a section transverse to the tunnel axis, far enough from the tunnel face. In these conditions, the shape of the green-field settlement trough at the ground surface may be described by a Gaussian distribution curve:

$$w = w_{\max} \cdot \exp\left(-\frac{x^2}{2i^2}\right) \tag{1}$$

where w is the vertical displacement at a distance x from the tunnel axis, $w_{\max}$ is the maximum vertical displacement and i is the distance of the point of inflection of the Gaussian curve from the tunnel axis. The maximum settlement $w_{\max}$ is obtained by integrating Equation 1 as:

$$w_{\max} = \frac{V_L \cdot V_0}{\sqrt{2\pi} \cdot i} \tag{2}$$

where D is the tunnel diameter and V_L is the volume loss, defined as the volume of the settlement trough at the surface divided by the nominal volume of the excavated tunnel V_0. For tunnels with a cover of at least one diameter, the value of i at the surface is proportional to the depth of the tunnel axis, z_0 (O'Reilly & New, 1982):

$$i = K \cdot z_0 \tag{3}$$

with a width parameter K depending essentially on the soil type. Equation 1 can be rearranged in the form:

$$\ln\left(\frac{w}{w_{\max}}\right) = -\frac{z_0{}^2}{2 \cdot i^2} \cdot \frac{x^2}{z_0{}^2} \tag{4}$$

which is a linear equation in the semi-log space $(x/z_0)^2$, $w/w_{\max}$. Equations (4), (3) and (2) can be used to calculate the observed values of i, K and V_L using the monitoring data, for a given depth of the tunnel z_0 that is about constant (= 24.9m) between the TBM launching pit and *San Giovanni* station. Figure 10 shows the settlement trough measured at the instrumented section MOM-06 averaging the daily readings taken in the time interval from 07/2/2017 to 15/02/2017. At that time, only the "even" tunnel was excavated, and the tunnel face was 20m ahead of section MOM-06, so that plane strain conditions can be assumed. The maximum settlement induced by excavation of the "even" tunnel was about 3 mm. Figure 10 also shows that best-fitting the measured settlements with a Gaussian curve yields a volume loss $V_L = 1.39\%$. This value is substantially larger than the ones observed during the closed-shield excavation of the large-diameter tunnels of the line in other sites along the route, where values of VL ≤ 0.5% have been typically measured. However, the high cover to diameter ratio $C/D = 7.8$ is such that the maximum settlement is small. Table 2 summarises the maximum settlements and the settlement trough characteristics evaluated for the 7 instrumented sections: maximum settlement $w_{\max}$ is ranges from 0.73 mm to 2.75 mm, the width of the gaussian curve i varies from 7.99 to 14.25m, while values of $K = 0.32$ to 0.57 were obtained; the computed volume loss ranges between 0.35% and 1.39%.

5.2 Excavation of the "odd" (Northern) tunnel

The net effect of the excavation of the "odd" tunnel was obtained by subtracting the settlement induced by the "even" tunnel to the values measured after the completion of both tunnels. The results are plotted in Figure 10 for section MOM-06, while Table 3 reports the

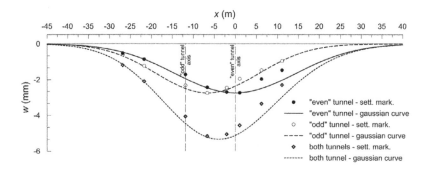

Figure 10.   Vertical displacement measured at section MOM-06 during mini-tunnels excavation.

Table 2. Parameters of the gaussian curves calculated from the measured settlements for the "*even*" tunnel.

Section ID	w_{max} (mm)	i (m)	K (-)	V_L (%)
MOM-01	-	-	-	-
MOM-02	0.73	13.76	0.55	0.35
MOM-03	1.55	9.72	0.39	0.54
MOM-04	-	-	-	-
MOM-05	1.88	12.50	0.50	0.83
MOM-06	2.75	14.25	0.57	1.39
MOM-07	2.48	7.99	0.32	0.70

Table 3. Parameters of the gaussian curves calculated from the measured settlements for the "*odd*" tunnel.

Section ID	w_{max} (mm)	i (m)	K (-)	V_L (%)
MOM-01	-	-	-	-
MOM-02	1.71	12.72	0.51	0.77
MOM-03	3.77	9.90	0.40	1.32
MOM-04	2.20	12.05	0.48	0.94
MOM-05	2.00	11.72	0.47	0.83
MOM-06	2.75	12.10	0.49	1.18
MOM-07	2.36	10.39	0.42	0.87

values of the parameters of the gaussian curve and the volume loss computed for all the monitoring arrays. The maximum settlement induced at the section MOM-06 by the second tunnel excavation is equal to the value measured after the completion of the first tunnel, while the volume loss is 15% smaller. The final value of w_{max} induced by both tunnels is about 5 mm. A similar behaviour is observed also for the MOM-05 and MOM-07 sections, with differences of w_{max} and V_L of about 6% and 31% with respect to the values reported in Table 1, respectively. Conversely, sensibly large increases of settlements and volume loss were recorded at MOM-02 and MOM-03, located near *San Giovanni* station, the values measured after completion of the second tunnel being more than twice those observed at the end of excavation of the first tunnel. Observed values of the width parameter K are in this case of 0.40 to 0.52.

5.3 *Effects of the ground improvement*

Figure 11 compares the vertical displacements measured at section MOM-03 at the end of the excavation of the two mini-tunnels with those induced by ground treatment: the maximum

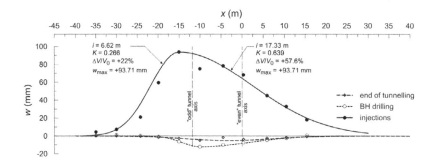

Figure 11. Vertical displacement measured at section MOM-03.

settlement measured after the completion of the two mini-tunnels is equal to about 5 mm; the radial boreholes drilling caused an increase of w_{max} to a final value of 12 mm which is about twice the value induced by tunnelling; grout injections caused instead a massive heave, as high as 93 mm. In fact, the two tunnels will be excavated in a fine-grained soil at the crown, and in a coarser soil at the invert. Therefore, the injections carried out at the invert resulted in grout permeation and reduction of the permeability of the coarser soil, while those performed at the crown mainly induced displacement of the fine-grained soil, with heave developing at the ground surface. Nevertheless, no sensible effects have been observed on the nearby buildings and, above all, on the ancient city wall, located at a distance (about 25 m) where the heave was negligible.

6 CONCLUSIONS

The surface settlement markers installed along 7 arrays transversal to the tunnels axis permitted to evaluate the ground settlements induced by construction of two 3-m-diameter tunnels excavated using a mini-TBM for a length of about 114 m, as well as the settlement induced by drilling of boreholes, radially to the tunnels lining and needed for installation of *tubes à manchettes* and subsequent soil treatment. The soil heave produced by low-pressure cement grouting was also monitored and seen to be much higher than the measured settlements.

Data from surface settlement markers measured after the completion of the first mini-tunnel showed values of the volume loss relatively high ($V_{L,max} = 1.39\%$), although the maximum settlement did not exceed 3 mm due to the high cover to tunnel diameter ratio ($C/D = 7.8$). The construction of the second tunnel caused the settlement to sensibly increase only in the two monitoring sections close to *San Giovanni* station.

Data from the excavation of the two mini-tunnels permitted to calibrate the empirical gaussian curves that were characterised by trough width parameters $K = 0.32$ to 0.57 for the "even" (southern) tunnel and $K = 0.40$ to 0.51 for the "odd" (northern) tunnel, these values being in the range of those evaluated for soils of similar grain size distribution.

Borehole drilling for installation of the *tubes à manchettes* produced a final maximum settlement of about 5 mm, thus almost doubling the settlement induced by tunnels excavation, while the subsequent low-pressure cement grouting produced a substantial heave, as large as about 12 mm at the ground surface. The above observations highlighted the importance of monitoring the installation effects of side construction stages that may have a larger impact than the one associated to the main excavation activities that are usually carried out under a more strict and accurate control. It is worth mentioning that for the case at hand, no sensible effects were detected on the ancient *Aurelian Walls* close to the tunnels.

REFERENCES

Colombo, A., Lunardi, P. and Pizzarotti, E.M. 1994. The Venezia Station of Milan Railway Link carried out by the Cellular Arch Method: water proofing, fire proofing and safety, ventilation. In *Proc. of the Int. Conf. "Underground openings for public use"*. Oslo.

Fantera, L., Rampello, S. and Masini, L. 2016. A mitigation technique to reduce ground settlements induced by tunnelling using diaphragm walls. In *Geotechnical engineering in multidisciplinary research: from microscale to regional scale CNRIG2016. Proc. VI Italian Conference of Researchers in Geotechnical Engineering*, Procedia Engineering, 158 (2016): 254–259, Elsevier DOI 10.10167j.proeng.2016.08.438

Masini, L., Rampello, S. and Soga, K. 2014. An approach to evaluate the efficiency of compensation grouting. *Journal of Geotechnical and Geo-environmental Engineering*, 140 (12), 04014073.

O'Reilly, M.P., New, B.M. 1982. Settlements above tunnels in the United Kingdom – Their magnitudes and prediction. In *Proc. Tunnelling '82 Symp.*, pp. 173–181. London.

Rampello, S., Callisto, L., Viggiani, G. and Soccodato, F.M. 2012. Evaluating the effects of tunnelling on historical buildings: the example of a new subway in Rome. *Geomechanics and Tunnelling*, 5(3): 254–262. DOI: 10.1002/geot.201200017

Tunnels and Underground Cities: Engineering and Innovation meet Archaeology,
Architecture and Art, Volume 12: Urban
Tunnels - Part 2 – Peila, Viggiani & Celestino (Eds)
© 2019 Taylor & Francis Group, London, ISBN 978-0-367-46900-9

A record of constructing a ramp section of an expressway by the underground widening excavation method

N. Matsukawa, S. Ikezoe, Y. Kono & N. Oshima
Hanshin Expressway Company Limited, Osaka, Japan

M. Watanabe
Kajima Corporation, Tokyo, Japan

ABSTRACT: Tokiwa work section in Hanshin Expressway Yamatogawa route, including 350m of main track and 500m of Tokiwa West exit ramp, was originally planned to be constructed using cut and cover tunneling method. However, non-open-cut methods were finally selected for the exit ramp to mitigate the inconvenience of local residents due to the restriction to the local road in front of their houses by retaining wall.

We adopted two non-open-cut methods. One was the rectangular-shaped earth pressure balanced(EPB) shield for the separate section of the ramp, which was first trial in Hanshin Expressway and enabled to keep the construction width in the right-of-way limit. Another was the underground widening excavation method to widen the turnout side from the main truck and the outline of this construction was reported in this paper.

1 INTRODUCTION

Yamatogawa Route is expressway of about 9.7 km connecting east and west with the coastal area and inland area (see Figure 1). Yamatogawa Route is a route forming part of "Osaka Urban Renaissance Loop Route" in the central part of Osaka. By constructing Yamatogawa Route, the traffic congestion of the general road in the southern part of Osaka is greatly alleviated and improves congestion relief and convenience of expressway use. Yamatogawa Route is an underground structure composed of Cut and cover tunnel and Shield tunnel. The connecting portion with other route and common street is a retaining wall structure (see Figure 2).

Tokiwa work section is construction work to 4 traffic lanes Expressway (Cut and cover tunnel), Tokiwa-Nishi Ramp and shaft for a shield machine to turn. In addition, in the densely populated residential areas where the first-class Nishi-Yoke river intervenes between them, the main body of the expressway with inner section dimensions with the height of 6.6 m, and the width of approximately 21.3–34.3 m is to be constructed under river with the soil covering approximately 8–9 m, The drilling would begin at a depth of 20 m and extend to 37 m at the deepest part of the shaft (see Figure 3). The plan is to temporarily fill up the first-class Nishi-Yoke river and perform construction using the open-cut method.

At the Tokiwa west exit and the widened part of the main line, as they are very close to residences, the non-open-cut method of construction was selected instead of the open-cut method to reduce burdens on residents. Where an exit alone is constructed, the rectangular-shaped earth pressure balanced(EPB) shield for the separate section of the ramp, which was first trial in Hanshin Expressway and enabled to keep the construction width in the right-of-way limit[1]. The underground widening excavation method applied from the open-cut part of the main line was used to construct the widened part of the main line (Figure 4). This paper reports the result of the underground widening excavation method used in a narrow underground space in an urban area at the Tokiwa work section.

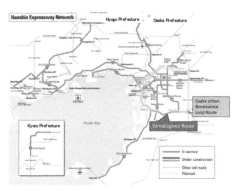

Figure 1.   Hanshin Expressway Network.

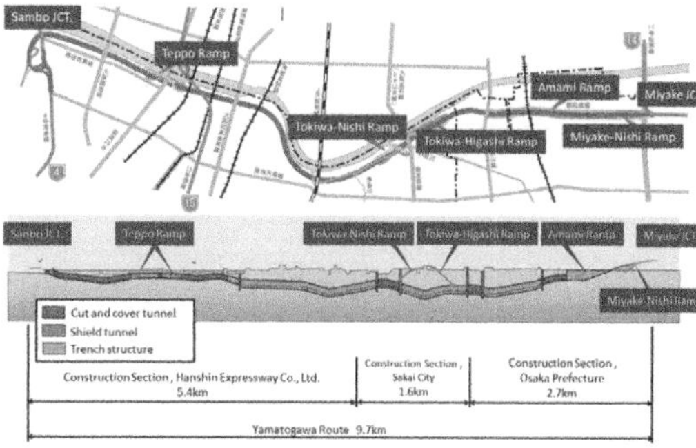

Figure 2.   Tunnel plan and profile of the Yamatogawa Route.

2   CONSTRUCTION SUMMARY

2.1   *Selection of the method*

The cross-section of the underground widening excavation part is shown in Figure 5. When a branch from the main line is constructed using the open-cut method, it is necessary to build an earth-retaining wall at the position of the secondary earth-retaining wall shown in the figure. However, it requires traffic blocking of traffic of a municipal road and will impose restrictions to vehicle access of neighboring resident. In response, a construction procedure specifically tuned to minimize the access restrictions was devised, which involves construction of the primary earth-retaining wall by maintaining traffic of at least one lane of the municipal road, The underground widening excavation part of the earth-retaining wall and the ground behind it from the underground excavated area of the main line, widening the excavated area toward the outside, and additionally constructing the structure. The construction method to realize the abovementioned procedure was selected by comparing its structural feasibility, stability, and past performance (Table 1).

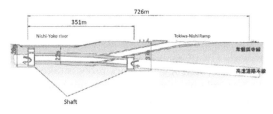

Figure 3.   Tokiwa work section sectional view.

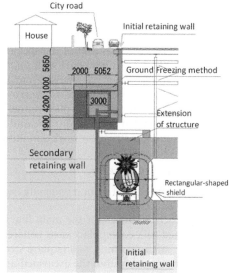

Figure 5.   Underground widening excavation method sectional view.

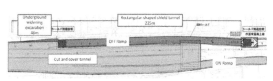

Figure 4.   Tokiwa work section plan view Route.

Table 1. Comparison table of construction method.

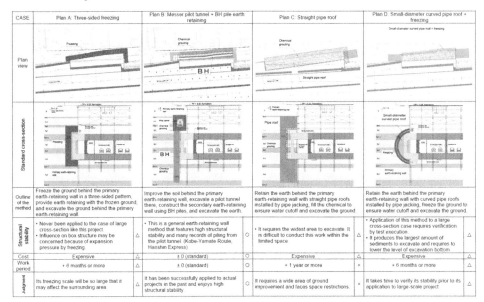

CASE	Plan A: Three-sided freezing	Plan B: Messer pilot tunnel + BH pile earth retaining	Plan C: Straight pipe roof	Plan D: Small-diameter curved pipe roof + freezing
Plan view				
Standard cross-section				
Outline of the method	Freeze the ground behind the primary earth-retaining wall in a three-sided pattern, provide earth retaining with the frozen ground, and excavate the ground behind the primary earth-retaining wall.	Improve the soil behind the primary earth-retaining wall, excavate a pilot tunnel there, construct the secondary earth-retaining wall using BH piles, and excavate the earth.	Retain the earth behind the primary earth-retaining wall with straight pipe roofs installed by pipe jacking, fill the chemical to ensure water cutoff and excavate the ground.	Retain the earth behind the primary earth-retaining wall with curved pipe roofs installed by pipe jacking, freeze the ground to ensure water cutoff and excavate the ground.
Structural stability	• Never been applied to the case of large cross-section like this project. • Influence on box structure may be concerned because of expansion pressure by freezing.	• This is a general earth-retaining wall method that features high structural stability and many records of piling from the pilot tunnel. (Kobe-Yamate Route, Hanshin Express) ○	• It requires the widest area to excavate. It is difficult to conduct this work within the limited space ○	• Application of this method to a large cross-section case requires verification by test execution. • It produces the largest amount of sediments to excavate and requires to lower the level of excavation bottom. △
Cost	Expensive	± 0 (standard) △	Expensive	Expensive △
			○	△
Work period	+ 6 months or more	± 0 (standard) △	+ 1 year or more	+ 6 months or more △
			○	×
Judgment	Its freezing scale will be so large that it may affect the surrounding area. △	It has been successfully applied to actual projects in the past and enjoys high structural stability. ○	It requires a wide area of ground improvement and faces space restrictions. ×	It takes time to verify its stability prior to its application to large-scale project. △

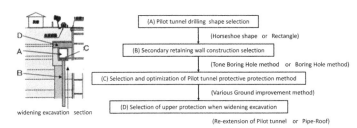

Figure 6. Flow of Temporary structure and Construction method.

2.2 Selection of the temporary structure

Figure 6 shows a flow chart for selection of the temporary structure at the underground widening excavation part and of the specific method. The procedure taken includes: (A) select the shape of the pilot tunnel part with safety, economic efficiency and work process taken into consideration, (B) select the kinds of earth-retaining pile method that can be applied within the space of the selected shape, (C) narrow down the applicable earth-retaining pile methods and select the specifications of pilot tunnel protection work, (D) conduct further optimization based on the updated field conditions, and (E) adjust the work timing to minimize disturbance to the main line construction work simultaneously under way.

2.3 Construction procedure

The procedure of underground widening excavation method is shown below: (Figures 7 and 8)

1) Construct the boxes for the main line in advance, backfill the widened part with liquefied stabilized soil, cast replacement concrete, construct temporary RC pillars, and set up temporary earth-retaining wall (square steel pipes) (Figure 9).
2) Protection work within the excavated area of the pilot tunnel (pilot tunnel entry and excavation arrival: Super Jet-grouting method; general pilot tunnel part: Chemical grouting method)
3) Apply freezing to the back of the primary earth-retaining wall for protection of the upper part of the pilot tunnel.

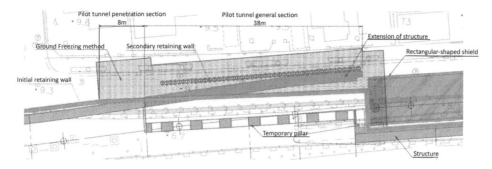

Figure 7. Underground widening excavation method plan view.

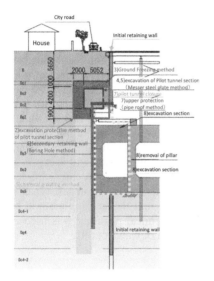

Figure 8. Underground widening excavation method sectional view.

Figure 9. Pre-built flow sectional view.

4) Excavate the area of the protection work for the primary earth-retaining wall and the entry part as 2) above to construct a pilot tunnel.

5) After excavating a 38-m long general part of the pilot tunnel (using the Messer method), apply chemical grouting to the back of the primary earth-retaining wall for water shut-off from inside the pilot tunnel.

6) Drive secondary earth-retaining piles in the extension area from inside the pilot tunnel (BH method).

7) After constructing the upper protection work in excavation of the extension (pipe roof method), the pilot tunnel was closed by filling liquefied stabilized soil and air mortar.

8) Between the secondary earth-retaining wall and the main line structure, excavate the ground behind the earth-retaining wall, remove the primary earth-retaining wall, construct the extension structure, and remove the temporary piles (wire sawing method) to complete the work.

3   CONSTRUCTION CONTENTS

Various methods are used in each type of work in a project where the underground widening excavation method is used. Three methods such as chemical grouting, jet grouting, and freezing, are used appropriately depending on the location at ground improvement for pilot tunnel excavation protection. In addition, the Messer method was used for pilot tunnel excavation, BH method for driving of secondary earth-retaining piles in the pilot tunnel, and pipe roof method for upper protection work. The details of each method are explained in the following:

3.1 Ground improvement - shifting to chemical grouting and freezing

Chemical grouting was used for the pilot tunnel protection work of the general part of the pilot tunnel excluding the upper part and for water cutoff in the back of the secondary earth-retaining wall.

Note that the upper part of the pilot tunnel protection work was the filled layer recovered after a past flood event and mixed with a large content of coarse gravely soil. The groundwater level in the neighborhood was continuously measured to find that the groundwater level showed seasonal fluctuation and that the level moved around the upper part of the pilot tunnel protection work. Under these ground conditions, the upper part of the pilot tunnel needed the strength of cohesive force (C) of 80 kN/m^2 or more and water cutoff work during pilot tunnel excavation to prevent face slip failure or ceiling fall (Figure 10).

In general, chemical grouting uniformly pushes out interstitial water among soil particles with the filler, permeate the soil voids with the filler and enhances the cohesive force of particles. At this construction site, trial was conducted at the site to verify if the method can ensure sufficient quality under these ground conditions. A solution type filler for smooth permeation and a suspended type filler for void filling were selected, and the filling ratio of 40.5% (filling ratio of 100%) and 58.5% (filling ratio of 130%) was chosen. After trial application of these fillers, no sufficient strength was generated from either filler. As a solution, the freezing method that can be horizontally applied from the primary earth-retaining side and is expected to generate high strength was included in a list of review, and room test with locally sampled soil was conducted. As result, for all samples and concludes that the method can satisfy sufficient required strength even though. Therefore, it was determined that the freezing method be used for the upper part of the pilot tunnel protection work. Chemical grouting was used in areas other than the upper part as they only required performance of water cutoff.

3.2 Ground improvement – Super Jet-grouting method

This method was used for protection of the ground at the back of the removed core material of the primary earth-retaining wall and for the entry of the pilot tunnel. It was also used at the gable part of the general part of the pilot tunnel for gable protection of the general part of the pilot tunnel and the extension of the structure (Figure 11).

Ground improvement work using, for instance, jet grouting is generally conducted vertically from the ground. However, since it was necessary to keep one lane of the municipal road available for traffic at the construction site, slanted application (slope of 23°) from within the occupation zone or horizontal application (depression angle of 7°) under the road would be also used in addition to conventional vertical application (Figure 12). When jet grouting is used in the slanted application, it is generally understood that the slope can be down to approximately 10° relative to the vertical direction mainly because of drilling precision or

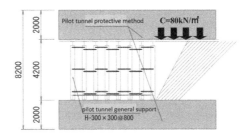

Figure 10.   Excavation of pilot tunnel section sectional view.

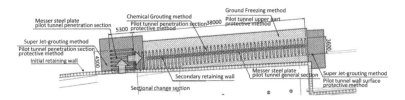

Figure 11. Pilot tunnel protective method and excavation to pilot tunnel section of plan view.

improvement diameter. As there are no sufficient data about slanted application of jet grouting, test application was conducted at the site to demonstrate its validity.

The test application results confirmed that slanted application successfully improved the ground without degrading the finish or affecting the ground surface. Therefore, slanted application of jet grouting was put to full-scale use. A total of 82 improved soil columns of 2.0 to 3.0 m in diameter were produced by slanted application, while 135 columns of 1.1 m in diameter by horizontal application. When the actual improved soil column was visually checked during pilot tunnel excavation, good cutoff performance and sufficient strength were confirmed.

3.3 Ground improvement - Freezing

Ground Freezing method adopted to brine type was adopted for its excellence in economic efficiency and safety. The calcium chloride brine was circulated in freezing pipes to freeze the ground around the pipes and unify frozen soil columns that grow in a tree-ring pattern. Finally, an impervious and completely frozen (water-shutting) wall was constructed.

The ground improvement work using this method intended to improve the ground of approx. 570 m^3 in the upper part of the pilot tunnel protection work using 85 horizontal freezing pipes and 154 wall-attached freezing pipes. The wall-attached freezing pipes were installed on the primary earth-retaining wall to ensure freeze-bonding between the frozen soil wall constructed by the horizontal freezing pipes and the earth-retaining wall (Figure 13).

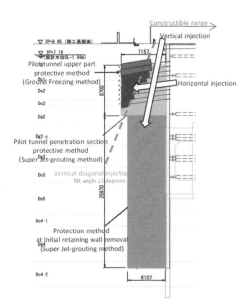

Figure 12. Vertical injection and Horizontal injection sectional view.

Figure 13. Wall-attached freezing pipes.

5891

The surrounding ground impact analysis indicated a ground settlement of approx. 20 mm. The settlement measurement during work showed the value below the former, meaning the settlement had no impact on the surrounding buildings.

3.4 Pilot tunnel excavation – Messer steel plate method

The shape of pilot tunnel excavation was examined, and various shapes were compared (Table 2).

As a result, since it was found that Plan A took time to erect the semi-circular shoring and required complicated structural system for the upper protection work, Plan B, which allows easy use of working space and upper protection work, was selected for the rectangular.

The Messer method involves surrounding a pilot tunnel section with several messer sheet piles, press-fitting sheet piles to the face side one by one, and excavating the face in the space surrounded by sheet piles with the press-fit sheet piles used as the primary lining. Behind the face, the method sets shoring with H beams and wooden sheet piles, repeats the processes of sheet pile press-fitting, excavation, and shoring, and creating a pilot tunnel.

In general, the Messer method constructs a pilot tunnel in one direction. At the construction site, the pilot tunnel was constructed in two directions. At the entry of the pilot tunnel, was dug toward the primary earth-retaining wall. The excavation direction turned to the 90° direction inside the entry part, was excavated as the general part of the pilot tunnel in parallel with the primary earth-retaining wall. The entry of the pilot tunnel was made bigger than the general part of the pilot tunnel so that excavation would re-start despite the change of excavation direction.

Since erection of shoring in the pilot tunnel had to be in a small space and materials could not be hoisted down from above, a handling machine fitted with an attachment that can hold steels was used (Figure 14).

Table 2. Comparison table of construction method

	Plan A: Horseshoe shape		Plan B: Rectangular	
Standard cross-section	3600 / 4100 / Ground improvement		3800 / 4200 / Ground improvement	
Timbering	H-200*200*6*12@800 (curved materials)		H-300*300*10*15@800	
Characteristics	1) Use of circular shape for the top of the pilot tunnel shoring will ensure rationalization of shoring members. 2) As the top shape is circular, the work space in the tunnel will become smaller than the case of a rectangular shape.		1) In the case of the rectangular shape, the work space inside the tunnel will be effectively used to enhance ease of operation. 2) We have experience of using this method in the similar scale project.	
Structural stability	The shoring is strong enough to sustain the total covering load, which demonstrates sufficient structural stability.	○	The shoring is strong enough to sustain the total covering load, which demonstrates sufficient structural stability.	○
Anomaly	The ground surface change will be small and pose no problem. (Approx. 1 mm in the public-private boundary)	○	The ground surface change will be small and pose no problem. (Approx. 1 mm in the public-private boundary)	○
Constructability — Advantages	1) Members to use are light in weight.		1) Either messer excavation, earth-retaining pile driving, or vertical chemical grouting in the back is easier to carry out than Plan A.	
Constructability — Drawbacks	1) When H beam splicing plates are installed using the erection derrick, there will be no room for lateral sliding. 2) It will pose restrictions to the length of the rod to conduct chemical grouting vertically into the rear earth. 3) As it requires the right-angle right turn from the entry of the pilot tunnel, a rectangular cross-section will be required for the pilot tunnel entry.		1) Members to use are heavier than Plan A.	
Work cost	1) Although the upper shoring may be downsized from H300 to H200, the H200 bending process will be a custom-made item and requires higher cost than H300.	○	1) The volume of earth to excavate at the corner will increase, but it will be only a little, or approx. 70 m².	○
Work period	3.5 months (It will take time to erect the semi-circular shoring.)	△	3 months	○
Auxiliary method	The structural system will be complicated.	△	It is composed of straight members and can flexibly cope with the situation.	○
Judgment		△		○
Remarks	Auxiliary method : (1) when another cut-and-open operation is necessary to protect the upper part of the initial cut-and-open work or (2) when horizontal soil improvement is conducted from the primary earth-retaining wall for protection of the upper part of the pilot tunnel.			

Table 3. Comparison table of construction method

		Plan A: TBH method		Plan B: BH method	
Characteristics	Outline of the method	• This method combining the conventional reverse method with the top drive method is designed to be applicable to large-diameter excavation in small and low-height working space. • Generally it is excellent in excavation of gravely layers and treatment of slime.		• BH drilling machine (improved type) is used to drill holes. • After forward circulation for drilling, another process, which consists of insertion of the suction pipe to the hole bottom, suction of the slurry in the hole with simultaneous feeding of the clean bentonite solution through the hole opening to replace the solution near the hole bottom with the clean solution, will be added.	
	Excavation method	Reverse circulation		Forward circulation	
	Quality	1) Drill holes will have no particular problem. 2) The contents inside the hole will be replaced by clean solution.		1) Gravel size is small and does not disturb drilling. 2) The contents inside the hole will be replaced by clean solution.	
	Machinery	It will require remodeling to match the conditions of smaller-scale pilot tunnels.		The general-purpose machine can be used.	
	Plant	Wide circulation solution plant will be necessary on the ground surface.		Smaller in scale than Plan A	
	Noise and vibration	Less noise or vibration		Less noise or vibration	
	Bearing capacity	• No problem (working load = 804.8 kN)		• It can support the vertical load produced by skin frictional force at the embedded part during excavation and will have no problem as supporting piles (allowable bearing capacity = 1,145 kN).	
Work cost		More costly than Plan B	△	Less costly than Plan A	○
Work period		It requires a lot of time to relocate equipment and materials although its drilling capability is high.	○	Although it allows shorter time for equipment relocation, its drilling capability is slightly poorer. In total, it is almost the same as Plan A.	○
Judgment		△		○	

Figure 14. Messer steel plate method situation.

Figure 15. Driving into hole Secondary retaining wall situation.

3.5 *Driving of secondary earth-retaining pile – Boring Hole method*

Methods to drive secondary earth-retaining piles were reviewed and compared, with machines applicable inside a pilot tunnel taken into consideration (Table 3).

The review concluded that both of them were applicable to the site conditions. However, Plant B "BH method" was chosen for its cost effectiveness.

The BH method uses a boring machine with powerful driving force. A bit attached to the front end of the boring rod is rotated to excavate the ground without casing. As the necessary machine is small in size, it can be applied to a small space like the inside of the pilot tunnel.

To begin with, chemical grouting was applied from inside the pilot tunnel to make the lower part of the tunnel water-tight to ensure water cutoff performance against the back of the secondary earth-retaining wall during cutting-and-opening. To construct the secondary earth-retaining wall, 44 H-beam cores (H - 414 x 405) measuring 24.5 m in length were driven into holes, each 700 mm in diameter, drilled by the BH method. After drilling with the BH method, short (2.0 m and 2.5 m) cores were hoisted down using erection derrick, mini crane, and chain block, and bolts were bonded. This operation was repeated to set the core steels to the prescribed height (Figure 15).

3.6 *Upper protection work - Pipe roof method*

Upper protection work plans were reviewed and compared as it was necessary to support the loads in the pilot tunnel area in order to remove the primary earth-retaining wall, excavate the earth behind the primary retaining wall, and cut and open the ground for structural extension in the space between the secondary earth-retaining wall and the main line structure (Table 4).

The review results evaluated five plans as follows: (1) Plan 1 has drawbacks including instability of the natural ground during excavation and structural extension and enormous

Table 4. Comparison table of construction method

	1) Re-freezing + pilot tunnel widening + steel placement	2) Temporary working piles + steel placement	3) Pipe roof	4) Core boring + φ 300 class steel pipe insertion
Schematic diagram				
Outline	• It widens the pilot tunnel at the side and installs steel materials as in the case of messer pilot tunnel excavation.	• It creates a work space with pipe roofs from inside the pilot tunnel and places steels there.	• The pipe roof of Plan 2) is used as the structural member.	• Its structural features are the same as Plan 3), with the pipe roof cost reduced.
Structure and specifications	• It supports the upper earth load with 300 mm high steels. • Re-freezing to ensure sufficient strength of the upper part will allow excavation of the work space without shoring.	• The structural features are the same as Plan 1).	• Pipe roof steel pipes, about 300 mm in diameter, are used to support the assembly.	• Steel pipes, about 300 mm in diameter, are used to support the assembly.
Constructibility	• It will have some problems in excavation in small space although it shares the same process as pilot tunnel excavation.	• It has a lot of work processes including pipe roof, excavation in a small area, or installation of steels.	• It requires provision of a sturdy frame to install pipe roofs on the main line structure. • The primary earth-retaining wall struts may disturb the work. Continuous drilling may be difficult.	• The clearance of primary earth-retaining wall core materials is 400 mm which is the upper limit of casing pipe installation space. • Provision of a simple scaffold at the main line structure side will allow this system to be installed. • Steel pipes are heavy.
Quality	• As freezing pipes are inserted slightly in a slanted position, it will remain some natural ground part near the back of the primary earth-retaining wall. It can cause some instability during pilot tunnel widening.	• It is most likely to make the natural ground above the work space loose.	• It needs some ideas to fix the circular pipe roofs to H beams on the secondary earth-retaining wall.	• Same as left
Safety and environment	• It can likely cause upheaval and settlement by freezing (possible disturbance to buildings).	• It has some safety problems such as flaking of sediments during fixation of pipe roof as it requires excavation of the ground that is once loosened.	• No problem in particular	• No problem in particular
Work period	3.5 months (More time will be separately needed for freezing operation.)	4 months	3.5 months	2.5 months
Cost	• An enormous amount of money will be required for re-installation and removal of equipment and operation of freezing equipment.	• It requires a variety of operations in a small space, which makes it a costly method.	• It requires a lot of cost to install pipe roofs.	Work is relatively easy.
Evaluation	×	×	△	○

cost of re-freezing; (2) Plan 2 faces safety problems including flaking of earth during fixation of the pipe roof; (3) Plan 3 requires installation of a sturdy and large work platform for installation of the pipe roof and high cost; (4) Plan 4 can drill holes through to the pilot tunnel with casing pipes, insert and install the pipe roof inside the hole, and carry out the work with a simple scaffold. Consequently, Plan 4, "core boring + insertion of φ 300 class steel pipes," was chosen.

After setting the simple scaffold, 53 casing pipes were installed by core boring, pipe roof steel pipes were inserted into the pipes, and the secondary earth-retaining piles were used to support them (Figures 16 and 17). After setting the pipe roof, the inside of the pilot tunnel was backfilled and closed to transfer the load of the upper part of the pilot tunnel to the pipe roof and the secondary earth-retaining wall. While liquefied stabilized soil was used for backfilling, no bleeding was conducted for the ceiling part, and the voids were filled with air mortar that can fill up voids and closed.

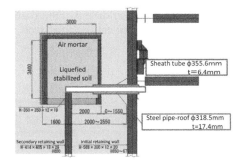

Figure 16. Upper protection work sectional view.

Figure 17. Pipe roof method completion situation.

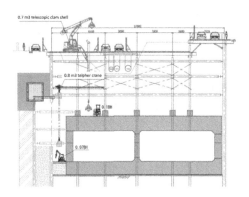

Figure 18.   Narrow part excavation sectional view.   Figure 19.   Removal of temporary pillar.

3.7   *Narrow part excavation - Open-cut method*

This method conducts, between the secondary earth-retaining wall and the main line structure, removal of the primary earth-retaining wall, excavation of the earth behind the primary earth-retaining wall, excavation of the antecedently backfilled liquefied stabilized soil, chipping of replacement concrete, removal of temporary earth-retaining (square steel pipes), installation of earth-retaining shoring, and construction of the extension structure according to the procedure of open-cut method.All these operations are conducted in a small area. Appropriate machines should be selected, and their appropriate layout should be developed. Since excavated earth or liquefied stabilized soil cannot be directly hoisted up on the ground, a 0.8 m³ telpher crane was installed under the road, and the earth was temporarily placed on the top slabs of the main line using the crane and lifted up on the ground with a 0.7 m³ telescopic clam shell(Figure 18).

3.8   *Removal of temporary pillar - Wire sawing method*

Prior to removal of temporary pillar s (a 1.5 x b 1.0 to 2.0 x h 6.7 m), a temporary receiving bent platform was assembled to remove the impact load that occurred during cutting of temporary pillar s and was imposed on the top slabs or transfer the load working on the temporary pillar s to the side wall. Then, the movable scaffold was assembled, and isolation drilling was conducted by core boring drilling (160 in dia.).Continuous 2 x 2 core boring drilling was conducted to pull out blocks. After the blocks were cut by the wire saw, a separated large block, weighing approx. 6 ton, was carried by a 10 ton class forklift (Figure 19).

4   CONCLUSION

The underground widening excavation method used to realize construction of an expressway ramp branch in a limited space under the narrow urban environment is highly expected to show its advantages as an option applicable to underground space construction in urban areas that go more and more complicated in the future. Various measures were taken to minimize the negative impacts on the neighboring residents, and ultimately the exit ramp at the main line widening section and in the non-open-cut part of the rectangular shield tunnel work was successfully completed. It is the authors' hope that this report will be used as reference data for similar kinds of projects since construction projects under space restrictions such as near residential areas or in urban areas, like our project, are expected to increase in number.

REFERENCES

Takahisa Fukushima, Michiharu Higashi, Takashi Kosaka, Naofumi Matsukawa, Mikihiro Watanabe: About non-Cut and cover tunnel method (Rectangular shield and Underground opencut method) in Hanshin Expressway Yamatogawa Route, WTC 2018 of ITA, Apr. 2018

Tunnels and Underground Cities: Engineering and Innovation meet Archaeology,
Architecture and Art, Volume 12: Urban
Tunnels - Part 2 – Peila, Viggiani & Celestino (Eds)
© 2019 Taylor & Francis Group, London, ISBN 978-0-367-46900-9

Brescia Driverless Metro Line – San Faustino Station: A joint work example between client, superintendence, designer and P&CM during the construction phase

A. Merlini
Brescia Infrastrutture, Brescia, Italy

A. Breda
Superintendence Archaelogy Fine Arts and Landscape for the Provinces of Bergamo and Brescia, Medieval Archaeology Official, Brescia, Italy

I. Carbone
Consultant and past P&CM Manager for Brescia Driverless Metro Line, Milan, Italy

M. Gatti
Rocksoil S.p.A., Milan, Italy

ABSTRACT: The archaeological investigation, requested by the "Brescia Superintendence" during the construction phase of the Brescia Driverless Metro Line, highlighted significant archaeological evidences at the areas of the San Faustino Station. The archaeological findings describe part of the historical access system of the city and the nineteenth-century bridge with a double arch that allowed crossing the protection moat of the Venetian Walls (thirteenth century). The Brescia Superintendence required to revise the approved project of the Station, given by the value of the archaeological evidences and make them visible to the station passengers in trans-it. The station, located at a depth of about 30 m below the surface level, has been opened in 2013. The paper highlights the importance to have a sensitive Client who drives all the stakeholders, specially the Designer and the P&CM, developing innovative solutions, in terms of architectural approach and construction technologies.

1 INTRODUCTION

During the 1980s, road congestion in the vicinity of Brescia rose dramatically, resulting in the City Council becoming interested in the adoption of a new mass transit platform to provide an alterative means of access around the city. Following studies of several mass transit systems, it was decided that the development of a light metro would be the most suitable option.

In 1989 Brescia City Council commissioned ASM Spa with the drafting of the tender. At a later stage, in 2003, Brescia Mobilità Spa awarded the construction work, via a Design and Build Procurement procedure, to Ati Metrobus, a joint-venture composed of Ansaldo Sts, Ansaldo Breda Spa and Astaldi Spa, which executed the works up to 2011. Brescia Infrastrutture S.r.l., a company fully owned by Comune di Brescia, was set up that year. The company brought the project to completion, tested it and started operation of the entire line in March 2013.

The Brescia Metro is a rapid transit network serving Brescia, Lombardy, Italy. Today, the network comprises a single line, having a length of 13.7 km and a total of 17 stations from Prealpino to Sant'Eufemia-Buffalora, located respectively at the north and southeast of Brescia (Figure 1).

All stations have been outfitted with automated doors along their platform edges. This is to act as a safeguard against passengers entering onto the track, accidentally or otherwise, while also increasing passenger comfort by reducing their exposure to wind and noise from train movements through the stations.

2 BRESCIA DRIVERLESS METRO PROJECT AND SAN FAUSTINO STATION

2.1 *Metro line*

Brescia's Automatic Light Metro was built in a heavily populated area and it stretches for about 13,7 km of double-track line divided as follows (Figure 1):

– approx. 3 km of double-track excavated using the Cut & Cover method;
– approx. 6 km of double-track bored tunnel, excavated using the TBM (∅ 9,15 m);
– 5 km of double-track line, including trenched, street-level and elevated sections (1,7 Km);
– 17 stations, of which 13 underground and 7 above surface;
– 13 underground shafts for safety access;
– 1 depot-workshop running along 5 km of track.

A first archaeological survey was conducted by contractor Ati Metrobus during the tender stage. The survey was based on historical documents that highlighted the existence of archaeological remains in the city fabric, particularly in the oldest part of the city centre. No preventive survey involving excavation and aiming to show what exactly was lying underground was conducted. Once the contract was awarded to Ati Metrobus, the archaeological activities got underway as follows:

– a risk assessment report relevant to the intended works was conducted prior to the opening of the sites;
– the Archaeological Heritage Office supervised and offered assistance to the earthmoving works from the time of the opening of the sites (starting in April 2006).

Brescia Infrastrutture S.r.l. worked in close collaboration with the Archaeological Heritage Office; this collaboration allowed to fully understand the importance of surveying and hands-on knowledge, of classifying individual findings within the historical and archaeological context, of properly assessing the repeatability or the uniqueness, as the case may be, of a given find so that it can be fully appreciated by users.

Figure 1.   Brescia Metro Line Layout.

2.2 *Archaeological activities related to the construction of San Faustino Station*

One of the main urban areas where has been planned an excavation with archaeological evidence is the one of San Faustino Station (Figure 2). The surveys conducted at the various stages have unearthed: *a)* the Venetian walls (the old city walls) similar in shape and structure to those visible in other points of the city – Porta Pile or Porta Sancti Faustini (1254) with its flooring (Figure 3.a), the gateway to the city from the north (including a nineteenth-century bridge which in 1818 replaced the drawbridge, see Figure 3.b, crossing the moat and composed of two almost intact arches); *b)* a section of the Visconti walls basement (datable to the second half of the fourteenth century) which was superimposed by *c)* the next curtain belonging to the Venetian era (early sixteenth century; today well preserved in elevation for almost 300 meters along the adjacent Fossa Bagni, corresponding to northern moat of the city walls, d) the double archivolt and bridge-channel system (1518–1522) through which the Garza torrent, coming from the north, first passed under the counterscarp bank of the moat, then crossed the moat and finally entered the city crossing the walls and the embankment behind it, e) the remains of various medieval and modern structures located within the urban belt downstream of the gate.

The figures 4 and 5 show the relationship between the surface area of the station (in red), and the excavation work and archaeological remains actually found (in green) or theoretically existing (in blue). The assessment of the relevance and significance of the findings from an archaeological point of view determined the modification of the station project design.

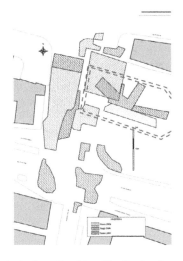

Figure 2. San Faustino Station: dashed red line identifies Station layout.

Figure 3. Porta Pile: a) Iseppo Cuman Map (1667, Brescia, Archivio di Stato, ASC 999); b) Stone bridge, built in 1818, found during metro line construction work (engraving of Fausto Pernici (1840–1850).

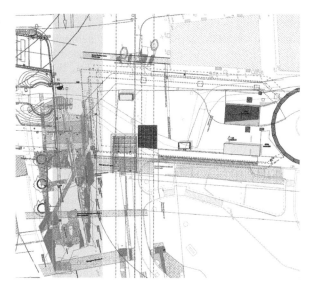

Figure 4. San Faustino Station (red dashed line), archaeological findings (green areas), historical structures at the ground level (light blue areas).

Figure 5. Archaeological excavation of the medieval gate and the stone bridge.

Further surveying was conducted in support of the excavation work until the completion of the project, as follows:

– in March-April 2009 excavation for the construction of the station took place in the moat alongside the Venetian walls;
– from November 2011 to September 2012 excavations took place to facilitate the rearrangement of the underground utilities layout and to build the access to the station;
– at the start of 2013 took place the final arrangement on the south side of the station.

3 SAN FAUSTINO AND THE ARCHAEOLOGICAL FINDINGS: AN INNOVATIVE APPROACH OF THE ARCHAEOLOGICAL PROJECT

The archaeological survey (Gallina, 2003) has been carried out along the entire route of the new metro line (13 km), in addition to the various and complex actions related to the urban impact assessment (at that time prescribed by D.P.R. 554/1999 on the organization of public works, which was equivalent to the current D.L. 50/2016 – Code of Public Contracts). The study has been commissioned by ATI to a professional archaeologist, qualified to the task and above all well aware of the history of the city constantly assisted by the Superintendence for Archaeological Heritage of Lombardy. The investigation has been based on two types of sources:

a) historiographic sources (published and unpublished written documents, iconographic documents);
b) the dense archaeological literature that dates back to the seventeenth century for Brescia and that since the fifties of the last century has grown exponentially, as a result of numerous emergency excavations and research that have accompanied the impetuous transformations of the contemporary city.

Obviously, the greatest attention has been paid to the urban districts of via Verdi, via Dante Piazzale Battisti, via San Faustino, where the excavation of the "Vittoria" and "San Faustino" stations has been designed, the two areas of the historic centre that would affect by the archaeological findings. The survey underlined a lot of critical issues for both areas, in particular, the certain presence of the archaeological findings related to the main gate of the city and the adjacent walls of the Renaissance has been highlighted for the San Faustino Station Area. In addition has been detected also the possibility of unearthing of some structures belonging to the fortified apparatus of the Middle Ages. The prevision is confirmed by the findings made during the various phases of the archaeological investigations carried out between 2006 and 2007, but extended with necessary assistance until the year 2012 (Figure 2 – Figure 4).

Brescia Infrastrutture and the agreed on the advisability of an *alternative project* in consequence of the high density of archaeological findings distributed in relatively small area (3000 sq). The new project should allows the construction of the station (not located nearby due to urban constraints and logistical opportunities) and, at the same time, the preservation of important evidence of the city's history.

A solution has been defined with the Institutions involvement (Archaeological Superintendence of Brescia, Cremona e Mantova; Superintendence for Archaeological Heritage of Lombardy; Directorate General Archaeology – Ministry for Cultural Heritage and Activities). At the end the station has been built in the predestined place, while ensuring the conservation of most of the ancient structures and the visibility of significant portions of the same within the station.

The station's structure originally should have been built through open cut excavation by building the station retaining walls first, followed by excavation and the total removal of soil and material in the designated area; this would allow free access to the TBM, as was the case with the other deep-level stations involving the excavation of a natural underground tunnel.

Given the importance of the archaeological layer along the north-south axis of Cesare Battisti Plaza and given the possibility to carry out a single open cut excavation in the area occupied by the Fossa Bagni old moat between the axis of the metro line and the existing underground car park, the station project was reviewed with substantial modifications to both the spatial distribution and the construction technique.

The entire distribution of the vertical connections was included in the openly excavated volume, while the platforms, maintained in their original position, were built by widening the underground tunnel (excavated with the TBM boring right through), allowing us to safeguard the archaeological layer above. The archaeological remains found in Cesare Battisti Plaza caused inevitable delays in the works, which in turn led to a series of variations in the San Faustino Station design, conceived to respond in the best possible way to the complications encountered. The revision of the station architectural design, was aimed to answer to the following specific requirements:

− first of all, to minimise the impact of the works on the archaeological findings;
− secondly, to follow the same principles that inspired the design of the other deep-level stations: namely, that the station should be perceived as a single volume without intermediate horizontal levels and that natural light should reach the platforms.

The Heritage Office requested that the archaeological findings be exploited and that solutions to allow the findings deemed most interesting to be in view within the station, compatibly with the station layout, be thought out. These findings include some sections of the Venetian walls and the counterscarp, as well as the system across the Garza stream and the arches of the nineteenth-century bridge. The station is embedded exactly along the width of the moat, which means that the longer sides of the station volume run along the Venetian walls and the counterscarp. The design principle with respect to the archaeological findings was therefore to make the Venetian walls, the moat and the counterscarp available for the public to view. This solution allows the public to admire sections of the ancient walls on the south side, a part of the bridge at the foot of the main stair access on the west side and the round arches at the point where the Garza stream crossed the counterscarp to reach a rectangular weir cut into the width of the moat (see Figure 6.a).

The Venetian walls, sectioned in that stretch by the ramps, have been left in full view in the cladding of the wall masonry that runs along the stairs leading to the upper concourse (south side). The cross section has been left visible to the side of the stairs (see Figure 6.b).

In addition, a long and particularly well-preserved section of the Venetian walls has been left visible along the south side. It can be admired from inside the station through an ad hoc 15-metre long glazed opening. The section made in the vicinity of the Venetian walls and the construction of the all-glass lift shaft let light into the station, while allowing us to respond to the design requirements of a single volume and natural light on the platforms. Due to the severe constraints imposed by the context and the restricted space available, it was only possible to meet the said requirements as far as the lower concourse level (Figure 7.a). A view of the bridge, whose depth has been left partly uncovered so that the structure can be appreciated, opens along the west side of the upper concourse. A photograph, on display beyond the perimeter wall masonry, offers a better understanding of the whole system (Figure 7.b).

Furthermore, a part of the front section of the counterscarp is visible from inside the station through an opening located on the north side of the upper concourse. The large glazed

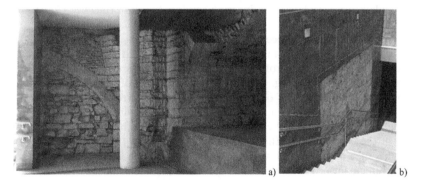

Figure 6.   a) Round arch of Garza stream; b) Venetian walls section visible to the side of the stairs.

Figure 7.   a) Large section of Venetian wall (sixteenth-century) visible along the south side of the station; b) nineteenth-century stone's bridge, upper atrium view.

Figure 8.   La grande finestrature che consentono al pubblico di percepire lo spazio della larga fossa che circondava le mura urbane: a) a sud verso le mura b) a nord verso la controscarpa.

opening on the south side (Figure 8.a) over the Venetian walls and the opening on the north side (Figure 8.b) over the counterscarp allow users to conjure up a mental picture of what the ancient system was like, giving them a sense of measure of the space that constituted the ancient moat.

4   THE CONTRIBUTION OF THE P&CM DURING PROJECT DEVELOPMENT

The new Metro Line of the city of Brescia and its metropolitan area marks a historical passage in the city life and determines a new way of relating with the urban fabric, the Contracting Authority decided to use technical engineering services for the Project & Construction Management (P&CM) of the works.

The international tendering procedure has been awarded by a group of engineering companies. with mandated Metropolitana Milanese S.p.A., today MM S.p.A., was the Authorized Representative. The P&CM is a professional service that uses specialized project management techniques to oversee the planning, design, and construction of a project, from its beginning to its end and the first months of operation. In the case of Brescia Metro Line, the P&CM went beyond the mere technical and regulation control by enlarging their operating area to the comprehension of the reasons behind the design, attuning themselves with the choices and prescription of the contracting authority. This Approach was fundamental for San Faustino Station and it is the normal way of work of MM S.p.A., which is the engineering Company of the Milan's Municipality with more than 60 years of experience in Metro Line realisation in Italy and abroad.

The Brescia Metro Line was a complex project, the P&CM decided to apply an organisational model which comply with the Contracting Authority needs and project requirements. The organization of P&CM includes a General Coordinator and a series of specialist and multidisciplinary managers who coordinate specific control work groups (cost, detailed and constructional design, Construction Supervision, safety coordination during the construction phase, for civil works and civil installations, railway installations and automation, rolling stock).

One of the main organisational choices of P&CM, which success has been later proved by results, was to employ most of the group engaged for the control phase of the project also for the Supervision during construction phase. The same choice has been made also for the railway system/automation and rolling stock, relatively to the phase of tests and trials, pre-operation and the first months of operation. The latter aspects are very tricky phase for a public transport system in general and, in particular for a Driverless Metro Line.

Particularly important were the interface and assistance activities to the Testing Committee and the Safety Committee. In particular, the representatives of P&CM worked from the outset within the Safety Committee – composed by representatives of the Transport Ministry and Contracting Authority to follow step by step all the various phases of the design – providing their contribution to the analysis of the documents submitted in progress by the Contractor and to the technical and contractual approval of the P&CM.

5 ENGINEERING: THE DESIGN AND CONSTRUCTION PHASE

The identification of archaeological finds required the modification of the architectural and structural design of the station, using underground excavation construction systems, without impact on the surface level and on the first thickness of the anthropic layer, where there were mainly placed the pre-existing to be preserved and enhanced.

The typological scheme of the stations located along the metro- underground section excavated by the TBMs referred to the cut & cover construction method. The San Faustino Station has been designed for cut&cover method only for a small section, the main longitudinal part of the station, containing the platforms of the station, has been located in a deep tunnel. The bottom-up shaft has been strategically located on the sides of archaeological finds, between the moat of Fossa Bagni and the car parking (Figure 9), by this way has been possible realise some openings on the retaining structures and make visible the archaeological finds. The portion of the shaft located along San Faustino Street has been connected with the deep tunnel. The shaft housed the vertical connections of the station, such as stairwells and elevators, technological and system components; the bottom part of the shaft includes the passageway to the two directions of travel of the trains, while most of the platforms have been placed inside the underground tunnel.

The civil works have been performed in a geotechnical context constituted by a 5,00 m thickness layer of natural and anthropic fill and by heterometric and polygenic gravel in a silty sand matrix, characterised by a non-cohesive behaviour. The geotechnical tests performed identified: a deposit with no cohesion and friction angle equal to 38°, while the Young's modulus (E) varying from 40MPa at the ground level to 220 MPa at the tunnel level. The water table height is 121,00 m above the sea level corresponding to a depth of 25–30m from the ground level and influences only partially the civil work. For the excavation of the station reinforced concrete diaphragm walls (open bottom-up method) executed by the use of hydro-mill were provided; this technology has been selected to guarantee the minimum verticality tolerances requested due to the reduced thickness available for both temporary and permanent structures. The use of the hydro-mill has also ensured better continuity and hydraulic seal of the joints. The diaphragm walls have been characterized by a thickness equal to 1,00 m and a width of 2,80 m, the diaphragm have been casted alternatively (primary and secondary

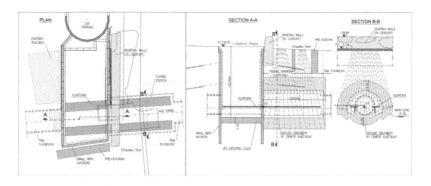

Figure 9.   Intervention overview: general plan and representative cross-section.

panel). The maximum excavation depth was 28,00–30,00 m, for this reason the depth of diaphragm walls was 43,00 m (embedded part equal to 12–13 m).

In some sections, the ones close to buildings and in correspondence to the existing car park, the construction of the diaphragms was brought forward by the installation of micropiles to protect the foundations of the buildings. The diaphragms walls affected by the passage of the TBM have been reinforced with fibreglass bars, according to the logic of *soft eye*.

In order to guarantee the seal of the bottom of the excavation with respect to the hydraulic pressure, it was decided to create a bottom plug by means of jet-grouting columns characterized by a nominal diameter of 1400 mm and a mesh of 1,05 m per 0,95 m. The execution of a field test led to calibrate the operating parameters of the treatments, with specific energy of 38–40 MJ/ml. During the excavation phase, the diaphragms have been contrasted by the installation of ground anchors (6–8 steel cable strands), arranged in numerical order of 2 per panel following a lowering excavation step of 5,00–8,00 m.

The design and execution of the station tunnel was a very demanding work. The ground improvement around the tunnel face has been performed by grouting activities carried out from the surface and localised in the areas identified as "free" as a result of the execution of archaeological surveys. The overburden in correspondence of the tunnel crown was about 15–18 m. The soil injection technology with cementitious mixture and integrative water-proofing mixtures by "tube-a-manchette" equipped with 3 valves per metre has been adopted. Since drilling had to be carried out from rather restricted areas, inclined drilling sequences have been defined, with the aim of obtaining 1,60 m by 0,80 m meshes at the bottom of the perforation, thus imagining a penetration radius around 90 cm. The station tunnel had an excavation diameter of 18,00 m, with a total excavation area of 168 m^2; a 4,5 m thickness around the tunnel crown was consolidated and about 3.5 m in correspondence of the invert, considering the modest hydrostatic head.

The construction phases of the works have been also modified with respect to the solution envisaged in the project and adopted for the other stations. The station excavation, originally, has been designed providing the excavation up to the level of the bottom slab, allowing the break-through of the TBM, i.e. through the station completely excavated. The construction of the metro line took place in sections from station to station; inside the excavated station the preparations for TBM starting were built, such as cradle and thrust frames, or slab for TBM arrival at the station after the tunnel excavated section. In the case of the San Faustino station, on the other hand, the TBM was forced to bored through the station, i.e. without having completed the excavation of the station; this was due to the accumulated delays, compared to the excavation schedule of the TBM, with the sets out of the archaeological variant. In order to allow the TBM tunnel to be maintained in safety, the excavation of the station was stopped at the necessary height to guarantee an overburden of 5,00 m in correspondence of the tunnel crown. The numerical analyses shown how this covering could guarantee of static stability conditions for the segmental lining ring; a reduced load on the crown would have decompressed the joints between the precast segments, thus not allowing the monolithic nature of the lining. In order to proceed as quickly as possible with the excavation of the station shaft, so as to start the construction of the civil works and the subsequent installation of the systems, excavations were carried out in the portion furthest from the transit way of the TBM, separating the two areas of the station, located at different levels of excavation, through the construction of a retaining wall by jet-grouting. Once the excavation by TBM had been completed for the entire metro line, it was possible to dismantle the service structure inside the tunnel serving the TBM (power lines, piping for supplies, etc.) in the sector inside the San Faustino station and proceed with the completion of the excavation of the station. For this reason, a set of rings was shoring up with steel arches and plates at the joints, by this way the excavation machineries could operate at the tunnel crown in safe conditions.

It was then possible to open a first window by demolishing some segments of the lining and proceed with the progressive filling of the tunnel, through the excavation resulting soil and its demolition until the excavation reaches the bottom side of the station. Once the excavation of the station between diaphragms was completed, up to the mechanized excavated tunnel, it was possible to proceed with the excavation of the tunnel station, enlarging the excavation already carried out by the passage of the TBM. The excavation proceeded from the station shaft

Figure 10.   Work Phases: the excavation of the station shaft and of the tunnel station.

towards the tunnel, thanks to the consolidation already carried out from the surface. The excavation was carried out for single excavation step of 0.90 m, by installing a sprayed concrete layer (30 cm thickness) reinforced by 2 IPN240 steel arch; the precast segmental lining was removed together with the ground of the tunnel front. The delays accumulated with respect to the original time schedule led to the speeding up of the internal structures construction through the use of precast elements; the structural system also had to adapt to the new architectonical layout of the station, providing large spans that could enhance, during the operation of the station, the view of the archaeological finds detected.

6   CONCLUSIONS

The Driverless Metro Line of Brescia and, in particular, San Faustino Station represent the opportunity to create a dialogue between past and present.

The fundamental choice of the Municipality of Brescia and of the Companies which have taken the leadership of the Project over the years, was to involve the Superintendence from the beginning and be supported by specialists (designers, P&CM and contractors) able to provide the maximum contribution to the design and implementation of innovative solutions, considering a global working method, which provides a constant cooperation between the different stakeholders involved in the project: Institution, Client, Contractor and Designer.

The city of Brescia has always been strongly oriented towards innovative solutions. It's enough mentioning the Brescia is one of the few Municipality in the world to have built a driverless metro in a metropolitan area of about 400,000 inhabitants.

The San Faustino Station is able to enhance the archaeological findings maintain its functionality, thanks to a perfect insertion into the pre-existing area; this result was achieved with a fully revised of the technical solution for the station, adapting the project to the findings and applying very demanding construction solution.

REFERENCES

Archivio di Stato Brescia, *ASC 999*
Gallina, D. 2003. *Brescia – Progetto Metrobus: Valutazione di impatto archeologico*, Ati Metrobus, Brescia.
Lunardi, P. 2008. *Design and construction of tunnels: Analysis of Controlled Deformation in Rock and Soils (ADECO-RS)*. Berlin: Springer.
Merlanti, P., Bellocchio, A., Sicilia, R. 2008. Analysis and prediction of subsidence phenomena induced by excavation employing an EPB-S in the works to construct the Brescia metro design predictions and back-analysis, in *Proceeding of AFTES International Congress 2008: "Le souterrain: espace d'avenir"*, 6th-8th October, Monaco, Germany.
Zanirato, L. 2018. *Metropolitana di Brescia – Una nuova concezione di mobilità. Brescia subway – Mobility redefined*, Massetti Rodella Editori, Roccafranca (BS), Italy.

*Tunnels and Underground Cities: Engineering and Innovation meet Archaeology,
Architecture and Art, Volume 12: Urban
Tunnels - Part 2 – Peila, Viggiani & Celestino (Eds)*
© 2019 Taylor & Francis Group, London, ISBN 978-0-367-46900-9

Tunneling under the Dutch capital with minimum overburden – the Victory Boogie Woogie Tunnel

J. Mignon
Combinatie Rotterdamsebaan, The Hague, The Netherlands

P. Janssen
Projectorganisatie Rotterdamsebaan, The Hague, The Netherlands

M. Potter
Combinatie Rotterdamsebaan, The Hague, The Netherlands

ABSTRACT: Currently in the south-east of the Dutch capital The Hague the Rotterdamse-baan road project is being tackled at a cost of some 272 million euros, almost ten years after the Hubertus Tunnel was opened. The section to be built runs from the Ypenburg interchange to the Central Ring, passing through the communities of Leidschendam-Voorburg, Rijswijk and The Hague. After a four-and-a-half-year period of construction the roughly 3.8 km long road link is due to be opened in July 2020. Apart from two ramp structures and a 650 m long tunnel built by cut-and-cover, passing beneath the A4 motorway, an exit and the connecting arc from the A4 and A13, there are also two urban tunnel tubes, called the "Victory Boogie Woogie Tunnels", to be constructed by TBM, each 1.65 km long. The tunnels with an internal diameter of 10.15 m will have a fire-resistant segmental lining. Between the two tunnels six cross-passages will have to be constructed using ground freezing method. The tunnel alignment is characterised by low overburden in the beginning and the end of the tunnels and a rather small clearance between the two tunnels. The tunnels underpass several residential buildings as wells as a harbour area. Due to the geological formation, predominantly comprising of sand, but also interstratified by silty clay and peat, a slurry TBM is used to construct the tunnels. Several innovative concepts to reduce energy-consumption are used in the project as for example tunnel vehicles with electrical drives.

1 GENERAL DESCRIPTION

1.1 *Construction project and client*

The project embraces the designing, production and servicing (Design, Build and Maintain) of the two-lane connecting route between the Ypenburg interchanger located in the south and The Hague's Central Ring (Figure 1).

The new road section to be created is about 3.8 km long and includes the following construction phases:

- Construction of a roughly 650 m long and 20 m wide tunnel by cut-and-cover, which under-passes the A4 motorway, the A4/A13 connecting road as well as the Laan van Hoornwijk.
- Construction of the 420 m long access ramp including the launching shaft at the Vlietzone, the subsequent operating building and a cross-passage between the two driving directions.
- Construction of two segmental lined tunnels, each having an inside diameter of 10.15 m and a length of 1.65 km. ("Victory Boogie Woogie Tunnel"), as well as six cross-passages between the tubes.

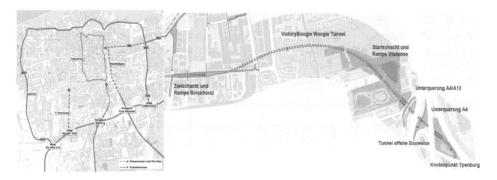

Figure 1. Project overview of the Rotterdamsebaan with on the left the alignment in green and on the right rotated by 90° and showing the individual sections.

- Construction of the 350 m long access ramp, including the arrival shaft at the Binckhors-tlaan, which will also have an operating building and two cross-passages.

This new connection is built on a turnkey basis, including all traffic and controlling installations. After commissioning, the contractor is responsible for the maintenance of the infrastructure for a period of 15 years, which is also an integrated part of the project.

The Rotterdamsebaan is a joint project involving the municipalities of Leidschendam-Voorburg, Rijswijk, the metropolitan region of Rotterdam-Den Haag and the Dutch Ministry of Environment and Infrastructure, represented by Rijkswaterstaat. It is part of a superordinate transport route plan, that is intended to improve daily commuter traffic in the center of The Hague.

1.2 Geological and hydrological conditions

The project is situated in the proximity of old marine deposits in The Hague's polder area. Before the tender phase, a soil investigation program was executed in order to provide a geotechnical longitudinal profile of the 3.8 km long stretch (Figure 2).

This investigation was the basis for tendering. Soil parameters (maximum and minimum values for soil density, friction angle, E-modulus, etc.) as well as groundwater levels (high and low ground water) were provided by the client. The layers of soil close tot the surface mainly consist of loosely bedded sand, interspersed with clay and silt lenses. Furthermore, larger settlement sensitive peat sections, especially in the vicinity of the site installation area, had to be taken into account. A roughly 6 to 8 m thick, densely bedded sand layer is located underneath. A clay layer of up to 5 m thickness with peat adhesions on its lower side is to be found below the sand layer. The level of the clay layer is just below the floor of the excavation for the launching and arrival shafts. For the dimensioning of the individual structural parts, the various soil layers must be considered, taking a variation of the given soil parameters (between favorable and unfavorable) into account. The groundwater level is to be found practically at

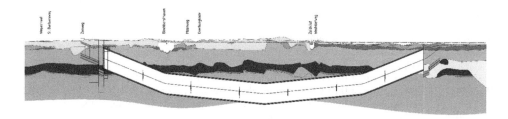

Figure 2. Geotechnical profile.

the surface as is common in the Netherlands. In addition, the tunnel underpasses the Binck-khorst port and some inland canals. A rise of the average groundwater level within the projected life time of 100 years was also to be taken into account, as well as measures for dealing with a groundwater rise during the construction period.

1.3 *Local residents, existing infrastructure and buildings along the alignment*

Because of the urban location, especially for the two shield-driven tunnels, the arrival shaft as well as the ramp structure in the Binckhorstlaan, special measures had to be taken for planning and execution. The tunnel tubes pass below many residential buildings with shallow foundations as well as under offices and other business premises founded on piles. The depths of these piles are not completely charted. The contractor had to carry out special surveys prior to the tunneling, so that eventual extra measures could be accomplished.

In the Binckhorstlaan a number of abandoned buildings were demolished prior to the start of the construction activities. These buildings had to be knocked down to create space for future constructions, as they didn't harmonize with the new concept that was adapted for this area. For the businesses and stores in operation, it had to be assured that they were accessible during the whole construction. It was essential for bidders to steer the construction phases to fulfill this requirement.

2 SCHEDULING AND CONTRACTUAL MILESTONES

The construction period for the Rotterdamsebaan project started in December 2015 and shall end in July 2020 at the latest, with the opening of the new road section. Apart from the production of the execution planning for all the civil constructions, the contractor also needs to produce all the documents necessary to obtain the commissioning from the Dutch authorities. Tabel 1 shows a summary of the construction schedule.

3 THE VICTORY BOOGIE WOOGIE TUNNEL

The two tunnels of each almost 1650 m length, which pass underneath numerous buildings, are driven by mechanical means and form the core of the Rotterdamsebaan. The tunnel will be named the Victory Boogie Woogie Tunnel, referring to the famous abstract art work of the Dutch painter Piet Mondriaan. The primary colors yellow, red and blue that Mondriaan used for his painting are predominantly present on the construction site.

3.1 *Choice of the Tunnel Boring Machine*

Already in the tender phase, the joint venture Combinatie Rotterdamsebaan had considered the reutilization of the TBM that bored the Sluiskiltunnel in the south of the Netherlands.

Tabel 1. Construction schedule.

Time period	Construction section / milestone
2016–2019	Construction of the launching shaft and the access ramp at the Vlietzone
2016–2018	Reconstruction of the Laan van Hoornwijk, construction of the underpasses of the connection A4/A13 and the A4 itself
2017–2019	Construction of the arrival shaft and the access ramp in the Binckhorstlaan
2018–2019	Construction of the two bored tunnel and the six cross-passages
2019	Finishing of the two tunnel tubes and access ramps and connection to the existing roads
2019–2020	Technical tunnel systems including testing
July 2020	Opening of the new connection

Figure 3.   TBM at the Herrenknecht facilities in Kehl, Germany.

JV-Partner Wayss & Freytag Ingenieurbau had taken over this TBM after conclusion of the project. As a result, a tunnel boring machine with a slurry-supported working face, having a boring diameter of 11.37 m and 3.6 bar design pressure is used to construct the Rotterdamsebaan.The entire TBM is about 75 m long and consist of a shield, a bridging structure and two back-up trailers. Apart from refurbishing measures due to wear, only slight project-specific adaptations had to be made.

The TBM was refitted at the Herrenknecht facilities in Kehl, Germany during the summer of 2017 (Figure 3). It was transported to the site in October 2017.

3.2   Tunnel design

The decision to reuse the Sluiskil TBM implied that there were certain fixed boundary conditions for the design of the concrete segmental lining. The necessary thickness of the lining of the Rotterdamsebaan is 40 cm, 5 cm less than on the Sluiskiltunnel. Other technical optimizations had to be undertaken, mostly due to the tunnel installations.

The segmental lining has to resist the Dutch Fire Curve. The elements have to withstand under load a fire impact of 120 minutes with a peak temperature of 1350 °C. The full-scale tests were executed at the MFPA in Leipzig, Germany in November 2016. After the successful conclusion of the tests, the production of PP-fiber reinforced elements (about 13000 pieces) started at the precast factory of Max Bögl in Hamminkeln, Germany (Figure 4).

Figure 4.   Segment production at Max Bögl in Hamminkeln, Germany.

The ring distribution as well as the ring width remained unchanged compared to the Sluiskil project at 7 + ½. However, the internal and external diameters were respectively adjusted to 10.15 m and 10.95 m. In order to improve installation, the opening angle of the key-stone, at half the length of a standard stone, was slightly reduced.

The reinforcement concept is based on a modular assembly of the reinforcement cages, that are made of matting and prefabricated ladders, which are welded to the connecting points.

Both static proofs based on strut and tie models, as well as with finite elements were demanded by the client. In particular the situation close tot the launching and arrival shafts as well as the very narrow space between the two tubes had to be taken into consideration.

3.3 Special measures for the TBM passage

Due to the specific geological conditions with peat and layers of clay, the tunnel alignment close to the surface and the urban environment, special measures were necessary to secure the stability of the tunnel face during construction as well as the stability of the lining in the operational phase. During the tender phase, the idea arose to increase the length of the tunnel towards the two access ramps. Given the fact that both tubes are already quite close to one another, an increase towards the ramps makes this situation even worse as the space decrease more and more. These conditions called for creativity in the choice of measures.

For the launching of the TBM in the Vlietzone, the soil was partly replaced. The layers of peat were removed and replaced by sand. To compensate the lack of overburden, a weight slab of sand was installed directly beside the shaft over a length of roughly 50 m. In order to assure tightness during the launching procedure of the TBM, a sealing block of low resistance mortar was placed in front of the start-up wall. In this area, the tubes at a distance of only 1.40 m.

3.4 Logistics and peripherical equipment

The separation plant (capacity 2500 m³/h), the centrifuges for the treatment of the fines, the compressor station, the neutralization unit and the badging plant were all installed on the surface foreseen for the subsequent road to the tunnel. As this area needed to be pre-consolidated anyhow, it was the best option to install all the peripherical equipment, as well as the stock for the segments here (Figure 5).

For the logistics vehicles on tyres were chosen because of the gradient of the access ramp (> 4%). The vehicles are entirely electrical and battery-operated, a particularly sustainable drive system that is used here for the first time (Figure 6).

3.5 Schedule

The TBM was transported from Kehl to The Hague by ship. It arrived in pieces in October 2017. The assembly took until Christmas 2017. The first tube was bored from January 2018 until June 2018. After having entered the arrival shaft, the shield was dismantled and transported back by boats on the canals (Figure 7).

The back-up trailers were pulled back to the arrival shaft through the tunnel.

The second drive started in September 2018 and ended just before Christmas 2018.

3.6 Cross-passages

Six cross-passages must be built to connect the two tubes and so create an escape route about every 250 m. Under protection of frozen ground, the tunnel lining will be opened. The cross-passages will be built in the frozen ground using conventional tunneling methods (Figure 8).

At each future cross-passage, some 20 freezing tubes will be drilled from the eastern tunnel (the first drive) to the western tunnel. The frozen area that is produced by each freezing tube is about 1.50 m in diameter. The freezing tubes are bored in a circular order, at a distance of

Figure 5. Peripherical equipment and segment stock.

Figure 6. Electrical tunnel vehicle.

1.00 m from each other. Next to the freezing tubes, some extra tubes for temperature monitoring and for drainage are driven into the soil.

As the two tunnel tubers are located very close to each other, only 3.60 m, the cross-passages are relatively short. The cross-passage itself is only 2.6 m long. The connecting concrete structure on each side is 50 cm long. They have a horseshoe shape with a footpath of 1.20 m and a free height of 2.30 m.

Figure 7.   Transport of the cutting wheel by boat on the canal.

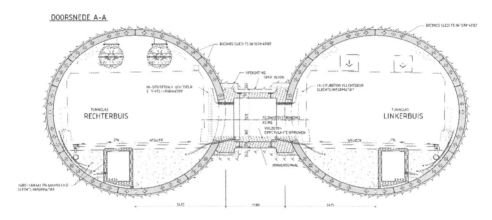

Figure 8.   Cross-passage.

After having cut in the frozen soil, a shell of 20 cm sprayed concrete is placed against the soil. A watertight membrane is then put against this shell. The cross-passage is finished with a reinforced inner concrete shell of 35 cm.

4   CONCLUSIONS

The two tunnel drives were performed as planned, with some delay on the first drive, but compensated by a higher advance rate on the second one. Although the small inter distance between the two tubes, the unfavorable soil conditions and the closeness of some foundations, settlements remained within the defined maximum values.

The tunneling process was an overall successful operation.

Tunnels and Underground Cities: Engineering and Innovation meet Archaeology,
Architecture and Art, Volume 12: Urban
Tunnels - Part 2 – Peila, Viggiani & Celestino (Eds)
© 2019 Taylor & Francis Group, London, ISBN 978-0-367-46900-9

Warsaw line II metro extension: TBM excavation under historical buildings in urban environment

M. Minno & V. Capata
S.G.S. Studio Geotecnico Strutturale S.r.l., Rome, Italy

T. Grosso
Astaldi S.p.a., Rome, Italy

ABSTRACT: The TBM excavation for the Eastern extension of the Warsaw line II metro has been completed in January 2018 by Astaldi/Gulermak joint venture. Excavation involved mainly fine and medium sands below water table which created a challenging environment for Earth Pressure Balance (EPB) TBM works. In order to reduce settlements at ground level for selected historical buildings, ground improvement was carried out through permeation grouting with the aim at reducing volume loss during excavation and inducing an initial heave of the buildings. Also, high-standard TBM performance was achieved with the use of bentonite and polymers, specific grout mix and controlled face and grout pressures. Extensive monitoring was carried out through manual and automatic measurements and results showed small movements of the buildings and very small damage. This paper is aimed at describing the challenges faced for the TBM excavation, the adopted solutions and the most relevant results.

1 INTRODUCTION

On March 2016, the Municipality of Warsaw, by the Investor Metro Warszawskie appointed Contractor Astaldi S.p.A for the North-East extension of the Metro line II, including the

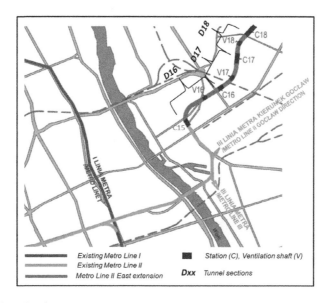

Figure 1. Tunnel works plan.

construction of 3 underground stations, an holding track and double twin tunnels on the area of Targowek-Praga districts. The section is the extension of the central part of the Metro line II, built by Astaldi and Gulermak joint venture and active since March 2015. The 6,3 km construction of the metro section was subdivided into 2x2,22 km long shield tunnel, and the rest on cut&cover technology for the station and ventilation shaft executions. Figure 1 shows a layout sketch of the whole project. The focus of this paper is the tunnel section D16 which is the most densely inhabitated. The 2 single-track tunnels have an inner diameter of 5,4m and the lining is made up of 5+1 precast segments with a thickness of 30cm. Classic steel reinforcement cage has been used and polypropylene fibers have been added to the concrete mix mostly for fire protection.

2   GROUND CONDITIONS

The tunnel route of interest is located within the area of regional geomorphological unit called Wisła River drainage basin. The morphology results from the glacial retreat occurred during the Central-Poland Glaciation and the subsequent fluvial processes which were continuously affecting the area during the Eemian Interglacial period, the Wisła River Glaciation, and the Holocene. The tunnel route encountered two geological formations (Pliocene and Quaternary deposits) other than anthropogenic fills. On a geotechnical point of view, the following units have been selected (from top to bottom):

- Fill (Holocene). It consists of uncontrolled and non-uniform anthropogenic fills of diversified origin (filled up old clay pits, cellars, old pavements and road sub-grades, mainly created in the course of war damages and post-war reconstructions) with a maximum thickness of 3 meters;
- Unit 4 (Quaternary). It consists of granular non-cohesive alluvial soils originating as fluvioglacial deposits, including mainly medium grain sands and coarse sands. Maximum thickness along the route is about 13 meters. N_{SPT} varying between 9 and 20 and a CPT cone penetration q_c of approx. 4MPa to 10MPa (with local peaks up to 15MPa);
- Unit 3 (Quaternary). It consists of granular non-cohesive alluvial soils originating as fluvioglacial deposits (Eemian interglaciation). These include mainly medium grain sands and coarse and dense sands. Maximum (verified) thickness along the route is about 15m. N_{SPT} varying between 16 and 40 and a q_c of approx. 15MPa to 45MPa;
- Unit 2 (Quaternary) – locally present. It consists of cohesive soils originating as marginal deposits of Odra glaciation. These include mainly clays and silty clays found in hard-plastic state. Maximum (verified) thickness along the whole tunnel route is about 13 meters. N_{SPT} varying between 14 and 17 and a q_c of approx. 2MPa;

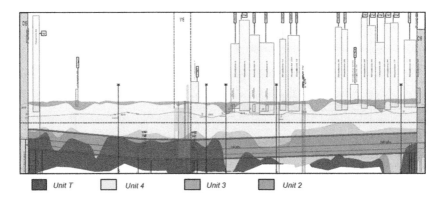

Figure 2.   Extract from the geotechnical profile along the D16 Right tunnel excavation (different scale in x and z).

– Unit T (Pliocene). It consists of cohesive soils originating from lacustrine deposits of Odra glaciation. These include mainly tills and silty tills. Maximum (verified) thickness along the route is about 10m. N_{SPT} varying between 9 and 15 and a q_c of approx 1–2MPa increasing with depth at about 1MPa/m.

In the area of interest, a single aquifer exists and it is related with the alluvial sand and gravel deposits. The water table level is found at approx. 2–3m below ground level and the aquifer is easily rechargeable due to the high permeability of the non-cohesive water-bearing strata.

The area along the tunnel route of interest is generally flat with the only exception of a railway embankment to be underpassed.

3 SITE CONDITIONS

While for the first 2 tunnel sections (D17 and D18) mostly green areas can be found at ground level, the area above the last tunnel section (D16) is densely built. Moreover, most of the buildings were built before the second World War and it is a almost a unique scenario in Warsaw since the city was highly destroyed and damaged during the war. These are generally 4–5 storeys masonry buildings with 1-level basement and founded on strip brick foundations; timber elements may also be present at the building roof or floor slabs. Many of them are in poor conditions and some of them show an extensive presence of cracks and spalling. Minor refurbishment and structural works were carried out in selected buildings during the years; just few of them have been completely re-built.

The 2 tunnels run along Strzelecka street directly below many of such buildings or in the close vicinity. In particular, the left tunnel is below one side of the buildings with the risk of inducing detrimental differential settlements. The soil overburden generally varied between approx. 7,7m and 12,1m.

4 TBM EXCAVATION

As in the Central part of the Line II metro, a mechanized excavation technology has been adopted for the tunnel excavation in order to minimize the impact on the city (reducing construction site areas, disturbance to public and traffic, damages to existing buildings and utilities, vibrations, etc.), improve safety of the workers and, at the same time, guarantee a high performance in terms of production and cost.

Two of the Earth Pressure Balance (EPB) TBMs successfully used in the heterogenous geological conditions of the Central part of the Line II were therefore renewed and adopted for the new excavation. At design stage it appeared already clear that excavation conditions

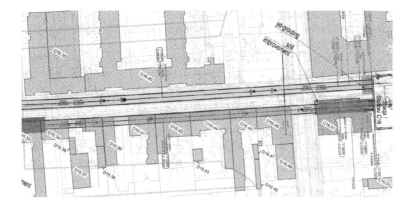

Figure 3.   Plan view of the building and tunnel location in the first part of D16 tunnel section.

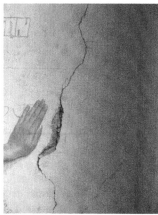

Figure 4.   One of the buildings that were underpassed by the TBM (existing conditions prior to excavation).

along this metro extension were challenging for several reasons: in some areas the ground conditions - clean sands and gravel below groundwater table often without any fine soils - were at the edge of the use of a EPB machine; excavation was generally shallow with several buildings and utilities above; buildings were often old, masonry-built and in poor conditions. Therefore, both prior to TBM launching and during the initial part of the excavation (on green areas), a series of improvements have been applied to minimize the volume loss, and therefore the subsidence at the ground level, and optimise the excavation process:

- Design pressure at the TBM front was increased by approx. 0.3bar comparing to values initially derived by standard calculations for face stability (Jancsecz & Steiner, 1994; Anagnostou & Kovári, 1996). Therefore face pressure was usually between 1.5bar and 1.7bar. Nevertheless, TBM was still able to excavate at good pace and without significant high thrust or torque;
- Although conditioning was used, due to the lack of fine fraction, the excavated soil was very liquid creating problem from a logistic point of view on handling the muck removal and affecting the control of the earth pressure at the excavation front. To face this problem, Contractor, Designer and Supplier decided to use, with very low dosage, a lubricating polymer able to decrease the water content of the muck so to better create a plug in the excavation chamber. This allowed the TBM operators to better manage the screw conveyor and hence the earth pressure at the TBM face;
- In selected sections bentonite was injected around the shield of the TBM during each advancement in order to have more plastic material around and at the tail of the shield;
- The bi-component grout quantity injected to the anular gap at the tail of the TBM has been generally higher than design value. Main parameters have been also improved; in particular, the gel time was decreased from 9 seconds to 7 seconds in order to have a more viscous grout. This allowed to achieve the design back pressures and, hence, to keep the cavity more stable. Design values of back-pressures were generally 0.3bar higher than face pressure.
- An appropriate specific configuration of cutting wheel dressing, with 14 disc cutters and additional rippers has been adopted; this was aimed at better protecting the excavation profile of the buckets from the gravels and pebbles encountered along the excavation and

Figure 5. Break-through of the Left TBM at C15 station.

allowed considerable savings in economic terms and time schedule avoiding any additional maintenance stop. This was decided following the arrival of the TBMs into the first station where a high wear of the cutter head tools was noted.

The TBM excavation works were successfully completed in January 2018. In the D16 tunnel, where high attention was paid due to the presence of historical buildings and utilities, measured Volume Loss was generally smaller than 0.3% with settlements of the buildings lower than 1.0cm and very minor damages recorded.

5 IMPROVEMENT WORKS

5.1 Ground movement and damage assessment

Ground movement assessment was carried out according to Burland methodology (Burland, 1997) with appropariate safety factors adopted to take into account the conditions of the buildings and their susceptibility to damage (vulnerability). A Volume Loss of 0.4% has been initially adopted and a trough width parameter as K=0.3 was generally selected for the excavation in granular materials (Mair et al., 1997). The outcome showed that improvement works were necessary to reduce the risk of excessive movements and damages, as already envisaged at tender stage. Therefore a series of interventions were carried out:

– Structural works aimed at improving the building capacity to resist to the induce ground movements, i.e. to reduce its vulnerability. This included steel anchors along the bearing walls, RC works to increase strength and stiffness of the foundation, replacement of some floor slabs (SGS only partially involved);
– Ground works aimed at reducing the magnitude of the settlements and differential settlements at building foundation level by reducing the Volume Loss during the TBM excavation and therefore minimizing the ground movements. This was done by permeation grouting works, as described in the paragraph below.

5.2 Permeation grouting

Permeation grouting technique adopts a low pressure injection system to improve the strength and reduce the permeability of granular soils. In the present works, 100mm diameter bores are drilled into the ground to the required depth with the use of external casing in granular materials. Due to the presence of the building above the tunnel the injections points were located along the street or the courtyard and pre-excavations were often carried out to identify the location of utilities. PVC guide-pipes were therefore used to host the sleeved pipe and avoid any damage to the surrounding utilities once the area was backfilled and working platform installed.

The TAMs (high resistance PVC pipes, 27/38 mm diameter with 3 valves per meter) have been then installed and the bore is filled with a weak sheath grout. The sheath grout acts to stabilize the hole and, when set, as a blocker to prevent injection migration vertically in the bore. Finally a sliding inner injection pipe with a double packer system is pushed down inside the TAM pipe and located at the required stage level. Injection can then start and will proceed sequentially in each valve in ascending stages. When the grout is pumped, the sheath is cracked and grout spreads in the surrounding strata. Injection pressure is low to allow the grout to penetrate the soil without significantly affecting the soil structure and avoiding fracturing or claquage. Only cement-grout injections were carried out in this project with fine cement (Blaine>5000cm^2/g), additives (anti-flocculant, dispersant, fluidificant), bentonite (b/c= 0.125). The design volume of grout to be injected (150 litres/valve) and Refusal Pressure (10–12 bar) were given. However, these parameters, along with the number of simultaneous injections occurring below the same building, had to be continuously reviewed and updated following the monitoring results.

A first aim of the permeation grouting is, of course, to reduce building settlements. In particular, during the TBM excavation works the improved soil helps in reducing the soil and groundwater movements towards the face of the tunnel and radially towards the tunnel lining behind the shield and therefore in reducing the Volume Loss and ground movements at ground level. However, there is a second aim which has been pursued with the permeation grouting and which can be seen as an '*advance compensation grouting*' (in relation to the standard compensation grouting, Mair R & Hight D, 1994). In fact, although the grout is injected at low pressure, some heave can be usually observed at ground level, especially for shallow works. This should be generally limited to a minimum in order to not give disturbance/ damage to the building above. However, the heave induced during grouting works can provide an initial upwards movement which will partially/completely compensate the future settlement induced by the TBM. This is a positive effect because it increases the amount of settlements that can be tolerated by the building and, at the same time, the building will experience a smaller residual settlement in a permanent condition.

FEM analyses were carried out in PLAXIS 2D to evaluate the positive effect of the grouting intervention. An area of improved soil was modelled around the tunnel and it was seen that a reduction of Volume Loss of approx. 30% of the greenfield value was achievable. This is in line with similar evaluations carried out for the Central Line (Lunardi et al., 2014). A damage

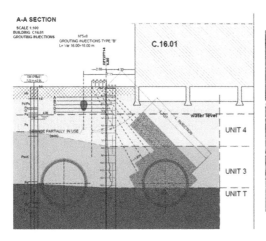

Figure 6.   One of the executed soil improvement works.

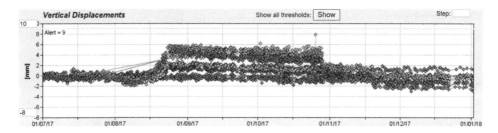

Figure 7. Vertical movements of Strzelecka Street 10 measured during injections (August 2017) and during TBM excavation (October-November 2017) through mini-prisms.

assessment showed that such reduction was sufficient to limit the expected damage within acceptable limits. Boreholes were carried out in the grouting area both before and after the grouting works. SPT tests, Le Franc tests and pressumeter tests were carried out. Comparisons of the results showed that improvement was mostly efficient in Unit 4. This could be seen: by visual inspection of the recovered soil cores; where loose sands are present, N_{SPT} showed a clear increase (N_{SPT} from 6 to 27); Menard modulus, fluage and limit pressures increased by at least 100%; permeability decreased by almost 2 orders of magnitude.

A comprehensive monitoring system was installed to verify the building response during injections and, later on, during the passage of the TBMs. This included manual readings (for ground pins, levelling pins on the building walls, crackmeters, clinometers), automatic readings (3D mini-prisms, electric crack-meters, electric clinometers) and visual observation. Therefore injections were tightly controlled in terms of grout quantity, max pressure and number of simultaneous injections in order to limit and control the building response. In certain situations, these were decreased down to 1 injection at a time and a maximum refusal pressure of 6MPa. A tight and successful collaboration was in place with the grouting Sub-Contractor (SCF).

5.3 Results

As discussed in section 4, the excavation in the D16 tunnel section, in particular below the historical buildings, has been successful. With regards to the buildings were soil improvement was carried out, results were satisfactory. In fact, it can be seen that for most of the cases the settlement induced by the TBMs was less than 1cm (therefore within expectations) and that the buildings raised up by the injections approximately returned to its initial level or decreased the fianl amuont of settlement. An example is the building located in 10, Strzelecka Street where monitoring results showed that it moved by the same amount (approx. 6mm) upwards during to injections and downward during the TBM excavation.

6 CASE OF TARGOWA 84 BUILDING

An interesting and delicate case is the building located at 84, Targowa Street. This is a 5-storey masonry corner building (plus basement) built in 1904. This already suffered some settlement from the excavation of C15 station (during the construction works of the Central part of Line II) and is directly above the Left Tunnel excavation. Not only, in order to safely enter the TBMs into the existing C15 station, a jet-grouting plug was foreseen to be installed adjacent to the station d-walls and hence in the immediate vicinity and even below Targowa 84 building. Therefore, a series of works have been carried out (see Figure 9):

– Jet-grouting works were carried out with extreme care with both vertical or slightly angled rods in order to improve the soil also below the building;

Figure 8. Targowa 84 building external view (left) and basement view (right).

– Permeation grouting works were carried out for the area of the building underpassed by the TBM and adjacent to the jet-grouting;
– A series of TAMs were installed and used in order to compensate any potential settlement caused by the jet-grouting works;
– A series of additional TAMs were also installed and left empty in order to have the possibility to compensate any potential settlement caused by the TBM excavation
– Strengthening of the building foundations was carried out by connecting new RC beams to the main bearing brick walls on 1-side or 2-sides of the existing footing (see Figure 10).

All these works were carried out with continuous and collaborative interface among SGS, Astaldi and sub-Contractor (Keller) since the building was very sensitive to such ground works. High frequency automatic measurements and inspections within the building were done to provide an immediate feedback and adjust the grouting works in order to achieve the scope of work without inducing damage to the building.

This work was successfully completed. A pattern of building raise and settlement (due to the improvement works and TBM excavation) is visible in Figure 11. A final settlement of

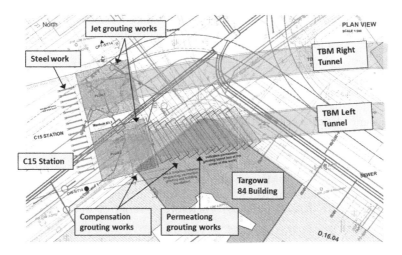

Figure 9. Plan view of the grouting works around Targowa 84 building.

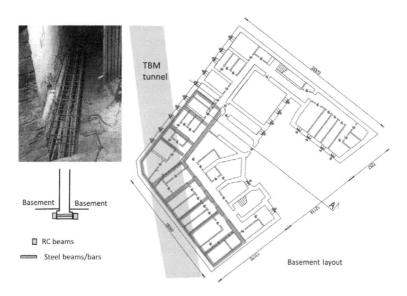

Figure 10. Strengthening works for Targowa 84 foundations.

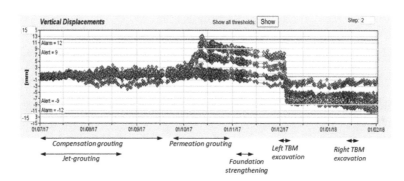

Figure 11. Measured vertical movements for Targowa 84 building.

approx. 1cm has been recorded comparing to initial status. Very minor damages with slight opening of existing cracks or formation of new small cracks have been noted within the building due to both raising and settling.

7 CONCLUSIONS

Contractor Astaldi S.p.A was appointed by the Municipality of Warsaw to execute the extension of the existing Metro Line II in Warsaw, Poland. Two tunnels were built between March 2017 and January 2018 through EPB TBMs connecting C18 and C15 stations on the area of Targowek-Praga districts. The excavation was particularly challenging due to the geological conditions (mostly sands and gravel below groundwater) and existing presence at ground level (historical buildings often in poor conditions) on D16 tunnel section. In particular for this section, the excavation was successfully completed in January 2018 with a measured Volume Loss of less than 0,3% and building settlements generally smaller than 1.0cm.

A series of improvements were done and contributed to such success. In particular, improvements were done for the TBM machine, for the advancement excavation parameters and for the materials used during the excavation At the same time in order to decrease the risk of excessive settlement and hence damage to the buildings, ground improvement was carried out in selected areas. This was done through permation grouting with the specific aim of increase the strength of the soil in front and above the TBM and also to provide an initial upwards movement to the building. This *advance compensation grouting* turned out to be very useful to limit the final settlement of the building. However, it should be remarked how an intensive monitoring system and a tight and responsive collaboration among Designer, Contractor and sub-Contractor are essential to the success of the work.

REFERENCES

Anagnostou G & Kovári, 1996. Face stability in slurry and EPB shield tunnelling. In Mair & Taylor (eds), *Geotechnical Aspects of Underground Construction in Soft Ground*. Rotterdam: Balkema.

Burland J.B., 1997. Assessment of risk of damage to buildings due to tunnelling and excavation. In Ishihara (ed.), *Earthquake Geotechnical Engineering*. Rotterdam: Balkema.

Jancsecz S. & Steiner W., 1994. Face Support for a Large Mix-Shield in Heterogeneous Ground Conditions. *Tunnelling 94*. London.

Lunardi G, Carriero F, Canzoneri & Carini M., 2014. Warsaw Metro's Improvements. *Tunnels and Tunnelling*: November, 31–37. London.

Mair R J & Hight D, 1994. Compensation grouting. *The Mining Journal*: November 7(8), 361–367.

Mair R J, Taylor R N & Burland J B, 1996. Prediction of ground movements and assessment of risk of building damage due to bored tunnelling. In Mair & Taylor (eds), *Geotechnical Aspects of Underground Construction in Soft Ground*. Rotterdam: Balkema.

Tunnels and Underground Cities: Engineering and Innovation meet Archaeology,
Architecture and Art, Volume 12: Urban
Tunnels - Part 2 – Peila, Viggiani & Celestino (Eds)
© 2019 Taylor & Francis Group, London, ISBN 978-0-367-46900-9

A case study of excavation induced displacements in dense urban area

M. Mitew-Czajewska & U. Tomczak
Institute of Roads and Bridges, Warsaw University of Technology, Warsaw, Poland

ABSTRACT: The paper presents a case study of a deep excavation constructed in the city center. The excavation is bounded on three sides by existing historic tenement houses and on the fourth side by a subway line. Detailed FEM 2D analysis of an impact of the designed deep excavation on the structures adjacent to it has been performed. The jet-grouting method of strengthening of the ground below the foundations of the tenement houses was proposed based on these analyses. The results of a second stage numerical FEM analysis including the soil reinforcement were compared with the measurement results of the vertical displacements of the tenement houses. Conclusions related to the impact of the deep excavation on the existing buildings are presented. Discussion on the technologies used in this context (i.e. the methods of supporting the walls of the excavation and reinforcing of the foundation) is also included in the summary section.

1 INTRODUCTION

Nowadays, due to a growing urbanization and rapid population growth in cities, one can see a significant increase in urban density. More and more structures are constructed in close proximity to the existing structures prone to settlement. According to the current regulations investors are obliged to provide appropriate number of parking spaces when offering new office or residential spaces. This requirement together with the limited space in cities causes the increase in the usage of the underground space. New structures usually have several underground storeys, which leads to the necessity of designing deep excavations. Proper excavation design means not only the design of safe stabilization of excavation walls but also verification of the influence of the excavation to surrounding structures Kotlicki & Wysokiński (2002), Siemińska-Lewandowska & Mitew-Czajewska (2009), (Łukasik et al. 2014), Mitew-Czajewska (2015), (Superczyńska et al. 2016), (Mitew-Czajewska, in-press) and geotechnical risk analysis Bogusz & Godlewski (2018a, 2018b), Kolic (2018).

The case described in the paper is a perfect representative of such investment. The 9-storey office building with 3 underground car park levels was built within the yard of the existing, historic tenement houses on a very limited construction site space. In addition, the construction site is limited from one side by the existing metro station, which is only 2,1m away from the excavation wall. The description of the case, results of the profound monitoring and numerical analysis of the influence of the excavation on surrounding structures are further presented in detail. In the discussion part the theoretical horizontal displacements of excavation walls and vertical displacements of foundations of tenement houses resulting from numerical analysis are compared to the appropriate measured values.

2 DESCRIPTION OF THE CASE

2.1 General information

The new investment, described in this paper, with 3 underground storeys is located in a small yard between old tenement houses as shown in Figure 1. From the forth side the investment is bounded by the metro station. The shortest distance between the excavation wall and the closest building was 47 cm. The excavation, which was 10.9 m deep was executed within 18.85 m deep and 80 cm thick diaphragm walls. The stability of the excavation walls was provided in temporary state by two underground slabs (-1 and -3) with round opening in the middle. Unfortunately due to the location of the entry ramp for cars, significant part of the excavation wall adjacent to the building was supported in the temporary state by three levels of steel tubular struts (instead of two slabs mentioned above).

The typical cross-section of the underground part including foundations of tenement houses is shown in Figure 2.

Basing on the preliminary calculations of the described excavation and its influence on the foundations of the old tenement houses it was decided that it is necessary to strengthen the soil below two strip foundations closest to the wall. The jet grouting method was chosen. The results of second stage, final analyses including the jet-grouting columns showed satisfying results in terms of expected displacements of tenement houses. The excavation and the whole new structure was constructed, with precise monitoring of displacements of surrounding houses and the metro station carried out continuously during construction.

The results of the monitoring are further compared to the results of the second stage numerical finite element analysis.

2.2 Geotechnical conditions

The analyzed case is situated in the downtown of Warsaw on the left bank of the Vistula river within the area of glacial plateau. The elevation of the surface of this area is approximately 35.5 m above the "0" level of Vistula river. According to the geological report directly below the surface, 2.2 m thick anthropogenic soils occur (Layer 1 - Fill). Below the layer of fills, approx. 10 m thick sandy formations occur in form of medium dense and dense fine fluvioglacial sands. This layer was divided into two layers distinguishing different parameters (Layer 2 and Layer 3).

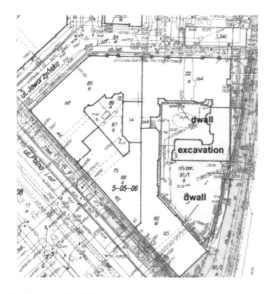

Figure 1.   Layout including designed building and existing infrastructure.

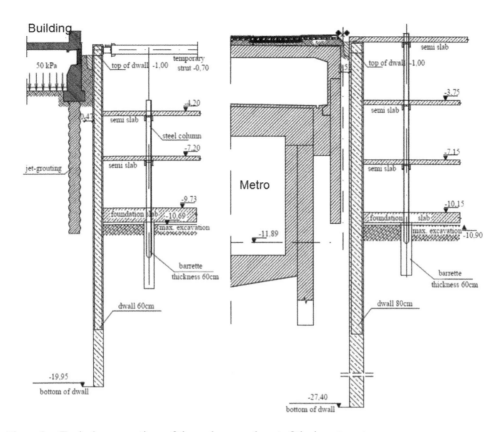

Figure 2.   Typical cross-sections of the underground part of the investment.

Table 1.   Geotechnical parameters.

Layer	γ [kN/m³]	φ' [°]	c' [kPa]	E [MPa]	Eur [MPa]
1	18,0	25	1	25	25
2	19,0	30,9	0	50	150
3	20,0	31,4	0	60	180
4	20,0	18	12	25	75
5	21,0	30	10	80	240

Below sand layer there are approx. 6 m thick stiff Tertiary clayey deposits (Layer 4) and down to the exploration depth stiff to very stiff sandy clays (Layer 5). Geotechnical conditions are shown in Figure 3, the parameters of each soil layer are given in Table 1.

The main groundwater level with free and locally slightly confined water table stabilizes at the level 31.35 m above the "0" level of the Vistula river (~4.15 m below the ground surface (bgs)).

2.3   Construction stages

Following construction stages were designed, executed and taken to numerical analysis:

- Stage 1 – Initial stage including existing neighboring structures
- Stage 2 – Construction of the diaphragm wall and strengthening of the soil below two strip foundations closest to the diaphragm wall by means of jet-grouting

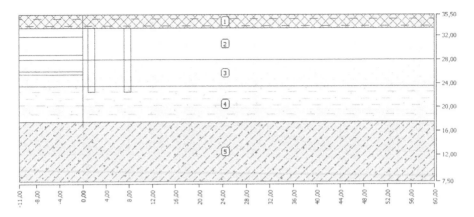

Figure 3.   Geotechnical conditions.

- Stage 3 – Installation of steel struts in the level of the 0 slab,
- Stage 4 – Excavation below the first underground slab (-1 slab), i.e. 3.7 m bgs,
- Stage 5 – Installation of steel struts in the level of the -1 slab,
- Stage 6 - Excavation below the second underground slab (-1 slab), i.e. 6.7 m bgs,
- Stage 7 - Installation of steel struts in the level of the -1 slab,
- Stage 8 – Final excavation to the level 25,50 m above Vistula river level, i.e. 10 m bgs.

The diaphragm walls on the whole periphery of the excavation were embedded into the impermeable clay layer, therefore the water table was lowered only within the excavation and no impact of dewatering on surrounding area was expected.

2.4   Description and results of displacements measurements

Detailed, continuous monitoring was made during construction. Following displacements were measured: horizontal displacements of excavation walls, vertical displacements of old tenement houses as well as vertical displacements of the track bed in the metro station.

The diaphragm wall displacements were measured by inclinometers installed inside the wall, in order to have a complete overview of the results obtained for each construction stage. The maximum displacement of the wall was 4.9 mm (Figure 4), occurred during the final excavation stage and was 3.6 mm smaller than the resulting displacement obtained in the numerical analysis. The shape of the wall displacement was similar for the measurements and the analysis. The differences between the values of measured and theoretical displacement for the following stages were as follows: 0.5 mm for stage 4, 1.3 mm for stage 6 and 3.6 mm for stage 8

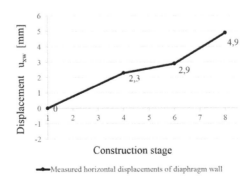

Figure 4.   Horizontal displacements of the diaphragm wall.

(final excavation depth). It could be noted that these differences fall within the range of the measurement error for the first stages.

The monitoring of the neighboring tenement buildings began a few months prior to the start of excavation works. The first vertical displacements were noted at benchmarks during the jet-grouting underpinning of foundations of existing buildings and the excavation of the diaphragm walls. The maximum "technological" settlement related to the strengthening of foundations and diaphragm wall excavation amounted to 7.9 mm for the measuring point no. 1004. The application of jet-grouting technology for the strengthening of the foundations was done with an intention to minimize the settlement of the building and to align it. While the maximum movement (settlement) of the tenement house wall closest to the excavation was 7.9 mm and eventually occurred during the underpinning of foundation and construction of the diaphragm wall. In the same time (stage) the most distant wall moved slightly upwards up to + 0.3 mm. The resulting differential settlement was 8.3 mm. The results of displacements measurements taken on several benchmarks located on the old tenement houses are presented in form of graphs in the following figures: on the wall closest to the excavation wall (Figure 5), on the wall in the yard located in the middle of the building (Figure 6) and on the furthest wall from the excavation (Figure 7).

Finally, because of the metro station proximity it was also required to take measurements of the track bed movements, which minimally settled during the first stage – construction of diaphragm walls (the values slightly exceeded the measurement error), and later on, during the stages of excavation went up to the maximum value of 2.5 mm. The following construction stages and reloading of the excavation has caused the track bed value to basically return to its

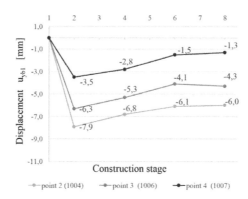

Figure 5.   Vertical displacements of the closest wall of the tenement houses.

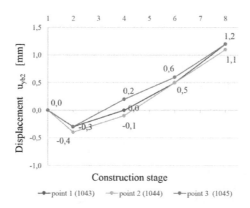

Figure 6.   Vertical displacements of the middle wall of the tenement houses.

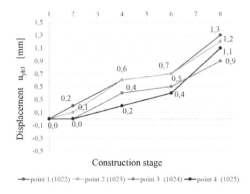

Figure 7. Vertical displacements of the most distant wall of the tenement houses.

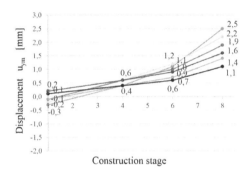

Figure 8. Vertical displacements of the track bed.

original state, recorded prior to the start of construction works (with a difference of only 0.3 mm). The displacements of the track bed are shown in Figure 8.

3 NUMERICAL ANALYSIS

3.1 *Assumptions for numerical analysis*

The analysis was made in 2D space, using GEO5 FEM software, which enables elastic-perfectly plastic analysis of two phase medium, assuming plain strain analysis, Fine (2018). The elastic-perfectly plastic constitutive soil model was used with the Coulomb-Mohr plasticity criterion, without isotropic and kinematic hardening. An assumption was made about the non-associated flow law. In reference to reinforced concrete structures, i.e. diaphragm walls, underground slabs and foundations of the neighboring buildings, a linear-elastic isotropic model was used.

The mesh was made of isoparametric, six-node triangular elements and interface-type contact elements (Figure 9). The diaphragm wall and the reinforced concrete structure of the building were modeled using beam elements assuming their appropriate stiffness. At the contact of the structure and the soil, contact, interface elements of zero thickness were used. They allowed modeling of adhesion and slip phenomena in accordance with the Coulomb-Mohr plasticity condition. The initial state was assumed to be geostatic stresses depending on the volumetric weight of soils and the loads of existing buildings.

Geotechnical conditions were modeled as described in Section 2.2, the cross-section is shown in Figure 3, the parameters are given in Table 1.

Construction stages considered in the analyses are described in Section 2.3.

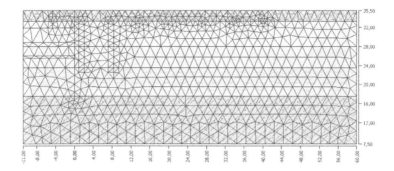

Figure 9.   Numerical analysis - finite element mesh.

3.2   *Results of finite element analysis*

The results of finite element analysis were obtained in all calculation stages including: displacements and stresses in the soil body, displacements of the diaphragm wall and settlements of foundations of tenement houses in the nearest vicinity. Below, the results of the analysis of three excavation stages (Stage 4, 6 and 8) are shown in form of maps of vertical displacements of the model (respectively in Figure 10, 11, 12).

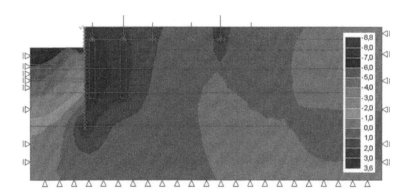

Figure 10.   The plot of vertical displacements of the model in excavation Stage 4.

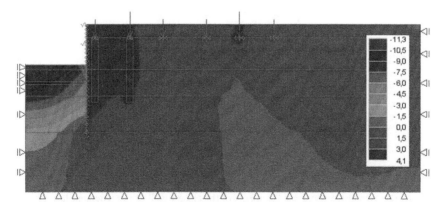

Figure 11.   The plot of vertical displacements of the model in excavation Stage 6.

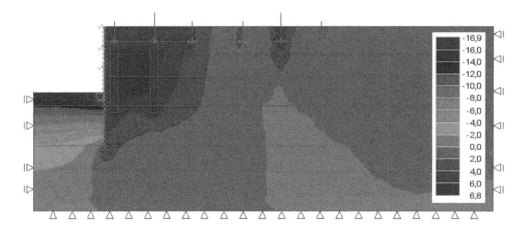

Figure 12.   The plot of vertical displacements of the model in excavation Stage 8.

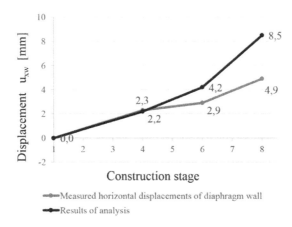

Figure 13.   Comparison of horizontal displacements of the diaphragm wall.

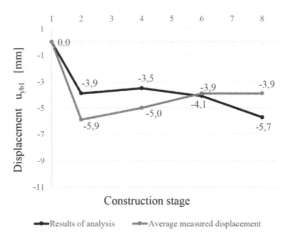

Figure 14.   Comparison of vertical displacements of the closest wall of the tenement houses.

In addition, theoretical, calculated horizontal displacements of diaphragm wall and settlements of foundations are compared to the average measured values for each excavation stage (Stage 4, 6 and 8). These results are shown in Figure 13 for the horizontal displacement of the diaphragm wall and Figure 14 for the vertical displacement of the nearest foundation of the tenement houses, respectively. Observed as well as theoretical vertical displacements of the middle and most distant walls of the tenement houses were negligible in the engineering point of view, amounted to 1 - 2 mm.

4   DISCUSSION AND CONCLUSIONS

The finite element analysis taking into consideration the strengthening of foundations of tenement houses by means of jet-grouting verified by comparing horizontal displacements of the diaphragm wall and settlements of foundations measured on site with appropriate theoretical values proved to be very precise and adequate. The differences in the maximum calculated displacement values didn't exceed 3.6 mm for the horizontal displacement of the wall in the last excavation stage, being in a very good accordance in intermediate calculation stages; 2,0 mm for the vertical displacement of the closest wall of tenement houses; up to 3 mm for the vertical displacement of the middle and the most distant walls of tenement houses.

The maximum settlement of the tenement house wall closest to the excavation was 7.9 mm during the underpinning of foundation and construction of the diaphragm wall, in the same time (stage) the most distant wall moved slightly upwards up to + 0.3 mm. The resulting differential settlement was less than 10 mm. This proved to be acceptable for the old structure of historical houses.

It can be stated that technologies chosen (the type of excavation wall, the method of excavation wall support, additional strengthening of the soil below two strip footings by means of jet-grouting) basing on the two stage numerical finite element analysis made prior to construction were suitable for execution of the excavation in such close vicinity to the existing old tenement houses.

REFERENCES

Amorosi A., Boldini D., de Felice G., Malena M., di Mucci G. 2014. Numerical modelling of the interaction between a deep excavation and an ancient masonry wall, in: Yoo, Park, Kim & Ban (Eds), *Geotechnical Aspects of Underground Construction in Soft Ground*: 245–250. Korean Geotechnical Society, Seoul, Korea.

Bogusz W., Godlewski T. 2018a. Predicting the impact of underground constructions on adjacent structures as an element of investment risk assessment. In: *Special issue: XVI DECGE 2018 Proceedings of the 16th Danube-European Conference on Geotechnical Engineering, ce/papers*, vol. 2, Issue 2–3: 281–286. Ernst&Sohn a Wiley Brand.

Bogusz W., Godlewski T. 2018b. Geotechnical interaction in underground space – theory and practice. In: C. Madryas, A. Kolonko, B. Nienartowicz & A. Szot (Eds), *Underground Infrastructure of Urban Areas 4*: 19–31. CRC Press/Balkema.

Fine Ltd. 2018. *GEO5 User's manual*. Fine Ltd. Prague.

Kolic D. 2018. Subway line optimization through risk management. In: C. Madryas, A. Kolonko, B. Nienartowicz & A. Szot (Eds), *Underground Infrastructure of Urban Areas 4*: 19–31. CRC Press/Balkema.

Kotlicki, W. & Wysokiński, L. 2002. *Protection of structures in the vicinity of deep excavations. Guide ITB no 376/2002*. Warsaw: Building Research Institiute. In polish.

Łukasik, S. & Godlewski, T. et al. 2014. *Technical requirements for designed and constructed investments, which could influence existing metro structures*. Warsaw: Building Research Institiute. In polish.

Mitew-Czajewska, M. 2015. Evaluation of deep excavation impact on surrounding structures a case study. In: C. Madryas, A. Kolonko et al. (eds), *Underground Infrastructure of Urban Areas 3*: 161–172. CRC Press/Taylor & Francis Group: Balkema.

Mitew-Czajewska, M. 2018. Displacements of structures in the vicinity of deep excavation, *Archives of Civil and Mechanical Engineering*. In press.

Siemińska-Lewandowska, A. & Mitew-Czajewska, M. 2009. The effect of deep excavation on surrounding ground and nearby structures. In Ng C.W.W., Huang H.W., Liu, G.B. (eds), *Geotechnical Aspects of Underground Construction in Soft* Ground - *Proceedings of the 6th International Symposium, (IS-Shanghai 2008)*, *Shanghai, China, 10–12 April 2008*: 201–206. CRC Press/Balkema.

Superczyńska, M. & Józefiak, K. & Zbiciak, A. 2016. Numerical analysis of diaphragm wall model executed in Poznań clay formation applying selected FEM codes. *Archives of Civil Engineering Vol. LXII, Issue 3*: 207–224.

*Tunnels and Underground Cities: Engineering and Innovation meet Archaeology,
Architecture and Art, Volume 12: Urban
Tunnels - Part 2 – Peila, Viggiani & Celestino (Eds)*
© 2019 Taylor & Francis Group, London, ISBN 978-0-367-46900-9

Conventional tunneling in urban areas

N. Munfah
AECOM, New York, New York, USA

V. Gall, W. Klary & T.M. O'Brien
Gall Zeidler Consultants, Ashburn, Virginia, USA

ABSTRACT: Conventional tunneling as defined by ITA's Working Group 19 is often referred to as New Austrian Tunneling Method (NATM); Sequential Excavation Method (SEM); or Sprayed Concrete Liner (SCL). It is being used more often in urban settings, in very soft ground, low cover, and overbuilt conditions for the creation of tunnels, station caverns and cross passages between previously TBM driven soft ground tunnels. To enable open face excavation ground improvement methods are used and include various methods of dewatering, grouting, and ground freezing. The paper examines criteria to be used for the implementation of conventional tunneling in urban areas, impact on existing facilities, ground improvement measures, and risks mitigation measures for conventional tunneling under such circumstances. The paper illustrates these aspects in two recent examples: Chinatown Metro Station in San Francisco and Sound Transit Bellevue Tunnel in Seattle, both very recently and successfully completed tunnels in urban settings in the US.

1 INTRODUCTION

Conventional tunneling has become a method of choice in urban areas to construct complex underground structures such as metro stations, multi-track metro lines, rail crossovers, short road tunnels, and underground road ramps in order to avoid cut and cover construction with its impacts on streets, utilities, traffic, businesses and the public. Under these conditions and where complex and challenging ground conditions exist, underground construction requires a flexible design that can be executed effectively and safely, while minimizing impacts to existing structures. This specifically includes tunneling in running and flowing ground, tunneling under high water pressure, encountering mixed face conditions, low ground cover, presence of sensitive building and structures within the influence zone of the excavation, presence of boulders, abandoned foundations or unchartered utilities and complex geometrical configurations. Conventional tunneling method minimizes impacts on traffic and utilities/services throughout construction, reducing disruption to everyday life. Urban settings provide a host of challenges that require a risk managed approach from the very beginning of the design process through construction. This paper evaluates the components of a risk mitigated project from design through construction in two recent examples of challenging conventional tunneling in urban settings projects: Chinatown Metro Station, San Francisco, CA and Sound Transit Bellevue Tunnel, Seattle, WA.

2 CHALLENGES OF TUNNELING IN URBAN AREAS

Conventional tunneling in a complex urban setting presents a number of unique challenges. A typical urban setting will include the presence of major roadways, potential shallow ground cover, soft ground conditions and potentially mixed ground, potentially existing or abandoned foundations and buried structures, and large intricate networks of utilities. Additionally, space constraints in urban settings magnify the challenge of implementing tunneling in such a manner as to avoid inducing displacements damaging to adjacent facilities, structures and utilities. Such challenges can be addressed with carefully designed excavation and support sequencing, including potential ground improvement and a robust instrumentation and monitoring program. With a risk mitigated approach during the design phase, conventional tunneling has proven successful in complex urban settings (Gall et al. 2016 & Gall et al. 2017).

3 GEOTECHNICAL AND HYDROLOGICAL CONSIDERATIONS

A thorough ground investigation program, including assessment of geotechnical and hydrogeological conditions is a key to the success of any conventional tunneling project. Such investigations facilitate the collection of information to assess the anticipated ground behavior during excavation. A clear understanding of the anticipated ground conditions along the tunnel alignment allows the implementation of ground improvement measures if needed, which could include dewatering, grouting or ground freezing. However, such a process needs to occur very early in a project so as to allow careful considerations with respect to tunnel design and the potential impact on existing structures and facilities along the tunnel alignment, which can limit the available techniques for ground improvement measures, such as dewatering, due to their potentially adverse impact on the structures. Identification of contaminated ground and groundwater and the presence of hazardous substances such as hydrocarbons, gases, and other hazardous materials allows for planning of special remedial/mitigation measures, or modification of the tunnel alignment if possible to avoid such ground conditions all together. Typical ground conditions that result in ground instability include: 1) fractured and decomposed rock in near surface conditions, 2) potential swelling mainly due to presence of swelling prone clay minerals, 3) mixed face conditions, 4) soft ground (cohesive and non-cohesive) and low ground cover, and 5) high ground water pressure.

4 GROUND IMPROVEMENT MEASURES

4.1 *Ground Improvement*

Ground improvement measures serve to improve strength and stiffness of the ground. With respect to conventional tunneling, ground improvement improves soil standup time during excavation and allows installation of optimized initial support while providing safe excavation (FHWA, 2009). Ground improvements also serve to control ground water, reduce ground loss and potential surface settlements and minimize the tunnel deformations during excavation. The variety of ground improvement techniques available are diverse and include dewatering, jet grouting, cementitious or chemical permeation grouting, compaction grouting, ground freezing, etc. In the scenario where settlement would potentially occurs, compensation grouting can be used as a remedial measure to overcome tunneling induced settlements. Instrumentation and monitoring are critical for detecting ground movement and implementing corrective measures.

4.2 Pre-Support Measures

Common methods of pre-support, which include spiling, pipe arch canopies and sub-horizontal jet grouting, act to improve the standup time of weak ground during and after excavation. In addition to minimizing risk during excavation, effective pre-support measures will minimize disturbance to in-situ ground during excavation, thereby limiting surface settlements. However, pre-support measures are only suitable for implementation when they have close contact with the ground. This is essential in order for ground and pre-support elements to work effectively as a reinforcement integrated into the ground.

5 EXCAVATION AND SUPPORT MEASURES

Conventional tunneling is an observational method that relies on the ground behavior and its interaction with the installed support system. Design of an effective excavation and support sequence is predicated on a comprehensive understanding of the anticipated ground conditions and behavior along the alignment, particularly with respect to weak soils/ground and local groundwater. Classification of the ground into different ground support classes provides flexibility during construction to implement pre-support, staged excavation and initial support measures consistent with the encountered ground conditions. The number of drifts, round length and sequencing of the excavation are based on the cross sectional size of the excavation, ground condition, and available cover; and they are critical for a successful tunneling in urban areas.

6 INSTRUMENTATION AND MONITORING

Instrumentation and monitoring is an integral part of conventional tunneling, allowing verification of design assumptions and the interaction between the ground and the support system during excavation (FHWA, 2009). The primary purpose of instrumentation in conventional tunneling is to monitor the initial lining deformation systematically as excavation progresses in comparison with anticipated deformations and to measure ground settlement at various depths above the excavation and at the surface. Displacement that exceeds critical threshold values triggers implementation of contingency measures such as the use of additional support, reduction of round length, or implementation of ground improvement. An extensive instrumentation program will also assess potential impacts on existing facilities, structures, and utilities during construction. This will permit the implementation of remedial measures in a timely manner if needed. Instrumentation usually includes internal extensometers, strain gages, total stationing, surface settlement markers, inclinometers, multiple point borehole extensometers, piezometers and shallow and deep settlement indicators.

7 IMPACT ON EXISTING FACILITIES

A robust design and a suitably designed staged excavation and support system for conventional tunneling serves to minimize the impact of construction on existing facilities by limiting settlement of the ground. Implementation of ground improvement measures and/or pre-support in the worst ground in conjunction with a compensation grouting program and a robust instrumentation and monitoring program will limit the impact of excavation on existing structures, while also having measures in place to immediately remediate and mitigate in the event of settlements in excess of allowable. A critical element to minimizing the impact on existing facilities, which is discussed in greater detail in the next section, is having experienced personnel that can recognize and react to deviations quickly and implement the necessary contingency measures to mitigate the issue.

8   RISK MANAGEMENT AND DESIGN ROBUSTNESS

All of the above-discussed considerations for a conventional tunneling project in urban areas, when thoroughly addressed, present the most effective manner of developing and executing a comprehensive risk managed tunneling program. Risk management begins during preliminary engineering with the identification of risks through the development of a risk register, which requires a thorough understanding of the project challenges, including existing structures and anticipated ground behavior. Through development of the risk register, preliminary engineering can address and mitigate the risks during the design by the development of ground improvement, pre-support measures, and a staged excavation with robust initial support.

During construction extensive geotechnical instrumentation and monitoring program as well as a strong Quality Assurance/Quality Control (QA/QC) plan are essential parts of risks mitigation. The effectiveness of a risk mitigated design program (design, monitoring, QA/QC) is also heavily dependent on the personnel chosen to execute it. A solid technical knowledge, suitable prior experience, and skills in assessing the ground behavior and interpretation of the monitoring program are required for successful execution of conventional tunneling. It is recommended to implement a pre-qualification process for the contractors including their key personnel to ensure requisite conventional tunneling capabilities. During construction daily communication and coordination between various project stake holders, including the contractor, tunneling crew(s), designer and owner's representative is very important for addressing challenges and mitigating risks as they materialize.

9   CASE HISTORIES

9.1   *Chinatown Station, San Francisco, California, USA*

9.1.1 *Background*
The San Francisco Central Subway is Phase 2 of the Third Street Light Rail Project and extends the existing Phase 1 initial operating segment from its current connection at Fourth and King Streets along Fourth Street to Market street, under the BART and Muni Metro tunnels and then north along Stockton Street to Chinatown, terminating in Chinatown Station (CTS). The project owner is San Francisco Municipal Transportation Agency. Station excavation has been completed and waterproofing and final lining construction are ongoing (Figure 1).

Chinatown Station was excavated as a mined cavern beneath Stockton Street, between Jackson Street and Clay Street, utilizing conventional tunneling. It was excavated after the completion of twin TBM tunnels passing through its location. The vicinity of Chinatown Station is one of the most densely populated areas in San Francisco, with many existing buildings and underground utilities as well as a large volume of bus and car traffic on the

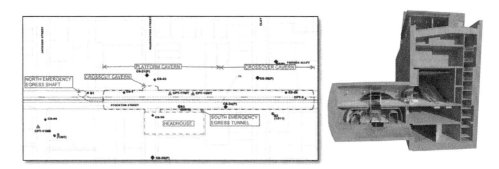

Figure 1.   Location of Chinatown Station and architectural rendering of cross cut and head-house shaft.

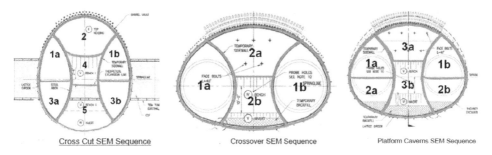

Cross Cut SEM Sequence   Crossover SEM Sequence   Platform Caverns SEM Sequence

Figure 2.   SEM excavation sequence for the caverns.

surface. The construction of Chinatown Station can be regarded as one of the most challenging tunneling projects in the US utilizing conventional tunneling method because of its exceptionally large size, limited access, presence of significant number of utilities, difficult ground and ground water regime, highly overbuilt neighborhood, and a very complex urban setting.

Several factors were of great concern during the construction of the station such as presence of numerous buildings, structures, and utilities in the area. Geology of the area is complex with varying soil and rock groups and high groundwater head along with restrictions on groundwater drawdown. Additional restrictions were also in place to ensure the major surface road remain open and unaffected by tunneling activities.

The major components of the station are the Crosscut Cavern, the Platform Cavern, the Crossover Cavern, the Head house, and two Emergency Egress Shafts (Figure 2). The Cross Cut cavern is approximately 13.1m wide, 16.1m high and 22.3m long. The Platform cavern is excavated in multiple drifts using a saw tooth profile to allow installation of pipe arch canopies. The cavern cross section is approximately 16.8m wide by 14m high. The Crossover cavern is also excavated in a similar saw tooth profile using multiple drifts. Its cross-sectional dimensions are 16.8m wide by 12m high. The overall length of the mined cavern is approximately 192m and at a depth from surface to the track level varies from 26.2m along the northern end of the station to 34.1m along the southern end of the station due to the change of the street level. All three caverns have similar structural support systems comprising of fiber reinforced shotcrete and lattice girders as the initial lining and cast in place final lining. The Crosscut Cavern has 450mm thick initial lining while the Crossover Cavern and the Platform Cavern has a 400mm thick initial lining.

9.1.2 *Geology*
The ground within the station area is grouped into two soil and rock groups. The soil group includes Colma Formation (Qc) and Colluvium (Qcol) and the rock group includes the Franciscan Complex Bedrock (KJf). The Colma formation (Qc) consists of dense to very dense sand or silty sand interbedded with stiff to very stiff clay and sandy/silty clay. The Colluvium (Qcol) consists of very dense, medium to fine brown sand with silt derived from complete weathering of the bedrock. The Franciscan formation (KJf) bedrock is highly variable in composition, degrees of fracturing, strength, hardness, and weathering. The rock mass is extensively sheared and a chaotic, heterogeneous mixture of small to large masses of different rock types, including sandstone, shale, siltstone, and various metamorphic rocks (such as metasandstone), surrounded by a matrix of pervasively crushed rock materials. The RQD values were consistently 0%. The rock/soil contact is locally undulating and irregular with an overall slope downward towards the east, and also towards the north.

9.1.3 *Ground Improvement*
The project also implemented dewatering and a complex compensation grouting scheme as ground improvement measures and building protection methods during construction. The

dewatering was chosen to provide a stable face during excavation while the compensation grouting was selected to restore any potential settlement of the surrounding buildings.

Dewatering of the Colma Formation (Qc) prior to tunneling was carried out by deep wells. Supplemental dewatering from within the excavation was also required in the Qc formation where pockets of perched groundwater are encountered that were not effectively dewatered by the deep well system. In these cases, a well-point dewatering system was provided as a backup system to reduce perched water pressures to maintain excavation stability and acceptable working conditions.

Dewatering of the Franciscan formation (KJf) material was accomplished with pre-drainage ahead of the excavation face or with well-point dewatering as needed. Additionally, pre-drainage of the face with gravity-flow well-points within the KJf rock units, and/or vacuum well-point dewatering to dewater local depressions or water-filled lenses that cannot be dewatered with the prescribed deep well system were used as additional contingency measure for groundwater control.

9.1.4 *Pre-Support & Excavation*
The caverns were excavated using a double sidewall drift excavation sequence. The design provided two side drifts and a center drift with multiple headings each (Figure 3a). Pre-support of the side and center drift excavations mainly consist of pipe arch canopies at the crown, to allow for micro-fine cement or chemical grouting of the surrounding ground mass.

Double rows of grouted pipe arch canopies were installed at the crown of the cross-cut cavern and a single row over the two side drifts of the cross-cut cavern; while a single row of pipes was installed at the crown for platform cavern and cross-over cavern (Figure 3b). Each pipe was 27m in length and 139mm in diameter; perforated steel pipes were installed at 300mm c/c spacing; GFRP pipes were used for the sidewall drifts to allow future easier removal when connecting the platform cavern with the cross-cut cavern. Pressure grouting was followed by the backfill grouting inside of the pipes. In addition to the pre-support, 12.2m long five to seven face bolts were used in the center drift top heading every 12.2m along the length of the drift depending on the cavern geometry.

9.1.5 *Risk Mitigation*
To protect the buildings and infrastructure near the station, a thorough instrumentation plan with monitoring details was developed. Existing buildings and structures in the excavation zone of influence were analyzed for impacts due to station construction, taking into account the proposed construction sequencing and excavation method. When the settlement of a major building across the excavation reached the initial threshold value, the compensation

a)   b)

Figure 3.   a) Crosscut Cavern showing top heading of the side and center drifts, b) Installation of Barrel Vaults for the Cross Cut Cavern.

grouting program was implemented arresting the settlement and partially restoring the building to its initial level.

An essential component of the daily conventional tunneling process is the use of the "Required Excavation and Support Sheet (RESS) Meeting". The project requires these meetings to be held every workday at a defined time, and conducted by the Senior Tunnel Engineer. These meetings are typically attended by the contractor's tunnel project manager, the design engineer, construction superintendent, project geologist, the geotechnical engineer, the surveyor, the quality control manager, the construction manager, and the owner's representatives. The RESS meetings provide an essential communication forum among the various parties and frequent and quasi-concurrent agreement on the tunneling process between the contractor's and the owner's representatives to reduce risk and improve tunneling performance.

9.2 Bellevue Tunnel, Seattle, Washington, USA

9.2.1 Background

Sound Transit's East Link Project is a 22.4 km (14-mile) long light rail transit (LRT) extension that will provide patrons a fast, frequent and reliable connection from Bellevue and Redmond, the largest population and employment centers east of Lake Washington, to downtown Seattle, Sea-Tac Airport and University of Washington (Figure 4).

While the majority of the East Link alignment will be at-grade or on elevated guideways, one of the most technically challenging components of the project is the Downtown Bellevue Tunnel. The Downtown Bellevue Tunnel is 740m long, as measured from the South Portal to the Bellevue Transit Center Station Interface, and runs under 110th Avenue Northeast through downtown Bellevue. The tunnel alignment is constrained by utilities and bounded on both sides by major buildings, including several high-rise structures and Bellevue City Hall (Figure 5).

Conventional tunneling, was chosen for the construction of the 605m long central portion of Downtown Bellevue Tunnel (DBT). The typical cross section is an 11.2m wide by 9.3m high ovoid, with a central fire separation wall. Near its mid-length, the tunnel cross section is enlarged to overall dimensions of 12.9m width by 11.5m height to provide space for an emergency ventilation fan room above the tracks. Maintenance access to the mid-tunnel ventilation fan room is provided through an access shaft (5.2m internal diameter and 15.5m depth) and a connecting adit, which was sequentially excavated from the enlarged tunnel towards the shaft.

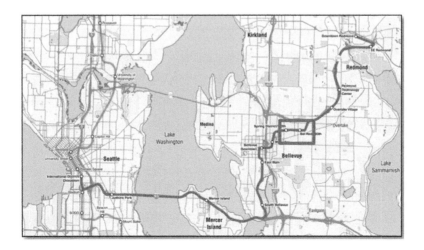

Figure 4.   Location of Downtown Bellevue Tunnel (Plan view Eastlink).

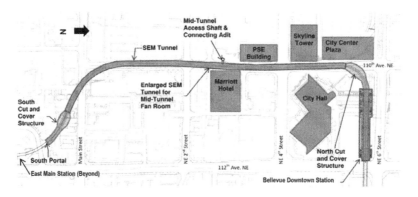

Figure 5.   Location of Downtown Bellevue Tunnel (*from* Wongkaew et al. 2018).

9.2.2 *Geology*

The geologic profile along the tunnel indicates glacial deposits consisting of glacially over-consolidated stratigraphic sequence that includes Vashon till, Vashon advance outwash deposits, and pre-Vashon glacio-lacustrine deposits (Figure 6). North of the enlarged section the profile indicates an "anomaly zone". During the design an extensive ground investigation program was executed but no conclusive geological model could be established for this zone. During excavation of the tunnel a change of the ground behavior has not been observed, however offsets in the stratigraphy have been encountered. The design groundwater table generally follows the top of the advance outwash which were expected to be encountered in the tunnel face during the second half of the tunnel. During excavation of the Tunnel the ground showed more favorable conditions than anticipated. In particular the groundwater table was much lower than expected and the planned dewatering measures which included dewatering with surface wells and vacuum dewatering from within the tunnel was not required. However perched ground water with a water inflow rate of approx. 0.75 l/min (0.2 gpm) from within sand layers in the till was encountered.

9.2.3 *Tunnel Support & Pre-Support Measures*

The DBT was designed as a single side-drift excavation with five Ground Support Classes and a round length from 1 to 1.5 m and systematic spiling (Figure 7a). From the South portal to the start of the enlarged section (approx. 50 % of the tunnel length) the Contractor suggested

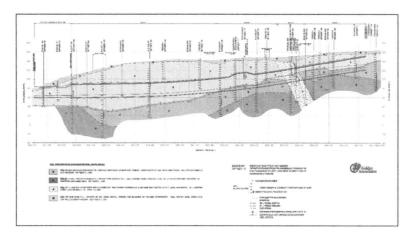

Figure 6.   Geologic profile along tunnel alignment.

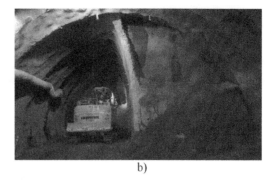

a)   b)

Figure 7.   a, b) Photos of tunnel excavation showing left and right drifts and short round excavation within each drift.

to change the excavation to a three heading sequence which included Top Heading, Bench, Invert (Figure 7b). From the enlarged section to the North portal a single side-drift excavation was implemented. Due to favorable ground conditions only, a small number of the designed spiles was installed. However, further investigation of utilities and basement of buildings showed that pre-treatment of the ground was required in the proximity of the Skyline Building and at the Intersection of 110th and 4th Street.

9.2.4 *Construction Challenges*

9.2.4.1 SKYLINE BUILDING

The Skyline Building is a high-rise building located on the west side of the tunnel. The design drawings showed that the basement (parking garage) of the building is as close as 1m next to the tunnel (Figure 8). It was required to install additional support on the garage wall from within the tunnel. During this work a water filled void behind the garage wall was encountered. Due to the close proximity to the tunnel and the risk of water causing further deterioration, the thin pillar was grouted extensively. Additional monitoring points were installed in the garage which included tiltmeters, strain gages and structural monitoring points. During excavation no water inflow was observed and the garage did not show any significant movement.

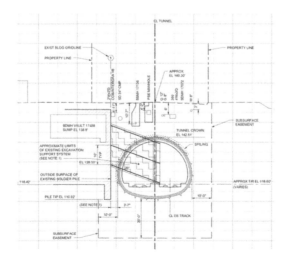

Figure 8.   Cross Section of tunnel relative to the Skyline Building.

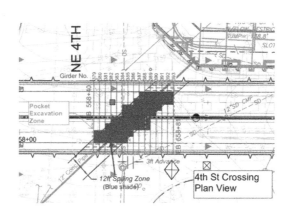

Figure 9. Plan view of 4th Street crossing showing utilities present.

9.2.4.2 INTERSECTION OF 110TH AND 4TH STREET

During the design phase it was identified that in the Intersection 110th Street/4th Street critical utilities such as storm drains, sewer pipes, water mains, high pressure gas lines, high voltage lines and fiberoptic cables are present (Figure 9). Further investigations of the storm drains revealed leakages in the storm drain. The proximity to the crown of the tunnel exhibited the risk of the leaking water to deteriorate the ground above or water inflow from sand layers in the Till. Since this would be very challenging to deal from within the tunnel the intersection was closed to traffic and the utilities exposed and the backfill replaced with controlled low-strength material (CLSM). As excavation below the leaking storm drain was not feasible the gravel below it was grouted instead. The excavation sequence and ground support during tunnel excavation was adjusted in this area. The advance length was limited to 1 m, spiles were installed and pocket excavation was utilized. Excavation of this area took place during the dry season.

10 CONCLUSIONS

Conventional tunneling is becoming highly effective method of tunneling in urban areas, in difficult ground, under high hydrostatic heads, and with limited cover. Properly implemented will avoid cut and cover construction and its associated impacts on traffic, utilities, businesses and the public. However, as was demonstrated for Chinatown Station, and Bellevue Tunnel, a robust design is required with detailed pre-support systems and ground improvement methods to mitigate potential risks. Excavation and support sequences were designed to address the anticipated ground behavior and limit the potential impact on existing facilities and structures. A comprehensive instrumentation and monitoring system with predetermined threshold limits and potential remedial measures is essential, along with pre-qualification of all involved parties. Effective communication and collaboration between the designer, contractor and owner's representatives is essential for successful implementation of conventional tunneling in challenging settings such as urban environments.

REFERENCES

Federal Highway Administration (FHWA). 2009. *Technical Manual for Design and Construction of Road Tunnels – Civil Element, Chapter 9*, FHWA-NHI-09-010, Washington, D.C.
Gall, V., Munfah, N. & Pyakurel, S. 2017. Conventional Tunneling in Difficult Ground Conditions, *In Proc. ITA-AITES World Tunnel Congress, Bergen, Norway, 9–15 June.*
Gall V., Munfah, N. 2016. Recent Trends in Conventional Tunneling (SEM/NATM) in the US, In *Proc. ITA-AITES World Tunnel Congress, San Francisco, 22–28 April.*

Munfah, N. 2014. Lessons learned from the first NATM Tunnel in California, the Devil's Slide tunnel, *In Proc. ITA-AITES World Tunnel Congress, Iguassu Falls, Brazil, 9–15 May.*

SFMTA (City and County of San Francisco Municipal Transportation Agency). 2011. *Geotechnical Baseline Report – Chinatown Station, Rev. 0.*

Wongkaew, M., Murray, M., Coibion, J., Frederick, C., & Leong, M. W. 2018. Design and Construction of the Downtown Bellevue Tunnel. In *Proc. North American Tunneling, Washington, D.C., 24–27 June.*

Tunnels and Underground Cities: Engineering and Innovation meet Archaeology,
Architecture and Art, Volume 12: Urban
Tunnels - Part 2 – Peila, Viggiani & Celestino (Eds)
© *2019 Taylor & Francis Group, London, ISBN 978-0-367-46900-9*

Effect of changing methodology of tunnel final lining on the construction duration of a metro project

M. Namli
Department of Civil Engineering, Faculty of Engineering and Naturel Sciences,, Istanbul Medeniyet University,
Uskudar, Istanbul, Turkey

I. Araz
Dudullu - Bostancı Metro Project, Arcadis & Tumas JV, Atasehir, Istanbul, Turkey

ABSTRACT: Uskudar - Umraniye - Cekmekoy Metro Line (UUCM) is the second metro project located at the Asian side of Istanbul. In this paper, effect of changing construction methodology of final lining of tunnels on construction duration and cost of Cakmak Metro Station is analyzed. Concourse of Cakmak Station was built by cut and cover method. The rest of the station was tunnel and built by NATM. Conventional cast concrete was used for the final lining of tunnels. Changing the final lining (secondary lining) of all tunnels in Cakmak Station from conventional cast concrete to sprayed concrete (shotcrete) is examined, since it was a very critical station in the middle of metro line.

Based on our analysis, using shotcrete as final lining of tunnel instead of conventional cast concrete could give the opportunity to complete the Cakmak Station about 6 months earlier and with 15 % less cost.

1 INTRODUCTION

Uskudar - Umraniye - Cekmekoy Metro Line (UUCMP) is a public transportation project consists of 20 km long double tube main line tunnel and 16 stations (Figure 1). The project owner is Istanbul Metropolitan Municipality (IMM) and contractor is Dogus Construction Corporation. UUCMP will integrate with Marmaray Undersea Metro Project at Uskudar station. The project started on March 2012 and has been taken completely into service in 2018.

The Peron tunnels have cross-section of 75.60 m^2. Geology in this section is composed of sandy clay and weathered sand stone, diabase andesite dykes are also rarely observed (Figure 2).

In UUCMP, while main line tunnels of UUCMP have been excavated by 4 tunnel boring machines, other tunnels have been excavated by the new Austrian Tunneling Method (NATM). However, using only NATM in weak rock and soil conditions is not always adequate for support. Therefore, to control the surface and subsurface deformations in weathered rocks and soils conditions, in addition to NATM, umbrella arch method (UAM) has been used for additional reinforcement.

In this study, using of shotcrete as secondary lining and cast concrete are compared in term of application time during the application of UAM and NATM.

Figure 1.   Uskudar - Umraniye - Cekmekoy Metro Line route plan.

1.1   The Geology of the Cakmak Station Area

UUCMP is carried out generally in Kurtkoy formation, Aydos formation, Gozdagi formation, Dolayoba formation, Kartal formation, Tuzla formation, and Trakya formation. This study is carried out in Cakmak station. There exists sand stone in the study area. But diabase and andesite dykes are also present. Some faults and geologic discontinuities are also exist. The overburden thickness above the tunnels varies between 20 and 30 m in the study area. Geological cross-section of the study area is presented in Figure 2.

1.2   Excavation and support of NATM

Station tunnels have 75.60 m^2 to 81.59 m^2 of cross-sectional excavation area depending on the support type. They are excavated in three sections; the upper bench, the lower bench and the invert. The upper bench is excavated before the lower bench which is 2 - 3 m behind the upper bench. Invert part is excavated after 10 - 15 m lower bench excavation (Figure 3). Excavators and hydraulic hammers are used in all tunnel faces.

Temporary tunnel support includes 4 m long rock bolts, wire mesh, lattice girder and shotcrete (Table 1). Excavation round changes between 0.4 m and 1.8 meters based on the face conditions. Depending on tunnel diameters, the final lining is undertaken with 35 - 45 cm thick in situ cast concrete (Ocak, et. al, 2015).

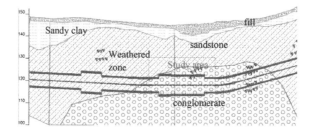

Figure 2.   Geological cross-section of the study area (IMM, 2012).

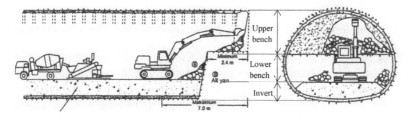

Figure 3.   Tunnel excavation steps (Ocak, et al. 2015).

Table 1. Types of support used in station platform tunnels (Ocak, et al. 2015).

Kind of support	A	B	C	D	E	F
Rock property	Very good	Good-medium	Medium-weak	Weak-very weak	Very weak	Very weak
RMR	>70	50 - 70	30 - 50	30 - 15	<15	<15
Shotcrete (cm)	25	25	30	30	30	30
Face Shotcrete (cm)	-	-	10	10	10	10
Wire mesh	1	2	2	2	2	2
Lattice girder (mm)	H=136	H=136	H=174	H=174	H=174	H=174
Pipe (L=4 m)	-	-	if needed	yes	-	-
Umbrella pipe(L=9 m)	-	-	-	-	4.8	4.0
Rock bolt	>1.8	0.8 - 1.2	0.6 - 0.8	0.4 - 0.6	0.8	0.8
Soil nail (L=6 m)	-	-	-	-	0.8	0.8
Excavation round	1.8	0.8 - 1.2	0.6 - 1.0	0.4 - 0.6	0.8	0.8

1.3 *Excavation and support of UAM*

In the umbrella arch method; steel pipes with 9 meters long are installed at upper part of the tunnel face. In order to install these pipes, the tunnel face is excavated 50 cm larger to have additional space for machines necessary to install the pipes. Steel pipes used for umbrella arch have diameter of 114 mm, length of 9 m and thickness of 6.3 mm and are installed 6°-8° upward direction. Umbrella pipes are installed into the holes with diameter 130 mm. The distance between the pipes is 30 cm and each section contains 25 steel pipes (Ocak, 2008).

The first 4.8 meters of the pipes are in the hole and 4.2 meters are used as overlapping. So, one part of the steel pipes that creates plastic zone is connected to lattice girders and other part stays in face, working as two-supported beams. This arrangement makes possible to support over stress in a safe manner. After the mounting of the steel pipes and completion of injection, tunnel excavation is started. Face is excavated in two phases in weak rocks and in three phases in soil formations. Each round is 80 cm. Thus 8 lattice girders are used in one period and excavation continues periodically (Ocak, et al. 2015).

2 EFFECT OF CHANGING THE METHODOLOGY OF TUNNEL FINAL LINING ON THE CONSTRUCTION DURATION OF A METRO PROJECT

Especially in the tunnels of metro projects, after the excavation support the final lining is mostly built with conventional cast concrete.

In this project (UUCMP), secondary final lining planned to be made with cast concrete. This study is carried out in Cakmak Station (Figure 4). Because, Cakmak Station is a very critical station which is in the middle of UUCMP line and completing this station may give opportunity to open metro line partly to service. Hence, Cakmak Station is chosen for making an analysis for making comparison of cast concrete tunnel lining and shotcrete tunnel lining, from view the point effect of changing construction methodology on the duration of project.

NATM work includes, excavation of tunnel, shotcrete application, mounting of lattice girder, wire mesh installation, mounting of connection steel and drilling and installation rock bolt and installation of steel pipes. UAM work includes excavation of tunnel, shotcrete application, mounting of lattice girder, wire mesh installation, mounting of connection steel and drilling and installation rock bolt, installation of umbrella pipes, installation of soil nails and application of face shotcrete. Recorded application times for NATM and UAM is given in Table 2.

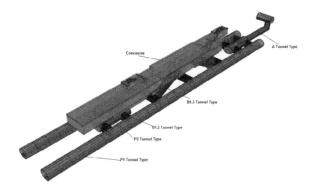

Figure 4. Cakmak station 3D view (IMM, 2013).

Table 2. Comparison of the application time (Selcuk, 2016).

	Quantity per meter		Average application time (min/m)	
Work item	NATM	UAM	NATM	UAM
Excavation (m³)	58.91	60.76	277	420
Shotcrete (m³)	5.70	5.74	68	60
Lattice girder (number)	1	1.25	65	90
Wire mesh (number)	5	9	40	45
Connection steel (number)	17	11	15	19
Rock bolt (number)	9	8	45	38
L steel (number)	6	6	4	4
Steel pipe (number)	6	-	41	-
Umbrella pipe (number)	-	6	-	399
Soil nail (number)	-	1	-	287
Face shotcrete (m³)	-	1.13	-	13
Total			555	1375

2.1 Cast Concrete Lining Method as Final Lining

The plan of tunnels of Cakmak station is in Figure 5. There are five types of tunnel cross sections in Cakmak Station (Figure 6, Figure 7, and Figure 8). After finishing tunnel excavations, construction of secondary cast concrete lining starts. The final linings of tunnels constructed in the station are made of classical cast concrete. The

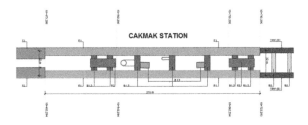

Figure 5. Plan view of Cakmak Station (IMM, 2013).

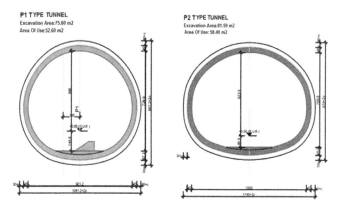

Figure 6.   P1, Peron tunnel, and P2 Ventilation fan tunnel cross sections (IMM, 2013).

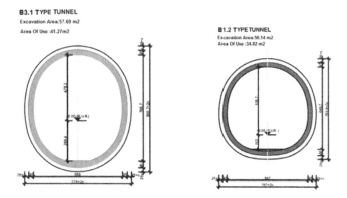

Figure 7.   B1.2 Ventilation fan connection tunnel, and B3.1 Ladder tunnel cross sections (IMM, 2013).

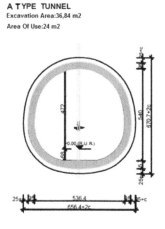

Figure 8.   A, Air conditioning tunnel cross sections (IMM, 2013).

length of the tunnels and duration of construction of the tunnels are given in Table 3. The cost of secondary final tunnel lining of Cakmak Station with cast concrete lining is given in Table 4. The stages of the cast concrete lining in the tunnels are shown in Figure 9.

2.2 *Using Shotcrete Method as Final Lining*

Taking into consideration the time of shotcrete application records at the jobsite, instead of cast concrete secondary lining, duration at the case of making final tunnel lining with

Table 3. Duration of Secondary final tunnel lining of Cakmak Station with cast concrete lining.

Tunnel types	Length (m)	Number	Time (month)
P1, Peron tunnel	270	2	8.5
P2, Ventilation fan tunnel	32	2	1.5
B1.2, Ventilation fan connection tunnel	20	7	2.5
B3.1, Ladder tunnel	350	2	1.5
A, Air conditioning tunnel	75.86	1	2
		Total time (month)	16

Table 4. Cost of secondary final tunnel lining of Cakmak Station with cast concrete lining.

Construction Name	Quantity	Unit price (€)	Amount (€)
Bar (ton)	1280.89	814.51	1,043,297.71
Concrete (m^3)	7981.25	49.45	394,672.81
Formwork (m^2)	20,140.55	19.55	393,747.65
Waterstop (m)	4784.14	8.26	39,516.96
Membrane (m^2)	20,160.45	7.95	160,275.59
		Total Amount	2,031,510.73

Figure 9. Stages of cast concrete lining.

Table 5. Duration of Secondary final tunnel lining of Cakmak Station with shotcrete lining.

Tunnel types	Length (m)	Number	Time (month)
P1, Peron tunnel	270	2	5
P2, Ventilation fan tunnel	32	2	1
B1.2, Ventilation fan connection tunnel	20	7	1.5
B3.1, Ladder tunnel	35	2	1
A, Air conditioning tunnel	75.86	1	1.5
		Total time (month)	10

Table 6. Cost of Secondary final tunnel lining of Cakmak Station with shotcrete lining.

Construction Name	Quantity	Unit price (€)	Amount (€)
Steel fibre (ton)	351.18	1350	474,086.25
Shotcrete (m^3)	8779.38	81.76	717,801.70
Waterstop (m)	4784.14	8.26	39,516.96
Sprayed membrane (m^2)	20,160.45	25	504,011.30
		Total Amount	1,735,416.21

shotcrete is summarized in Table 5 and the cost of Secondary final tunnel lining of Cakmak Station with shotcrete lining is given in Table 6.

3 CONCLUSION

Cakmak Station is a very critical station in UUC project which is in the middle of metro line. Hence, changing the final lining (secondary lining) of peron tunnels, ventilation fan tunnels, ventilation fan connection and ladder tunnels from conventional cast concrete to sprayed concrete (shotcrete) is analyzed.

Conventional cast concrete was used for the final lining of tunnels. Concourse of Cakmak Station was built by cut and cover method. The rest of the station is tunnel and built by NATM. Under this circumstances, construction of tunnels are completed approximately within 16 months and 2,031,510.73 euro cost. On the other hand, it is calculated that if shotcrete was preferred as final tunnel lining it might have taken approximately 10 months and with 1,735,416.21 euro cost.

Based on our analysis, using shotcrete as final lining of tunnel instead of conventional cast concrete could give the opportunity to complete the Cakmak Station about 6 months earlier with 15 % less cost.

ACKNOWLEDGEMENT

The author thanks to the representatives of Istanbul Metropolitan Municipality (IMM).

REFERENCES

Ocak, I. 2008. Control of Surface Settlements with Umbrella Arch Method in Second Stage Excavations of Istanbul Metro, Tunneling and Underground Space Technology, 23(6): 674–681.

Ocak, I., Selcuk, E., Eker H., Namli M. May 22-28. 2015. Cost Comparison of NATM and Umbrella Arch Method, ITA WTC 2015 Congress and 41st General Assembly.

Istanbul Metropolitan Municipality (IMM), 2012. Unpublished Technical Report (in Turkish).

Istanbul Metropolitan Municipality (IMM), 2013. Uskudar-Umraniye-Cekmekoy Metro Project, Cakmak station geotechnics report (in Turkish).

Selcuk, E. 2016. Comparing of NATM and Umbrella Arch Method in terms of Times, Cost and Deformation, Istanbul University Master's thesis (in Turkish).

Tunnels and Underground Cities: Engineering and Innovation meet Archaeology,
Architecture and Art, Volume 12: Urban
Tunnels - Part 2 – Peila, Viggiani & Celestino (Eds)
© 2019 Taylor & Francis Group, London, ISBN 978-0-367-46900-9

SCL vs squareworks – timberless tunnelling in future LU Station upgrade projects?

A. Nasekhian & A. Onisie-Moldovan
Dr. Sauer and Partners Ltd., London, UK

J. Ares & D. Kelly
Dragados SA., London, UK

P. Dryden
London Underground, London, UK

ABSTRACT: Two new adits, which are located between the two existing Northern line platform tunnels at Bank Station, would traditionally be approached and constructed using hand mining techniques. An innovative mechanised sprayed concrete lining (SCL) solution, with a lean support system, has provided greater working space and minimised the amount of hand excavation and manual handling, whilst simultaneously keeping the railway operational and reducing the construction programme duration and associated costs.

1   INTRODUCTION

Bank Station is a London Underground (LU) station located in the City of London financial district. It is a key interchange served by five lines, namely the Northern (NL), Central (CL), Waterloo & City (W&C), and the Docklands Light Railway (DLR), and at the Monument end of the same station complex, the District and Circle lines (D&C). Currently Bank Station suffers from heavy passenger congestion during peak hours for boarding, alighting and interchanging between the different lines. The Bank Station Capacity Upgrade (BSCU) is a large scale underground station expansion project intending to increase capacity and to account for future forecast demand at Bank/Monument Station, maximise savings in journey times, provide Step-Free Access to the underground lines, and provide emergency fire and evacuation protection measures for the station. The upgrade includes the provision of a new ticket hall at King William Street and a series of new tunnels of a total length of approximately 1330 m, including a new running and platform tunnel for the Northern line, concourse tunnels, cross passages and escalator barrels Figure 1. The complexity of the project has demanded state-of-the-art design solutions where meeting tunnels that are already over 100 years old.

Typically on upgrade projects in London, new SCL tunnels are connected to existing assets using the traditional timbering system (Mackenzie, 2014) called a squarework construction method (SQW). This method had also been considered in the initial design stage of the BSCU. Since the beginning of the project there was an aim to replace, wherever possible, the hand mined sections of work with mechanised sprayed concrete lined tunnels and to minimise the amount of manual excavation in a confined space involved in SQW construction; a slow construction process that involves health and safety issues such as high risk of hand-arm vibration syndrome (HAVS) and injuries due to manual handling.

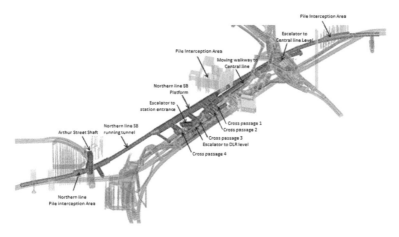

Figure 1. Bank Station Capacity Upgrade project layout – the dark blue illustrates new tunnelling works.

This paper explains the process followed and the considerations taken to implement this innovative engineering proposal, and how a collaborative design-build environment allowed for the development of an ambitious SCL design to substitute the SQW methodology in order to reduce the health and safety risk profile on site while also providing cost and programme benefits to the project.

2 INITIAL DESIGN IN SQUAREWORKS

In the BSCU scheme, there are several locations at the Central line (CL), NL and Docklands Light Railways (DLR) station levels where the new SCL tunnel structures are connected to the existing Bank Station tunnel assets.

One of the problems for the project was how to safely build new adits between the two existing NL platforms (NL-PT) whilst keeping the railway operational. This paper focuses on the temporary works design for NL Adit No.3 (NL-A3) and NL Cross Passage No.4 (NL-CP4) connection Figure 2.

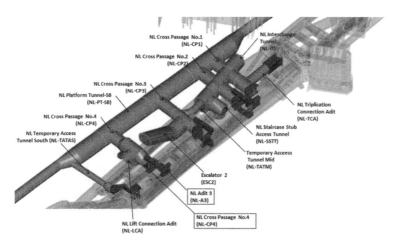

Figure 2. Early stage of the design scheme, SQWs (in red) connecting SCL tunnels (in blue) to the existing LU assets (in grey). NL-A3 and NL-CP4 are highlighted in red.

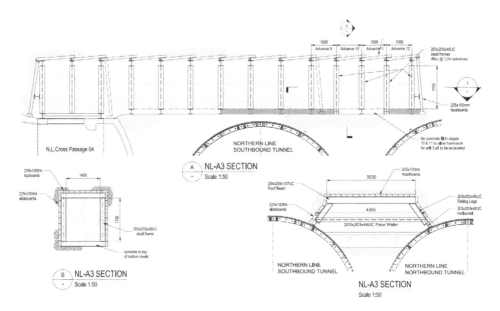

Figure 3. Section from the SQW Design proposal for adit access.

An early design for the temporary works enabling the construction of the new adits between the NL Platform Tunnels employed timber and steel squareworks tunnels using traditional hand mining techniques – a solution that has been proven through over a century of construction on the Underground network. A small hand-mined tunnel would be driven from the crown of the SCL tunnel from the new platform (NL-CP4) over the existing southbound platform (SB-PT) tunnel. The tunnel would be supported with steel and timber and have internal space of approximately 1.7x1.4m (Figure 3). From this access tunnel, hand excavation between the platform tunnels would continue to create the space needed for the permanent works.

Traditional SQW construction employs handheld pneumatic clay-spades to excavate the material by hand; this requires the workforce to be at the face working in confined spaces that are frequently difficult to access. This is followed by support using timber and steel sections that are often lifted and installed by the team on site. While proven successful for construction of tunnels in London Clay in close proximity to existing assets, there are a number of inherent issues with this technique; including high exposure to Hand Arm Vibration Syndrome (HAVS), limited means of access and rescue, lifting of large structural members in small spaces, significant manual handling and excavating overhead that mean that it should only be employed when it can be demonstrated to reduce overall risks to ALARP.

3 ALTERNATIVE DESIGN IN SCL

During design development significant steps had been taken to minimise hand mined sections of work by maximising the length of mechanised SCL tunnels to as short a distance from the existing assets as considered practicable at that point. As the construction works on site progressed, and the understanding of the behaviour of existing assets when tunnelling adjacent to them improved, a review of the square works methodology suggested that replacement of some further sections of the SQW would simplify and potentially de-risk the construction of the adits. The aim behind changing the design and therefore, where possible, using SCL in lieu of Squareworks techniques for temporary works was the possibility of:

- Maximising mechanised excavation
- Providing better and larger access with increased work space
- Minimising Manual Handling required
- Facilitating easier installation of lintel and jamb frames (large structural steel beams)
- Continuing the SCL Safe System of Works (SCL) that was working well
- Improving waterproofing details, by allowing for installation of a waterproofing membrane
- Designing permanent and temporary works in concert using the same designer
- Minimising disruption to existing assets and station operation by removing pan stiffeners (required in the squareworks design) where possible
- All of the above are CDM mitigations that reduce workforce exposure to risk, increasing the maximum of one of the BSCU main objectives, CDM – Safe by Design.

Consideration was given to the following parameters so an informed decision could be made.

In order for the proposal to be viable, Dr Sauer and Partners undertook a feasibility study; looking at buildability, space-proofing and the impact on the existing cast iron tunnel. Through the employment of a detailed staged 3D finite element modelling (3D FEM), the consultant provided both Dragados and London Underground with the

Table 1. SLC vs SQW comparison.

	Initial design – SQW		Alternative design - SCL	
	Disadvantage	Advantage	Disadvantage	Advantage
Health & Safety	– A lot of hand excavation therefore high exposure to HAVS (–) – Manual handling of large steel and timber sections with an elevated risk of injury (–) – Confined work space (–) – Difficult work space for installation of lintels, jamb frames and formwork (–) – Poor access/rescue space (–) – Hot Works (–)	– Smaller excavation in proximity of platforms in use (+)	– Larger excavation volume in proximity to platforms in use (–) – Increased risk of physical damage to aged structures by mechanical excavation in direct proximity (–)	– Better and larger access, increased work space (+) – Minimum manual handling required through mechanical excavation and support (+) – Easier installation of lintel and jamb frames (+) – - Continuation of the SCL SSOW (+)
Quality	The SCL design provided a more open workspace and consequentially simplified construction of the new structure within it. The waterproofing detail is easier using SCL compared to squareworks, as a membrane can be installed between linings			
Programme (est.)	88 days		65 days	
Cost (est.)	24% savings in SCL construction			
Impact on Existing LU Station	Further to the ground movement assessment, movements on the cast iron in SCL were expected to remain similar to movements in SQW.			
Impact on other stakeholder assets	No net difference on buildings, highways and utilities was assumed. For assessment purposes, the project has adopted moderately conservative values of volume loss of 2% for SQW excavation and 1.5% for SCL.			

Figure 4.   SCL versus Squareworks. SCL Access Tunnel at BSCU (left), Typical SQW Heading (right).

confidence to support the use of SCL method, particularly addressing the concerns on the impact on existing assets and demonstrating validation against examples of previous work of a similar nature.

Mechanical excavation and robotic spraying of SCL improved the construction process significantly by reducing the risk of long term injuries and conditions associated with HAVS and musculoskeletal disorders (MSDs). The replacement of the SQW method here allowed for full height tunnels with improved access and egress, and the employment of exclusion zones keeping the workforce away from the excavated face and risks of falling ground (See Figure 4).

Additionally, the use of SCL was initially predicted to reduce the construction time by 23 days (however, the actual construction of NL-A3 was finished a further week ahead of programme), resulting, in less programme pressure on the tunnelling workforce. The overall effect is that there is less stress across the project while allowing the delivery to remain on programme and positively impacting the long-term wellbeing of the workforce.

Traditionally with SQW water tightness is achieved by controlling crack widths which can prove difficult to achieve both in initial construction and subsequent operation. With the SCL method, the waterproofing system includes a continuous spray applied waterproof membrane extending onto the existing cast iron (CI) segments with hydrophilic strips and re-groutable hoses as an additional measure to provide a water tanked space outside of the permanent works. Providing the space for the temporary works using a fibre reinforced SCL shell means that risk of expensive maintenance in the tunnels is significantly reduced, driving down the lifetime cost of the project.

In essence, this approach brought significant benefit, not only in respect of workforce safety and safety by design principles, but also in terms of cost, programme and quality; so a decision was made to proceed with the alternative design approach in SCL.

4   GROUND CONDITIONS

The new SCL structures lie within the typical London clay underlying eight metres of River Terrace Deposits and Made Ground. Axis of the new tunnels at CL, NL and the DLR levels is nearly 20, 30 and 40m under the subsoil surface, respectively. The groundwater table is assumed 8 metres below the ground surface. More details in association with the ground condition and geotechnical parameters of subsoil layers can be found in Nasekhian and Spyridis 2017.

5   DESIGN

Figure 5 illustrates a cut through the Building Information Modelling (BIM) model made for modelling the temporary SCL tunnels required for access to the NL-A3 shaft between the existing platform tunnels which stem from the end of the NL-CP4 tunnel. The new design had NL-CP4 excavated from the new NL Southbound Platform tunnel (NL-PT) to a temporary headwall 3.0m behind the extrados of the existing NL Southbound Operational Platform. From this headwall a temporary inclined SCL Access Tunnel was excavated directly over the

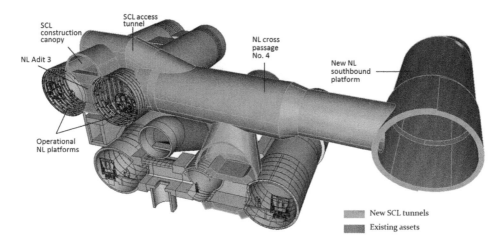

Figure 5.   Tunnel layout showing Cross Passage 4 and the temporary SCL Access Tunnel and SCL Construction Canopy which provided access to construct NL Adit 3 between the operation NL Platforms.

live platform tunnel, followed by a temporary SCL Construction Canopy, with 3m headroom, constructed above the two existing platform tunnels which provided access for the new permanent adit (NL Adit 3) to be excavated downwards between the platform tunnels. The temporary SCL tunnels were built with the following construction sequence that has been shown schematically in Figure 6 and Figure 7:

- Stage 1 – Excavate and spray primary lining of NL-CP4 up to temporary headwall. [Excavation size: Height 6.8m, Width 6.5m]
- Stage 2 – Breakout from temporary headwall, excavate and spray primary lining of NL Access Tunnel No.1 (NL-AT1). [H 3.8m, W 3.3m]
- Stage 3 – Excavate and spray primary lining of NL Breakout Chamber No.1 (NL-BC1). [H 4.2m, W 5.1m]
- Stage 4 – Breakout, excavate and spray primary lining of NL Adit 3 Canopy (NL-A3C). [H 4.9m, W 3.7m]
- Stage 5 – Excavate NL Adit 3 Shaft (NL-A3S), install waterproofing and reinforced cast-in-place opening frame NL-A3. [length 6.4m, Depth 5m, narrowest point between platforms nearly 0.9m]
- Stage 6 – Install foam concrete backfill to temporary tunnels NL-A3C, NL-BC1, NL-AT1
- Stage 7 – Breakout temporary headwall, excavate and spray primary lining of NL-CP4. Install waterproofing and NL-CP4 opening (NL-CP4-OF) frame permanent works - reinforced cast-in-place concrete structure.

The SCL tunnel primary linings were constructed with fibre reinforced concrete and contained no bar reinforcement. Temporary access tunnels are SCL tunnels with primary lining only. To allow for the breakout of the invert of the canopy to excavate the NL Adit 3 Shaft (NL-A3S) a steel frame was installed to provide temporary support to the canopy NL-A3C (see Figure 8). Due to space constraints between the existing platforms,

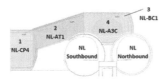

Figure 6.   Construction sequence for NL-A3 and NL-CP4 – Steps 1 to 4.

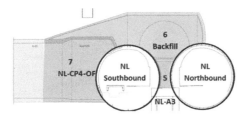

Figure 7.   Construction sequence for NL-A3 and NL-CP4-OF – Steps 5 to 7.

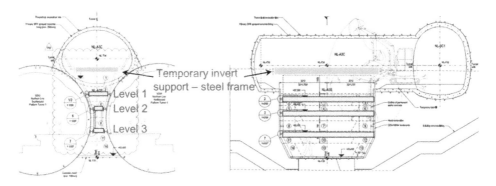

Figure 8.   Temporary steel frames for Adit 3 construction, a) cross section b) long section.

which are just 0.9m apart, it was not possible to use mechanised excavation for the construction of the NL Adit 3 shaft (step 5 above), therefore traditional hand mining techniques with timber and steel temporary support were used for NL-A3S. When excavating the NL-A3S in-between the existing platform tunnels, three levels of steel frames were installed to provide temporary support to the existing NL-PT as shown in Figure 8. The permanent works NL Adit 3 opening frames into the existing NL-PT were constructed in reinforced cast-in-place concrete. The NL-CP4-OF opening width in the existing southbound cast iron (CI) lining is 3.3m (i.e. 7-ring wide opening) and the opening width for NL-A3 is 4.3m equivalent to removal of 9.5 CI rings. Lining thickness of the SCL structures varies from 150 to 300mm.

6   FINITE ELEMENT MODELLING

A detailed 3D FEM was developed in order to design the SCL structures and assess the impact of the temporary works on the existing LU civil assets. The program used for the analysis is Abaqus 6–13 Dassault Systèmes. All of the excavation sequences of the temporary works was simulated. The SCL tunnels' simulation in the model commences after simplified modelling of the existing platforms which included using a so-called wished in place method. The SCL tunnels and existing assets were modelled using triangular shell elements while for the cast-in-place concrete opening frames volume elements were employed. Segmental lining properties of existing platforms are transformed to uniform ring properties based on the method proposed in Muir Wood (1975). Nasekhian and Spyridis (2017) give further details regarding the methodology of FEM used for the BSCU project.

Results from the temporary works finite element analysis in conjunction with contemporary closed-form methods showed the risk of damage to the existing platforms to be manageable. In light of this, the requirement for the strengthening of the existing cast iron tunnel lining by

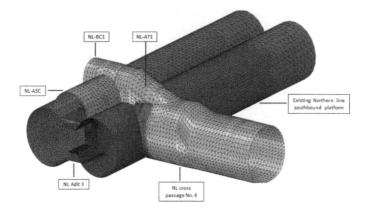

Figure 9.   A section through NL-Adit 3 in the 3D FEM.

the installation of stiffener plates was eliminated, which had significant cost, programme and operational benefits. A good estimation of tensile stresses in the CI segments before and after the construction was crucial in order to make the decision to remove these strengthening works (pan stiffeners) required in the original design.

7   CONSTRUCTION

The SCL temporary access tunnels were successfully constructed in a short time relative to the previously planned squareworks, without affecting the station operations. These tunnels provided a large safe work space for the construction team to excavate and construct the NL Adit 3 Shaft (NL-A3S). A monorail beam was used for moving large/heavy structural members and as a muck-away system during excavation works. Figure 10 and Figure 11 depict the provided spaces and logistics for construction of NL-A3S which enhanced significantly the work environment in terms of health and safety.

As the temporary tunnels were constructed directly over the existing tunnel lining, which continued running as a platform throughout, caution during excavation was employed; probe excavations were used to determine the exact location, and care taken over selection of equipment to protect the tunnel lining at all times.

The support system for excavation between the back of the platforms made use of relatively small sections of steel (Figure 12). The system required no bolting into the cast iron linings

Figure 10.   SCL Access Tunnel (NL-AT1).

Figure 11. SCL Construction Canopy (NL-A3C) with steel frame installed to allow for breakout of invert.

Figure 12. NL A3 Shaft Construction Works: Access area with monorail beam (LHS). Installation of steel props between existing platform tunnel at spring line (Centre). Spray applied waterproofing (RHS).

and maximised the limited space between back of the platforms. At the springline the gap was less than 1.0m.

The temporary works for NL-A3 were completed on March 16, 2018, almost a week ahead of the already reduced programme, without any issues or impact to railway operations. LU were satisfied with the design and execution of the temporary works, considering it a success and an important case study in the planning of both other BSCU works and on future station upgrade projects.

For the construction Cross Passage 4 Opening Frame (NL-CP4-OF) permanent works, at the junction between the existing Platform Tunnel and new Cross Passage 4, the use of SCL as a Primary Lining allowed for the installation of a continuous spray applied waterproof membrane extending onto the existing cast iron (CI) segments as shown in Figure 13. The junction between new and old structures is often a problematic area to waterproof, and the use of a fully tanked waterproof membrane greatly improves the water tightness in this area. The large working area provided by the SCL solution made the installation of the waterproofing a simple and safe operation while also improving the quality control.

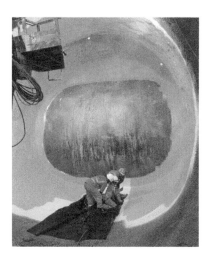

Figure 13. Waterproofing membrane at NL-CP04-OF.

8 IMPACT ON EXISTING PLATFORMS AND RAILWAY OPERATION

Mining in close proximity to the existing platform tunnels (partial removal of ground support of the tunnels) may induce large transverse and longitudinal distortion in the assets which potentially impacts both serviceability and operability of the live platform and structural capacity of the linings.

The existing NL-PT has 6.7m external diameter and the lining is comprised of 12 CI segments (with one rotated key) per ring connected circumferentially with 83 bolts. The CI segments have good strength in compression but are weak in tension (nearly a quarter). A comprehensive structural assessment using a risk-based approach was carried out on these assets to ensure that structural integrity and serviceability of the assets are maintained. FEA predicted a convergence of 15 to 20mm in the NL Southbound and slight movements in the NL Northbound. Structural assessment of the lining revealed that tensile stress in the flange of segments near the opening areas will slightly exceed the permissible tensile capacity of the cast iron in a limited area. A concession was approved by the client for those segments that would not be compliant with the LU requirements in terms of structural capacity since tunnel safety was demonstrated to be maintained through a combination of monitoring, inspection and provision of protection to passengers

The most important design requirement was that the NL-PT shall remain operational, as any interruption in railway traffic would be of significant impact to the project. Thus, a precise monitoring regime was put in place, not only to observe the longitudinal and transverse movement of the segmental lining but also to measure the track in terms of settlement/heave, cant/twist and clearances for passing trains. Monitoring sections comprising 5 targets were installed at two and a half metre centres in the critical zone and read automatically using total stations installed in the tunnel crown. A traffic light trigger system was applied, with the amber level the prediction from the FE model and higher triggers indicating behaviour beyond expectations but prior to critical limits. During construction of the NL-Adit 3 and CP4, the existing NL-SB-PT responded to ground movements but did not breach amber triggers at any point. Figure 14 depicts where the maximum convergence happened during the main construction stages. A settlement of 2mm was recorded on the southbound track after the construction of NL Adit 3 Shaft, and 5 mm heave when the NL-CP4 was completed.

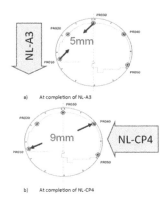

Figure 14.   Maximum convergence in the NL-SB-PT.

During these works the platforms remained operational and the existing linings behaved as expected with deformation magnitudes lower than predictions. No damage/cracks were reported in the flange area of the CI segments.

9   CONCLUSIONS

The temporary works solution provided for the construction of some of the adits at BSCU employs SCL lining in a situation where timber squareworks would traditionally have been used. This solution paves the way for fully mechanised excavation and support in and around existing LU assets, virtually eliminating many risks associated with the traditional hand mining method.

Our solution improved safety to the workforce during execution, by removing handworks with open faces (minimising HAVS), and reducing manual handling until the narrow shaft works where it was unavoidable. It also created a more comfortable working environment, where people and plant could easily pass each other, and material could be safely moved via a monorail beam. The removal of structural interfaces and the robust waterproofing provided by SCL works, create a better permanent works solution with much reduced maintenance issues, this enhanced sustainability in the design.

Our main challenge was to ensure that the railway remained operational at all times in light of the larger excavation faces, reduced temporary support and close operation of mechanised plant. The staged and calibrated 3D FEM showed and gave confidence that the existing platform tunnels would be safe to operate, and not require extensive enabling works to stiffen up the cast iron segments. This was confirmed with observation and in situ monitoring and no incident was reported.

The change in construction methodology has not only allowed for streamlined temporary works, but also allows for a better permanent solution. Moreover, this design is quicker and safer, proving that productivity and safety can be married through innovative design.

10   ACKNOWLEDGEMENT

The authors truly appreciate BSCU project Executive for review and granting permission to publish this article. The authors also are grateful to Bethan Haig and Viki James for their efforts to review and provide constructive comments on the paper.

REFERENCES

Mackenzie, C. 2014. Traditional timbering in soft ground tunnelling - a historical review, British Tunnelling Society.

Nasekhian, A. & Spyridis, P. 2017. Finite Element modelling for the London Undergroun

Bank Station Capacity Upgrade SCL design and deep tube tunnels assessment. Proc. of IV Int. Conf. on Computational Methods in Tunnelling and Subsurface Eng. (EURO:TUN).

Wood, A.M. 1975. The circular tunnel in elastic ground. Geotechnique 25 (1).

Tunnels and Underground Cities: Engineering and Innovation meet Archaeology,
Architecture and Art, Volume 12: Urban
Tunnels - Part 2 – Peila, Viggiani & Celestino (Eds)
© 2019 Taylor & Francis Group, London, ISBN 978-0-367-46900-9

Performance and modelling of Fort-d'Issy-Vanves-Clamart metro station: A 32 m deep excavation of the Grand Paris project

K. Nejjar
Terrasol, Paris, France

D. Dias
3SR Laboratory, Grenoble Alpes University, Grenoble, France

F. Cuira, H. Le Bissonnais & G. Chapron
Terrasol, Paris, France

ABSTRACT: As part of the Grand Paris project, Fort-d'Issy-Vanves-Clamart metro station is a 32 m deep excavation supported by 41 m deep retaining wall. The use of the classical subgrade reaction method to design the retaining wall shows its limits especially in terms of underestimating loads in the support elements. Arching effect undergone by retaining soil seems responsible of overloading upper struts. Finite element modelling is necessary to grasp the soil behavior and highlight the load transfer. A comparative study was carried out between finite element modelling and the subgrade reaction method. In spite of similar results in terms of wall displacement, loads in support elements are significantly different and wide discrepancies are noticed in the earth pressure diagram. The results are compared to the real performance of the retaining wall through measurement of wall deflection, strut loads and earth pressure with fiber optic, inclinometers, strain gauges and pressure cells.

1 INTRODUCTION

As part of the Grand Paris Express project, many deep excavations will be dug in dense urban areas and often nearby sensitive structures as high-speed rail lines or tower blocks. Hence, numerical modelling becomes indispensable since it takes into account soil structure interactions and assesses accurately settlements.

Moreover, the first studies to design the underpinning system of the future metro stations revealed important discrepancies between subgrade reaction method and finite element method. The first metro station under construction Fort-d'Issy-Vanves-Clamart was ideal to be equipped with advanced instrumentation in order to find out the origin of the differences between the two design methods and confirm the real behavior of the retaining wall. Such interesting feed-back could allow optimizing the design of further metro station of the Grand Paris Express project.

2 FORT-D'ISSY-VANVES-CLAMART METRO STATION

Fort-d'Issy-Vanves-Clamart station is part of the new metro line 15 under construction since spring 2016 within the Grand Paris Express project. The station enclosure consists of a 1.2 m thick slurry walls to a depth of 41 m. Its dimensions are 110 m long by 23 m large and 32 m deep. The present paper focuses on assessing the performance of the excavation to 20 m deep

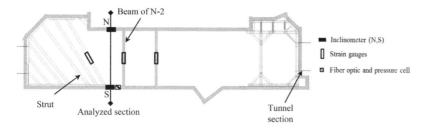

Figure 1. Top view of the station and instrumentation location.

only regarding work's progress. Up to this depth, support elements of the wall are composed of the cover slab, 2 inclined strut levels (B1 and B2) and 2 floor levels (N-1 and N-2).

The analyzed section is shown in the Figure 1 with the instrumentation around. The latter is composed of two inclinometers (N = north, S = south), 3 strain gauges in each instrumented strut in both levels B1 and B2, 2 strain gauges in each instrumented beam of the floor level N-2, 100 m of fiber optic installed in one wall panel and a pressure cell at the interface soil/wall in the excavation side at a depth of 34.5 m. Wall displacement is measured with inclinometers, forces in support element (strut and beam) are measured with strain gauges, wall bending moment is evaluated through fiber optic and earth pressure under excavation side is evaluated with the pressure cell. The fiber optic is set in the form of a "U" in order to grasp axial strains of compressed and tensed sides which difference divided by width of the "U" and multiplied by inertia product of the wall gives the bending moment (Nejjar et al. 2018).

The construction phases are detailed herein:

- Initialization of the stress state taking into account a surcharge loads of 30kPa due to nearby traffic
- Excavation with slope and installation of the slurry walls at 7.5m depth
- Extend slurry walls upward to ground level (0 m) and backfill the slope
- Set up cover slab and strut level B1 at 5.8 m prestressed at 79 kN/lm
- Excavation at 10m depth
- Set up N-1 floor level at 9 m
- Excavation at 14.2 m depth
- Set up strut level B2 at 13 m prestressed at 300 kN/lm
- Excavation at 16.5 m depth
- Set up N-2 floor level at 15 m
- Excavation at 20 m

3 GROUND CONDITIONS

The station is located in the geological context of the Parisian sedimentary basin. The studied section is composed of a backfill of 11m height, 10m of damaged Hard limestones (middle Eocene), 8m of Plastic clays (lower Eocene), 10m of Meudon marls (Paleocene) and Chalk (Cretaceous). The retaining wall is embedded in the Chalk over 2m.

A wide geotechnical soil investigation provided the soil identification. Figure 2 presents the soil stratigraphy through the dynamic shear modulus G_0 measured with a cross hole seismic testing. Nineteen boreholes were drilled for pressuremeter tests along 40m depth with a maximum step of 2m. Statistical analysis of pressuremeter modulus E_M and the consideration of the rheological coefficient α recommended by Ménard (1968) leads to the summarized values given in Table 1. The effective friction angle φ' and cohesion c' comes from 69 Consolidated Undrained triaxial tests with pore pressure assessment (CU+u) and 35 Consolidated Drained triaxial tests (CD). The earth pressure coefficient at rest K_0 is evaluated using Mayne and Kulhawy (1982) correlation with friction angle and overconsolidation ratio OCR presented is the equation 1.

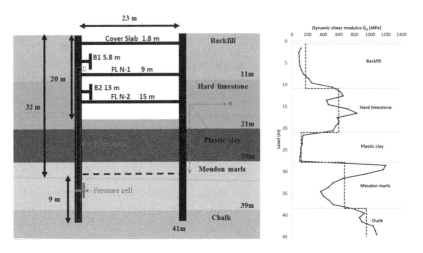

Figure 2.   Soil stratigraphy and dynamic shear modulus from cross hole seismic testing.

Table 1.   Geotechnical properties of soil layers.

	G_0 (MPa)	E_M (MPa)	α	φ' (°)	c' (kPa)	OCR	K_0
Backfill	175	6	0.5	29	0	1	0.515
Hard limestone	600	25	0.5	35	40	1	0.426
Plastic clay	156	40	1	18	10	1.5	0.85
Meudon marls	670	100	0.67	25	30	1	0.577
Chalk	950	170	0.5	35	40	1	0.426

$$K_0 = (1 - \sin(\varphi'))\sqrt{OCR} \tag{1}$$

The groundwater table is located in the Hard limestone at 16 m depth. The steady state water pressure profile is presented in blue in the Figure 2, the low permeability of Plastic clay is responsible of the decreasing pressure to zero.

4   FINITE ELEMENT AND SUBGRADE REACTION MODEL

Both finite element method (FEM) and subgrade reaction method (SRM) are applied to reproduce the behavior of the deep excavation. The first is carried out with Plaxis 2D v.2017 and the second with K-Réa v.4.

For FEM, the advanced soil model HSS (Plaxis manual) is used since it distinguishes between loading and unloading by including two different Young modulus and the yield surfaces with plasticity formulation provide an hyperbolic stress-strain curve for triaxial shearing similar to laboratory observation. In addition, it takes into account the small-strain stiffness by including a shear modulus degradation curve following Hardin & Drnevich (1972) model that involves two parameters G_0 and $\gamma_{0.7}$ (Eq. 2). $G_0 G_0$ corresponds to the dynamic shear modulus and $\gamma_{0.7}$ is the distortion reached at 70% of modulus reduction.

$$G = \frac{G_0}{1 + 0.385\frac{\gamma}{\gamma_{0.7}}} \tag{2}$$

Table 2. Inputs of HSS soil model.

		Backfill	Hard limestone	Plastic Clay	Meudon marls	Chalk
HSS	E_{50}(MPa)	30	125	100	375	850
	E_{oed}(MPa)	30	125	100	375	850
	E_{ur}(MPa)	90	250	200	750	1700
	v_{ur}	0.2	0.2	0.2	0.2	0.2
	m	0	0	0	0	0
	φ' (°)	29	35	18	25	35
	c' (kPa)	0	40	10	30	40
	K_0	0.515	0.426	0.85	0.577	0.426
	R_{inter}	0.66	0.66	0.33	0.66	0.66
	G_0(MPa)	175	600	156	670	950
	$\gamma_{0.7}$	9.5E-5	9.0E-5	1.04E-4	1.5E-4	1.3E-4

Table 3. Stiffness of support elements.

	Stiffness (MN/m/lm)
Cover slab	2760
B1	59.87
N-1	2800
B2	122.2
N-2	1600

For normally consolidated soil, Benz (2007) presents the Eq. (3) to assess $\gamma_{0.7}$, which will be used for Backfill, Hard limestone, Meudon marls and Chalk soil layers. The principal stress $\sigma'_1 \sigma'_1$ is equal to the vertical stress at the middle of the layer.

$$\gamma_{0.7} = \frac{3}{28 G_0} \left(2c'(1 + \cos(2\varphi')) + \sigma'_1(1 + K_0)\sin(2\varphi')\right) \tag{3}$$

For the overconsolidated layer of Plastic Clay, the $\gamma_{0.7}$ is determined from the fitting of the Hardin & Drnevich (1972) model with the measured $(E_{50}, y_{E_{50}})$ from Consolidated drained triaxial tests.

The Table 2 summarizes the inputs of HSS soil model. By default, E_{50} is taken equal to E_{oed} and E_{ur} is the double except for Backfill which is taken the triple. In order to fit the observed wall displacements, the E_{50} is considered as a multiple of the measured E_M/α. The multiplying factor is found equal de 2.5.

For SRM, the reaction coefficient k_h is computed from the Schmitt (1995) formula involving E_M/α and wall inertia. However, in order to fit the observed wall displacements, the E_M/α need to be heightened compared to the measured value by a factor of 1.5.

The retaining walls are modeled using linear elastic plates. The slab, floor and strut levels are modeled using linear elastic anchors. The Table 3 summarizes the stiffness of support elements per linear meter for 2D modelling. The wall has an inertia product of EI = 3456 MN.m²/lm.

5 PERFORMANCE OF THE RETAINING WALL

5.1 Wall displacement and moment

The Figure 3 presents a comparison between displacements and bending moments derived from both SRM and FEM models and the measured profiles coming from inclinometers

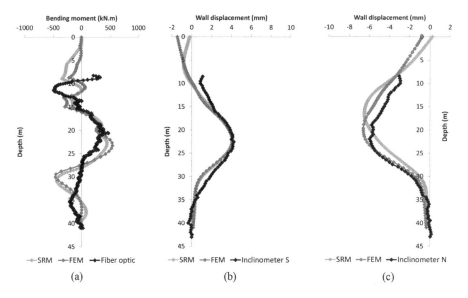

Figure 3.   Comparison between SRM/FEM models and inclinometers/fiber optic measurements.

and fiber optic. Both models give concordant results and satisfactory reproduce the wall performance. However, the north wall has a lower deflection than the computed one and the FEM shape approaches more the inclinometer than the SRM. The cause could be a higher Hard limestone modulus in this area which is confirmed at worksite by the observation of non-continuous limestones banks during excavation. The shape of bending moment do not fit perfectly the fiber optic measurements but at least one can retain that the maximum bending moment has the same amplitude as computed and almost the same location.

5.2   Strut and concrete beam

Figure 4 presents for struts B1 and B2 the increment of effort above the prestress and for floor N-2 the total effort after casting. The strain gauges in N-2 are completely encased in concrete, the modulus used to derive the force is 30GPa. It corresponds to the concrete class C35/45 and could be much higher.

The graphs show clearly the large difference between SRM and FEM models. Measured forces in B2 and N-2 are close to FEM model whereas SRM model underestimates substantially the support forces much more for upper one B2 than N-2. For B1, both models present a lower value compared to the measured one, the origin of that discrepancy could be a

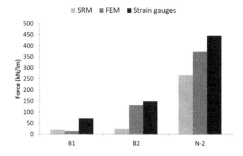

Figure 4.   Comparison between SRM/FEM models and strain gauges measurements.

difference between the prestress set up in the strut and the real prestress perceived by the ana-lyzed 2D section.

5.3 Lateral earth pressure

Figure 5 presents the earth pressure in front of the wall at 34.5 m depth derived from modelling and measured from the pressure cell installed at the interface soil/wall in excavated side. FEM gives a pressure closer to the measure compared to SRM, but still lower. FEM is 10% to 28% lower and SRM is 26% to 50%. Knowing that the pressure cell has an accuracy of 15%, FEM results could be acceptable but SRM underestimate largely this mobilized passive earth pressure. Consequently, the ratio between limiting and mobilized passive earth pressure is overestimated, which is equivalent to overestimate the safety margin available.

Figure 6 (a) shows the earth pressure of both excavation and soil sides in order to understand the origin of this greater mobilization of passive pressure. Indeed, FEM mobilizes a greater active pressure behind the wall, hence soil in excavated side needs to mobilize larger passive pressure to reach equilibrium. The origin of this greater active pressure behind the wall is due to arching effect. Indeed, Figure 6 (b) presents the incremental stress behind the wall at the last excavation phase (the difference between the earth pressure diagram of final phase and the previous one), the volume of excavated soil is indicated between dashed lines.

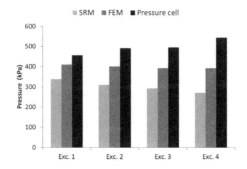

Figure 5. Comparison between SRM/FEM models and pressure cell.

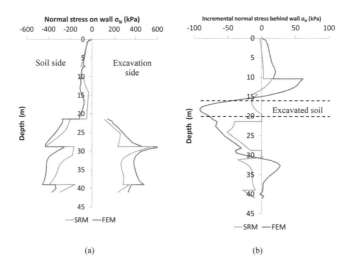

Figure 6. Comparison of earth pressure between SRM and FEM, (a) at final excavation phase, (b) incremental normal stress at final excavation phase behind the wall.

FEW incremental diagram depicts a clear load transfer mechanism. It consists of a significant unloading (-100 kPa at 18m) at the excavated zone and simultaneous loading at the upper part (+50 kPa at 11m) and the lower part (+25 kPa at 32 m) of the wall, which highlights a manifest classical arching effect. The latter is hence directly responsible of the greater earth pressure behind the wall and consequently the greater mobilized passive pressure in front of the wall.

SRM incremental diagram shows only an unloading of the soil behind the wall systematically and a minor loading at 11 m certainly due to a movement of the wall toward the soil. Thus, SRM could not reproduce arching effect since it involves independent reaction coefficients k_h responding only horizontally whereas FEM considers the equilibrium of a continuum media including shearing stresses in addition to vertical and horizontal stresses.

6 CONCLUSION

Fort-d'Issy-Vanves-Clamart metro station represents an interesting well documented deep excavation owing to its advanced instrumentation. The underpinning system in addition to the specific stratigraphy and excavation geometry form a suitable configuration to arching effect development. Both SRM and FEM models were used to fit the wall displacements. The best fitting was reached with an amplification of E_M/α ratio by 1.5 for SRM and a secant modulus $E_{50}E_{50}$ equal to 2.5 E_M/α for FEM. However, the forces in support elements were significantly different between the two methods. FEM approaches the most measurements whereas SRM underestimates significantly forces in strut and concrete beams. It underestimates also the mobilized passive earth pressure which overestimates critically the safety margin available. The analysis of earth pressure diagram confirms the responsibility of arching effect to explain these discrepancies between FEM and SRM in spite of the perfect concordance of wall displacements and bending moment.

The further excavation phases up to the final bottom level of the dig (32 m) will provide a better enlightenment about the persistence of arching effect with wider differences between FEM and SRM outputs. The modulus choice fixed in the present modelling to fit wall displacement will also be tested for the following excavation phases.

ACKNOWLEDGMENTS

The writers would like to thank the Société du Grand Paris for financial support of the installation of advanced instrumentation. We would like to warmly thank our partners in this work namely Lerm, Gexpertise, Ifsttar, Geokon and Dimione. We express our gratitude for the precious support of worksite teams of both Soletanche Bachy and Bouygues TP.

REFERENCES

Benz, T. 2007. Small-Strain Stiffness of soils and its numerical consequences. Universität Stuttgart.
Bjerrum, L., Clausen, C.J.F. & Duncan, J.M. 1972. Earth pressures on flexibles structures – a state of the art report. Proceedings of the 5th European Conference on Soil Mechanics and Foundation Engineering, Madrid, pp. 169–193
Handy, R.L. 1985. The arch in soil arching, Journal of Geotechnical Engineering. Vol. 111, Issue 3
Hardin, B.O. & Drnevich, V.P. 1972. Shear modulus and damping in soils: Design equations and curves. Proc. ASCE: Journal of thr Soil Mechanics and Foundations Division, 98(SM7), 667–692
Hashash, Y.M.A. & Whittle, A.J. 2002. Mechanisms of load transfer and arching for braced excavations in clay. Journal of geotechnical and geoenvironmental engineering, Vol. 128, N°3, pp 187–197
Mayne, P.W. & Kulhawy, F.H. 1982. K0-OCR relationships in soil, Journal of the Geotechnical Engineering Division. Proceedings of ASCE, Vol 108, N° GT6, pp 851–869
Ménard, L. 1968. Règles d'exploitation des techniques pressiométriques et d'exploitation des résultats obtenus pour le calcul des fondations. Sols-Soils, N°26 Paris

Nejjar, K., Dias, D., Chapron, G., Cuira F., Le Bissonnais, H. & Fluteaux, V. 2018. Advanced instrumentation for the retaining wall of the metro station Fort d'Issy Vanves Clamart. Journées Nationales de Géotechnique et de Géologie de l'Ingénieur, Champs-sur-Marne

Nadukuru, S.S. & Michalowski, R.L. 2012. Arching in distribution of active load on retaining walls. Journal of Geotechnical and Geoenvironmental Engineering, Vol. 138, No.5, May 1, pp. 575–584

Schmitt, P. 1995. Méthode empirique d'évaluation du coefficient de reaction du sol vis à vis des ouvrages de soutènement souples. Revue française de Géotehnique, 71, pp. 3–10

Zghondi, J. 2010. Modélisation avancée des excavations multi-supportées en site urbain. Mémoire de thèse, Institut National des Sciences Appliquées de Lyon

Tunnels and Underground Cities: Engineering and Innovation meet Archaeology, Architecture and Art, Volume 12: Urban Tunnels - Part 2 – Peila, Viggiani & Celestino (Eds)
© 2020 Taylor & Francis Group, London, ISBN 978-0-367-46900-9

Design and excavation of Camlica Tunnels in Istanbul

H. Nurnur, M. Kucukoglu & F. Efe
EMAY International Engineering and Consultancy Inc., Istanbul, Turkey

ABSTRACT: A tunnel network consisting of three single tube tunnels which have been projected based on New Austrian Tunneling Method (NATM) in Istanbul. The tunnels were excavated in Sandstone (very poor - poor strength), Quartzite (fair - good strength) and Arkose (very poor - poor strength). Tunnels are passing very shallow under residential areas especially at the entrance and exit parts of the tubes. In this paper, the main points of the tunnel design and construction will be explained and the measurements made at the site during the construction will be shown. These monitoring data will be compared with the numerical analysis results obtained during the design phase.

1 INTRODUCTION

An urban tunnel network project has been planned within the scope of highway transportation projects of Camlica Mosque, capable of accommodating fifty thousand people, which is under construction at Camlica Hill, Istanbul.

With the project, it is aimed to reduce the traffic load in Camlica and to provide more convenient and comfortable transportation in a shorter time as well as reducing the fuel loss and damage given to the environment.

The tunnel network, within the excavation and support work is completed and the electromechanical projects are continuing consists of three single tube tunnels. The T1 and T2 Tunnels, which have entrance and exit portals on the Libadiye side of the route, start side by side at the first 400 m. Then they are separated from each other and goes different sides of the mosque. In order to be able to regulate the traffic ring around the mosque and provide the passage between the two tubes, a third tube (T3 Tunnel) was built around the mosque. The length of the tunnels are given in Table 1 and the plan view of the tunnel network is shown in Figure 1.

The Camlica Tunnels have been projected according to New Austrian Tunneling Method (NATM) principles. The tunnels were excavated mostly in poor strength rock formations and are passing very shallow under residential areas especially at the entrance and exit parts of the tubes. Measurements have been made at the site for the tunnels and the surrounding buildings.

In this paper, the tunnel design and construction works will be explained and the measurement data will be shown. The numerical analysis results will be compared with the monitoring data.

Table 1. The length of the tunnels.

Tunnel	Start	Finish	Length (m)
T1 Tunnel	0+190	1+395	1,205
T2 Tunnel	0+220	1+620	1,400
T3 Tunnel	0+668	1+240	572

2 GEOLOGICAL AND GEOTECHNICAL CONDITIONS

2.1 Geological situation

There are three different geological formations in Camlica region. These formations are in order from oldest to youngest as Kurtkoy Formation (Lower Ordovician), Aydos Formation (Lower Ordovician) and Yayalar Formation (Upper Ordovician - Lower Silurian). These formations are mainly composed of Arkose (Kurtkoy Formation), Quartzite (Aydos Formation) and Sandstone (Yayalar Formation) units (Ozgul, 2011). The surface geology map of the area can be seen from Figure 2.

2.2 Geotechnical conditions

The tunnels were excavated in Sandstone (very poor - poor strength), Quartzite (fair - good strength) and Arkose (very poor - poor strength). Based on the engineering geology characteristics of the mentioned units for Camlica Tunnels, homogeneous zoning was performed as a result of laboratory experiments and empirical calculations. During the homogenous zoning

Figure 1.   Plan view of the Camlica Tunnels.

Figure 2.   Surface geology map of the Camlica Tunnels.

Table 2. The rock support classes of the tunnels.

Tunnel	B3 Class	C2 Class	C3 Class	Total Length
T1 Tunnel	664 m	240 m	301 m	1,205 m
T2 Tunnel	543 m	537 m	320 m	1,400 m
T3 Tunnel	255 m	277 m	40 m	572 m

studies, portal zones, areas with shallow cover thickness with poor rock conditions, rock units with potential for separation, water bearing zones, lithologies with aquifer properties, fault zones, swelling potential levels and deep sections are considered.

For each of the determined homogeneous zones, Q and RMR rock mass classifications were made. Based on the classifications aimed at determining the quality of rock mass, the rock support classes according to the ÖNORM B2203 standard have been calculated. These classes and their quantity are given in Table 2.

3 DESIGN OF THE TUNNEL CROSS-SECTIONS AND SUPPORT SYSTEMS

3.1 Tunnel cross-sections

The Camlica Tunnels have been constructed as single tube with two lanes and a safety lane, the clearance of the tunnel is 9.50 x 5.00 meters. Approximately the maximum excavation height of the tunnels is 8.50 m and the width is 14.50-15.00 meters. The typical cross-section of the tunnels is shown in Figure 3.

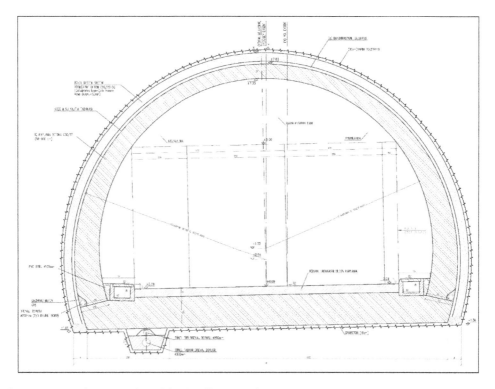

Figure 3. Tunnel cross-section of the Camlica Tunnels.

3.2 Tunnel excavation and support systems

For each NATM rock support classes, excavation and support type sections were established according to the Technical Specifications of Highways. These support systems can be seen from Table 3 and the Figure 4.

Table 3. The NATM support systems of the tunnels.

Support Type	B3 Class	C2 Class	C3 Class
Forepole	1.5" - 6m	2.5" - 6m	4.0" (Umbrella) - 6m
Shotcrete	C25/30 - 25cm	C25/30 - 30cm	C25/30 - 35cm
Wire mesh	Q221/221 x2	Q221/221 x2	Q295/295 x2
Steel beam	IPN 160	–	–
Lattice girder	–	2xϕ26 + ϕ32	2xϕ26 + ϕ32
Bolt	PG - 28mm - 4-6m	PG/IBO - 28mm - 4-6m	PG/IBO - 32mm - 4-6m

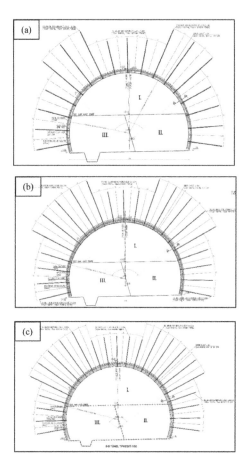

Figure 4. Excavation and support systems of the Camlica Tunnels a) B3 Class, b) C2 Class, c) C3 Class.

4 NUMERICAL MODELLING

Numerical analyses were carried out in order to test the reliability of the excavation stages and supporting systems determined for Camlica Tunnels. Tunnel stability analyses were made with the finite element method using PLAXIS 2D software.

Mohr-Coulomb material model was used for the analyses performed with PLAXIS 2D. In this model, the unit volume weight (γ), cohesion (c), friction angle (ϕ), Poisson's ratio (v) and Young's modulus (E) data are entered in order to define the geomechanical properties of the ground. The geomechanical parameters of the material which is used at the selected section are represented in Table 4.

Stability analyses for Camlica Tunnels were made for each section determined by the geotechnical evaluation. However, this paper focused on the Libadiye side of the route. This portal region is the most critical part of the project because the tunnels pass under the buildings in the shallowest section and the single tubes join at this area and become double tubes. The analysis model can be seen from Figure 5.

In the model, a spread load of 10 kN/m/m is defined for each storey, where the buildings are located. All the excavation stages are modeled in the analyses. At the end of the analyses,

Table 4. The geomechanical parameters of the Sandstone.

Lithology	Sandstone
Unit volume weight (kN/m^3)	26.00
Cohesion (MPa)	0.107
Friction angle (°)	52.10
Poisson's ratio (–)	0.27
Young's modulus (MPa)	556.00

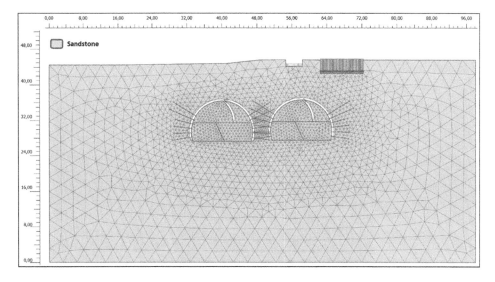

Figure 5. Analysis model of the selected cross-section.

the displacements in the ground and support system, forces acting on the support system, stresses in the ground and failure zones were obtained for all excavation stages.

As a result of the analyses, it is seen that the displacements around the tunnel shown in Figure 6 are within the tolerances of deformation determined in the tunnel design. In addition, normal forces, bending moments and shear forces in tunnel linings and normal forces acting on the bolts are given in Figure 7.

The most sensitive subject to the area concerned is to determine how much the buildings will be affected by surface settlements. Figure 8 shows the vertical displacements

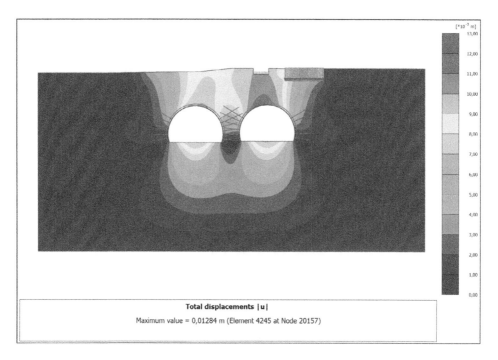

Total displacements |u|

Maximum value = 0,01284 m (Element 4245 at Node 20157)

Figure 6. Total displacements around the tunnel.

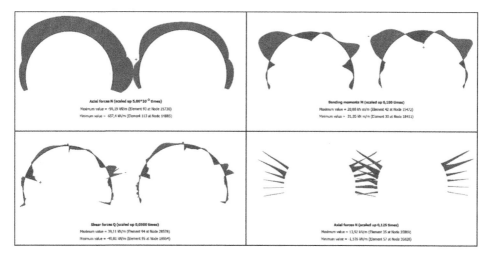

Figure 7. Forces acting on the linings and bolts.

under the buildings, from which it is seen that the maximum value is 10.3 mm. The vertical displacements occurred under the building started to be formed from the first excavation stage and proceeded to increase. The surface settlements according to tunnel excavation stages are shown in Figure 9.

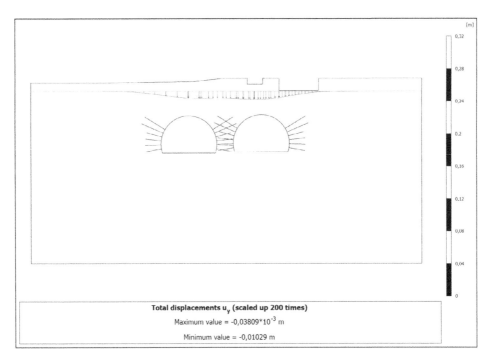

Figure 8.   Vertical displacements under the buildings.

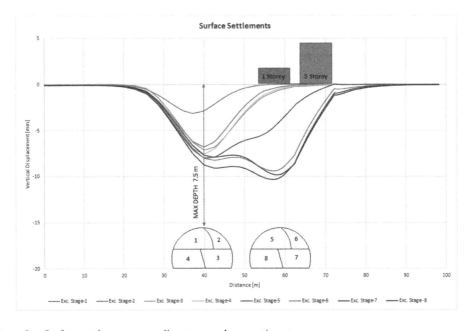

Figure 9.   Surface settlements according to tunnel excavation stages.

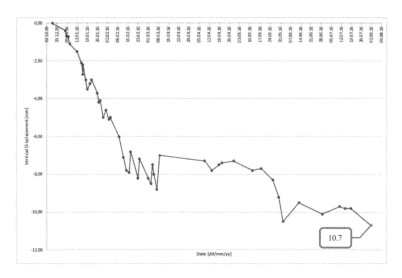

Figure 10. Surface settlements according to daily monitoring results.

5   MONITORING SYSTEMS

The basic rule for the supervision of the workings in the tunnel projects and the demonstration of the adequacy and quality of the construction is to observe the stresses and deformations (Vardar, Kocak & Tokgozoglu, 1998). One of the biggest problems encountered in tunneling is the surface settlement that occurs in the excavated land. If no precautions are taken, surface settlements formed by the construction of the tunnel can lead to structural damage in very serious levels in the upper structures.

Optical methods and convergence measurements were made in the tunnels and at the surface in order to continuously monitor the displacements increase. Readings were followed at frequent intervals and displacement - time curves were created after each reading. Figure 10 shows the measurements taken periodically from the relevant point, it has been seen that the surface settlement values continued increasing at every excavation stage and the maximum value is 10.7 mm.

6   CONCLUSIONS

In this paper, the interaction between the Camlica Tunnels and the buildings has been studied. The tunnel design and construction works have been explained and the measurement data has been shown. Finally, the numerical analysis results will be compared with the monitoring data.

When the amounts of the settlements formed under the buildings are examined, it is estimated that there will be 10.3 mm settlement on the surface will occur in worst case. As a result of the geotechnical measurements made in the field, the highest value is 10.7 mm.

The compliance of results of the calculations and the field measurements showed that the numerical modeling with the finite element method was performed successfully.

REFERENCES

EMAY International Engineering and Consultancy Inc. 2015. Design report of Camlica T1-T2 Tunnel.
EMAY International Engineering and Consultancy Inc. 2015. Geological and geotechnical investigation report of Camlica T1-T2 Tunnel.

General Directorate of Highways. 2013. Highway Technical Specifications.

Ozgul, N. et al. 2011. Geology of Istanbul province area. Istanbul Metropolitan Municipality Department of Earthquake and Soil Research. Istanbul.

PLAXIS BV. 2016. PLAXIS 2D User Manuals.

Vardar, M., Kocak, C. & Tokgozoglu, F. 1998. Examples of the origin and development of time-dependent deformations in Bolu Tunnel. 4th National Rock Mechanics Symposium Book: 173-185.

Tunnels and Underground Cities: Engineering and Innovation meet Archaeology,
Architecture and Art, Volume 12: Urban
Tunnels - Part 2 – Peila, Viggiani & Celestino (Eds)
© 2019 Taylor & Francis Group, London, ISBN 978-0-367-46900-9

Design and construction of large span tunnels and caverns in Sydney, Australia

D.A.F. Oliveira

Jacobs Engineering Group and University of Wollongong, Australia

ABSTRACT: Increased demand to future-proof tunnel projects with respect to traffic has led to the proposal of some very large tunnel spans in recent road tunnel projects in Sydney. For example, four lane tunnels are currently under construction in Sydney with mined spans of approximately 20 m and Y-junction caverns exceeding 30 m spans, all with a requirement for 100-year design life. As these spans are unprecedented in Australian civil tunnels, a direct comparison with local experience is not possible and simple extrapolation of precedent designs, not necessarily adequate. This paper intends to present and discuss how recent designs that focus on first principles and the basic objectives of rock reinforcement have overcome the challenges of these designs, satisfied codes and standards requirements but at the same time provided savings with respect to ground support. The key to the design involved understanding the key failure mechanisms that needs to be addressed, its relationship with the different actions of rock bolting, i.e. suspension/anchorage and/or rock reinforcement and what could be considered acceptable for design.

1 INTRODUCTION

With rapid development of cities, it is crucial that the use of the underground space is made efficiently with projects that can cater for the needs of the population for several decades. This has led to an increased demand to future-proof tunnel projects with respect to traffic which resulted in the proposal of some very large spans in recent road tunnel projects currently in construction in Australia all with the requirement for 100-year design life.

For example, several kilometers of four lane tunnels are currently under construction in Sydney with mined spans of approximately 20 m. Such spans had only been experienced in localized excavations in widened sections such as breakdown bays and Y-junction caverns but not for long lengths of tunneling. In addition, the Y-junction caverns now required for these tunnels are also unprecedented for road tunnels in Australia with spans reaching 31 m and exceeding experience in Australia which include the Eastern Distributor in Sydney (24 m) and both Kedron (26 m) and Lutwyche (27 m) caverns of the Airport Link tunnels in Brisbane.

These large span road tunnels are currently in construction for the New M5 and M4-M5 Link tunnels part of the infrastructure project known as WestConnex. The WestConnex project is a 33-kilometre underground motorway currently being constructed in Sydney's Inner West (Figure 1).

As these excavation spans are unprecedented in Australian civil tunneling, a direct comparison with local experience is not possible particularly considering the semi-flat roof tunnels typically excavated in Sydney. Although simple extrapolation of precedent designs could potentially provide a solution, two risks arise: (1) the extrapolation based on different excavation shapes that do not necessarily target the actual failure mechanisms involved in the excavation of such larger spans; and (2) provide uneconomical solutions. International experience could certainly be used but adequate design justifications and analysis would still have to be provided to verify its application locally.

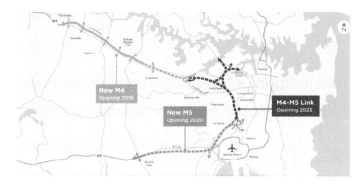

Figure 1.   WestConnex Motorway.

The search for solutions to new problems often target innovation. However, considering that the new challenge described above in fact involves an "old" problem but at a larger scale, it is considered appropriate to review the basic design assumptions to find robust solutions in more fundamental design principles. As a result, this paper intends to present and discuss how a design that focus on the basic objectives of rock reinforcement may allow for a better under-standing of the design requirements and still provide savings with respect to ground support. The key to the design involves understanding the failure mechanism that needs to be addressed, its relationship with the different actions of rock bolting, i.e. suspension/anchorage and/or rock reinforcement and what could be acceptable.

2   PRECEDENT DESIGN

Figure 2 presents a comparison between span and bolt length for several projects in Australia.

Significantly experience is observed for tunnels under 20 m span with most of the projects in Sydney, i.e. within similar geology (Hakwesbury Sandstone). In addition, a reasonable number are road tunnels with similar construction methodology, i.e. roadheaders, and excavation profile. For these smaller span tunnels, the bolt length typically follows a power curve with respect to the span that if assumed for tunnels of approximately 20 m span, a bolts length of approximately 6 m would correspond to the precedent design.

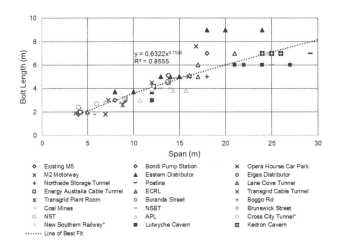

Figure 2.   Span versus bolt length for several projects in Australia.

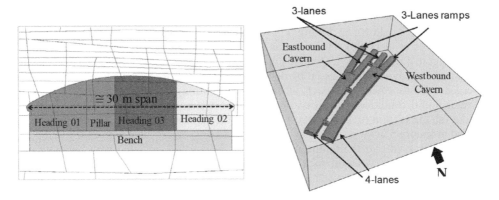

Figure 3.   Typical cavern geometry proposed in Westconnex.

For caverns approaching 30 m span, the precedent bolt length would seem to be approximately 8 m long. However, it is important to note that very few data correspond to tunnels in Sydney beyond approximately 18 m span, i.e. only the Eastern Distributor cavern with 9 m long bolts up to a span of 24 m and Lane Cove Tunnel with 7 m bolts for a 22 m span excavation. The other projects are essentially outside Sydney, namely the Poatina Power Station cavern in Tasmania with 7 m bolts for an approximately 29m span excavation, the Kedron cavern in Brisbane with 7 m bolts for a 26 m span and the Lutwyche cavern with 6 m bolts for an approximately 26 m excavation.

The main difference between these larger excavations outside Sydney and the large span tunnels excavated in Sydney is that they are all fully arched structures where the tunnels in Sydney are semi-flat roof tunnels, i.e. very low arch tunnels (Figure 3). Stress arching and combined use of passive shotcrete support in arched excavations may allow for shorter bolts than would have been used in semi-flat-roofed tunnels in Sydney. One of the reasons is also associated with the precedent design and underlying philosophy for flat-roofed tunnels in Sydney which would generally involve selecting a roof beam thickness based on a level of deflection deemed acceptable, typically between 15 mm and 20 mm, which would in turn dictate the rock bolt length (Bertuzzi and Pells 2002). As a result, this design approach typically results in longer rock bolts due to the larger deflections expected in flat-roofed tunnels than in arched excavations.

As a result, the use of the above precedent design in Australia for the design of the current large tunnels needs to be assessed with care.

3   FAILURE MECHANISM

A fundamental step in the design of any ground support is the understanding of the geomechanical behavior of the structure and how it would ultimately collapse. In other words, the failure mechanism ultimately dictates the required ground support.

An analogy can be made, for instance, with concrete columns that are subjected to both axial and flexural loads. An increase in the amount steel reinforcement would provide small benefit if the primary mode of failure is compression due to high axial loads whereas an increase in concrete strength or cross-sectional area would likely be required. On the other hand, an increase in steel reinforcement would likely be the required solution if the main failure mechanism is flexural.

To confirm the likely failure mechanism associated with such large span excavations, the construction of the proposed large tunnels in Sydney was simulated using a Distinct Element Method code, UDEC (ITASCA), and assuming an unsupported full-face excavation. A typical ground condition observed in Sydney Hawkesbury Sandstone is assumed with Sandstone

Class I/II (SS-I/II) but with relatively adverse thin laminations (SS-II/III) present in the immediate tunnel roof as shown in Figure 4. For the Sydney Rock Classification, the reader is referred to Pells et al (1998). However, as an immediate simple comparison, the condition shown in Figure 4 would have a Q-value greater than 4.

The ground behavior was assessed by developing a Ground Reaction curve similarly as in a Convergence-Confinement approach where the excavation perimeter tractions are reduced from their initial condition (no excavation or support stress equivalent to the in-situ condition) to a full excavation (0% support). The failure mechanism and ground reaction curves are presented in Figure 5.

It can be observed that the failure mechanism is primarily associated with buckling of the rock beds similar to a "voussoir rock beam" mechanism as often considered in the design of tunnel support in Sydney (Bertuzzi and Pells 2002). The ground reaction curve indicates that at a stress level equivalent to approximately 12–14% of the original in-situ stress the rock mass behavior is within its non-linear range primarily due to yielding along the bedding partings. At a stress equivalent to approximately 9–10%, the main collapse mechanism is initiated, and crown sagging becomes more pronounced and asymptotic. The collapse mechanism develops approximately the following process:

• Partial loss of abutment support due to corner wedges formed by arched roof
• gradual development of bedding plane shearing and slip with inward movements from abutments
• bedding separation in the mid-span
• coalescence of bedding slip from abutment towards central zone above crown
• buckling of rock beds with propagation towards stiffer and more competent beds (if available) potentially leading to formation of a chimney type collapse.

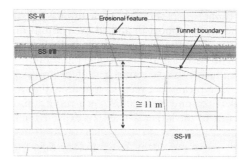

Figure 4. Ground conditions assumed.

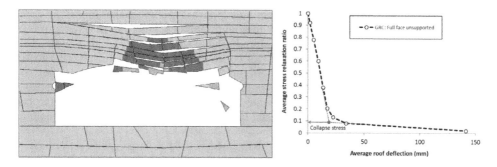

Figure 5. Collapse of unsupported cavern and associated Ground Reaction Curve.

4 DESIGN PHYLOSOPHY AND SUPPORT STRATEGY

Once the main failure mechanism has been confirmed, the appropriate support interaction needs to be selected. In general, ground support in underground excavation can be divided into three main primary functions as illustrated in Figure 6: (a) rock reinforcement; (b) rock suspension or hold action and (c) surface retention/support.

Rock reinforcement is a means of conserving or improving the overall rock mass properties from within the rock mass by techniques such as rock bolts and cable bolts. The primary objective of the reinforcement is to resist the actions induced by stress redistribution within the rock mass as a result of excavation, thus, reducing shear slip and tensile separation along rock discontinuities. Rock suspension or "hold" action basically involves the use of rock bolts and cable bolts to anchor unstable rock blocks that would otherwise fall-out under gravity loads, thus, addressing local failure mechanisms. Surface retention/support is the application of a reactive force to the surface of an excavation and includes techniques and devices such as timber, fill, shotcrete, mesh and steel or concrete sets or liners.

Although tunnel support may involve all three types of support mechanisms concomitantly, according to Brady and Brown (2004), potential slip on bedding planes is generally the main problem in the design of a stratified rock mass. The extent is related to the virgin in-situ stress field and the shape of the excavation. As a result, the primary support mechanism required in the case of stratified rocks such as the horizontally bedded Hawkesbury Sandstone typically involves rock reinforcement, and this has certainly be confirmed above.

As a general rule, cases where the span/bed thickness ratio (s/t) is low will be subject to slip mainly in the haunch area. This may be expressed in the rock mass as the opening of cracks

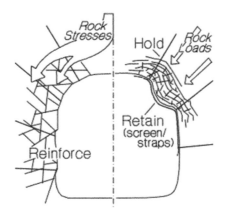

Figure 6.   Primary support functions (after Kaiser et al, 1995).

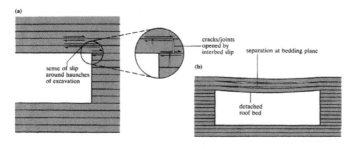

Figure 7.   The effects of slip and separation on excavation peripheral rock (after Brady and Brown, 2004).

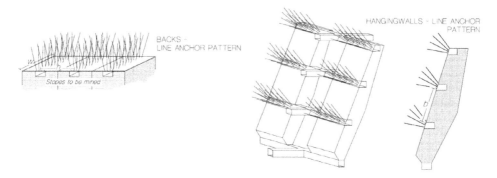

Figure 8. Sub-span design approach (after Hutchinson and Diederichs, 1996).

sub-perpendicular to bedding, perhaps coincident with any cross joints in the medium, as illustrated in Figure 7a. For a configuration in which the s/t ratio is high (i.e. beds relatively thin compared with excavation span), the zone of slip may extend further along the span of the immediate roof. Since the sense of slip on bedding is such as to cause inward displacement towards the span centreline of beds, the tendency is for isolation of the lower bed, at its centre, from the one immediately above it. Separation of a roof bed from its uppermost neighbour is highly significant because it implies loss of support of the roof by the overlying beds, as can be appreciated from Figure 7b. Prior to decoupling of the roof layer, its gravitational load is carried in part by the more extensive volume of rock in which the layer is embedded. After detachment of the roof, the bed itself must support its full gravitational load.

The above mechanism is consistent with the failure process described in the previous section and that of a "voussoir beam analogy". Based on this mechanism, the primary objective of the ground support is to provide a reinforcement effect near the abutments to reduce bedding slip and consequently bedding separation near the mid-span. When possible, rebar bolts are to be used near the abutments considering their better performance in shear reinforcement.

On the other hand, the tendency of the central zone is to separate from the upper roof beds once pronounced bedding slip develops near the abutments. As a result, the ground support within the central zone above crown should focus on a suspension action. In this case the bolts generally need to be longer than the bolts in the abutment zones bolts in order to achieve a better anchoring effect above the zone of more significant loss of confinement and roof movements. As a result, such an effect is often better achieved using cable bolts.

The concept of a suspension type support in central span of the cavern also provides an added benefit. The cable bolt reinforced central zone not only assists in limiting crown deflection and dilation but act as an "artificial" abutment for the adjacent spans (Fuller, 1983). For example, Figure 8 illustrates how that the "line-anchor pattern" provides an "artificial" abutment such that secondary smaller spans develop as marked by dimensions "b" and "h". This support strategy is referred to as sub-span design and is often applied in very large mining openings. It considers the global span to be supported (for the purpose of mining global stability) although the sub-spans are unsupported. This support strategy is only effective for rock masses that are primarily dominated by laminations parallel to the excavation face being supported which is consistent with the horizontally bedded Hawkesbury Sandstone.

5 SUPPORT OPTMIZATION

Based on the assessment of failure discussed in Section 3, the minimum required reinforcement zone seemed to be approximately a 5 m thick into the tunnel roof. In contrast to previous designs in Sydney, which used deflection the governing factor that dictates rock beam thickness, Oliveira and Paramaguru (2016) presented an approach for the design of rock bolts for laminated rock beams where the main focus is on satisfying the development of the compressive arch of an equivalent thicker voussoir beam analysis. The rock bolt reinforcement is designed to provide the necessary capacity to overcome the excess shear stresses in the bedding partings, thus stitching the thinner beams together into an equivalent thicker rock beam.

The only modification to the proposal by Oliveira and Paramaguru (2016) was the inclusion of the suspension effect of the longer bolts within the middle zone as targeted in the previous section. This was done by considering the effect of the "extra-length" of the bolts within the mid-section of the roof acting as a parabolic pressure as proposed by Diederichs and Kaiser (1999).

The above approach was written in a Mathcad sheet allowing for conveniently verifying several ground conditions and support spacing to confirm suitability of the proposed support strategy before more sophisticated numerical modelling.

The effect of ground support was also investigated numerically with discontinuum models using the Finite Element code RS2 from Rocscience to run a large number of models and different ground conditions and for a few cases using the DEM code UDEC of Itasca. The support was assessed by including both rock bolt support at the abutment and cable bolt support in the central zone.

For example, Figure 11 shows an example considering the ground conditions assumed in the failure mechanism model, with an unfavorable layer of SS-II/III in the tunnel crown. Although, a tighter bolt spacing would typically be specified for this ground conditions to control deflections, to assess the potential risk of misclassification during construction, a wider rock bolt spacing has been modelled. The bolt spacing was adopted at 1.75 m c/c spacing with a 50 mm shotcrete. The rock bolts are 5 m long at the abutments and cable bolts 7.5 m long in the central zone. As a result, some larger roof deflections could be anticipated.

Both multiple heading and full-face excavation sequences have been analyzed but only the full face sequence discussed. The latter was adopted for comparison with the failure mechanism analysis discussed above.

As observed in Figure 11, the rock support can control the failure mechanism. The predicted crown deflections vary from approximately 35 mm to about 45 mm within the central cable bolted zone which results in deflection to span ratios of approximately 0.15%. This could be considered reasonable for such large spans and comparable to many large concrete structures spanning such distances, for example in deep excavations where deflections between 0.1% to 0.2% are often observed in stiff fully propped or anchored retaining walls.

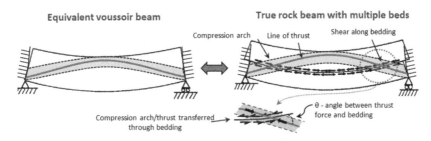

Figure 9. Rock reinforcement effect for equivalent rock bolt stitched rock beam (after Oliveira and Paramaguru, 2016).

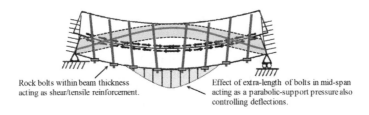

Figure 10.   Rock reinforcement concept adopted for large span caverns.

Figure 11. Vertical displacements with a supported full-face excavation using 5 m bolts near the abutments and 7.5 m high strength cable bolts within the middle-section both spaced at 1.75 m c/c.

Figure 12. View of 4-lane tunnel in construction (left) and Y-junction cavern (right).

Similar exercise was also carried out for 4-lane tunnels with spans of approximately 20 m. The main difference between the 4-lane tunnels and the caverns was that the former did not require the use of the longer cables in the middle zone with only 5 m long rock bolts used throughout the span but still 1 m shorter than the precedent design.

6 CONSTRUCTION

As previously discussed, these large tunnels are currently in construction. However, the available monitoring data from 3D prism convergence (survey), multipoint extensometers and instrumented rock bolts have generally confirmed the predicted tunnel performance using the proposed approach. Most of the large span roof deflections have been observed to be between 10 mm to 50 mm, depending on several factors such as rock quality, excavation sequence, insitu stress magnitude and orientation etc.

7 CONCLUSIONS

This paper presented a discussion on the design of large span tunnels in Sydney Australia. Some precedent design was presented and demonstrated that, although very valuable, past projects are not the only single source of design.

Design efforts that focus on first principles and the basic objectives of rock reinforcement have overcome the challenges of these designs, satisfied codes and standards requirements but at the same time provided savings with respect to ground support. For example, tunnels spans of approximately 20 m were demonstrated to be satisfactory with 5 m long cables in contrast to a precedent design value of 6 m long bolts. Large span caverns also have the maximum bolt length of 7.5 m in its mid-span with shorter bolts at the abutments compared to an precedent design of 9 m long bolts.

The key to the design involved understanding the key failure mechanisms that needs to be addressed, its relationship with the different actions of rock bolting, i.e. suspension/anchorage and/or rock reinforcement and what could be considered acceptable for design.

REFERENCES

Bertuzzi, R. and Pells, P.J.N. (2002). Design of rock bolt and shotcrete support of tunnel roofs in Sydney sandstone. Australian Geomechanics, 37(3).

Brady, B. H. G., and Brown, E. T. 2004. Rock Mechanics for Underground Mining. Dordrecht: Kluwer Academic Publishers.

Diederichs, M S and Kaiser P. K. (1999): Stability Guidelines for Excavations in Laminated Ground - The Voussoir Analogue Revisited, Int. J. Rock Mech. & Min. Sci.; 36, pp 97–118.

Fuller, P.G. (1983) Cable support in mining: A keynote lecture. Rock Bolting, Rotterdam: A.A. Balkem, 511–522.

Hutchinson, D.J. and Diederichs, M.S. (1996). Cablebolting in Underground Mines. Bitech Publishers Ltd., Canada.

Kaiser, P.K., McCreath, D.R. and Tannant, D.D. (1995). Rockburst support handbook. Geomchanics Research Centre, Sudbury.

Oliveira D. and Paramaguru, L. (2016). Laminated rock beam design for tunnel support. Australian Geomechanics Vol 51(4),pp.1–17, December.

Pells, P.J.N., Mostyn, G., Walker, B.F. (1998). Foundations on sandstone and shale in the Sydney region. Australian Geomechanics; Vol 33(3):17–29.

Tunnels and Underground Cities: Engineering and Innovation meet Archaeology,
Architecture and Art, Volume 12: Urban
Tunnels - Part 2 – Peila, Viggiani & Celestino (Eds)
© 2019 Taylor & Francis Group, London, ISBN 978-0-367-46900-9

An investigation of the occurrence mechanism of centralized crack on twin tunnel

T. Ono, H. Hayashi & M Shinji
Yamaguchi University, Ube, Japan

A. Kitamura
Former Yamaguchi University, Ube, Japan

S. Morimoto
DOBOCREATE CORPORATION, Ube, Japan

ABSTRACT: A twin tunnel is constructed in a challenging construction environment such as shallow tunnel and neighboring construction. Therefore, it is considered that the occurrence mechanism of crack in tunnel lining concrete of twin tunnels is more complicated than single tunnel. In the twin tunnel (A tunnel) which target of this study, crack occurrence in tunnel lining concrete was increased after the twin tunnel completed, because of difference in construction of two tunnels (first tunnel and second tunnel). In this research, we carried out excavation simulation of the second tunnel by using 3D-numerical analysis. In addition, we assumed the occurrence mechanism of crack that occurred after the twin tunnel completed by considering load that acts on the tunnel lining concrete after the twin tunnels are completed.

1   INTRODUCTION

Many tunnels in urban areas are constructed in a special construction environment such as shallow tunnel and neighboring construction. A twin tunnel is used in such construction environment. This tunnel is constructed in close proximity to each other, and these share a center pillar (CP). Therefore, twin tunnel is easy to be affected by construction. Table 1 shows the outline of the twin tunnel (A tunnel) which target of this study. From this table, it can be seen that, the second tunnel was excavated about 16 years after the first tunnel was completed. Therefore, it is also a major feature that the two tunnels (first tunnel and second tunnel) are constructed with different excavation methods. From such a special construction environment, as shown in Figure 1, cracks have occurred on one side of the lining in first tunnel after completion of the second tunnel.

Therefore, the purpose of this study is to clear the factors of occurrence of biased crack distribution. In this study, it's called centralized crack. Procedure of this study is as follows:

1) To reproduce the behavior of the ground and the tunnel lining stress by using the construction data obtained at the time of the second tunnel construction.
2) To assume the occurrence of crack that occurred after the twin tunnel completed by considering load that act on the tunnel lining concrete after the twin tunnel completed.

Both 1) and 2) are considered by using 3D-numerical analysis.

Table 1. A tunnel Overview.

	Construction	Completion		Length	Overburden (max)	Overburden (min)
	year	year	Method	m	m	m
First	1983	1984	Sheet pile	87	20	5
Second	1999	2000	NATM	84	20	5

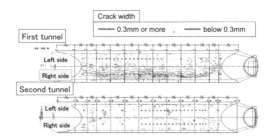

Figure 1. Crack map of A tunnel (2015 checked).

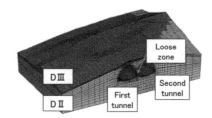

Figure 2. Ground model of A tunnel.

2 REPRODUCTION DURING SECOND TUNNEL EXCAVATION

2.1 Analysis conditions

In this research, we used 3D FDM code FLAC3D. Figure 2 shows the ground model. As can be seen from this figure, this ground model reproduces the geology and inclination of the site by creating a 3D model based on geological report and contour line. When doing analysis of tunnel excavation, the second tunnel was excavated 1m at a time, and the first tunnel was done 70m at a time on analysis. The numerical boundary condition was fixed on five sides other than the top side. The length of analysis model of A tunnel is 70m.

2.2 Setting of the material properties

In this study, back analysis was performed by using construction data (subsidence of ground surface, crown and CP) in order to identify material properties. As the procedure, by repeating excavation analysis of the second tunnel so as to approximate the analysis result to the construction data, we tried to reproduce the behavior of the ground and tunnel lining concrete and calculate appropriate parameter of the ground. Figure 3 and Table 3 shows the comparison between the construction data and result of the back analysis. From this figure, it is understood that the construction data can be accurately reproduced by numerical analysis. Table 2 shows the material properties determined by back analysis. In this study, we thought that ground beneath CP was weakened because of special construction environment. Therefore, the loose zone is created on ground beneath CP (shown Figure 2). By assuming this area,

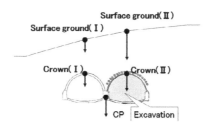

Figure 3.   Construction data and result of back analysis.

Table 2.   Material properties.

	Young`s modulus	Poisson`s ratio	Unit weight	Cohesion	Internal friction
	MPa	MPa	kg/m³	MPa	°
DIII	8	0.35	2000	0.08	30
DII	65	0.40	1850	0.05	30
Loosen zone	12	0.40	2000	0.08	30
Concrete	2000	0.30	2400	2000	45
Pipe roof	1000	0.30	1900	0.04	35
Shotcrete	9235	0.25	2400	2000	45

Table 3.   Construction data and result of back analysis.

	Subsidence based on construction data	Subsidence based on analysis data
	mm	mm
Surface ground(I)	10.8	9.1
Surface ground(II)	25.1	25.9
Crown	7.3	9.7
Crown	14.2	14.5
CP	12.9	14.5

result of the back analysis well agreed with the construction data. The material properties of the loose zone are as shown in Table 2.

2.3   Behavior of the tunnel lining after construction of second tunnel

The displacement and stress of the first tunnel's lining by using the material properties calculated in the previous section. Figure 4 shows the displacement and volume of subsidence at first tunnel after the construction of second tunnel, and Figure 5 shows the tunnel lining stress (maximum principal stress) at the first tunnel after the construction of the second tunnel. From Figure 4, it can be seen that the whole first tunnel was pulled toward side of the second tunnel due to construction of the second tunnel. This is considered to be due to the CP subsidence caused by the loose zone beneath CP. Moreover, it can be seen from Figure 5 that a tensile stress is generated on the right side of the first tunnel lining. It is considered that tensile stress was caused by pulling the first tunnel to side of the second tunnel. From these facts, it can be considered that tensile stress acted on the right side of the first tunnel lining because of the occurrence of loose zone beneath CP. From this result, it is considered that the occurrence of the loose zone beneath CP is factor of the occurrence of the centralized crack on the first tunnel lining.

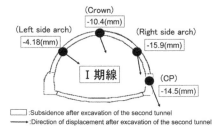

Figure 4. Displacement and volume of subsidence at the first tunnel after construction of second tunnel.

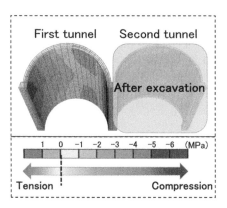

Figure 5. Tunnel lining stress (maximum principal stress) at the first tunnel.

2.4 Behavior of ground lining after construction of second tunnel

Figure 6 shows the vertical displacement of the ground and tunnel lining after construction of the second tunnel. From this figure, it can be seen that ground above the tunnel subsided due to excavation of the second tunnel. Also, ground above the second tunnel subsided uniformly from the crown to the ground surface. This is because the overburden of A tunnel is extremely small, so it is thought that the influence of excavation extends to the ground surface.

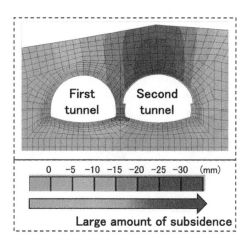

Figure 6. Vertical displacement of the ground and tunnel lining.

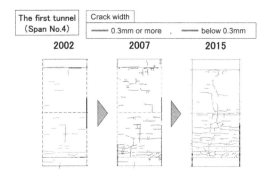

Figure 7.   Aging change of crack map of first tunnel.

Figure 7 shows the changing in crack of the first tunnel due to age. From this figure, it can be seen that the cracks occurring on the tunnel lining of the first tunnel increased over time after the twin tunnel was completed. From the above, in this study, we assumed that ground above the twin tunnel acts as a loose load after the twin tunnel was completed, and reproduced this loose load by numerical analysis.

3   REPRODUCTION OF LOOSE LOAD AFTER COMPLETION OF TWIN TUNNEL

3.1   *Analysis model and method of loose load loading*

Figure 8 shows the analysis model used in this section. This model was created without considering the ground surrounding tunnel in order to clarify the relationship between the ground under the tunnel and tunnel lining concrete. The material properties were the same as those in the previous section. Loose load was reproduced by loading forced load on the tunnel lining concrete of the Figure 8. As shown in Figure 9, the load applied the overburden pressure of the ground between the centers of both tunnels as stress to the outer element of the tunnel lining concrete. Moreover, the horizontal load was calculated by using the Poisson's ratio of the ground. In this study, tunnel lining concrete was modeled as elasto plastic body.

3.2   *Result of the loose load loading*

Figure 10 shows the vertical displacement of tunnel lining concrete after loose load loading. From this figure, it can be seen that the volume of subsidence increased as getting closer to side of the second tunnel in the first tunnel lining concrete. Especially CP was greatly settled, and it was estimated that the volume of subsidence of CP was increasing because of acting

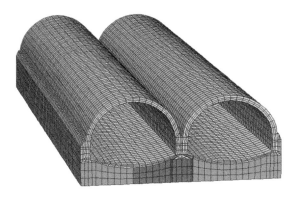

Figure 8.   Analysis model using loose load loading.

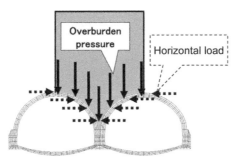

Figure 9.　Method of loose load loading.

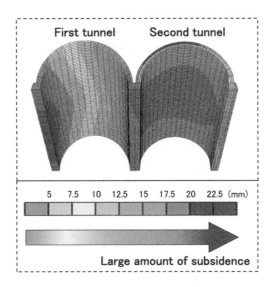

Figure 10.　Vertical displacement of tunnel lining concrete.

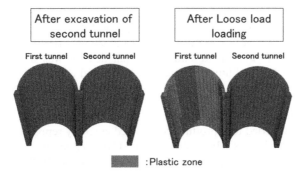

Figure 11.　Plastic zone of tunnel lining concrete.

overburden pressure. Figure 11 shows the plastic zone of tunnel lining concrete after loose load loading. From this figure, it was found that a plastic zone was generated at the side wall and right side of tunnel arch of the lining concrete due to the loose load loading. In addition, we compared the plastic zone with the crack map of Figure 7. From this comparison, it was found that the plastic zone coincided with location of occurrence of crack.

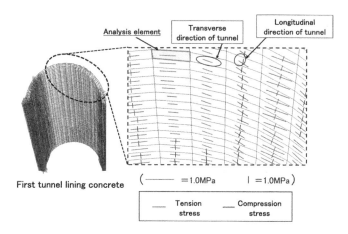

Figure 12. Occurrence of stress generated on the tunnel lining concrete.

Figure 12 shows the occurrence of stress generated on the first tunnel lining concrete. It can be seen from this figure, tensile stress was generated on the first tunnel lining concrete after loose load loading. In particular, it can be seen that tensile stress of transverse direction of the tunnel was generated on side of the second tunnel in the first tunnel. In other words, it seems that the longitudinal direction cracks were caused by the transverse tensile stress on this zone. Looking at the crack map of Figure 7, it can be seen that longitudinal direction cracks was actually occurred on this zone. From this, it is considered that the analysis result by loose load loading was consistent with the actual phenomenon. From this result, loose load loading was considered to be one of the factors of the occurrence of centralized crack on the first tunnel lining concrete.

4 CONCLUSION

In this research, 3D numerical analysis was carried out for A tunnel which is a twin tunnel constructed in a special construction environment, and the occurrence factors of centralized crack on the first tunnel lining concrete. In the second tunnel excavation, since it was a special construction environment, the loose zone was created due to consider the weakness of the ground beneath CP. The loose zone caused the subsidence of CP and generation of tensile stress on the first tunnel lining concrete, so it is considered that, the weakening of the ground beneath CP was one of the factors of the centralized crack occurred on the first tunnel lining concrete. In the analysis of loose load loading, the plastic zone was shown on the side of the second tunnel in the first tunnel lining concrete by loose load loading. Moreover, from checking the direction of tensile stress, it was found that the analysis result agreed with the actual phenomenon. From this result, it is considered that loose load loading was one of the factors of occurrence of centralized crack on the first tunnel lining concrete.

Thus, in this study, we tried to reproduce centralized crack by reproducing the actual tunnel construction on analytical basis. However, other factors such as the action of external force other than loose load cannot be reproduced, so we will set them as future issues.

REFERENCES

Kitamura, A., Morimoto, S., Shinji, M. 2017. Evaluation of maintenance priority point for road tunnel by using Tunnel lining Crack Index (TCI). *Proceedings of tunnel engineering, JSCE* 27(I–1):1–6
Aoki, K., Wakasa, M., Kamimura, M., Shinji, M., Nakagawa, K. 2002. Ground behavior and performance of eye-grass-frame tunnels based on measurements. *Proceedings of tunnel engineering, JSCE* 12: 371–376
Japan society of Civil Engineers. 2003. *Tunnel Deformation Mechanism.* Tokyo: Japan society of Civil Engineers Rock Mechanics Committee.

*Tunnels and Underground Cities: Engineering and Innovation meet Archaeology,
Architecture and Art, Volume 12: Urban
Tunnels - Part 2 – Peila, Viggiani & Celestino (Eds)*
© 2019 Taylor & Francis Group, London, ISBN 978-0-367-46900-9

Challenges in tunneling underneath houses in Contract T217 of Thomson East Coast Line of Singapore

C.N. Ow, R. Nair, A. Jadhav & M. Shelke
Land Transport Authority, Singapore

ABSTRACT: Alignment of twin bored tunnels in Contract T217 of Thomson East Coast Line (TEL) runs through challenging ground of Bukit Timah Granite formation consisting of mixed faces and full face granite. One of the challenges in tunneling was to under cross several residential buildings in mixed face conditions. Parallel seismic method and borehole radar survey was employed for investigation of foundations of some of the buildings to cross verify foundation details with as built foundation drawings. Careful control over parameters of Tunnel Boring Machine (TBM) was exercised to mitigate risk of over excavation. In cases of over excavation where access from surface for grouting of void was not available, especially when tunneling underneath buildings, the procedure for grouting from the TBM was developed. This paper describes challenges during tunneling in mixed face conditions underneath several houses.

1 INTRODUCTION

1.1 *Tunnel alignment in contract T217*

Figure 1 shows alignment of T217 twin bored tunnels from launch shaft in Napier Station (T217) to Orchard Boulevard Station (T218). Alignment passes through Bukit Timah Granite formation of Singapore. Approximate tunnel route for each drive is of 750m. One slurry tunnel boring machine was employed to mine two tunnel drives totaling 1500m. The slurry tunnel boring machine was reused for second drive. As shown in figure 1, these tunnel drives under cross several buildings. Initially in stacked fashion, these two tunnel drives are located at 20m to 48m below ground. These two tunnel drives posed challenge of undercrossing number of buildings in mixed face conditions. About 30% of the alignment of these two tunnel drives is in mixed face of Bukit Timah granite formation.

Figure 1. Buildings on T217 tunnel alignment.

Table 1. Buildings to be undercrossed on T217 Tunnel alignment.

Building/Structure to be undercrossed by tunnel drives	Foundation Type
Building A, Sherwood Road	Shallow foundation
Building B, Tanglin Road	Shallow foundation
Building C, Tanglin Road	Shallow foundation
Building D, Tanglin Road	Tantalized Piles
Building E, Rochalie Drive	Tantalized Piles
Building F, Rochalie Drive	Shallow foundation
Building G, Rochalie Drive	Shallow foundation
Building H, Rochalie Drive	Shallow foundation
Building I, Rochalie Drive	RC Piles

2 TUNNELING UNDER BUILDINGS IN CONTRACT T217

2.1 *Buildings to be under crossed on tunnel alignment*

Both tunnel drives were required to under cross several buildings. Table below shows details of buildings to be undercrossed by TBM. Type of foundation is also shown in table. Contractor procured as built foundation drawings for all buildings on tunnel alignment and conducted trial trenches to verify shallow foundations.

Also, pile investigation was carried for building I, Rochalie Drive to cross verify as built foundation details. The details of parallel seismic logging and borehole radar methods of pile investigation for building I, Rochalie Drive are discussed in section 3.

3 PARALLEL SEISMIC LOGGING AND BOREHOLE RADAR METHODS FOR PILE INVESTIGATION OF BUILDING I, ROCHALIE DRIVE

3.1 *The Borehole Radar survey and Parallel Seismic Logging method*

The Borehole Radar survey and Parallel Seismic Logging method was used to estimate the depth of the underground RC pile at building I, Rochalie Drive. Both methods are explained in detail below. The Borehole Radar Survey was carried out using two (2) boreholes drilled next to the concerned RC piles. The soil information obtained from the nearest bore log was used for interpretation of survey results.

Layout of two boreholes to be used to investigate the toe level of RC pile in question are shown in figure 2 by two red dots. 1-D radar system was used to carry out the survey. The antenna which acts as a both the transmitter and receiver, was lowered into pipe casing from

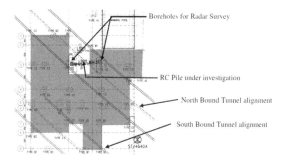

Figure 2. Layout of building I with boreholes for BH Radar Survey.

the ground surface. Radar signals were transmitted 360 degrees from the antenna and any reflection due to dielectric constant changes (density changes) were received by the same antenna. Method is illustrated in figure 3 below.

The survey data together with supporting information from soil bore log, the toe depth of an anomaly that was interpreted as the existing RC piles, which is 32.50m for BH-01 and at 35.30m for BH-02.

Parallel Seismic Logging uses principal of sound waves travelling in different media with different velocities. By the contrast of these velocities, we can estimate the change in material; hence the intersection of these velocities boundary will be interpreted as soil-pile interface. In this method, the wall of the building which is connected to the foundation is impacted by an impulsive hammer to generate compressional or shear waves, which travel down the foundation and are refracted to the surrounding soil. The refracted wave arrival is tracked at regular intervals by a three-component hydrophone receiver in a cased borehole. A schematic diagram showing the principal of parallel seismic logging is shown in figure 4 above. With the Parallel Seismic survey, it was concluded that the depth of existing RC pile at BH-01 is 18m while 20m for BH-02.

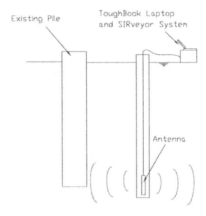

Figure 3.   Borehole Radar Survey Method.

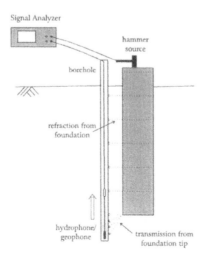

Figure 4.   Parallel Seismic Method.

3.2 *Interpretation of results*

As-built foundation record of the building has indicated the pile length of 19.5m and 17.5m near BH-01 and BH-02 respectively. Results validate the estimated pile depth calculated using Parallel Seismic Logging. With these estimated depths of the piles TBM would have clearance of 4.51m and 1.91m at piles near BH-1 & BH-1 respectively. No pile was encountered during both TBM drives.

4 TUNNELING UNDER BUILDINGS IN MIXED FACE CONDITIONS

4.1 *Tunnelling in mixed face conditions in Bukit Timah Granite formation*

North Bound tunnel drive, being at deeper level, was launched first from T217 launch shaft to T218 station. As shown in figure 5 below, North Bound tunnel drive undercrossed buildings B, C and D at Tanglin Road. Tunneling in mixed face was carried out under Tanglin Road and buildings shown in figure 5.

South Bound tunnel drive was launched after completion of North Bound tunnel drive. TBM was re used for this drive from T217 launch shaft to T218 station. As shown in figure 5 above, South Bound tunnel drive undercrossed buildings B, C and D at Tanglin Road. Tunneling in mixed face was carried out under Tanglin Road and buildings shown in figure 5, for this tunnel drive. Cross section AA and cross section BB (on plan in figure 5) are shown in figure 6 below. Foundation details of building D, clearance of tunnel crown from pile toe and mixed face geological profile for two tunnel drives are shown in figure 6.

Detailed geological profile of mixed face for South Bound Tunnel drive is shown in figure 7. Geological profile for North Bound tunnel was similar at this location. This profile is based on

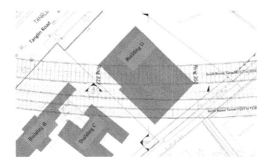

Figure 5.   Tunnel alignment in mixed face underneath buildings.

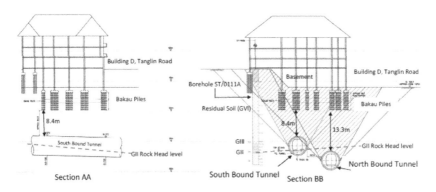

Figure 6.   Cross section of building D showing mixed face underneath.

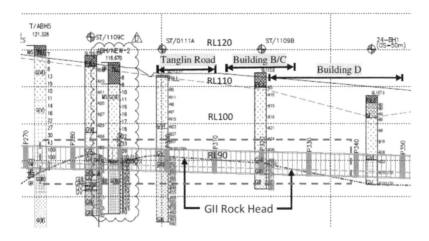

Figure 7. Geological profile of mixed face under buildings on tunnel alignment.

soil investigation boreholes at tender stage and additional boreholes carried out by the contractor. Mining in mixed face was carried from ring number 275 to ring number 335, shown by dotted red colored box in figure 7.

4.2 Standard Operating Procedure (SOP) for TBM undercrossing of buildings

Standard Operating Procedure (SOP) was formulated for each building to be undercrossed by TBM. Some of the salient points of these SOPs are discussed below. For each building a table indicating foundation details of building was created. For all the piles above tunnel alignment, high risk piles were identified. These piles were having clearance of less than 3m from pile toe and TBM cutter head. Chainages of these piles were marked on tunnel alignment drawing. All these details were displayed in TBM operator's cabin. Tunneling under these high risk piles was controlled in such a way that TBMs speed was limited to 10mm/min and torque was limited to 1500kNm. Excavation was to be stopped if sudden surge in thrust force and torque was observed. Slurry Treatment Plant was also monitored to pick up any signs of concrete debris. Experienced slurry TBM operators were deployed.

5 INSTRUMENTATION AND MONITORING

5.1 Instrumentation and monitoring of building D

Maximum consolidated volume loss for both drives from ring number 275 to ring number 335, in mixed face ground, was 0.8%. Building instrumentation results for one of the building D at Tanglin Road, are presented below. Building instrumentation plan for building D is shown in figure 8.

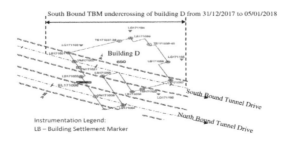

Figure 8. Building Instrumentation plan of building D.

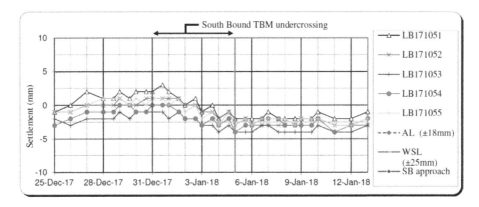

Figure 9. Building settlement of building D during undercrossing of South Bound TBM.

It can be seen from figure 9 that for both tunnel drives maximum building settlement was less than 5mm, well within alert level.

6 PROCEDURE FOR GROUTING THE VOID FROM TBM'S PROBE DRILLING HOLES

6.1 Grouting procedure from TBM

Wherever access from surface was not available to carry out grouting to fill up the voids, especially under buildings, comprehensive procedure to fill the over excavation voids from TBM was developed. Probe drilling holes in the TBMs middle shield were to be utilized, if needed. Figure 10 shows assembly of grouting drill with blow out preventer, rotary joint and drilling machine. The drill bit acquired could simultaneously drill and grout.

Figure 11 shows cross section through TBM to show longitudinal profile of probe drilling (for grouting) at 10 degrees. Target areas to fill the void with grouting, ahead and above cutter head are shown in figure.

6.2 Grouting procedure

The procedure for grouting the void from TBM is described below. If over excavation is detected, then stop the advance of the TBM. Inject chemical catalyst in excavation chamber to solidify slurry. The dosage of chemical catalyst should be 4kg/m3. Slowly rotate cutter head until bentonite slurry forms gel. Start filling up the void with cement grout with pressure lesser than face pressure. Wait for 6 hours while slowly rotate cutter head at 0.5RPM. Again, inject

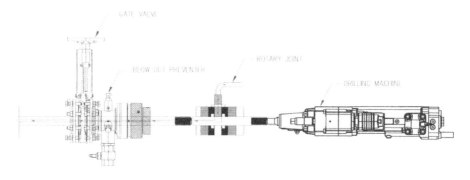

Figure 10. Assembly of Grouting Drill (for grouting voids from TBM).

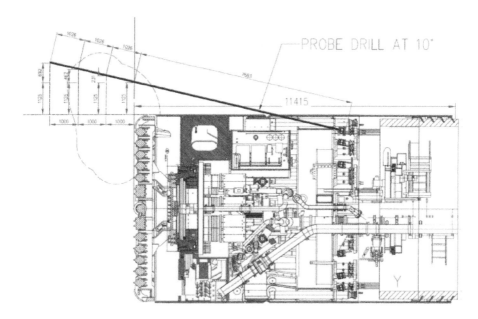

Figure 11.   Cross section of Probe Drilling through TBM (for grouting).

chemical catalyst in excavation chamber with dosage of 6kg/m3. Rotate cutter head until bentonite disperses. Replace bentonite with fresh slurry.

Once TBM starts advancing, inject primary grout as much as possible at location of the void. Once ring is built at over excavation location, carry out secondary grouting. Also, possibility of filling up excavation chamber with clay shock before carrying out grouting from TBM was considered. Throughout the tunneling underneath the buildings, grouting from TBM was not necessary as tunneling was carried out diligently as not to have over excavation in first place.

7   CONCLUSIONS

Investigation of foundation of buildings, well in advance, is of utmost importance for proper planning and execution of tunneling underneath buildings. This paper has demonstrated that two methods employed to cross verify foundation details were useful. For tunneling underneath buildings, especially in mixed face conditions, it is important to control vital TBM parameters. It is prudent to prepare for likely scenarios such as filling the over excavation void from TBM. Also, proper SOPs need to be in place during tunneling underneath houses and important structures.

Tunnels and Underground Cities: Engineering and Innovation meet Archaeology,
Architecture and Art, Volume 12: Urban
Tunnels - Part 2 – Peila, Viggiani & Celestino (Eds)
© 2019 Taylor & Francis Group, London, ISBN 978-0-367-46900-9

Integrated functional metro station design using BIM tools

C. Pallaria, D. Vercellino & A. Bolzonello
Geodata Engineering S.p.A., Turin, Italy

ABSTRACT: The continuous expansion of the Istanbul metropolis, the peculiarities of the city with the complicated urban texture characterized by new and historical infrastructures and buildings, the presence of the Bosporus that divides the city into the Asiatic and European sec-tors, are all factors to be considered in the mobility development plan of Istanbul. Local and national administration radically faced the situation by adopting the decision to realize a vast network of subways, large more than 500km, to be built within 2025. The current projects are part of this large transit development plan. The whole projects have been designed with a "full BIM 360" approach. Project management, coordination between disciplines, Bill of Quantities and 4D&5D simulations have been implemented through BIM approach, by means of developing specific codes for Revit and the other tools. The approach followed, and the main results obtained are described in the paper.

1 INTRODUCTION

Istanbul is a megalopolis with over 14 million of habitants, characterized by a mixture of historical and modern urbanization, often uncontrolled, and natural constraints. This poses significant challenges in infrastructure planning, considering also the continuous growth of the city.

In last ten years, the Municipality of Istanbul (İstanbul Büyükşehir Belediyesi - IBB) has financed a large investment plan to face the growing and urgent demand for public connectivity. It consists of building, by 2023 (100th celebration of the founding of the Turkish Republic), a new metropolitan railway network, covering the metropolitan area of about 500 km. The geography of Istanbul, heavily hilly and with the Bosphorus that separates the Eastern and Western sides, adds greater complexity to these challenges.

Within this ambitious investment plan, Geodata has been involved in the development of four subway projects:

- Dudullu-Bostanci Metro Line (14 km, 13 underground stations and 1 depot). Fi-nal and Construction Design.
- Ümraniye-Ataşehir-Göztepe Metro Line (13 km, 11 underground stations and 1 depot). Final Design
- Kirazli-Halkali Metro Line (9,6 km, 8 underground stations, 1 aboveground sta-tion and 1 depot). Final and Detail Design.
- Gayrettepe Metro Line– İstanbul Yeni Havalimani, line that will connect the city centre with the new international airport under construction. (37 km, 10 under-ground stations). Detail design revision and Construction Design.

All abovementioned projects have been done with a "full BIM approach", from the alignment up to stations design, considering also temporary and permanent structures. In this paper the approach followed for the different stages of the projects will be described.

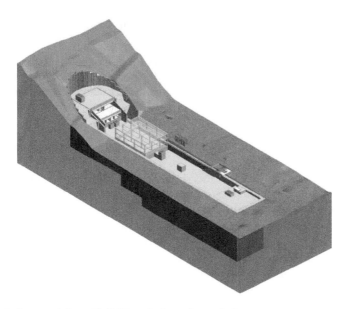

Figure 1. Terrain imported from Civil 3D to design urban solutions.

2 THE ALIGNMENT IN CIVIL 3D

Corridor definition and station positioning is the first step in the design of a Metro line.

In this step, the use of Civil3D software allowed to compare multiple alternatives, giving to the client the possibility to speed up the decision-making process. The digital model of the terrain was created starting from graphic elements (points and lines) of the photogrammetric survey performed on the whole urban area of Istanbul. During the design, the general model was integrated locally with detailed surveys corresponding to the major buildings (stations and shafts). The software allowed to immediately modify the surfaces and to integrate them eliminating the overlapping points.

Subsequently, after the plan-profile definition, the model of the tracks and a macro-model of the tunnel was created, useful for other discipline studies (structural/geotechnical). Moreover, the earth movements have been automatically calculated, especially for the optimization of the depot level in relation to the hard morphology of the ground.

Autodesk's Subassembly Composer has been used to define typical section types with the aim both to generate custom parametric sections by setting "ad hoc" parameters, and to extract quantities of each component for the BoQ.

During the detailed design, using Civil 3D and Excel it was possible to generate the XYZ coordinates of the center of the tunnel and model the entire TBM tunnel inside Revit using a specific Dynamo code developed internally (see chapter 6.2).

3 ARCHITECTURAL AND FUNCTIONAL DESIGN USING REVIT

The functional spaces definition of the stations is at the base of any metro line project. The interaction and the interface with all the other related disciplines, structural design, MEP design, architectural finishes, is a must. The greater the interaction between the disciplines and the minors are the negative impacts during the construction phase.

From the first day it was decided to work globally in the BIM environment, using a common language defined in the BIM Execution Plan (BEP) in which the Contractors and the Customer were fully involved.

The BEP defined the steps and the settings to generate a federated model, that consists of links of the different disciplines, in particular:

- Functional/Architectural model, which welcomes all disciplines: it is the basis for the division of spaces, each of which belongs to a specific category. The various rooms are marked by a unique coding system and thanks to the modelling detail it is possible to extrapolate surfaces and volumes useful for MEP sizing. In this discipline, moreover, the complex of the finishes has been managed, from the plasters to all the supporting structures for the panels, false ceilings and floating floors.
- The Structural model which include: formworks models, analytical models, supports for provisional structures
- MEP models: each single model of this discipline includes the management of the clearances and provides the preliminary dimensioning values, including also the actual data coming from the calculation reports.

The BIM approach made possible to manage the material schedules individually, allowing the verification and the updating of the BoQ simultaneously. By specific parameters applied to families and materials, it was possible to link Revit's internal materials to the technical specifications and unit prices, in accordance with the local legislation, through an exchange of information with the company, which periodically updated the BoQ. This procedure of parameterization and population of the families also includes the WBS codes useful for the management of 4D and 5D, developed in a post-modelling phase.

By the import of the terrain data deriving from Civil 3D it was possible to evaluate all the superficial impacts of this underground infrastructure, managing with precision the surface landings of the accesses, the emergency exits and the ventilation grids, minimizing urban impacts (please see Figure 1: Terrain imported from Civil 3D to design urban solutions.).

The use of the BIM approach has improved both internal management within Geodata and towards other Stakeholders. In particularly, it has been extremely useful for sharing the same model simultaneously with structural, architectural and MEP designers. This made possible to eliminate inconsistencies between disciplines and to deliver to the client an integral and congruent product.

Figure 2.   Internal view of a station with all construction details.

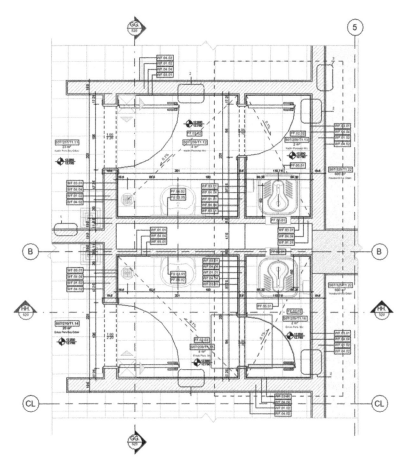

Figure 3.   Detailed plan of toilets (Construction Phase).

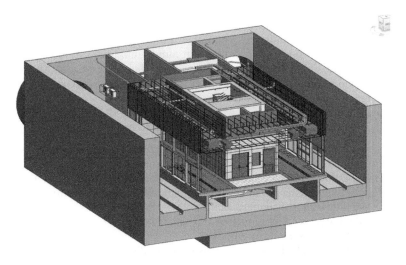

Figure 4.   Structural, Architectural and MEP models linked into a master central file.

The clash detection activity has been carried out periodically by the various BIM coordinators both within the single discipline and in a multidisciplinary environment with the use of Navisworks, which becomes a key tool in the management of the project. As it is stated in chapter 5, every week a coordination meeting gave an "issues and changes report" to correct the project day by day for all disciplines.

4   AUTODESK BIM360 FOR WORLDWIDE COLLABORATION DURING CONSTRUCTION

Projects of this dimension at different design levels, from final design to construction, can only be carried out efficiently if there is good coordination between all the stakeholders involved. In this regard, Autodesk BIM360 was used at the construction level, in order to give a full access of the design documentation to all involved person, with different access rules according to the user. Everyone was able to insert comments, even uploading images directly from the station sites. All this allowed a simultaneous coordination between the various figures involved, especially between the Construction Supervision, the Contractor and the designers.

5   BCF (BIM COLLABORATION FORMAT) TO REALIZE ISSUES REPORTS

For the design phases, the BCF format was used to check errors, to make "issues and changes report" and to keep track of all the changes requested, made or rejected. The type of BCF format also allowed to involve non-BIM professional figures who could read and comment on the project, since the BCF format can also be managed through stand-alone software. Every Friday a general coordination meeting was held between the various disciplines and a specific report of issues and changes in BCF format was re-issued.

6   INVENTOR, DYNAMO AND COMPUTATIONAL DESIGN

Computational design consists of applying computational strategies to conventional design processes. While in traditional processes the designers rely on intuition and experience to solve the design problems, the computational design aims to improve and support this process, codifying "the experience" in machine language.

Generative Design is an evolution of computational design in which algorithms that mimic natural evolution are used. Designers insert the expected results together with a series of boundary conditions such as materials, manufacturing methods, costs and physical constraints. Subsequently, it is possible to explore an almost unlimited number of alternatives, deriving from all the possible permutations generated by the defined algorithm. With generative design there is no single solution but, potentially, thousands of excellent solutions. It will be up to the designer to identify the one that best meets his needs.

Computational and generative design were the basis of the BIM design process in the Istanbul projects, which led to the creation of specific tools to perform:

- Automatic filling of title blocks by reading the parameters from the processed list of drawings
- Automatic filling of stations room names according to the coding for each room
- Automatic numbering of model elements (rooms, walls, pillars, slabs, etc . . .)
- Automated writing of specific parameters extrapolated from external database
- Automatic modelling of TBM tunnels
- Automatic modelling of NATM tunnels
- Automatic modelling of excavation supports
- Automatic modelling of TBM segments by using a specific tool developed for Inventor

Autodesk Dynamo was the platform for creating computational design scripts, to aid parametric modelling.

C# was used to create the specific tool working within Autodesk Inventor for the design of the rings of the TBM rings.

Python has been used to create custom Dynamo nodes that don't exist in the default database, reading the Revit API and breaking down all limits that the software seems to have during the first approach.

The new frontiers of generative design now include the use of algorithms of artificial intelligence, which take advantage of machine learning to develop solutions that improve and get closer and closer to the optimal solution. This new approach will be a must with the Big Data management in next years.

6.1 *Dynamo codes and internal tools to speed up architectural design*

Dynamo codes were used for automatic compilation of the parameters, importing an Excel file with all data of rooms, walls, materials, etc...

The first column identifies the code of the room or the element and writes inside it all parameters included in the specific row. This code saved several man-hours for repetitive work that would not only alienate the resources, but also made it lose several hours to do what the code realizes in a few seconds, with a margin of null error due to the elimination of the human factor.

Particularly, the writing of specific parameters extrapolated from an Excel Database turned out to be very effective, since every change within the database was reflected in automatic form within the models, so that the only objective was to maintain the database updated properly, and then apply it to all models.

6.2 *Dynamo for TBM tracking*

The Dynamo code imports an Excel, XML or JSON file with the XYZ coordinates of the tunnel's central axis coming from Civil3D, subsequently elaborated in Dynamo to create a 3D curved polyline (NURBS). This curve is then discretized to identify points at regular intervals, equal to the length of the single ring; these points will then define the insertion point of the ring itself. Computational design largely relies on the rules of vector geometry for the

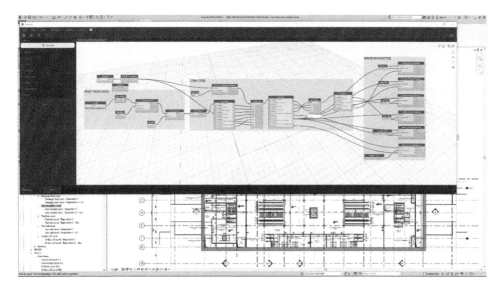

Figure 5. Example of Dynamo code used to write parameters reading an excel database file.

calculation of tangents and normal for curves. Once tangents and insertion points have been identified, this technique has been used for the positioning of all the other line elements such as pipes and various MEP elements. The results obtained, in terms of accuracy and reduction of hours/man, for the production of all the tunnels foreseen in the "scope of work", would not have been achieved with traditional modelling techniques.

6.3 Dynamo for NATM Tunnel modelling

A very similar technique has been adopted regarding the modelling of NATM tunnels, modelling the excavation progress. The particularity of the code is that it is possible to de-fine the length of the discretization sections, so that the individual pieces of tunnel progress can be modelled automatically simply using the coordinates of the alignment and the cross-section profile of the tunnel. To each portion have been assigned parameters containing various information, including the WBS codes of construction planning in order to develop a 4D and a 5D in an automated process directly in Navisworks.

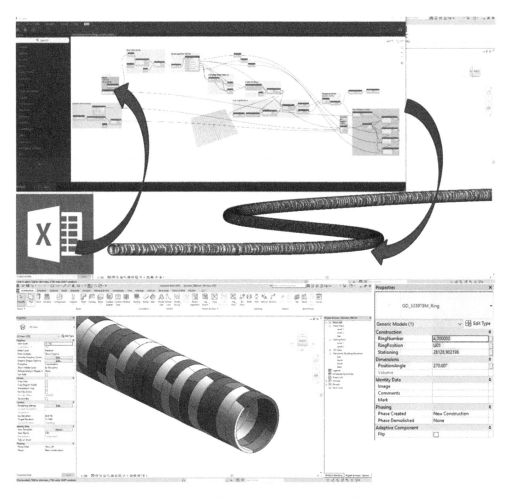

Figure 6.   Dynamo code used to model the TBM tunnel using XYZ coordinates from Civil 3D (or directly from TBM machine, depending on design stage. The final result is a model with the right ring rotation and all parameters needed written inside each element (such as stationing, ring number, ring rotation, etc...).

The modelling was carried out with Revit, Civil 3D and Dynamo in order to realize all NATM tunnels. The purpose of the model was gathering and organize information in order to develop an accurate and consistent 5D model for the quantity take off. Once defined the support classes, every tunnel has been developed by specific adaptive families able to include high level of information within a basic geometry.

A dynamo code has been used to re-create alignments by a JSON file (extracted from Civil 3d) and repeat every adaptive family along that curves.

Then every tunnel has been linked in a Revit Master file and, by another specific algorithm, populated with all parameters needed for quantification inside Navisworks.

6.4 *Dynamo for supports in tunnel construction*

The Dynamo code used for this purpose allows to apply to all kind of surfaces (flat or curves) the supports for temporary structures during excavation phase. It is possible to de-fine the steel ribs characteristics, diameter, length, wheelbase of bolts, as well as write spe-cific parameters in order to identify the type of support, especially in the BoQ phase.

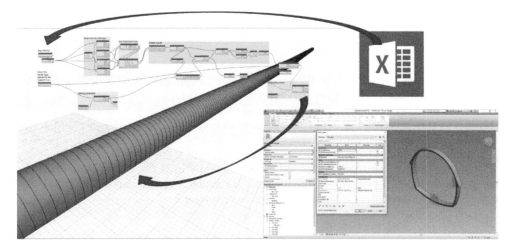

Figure 7.   Dynamo code used to model the NATM tunnels using XYZ coordinates from Civil 3D.

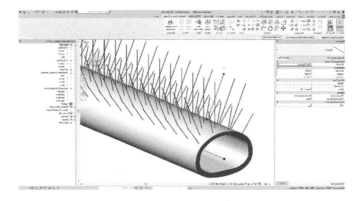

Figure 8.   Supports installation realized by a Dynamo code run by a "user friendly" interface of Dynamo Player.

The code divides the surface into portions with a step defined in the input data, identifies the coordinates of the points, finds the normal point to the surface having distance set in the input data and put on those points an adaptive family in which further parameters are written, including the diameter and characteristics of the support.

6.5 *The automatic design of TBM segments with inventor*

The Istanbul metro project has allowed the development of a new internal work methodology based on the release of a platform developed and structured within the Autodesk PLM software (Inventor) dedicated to the design of any type of prefabricated TBM segments, modelling the complex geometry.

The tool developed internally and called "The Ring", allows the user to act in real time on a parametric model managing the customization of each segment, starting from the definition of the basic geometry up to the details. By acting on the numerous parameters (almost 200 per segment) it is possible to define the ring radius and the minimum curvature of the tunnel in the specific project, but also select and customize the type, dimensions and orientation of:

- Pocket
- Anchor socket
- Gasket
- Guide bar
- Conex.

Moreover, thanks to the access to the database of all coordinates of the various segments, it is already possible during data entry, to perform real time checks of the insertion of the key segment and of the extra-stroke in the TBM thrust jacks.

Thanks to this procedure it is possible to immediately verify if the design choices made reflect the design and verification requirements, all within a short time and with a precision typical of mechanical products.

The model, once created and completed, is subsequently used for the production of two-dimensional drawings. Furthermore, the Inventor 3D model is imported into Revit (thanks to the use of Dynamo) with the technique described in chapter 6.2.

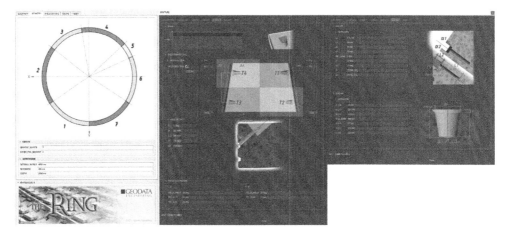

Figure 9.   Graphical User Interface of "The Ring", a custom tool developed by Geodata for TBM segment design.

7 CONCLUSIONS

We are always used to see the future as a space open to more forward-looking challenges, both today and 6 years ago, when BIM applied to infrastructure was almost a mirage. Today, unfortunately, although the technology is very developed and ready to be applied to the construction industry, it is not ready for infrastructure projects and we must always think of "how to solve the problem" that the software is not able to solve. This means that, in some specific cases, the work required to achieve the goal is very large and the efforts do not justify the adoption of BIM for such situations yet.

For this reason, one of the future objectives that we are setting and on which we are constantly working, is to create a large library of tools implemented internally in Dynamo and C #, which can be useful in order to speed up all processes that now seem too much difficult and make us say "Let's do it in traditional CAD technology".

Tunnels and Underground Cities: Engineering and Innovation meet Archaeology,
Architecture and Art, Volume 12: Urban
Tunnels - Part 2 – Peila, Viggiani & Celestino (Eds)
© 2019 Taylor & Francis Group, London, ISBN 978-0-367-46900-9

Limestone cavern influence to primary lining analysis for tunnel Soroška, Slovakia

P. Paločko, M. Bakoš & J. Ortuta
Amberg Engineering Slovakia, Bratislava, Slovakia

ABSTRACT: Motorway R2 will be a part of the Slovakia southern corridor connecting west part of country to east. The tunnel Soroška 4.2 km long will be a part of this motorway. It is placed in Slovak karst - listed as UNESCO protected area. Slovak karst is known for a lot of caverns and caves. Cavern in the rock environment can be filled by gas or water. This causes next change in stress state. The next parameter coming to an equation is a permeability of cavern boundary and difference between pore pressure in rock environment and pressure in cavern. The authors present an idea of pseudo-discrete continuum model of the caverns or collapsed caves. This model conducted study the behavior of caves in the primary failure mode that is associated with initial breakage of the rubble skeleton. The model is a combination of the discrete particle assembly generation and the finite element analysis of this assembly.

1   INTRODUCTION

One of the main goals for the Slovak Republic is to connect the local transport network to European transport network. The proposed motorway R2 lot Rožňava – Jablonov nad Turňou is in accordance with the development program of the Slovak national motorway network.

Proposed tunnel Soroška is located between the villages Lipovník and Jablonov nad Turňou. This area is a part of Slovenský kras national park and it is considered as protected area. The terrain creates steep slopes up to 15 %.

Slovenský kras consists of limestone and limestone tableland divided by deep valleys with many surface and underground karst phenomena. About 1300 caves are located in this area (Figure 1).

The alignment of the tunnel Soroška crosses several protected areas. This influences the technology of driving and other building works.

The areas of impact are:

- Slovenský kras national park and associated protected areas
- Protected area NATURA 2000 – protected bird area SKCHVU027 Slovenský kras
- Area NATURA 2000 - habitat of European importance Hrušovská forrest-steppe
- UNESCO objects – national nature monument Hrušovská cave and Krásnohorská cave – listed in UNESCO heritage

2   GEOLOGICAL AND HYDROGEOLOGICAL CONDITIONS

The site of planned tunnel belongs to the most interesting members of table karst in the Slovakia. All possible karst phenomena, such as caves, abyss, sinkholes, karst ponds, karst springs and others can be found.

Figure 1. Hrušovská cave (left) is UNESCO listed heritage. Extent of this cave is unknown. Uncovered karst pit (right) created on the boundary of two tectonic structures.

According to the geological investigation the geological structure of the area is simple. Young Triassic limestone covers the main area in the center. The bedrock consists of lower Triassic layers mainly slate, limestone and dolomite.

From geotechnical point of view, the structure is more complicated. Rocks of Silická plane are highly tectonic disrupted by steep faults of NW-SE orientation.

The next problems are karst phenomena. The measurement shows deep low resistance zone. This zone crosses the tunnel in 250 m section.

Detailed engineering geological investigation was performed in 2017 (Ortuta 2017a). This investigation found several new caverns 8-9 m high (Figure 2) and several large hollow volumes (height about 23-25 m).

These features resulted into a need of use several auxiliary measures to protect the tunnel during excavation.

Slovenský kras is a unique nature complex due to the presence of groundwater. It is characterized as extreme rich groundwater source and it is important source for water management.

Figure 2. Unique decoration of underground space towards tunnel route.

Because of this, a part of national park is declared as protected zone of natural accumulation of the water.

The flow of underground water in rock mass is quite complex. Infiltrated water keeps vertical direction and later it is changed to horizontal direction, called siphon flow.

Flow velocity of karst water is important hydrogeological characteristics. Dueto high permeability on the surface, wateris accumulated in deeper carbonate rock mass.

3 CONSTRUCTION OF THE TUNNEL LINING NUMERICAL MODEL

The problem of mathematical modeling is the assumption including a wide spectrum of geotechnical problems.

While designing the tunnel designer must keep in mind whole context and possible complication of future work and must deliver technically suitable and economically feasible solution.

Manmade constructions are defined more or less exact and it is a good base for static calculation.

However, rock environment is for geotechnical engineer a construction material and it creates a basic element of bearing system lining-rock. Parameters of this rock environment are very limited and geotechnical engineer must carefully decide on input parameters for static calculation (Ortuta 2017b).

The results of this static calculation directly influence the safety and price of the construction. The second is quite high in case of infrastructure projects.

While gravitational stress can be calculated, other parts of geotechnical state – residual and tectonic stress, can be only estimated. The wrong interpretation will lead into bad assumptions and inaccurate calculation.

4 INFLUENCE OF FAULT ZONES TO STRESS CALCULATION

To evaluate the influence of rock mass faults, the assumption that contact stress is lower in fault/crack area was applied. This stress can vary according to fault/crack size and filling (water or air).

There is however a question how to evaluate vertical deformations according to different compressibility of water and air. In this case, we can work with the assumption that the geology age is reached equilibrium stage and the deformation is zero (Ortuta 2018).

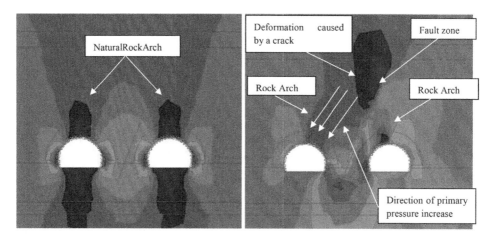

Figure 3. Equivalent relative deformation around the tunnel (left: without influence of cracks, right: with influence of cracks).

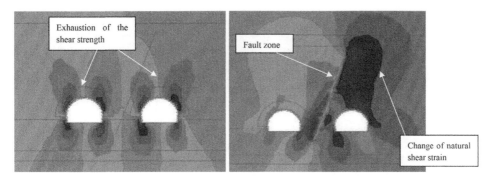

Figure 4. Exhausting of shear strength around tunnel (left: without influences of cracks, right: with influence of cracks).

This way, we can limit the number of variables influencing primary stress state (Figure 3 and 4).

5 INFLUENCE OF CAVES AND HOLLOW SPACES TO STRESS CALCULATION

The next factor influencing primary stresses for tunnel driving is presence of caves and hollow spaces. Risk raising from those spaces cannot be exactly determined. However, it needs to be taken in account in the initial phase of calculation.

Stress state of rock mass was changed during geological ages and it is in a certain equilibrium. This equilibrium will be disrupted during excavation and this will lead to changes of deformation. The load on the primary lining is automatically changed in the calculation. The result is needed for additional support – the change of primary lining.

Figure 5 shows simulation of collapsed system and induced deformations of rock mass.

This calculation represents the first phase of stress calculation using finite element method.

In this phase stress will be distracted and tunnel will be loaded in dependence on character and orientation of weakening. The weakening will be activated by excavation and rock mass will reach new state by change of individual stresses.

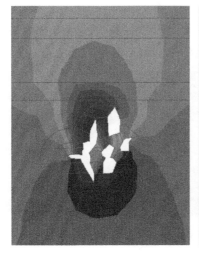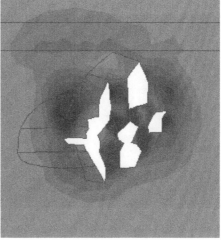

Figure 5. Simulation of the fallen cave system (left: vertical deformations and formation of rock arch; right: plastic deformation and weakening zones).

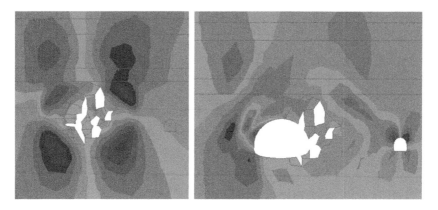

Figure 6.   Change of shear stress orientation for unsupported excavation.

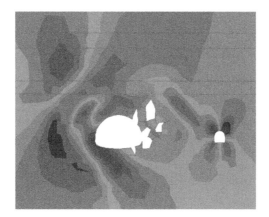

Figure 7.   Shear stresses in rock mass after tunnel construction.

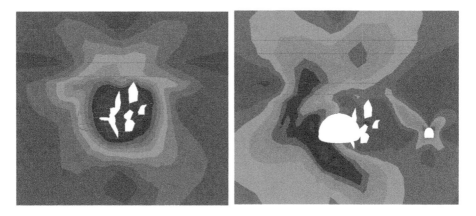

Figure 8.   Movement of intermediate principal stress in the rock mass.

Described process is reflected/considered in the next phases of calculation.

Figure 6 shows unsupported excavation and change of shear stress in surrounding of excavated tunnel.

Stresses are moving and with excavation and support is reached equilibrium state with newly built tunnel (Figure 7).

Movement of intermediate principal stress also completes the idea about movement of forces acting on primary lining of tunnel.

6 CONCLUSIONS

The route of the tunnel Soroška crosses massif of Slovak karst. Geological environment consists of upper Mesozoic and quarter sediments.

Geoelectrical measurements in high resistance carbonate rocks give clear results.The interpretation is based on homogeneity of rock types and drop of resistance marks worse geological conditions in tunnel driving. Higher value of resistance marks more intact rocks. Rocks on sinské layers have lower geoelectric resistance which induces lower strength. Important information is locality of the contact with wetterstein limestones located in km 8,260. Appearance of the karst phenomena can be expected in those rocks.

The design of primary lining is connected with two basic questions needed to be asked at the start on every design of numerical model:

What will be modelled and what kind of results are expected from the calculation?

The first question is connected with the model. There is no need to build a complex model without feedback at the beginning. The second problem is connected with mathematical analysis. We need to know what to expect from the calculation. The different approach is used for evaluation of deformations caused by anthropogenic impact and different for examination of internal forces in structure.

The decision on the effectiveness of constitutional relation to the calculation must be made based on the information about modelled area.

This fact emphasizes the importance of geological investigation and mapping before start of the design process.

Based on the results of this investigation geotechnical engineer can choose the most suitable numerical model for the current conditions to provide economic efficiency and structurally safe design. This plays a decisive role for tunnel design.

REFERENCES

Ortuta, J. & Paločko, P. 2017a. Vplyv seizmického zaťaženia na geotechnické konštrukcie. (The Effect of Seismic Loading on a Geotechnical Constructions), *Aktuálne geotechnické riešenia a ich verifikácia. Bratislava 5-6 Jun 2017(in Slovak)*

Ortuta, J. Hlaváč, M. & Chabroňová, J. 2017b. Materiálové modely a ich vplyv na výpočet primárneho ostenia seizmického zaťaženie na geotechnické konštrukcie. (The Material Model and its Influence for the Design Primary Linning), *Aktuálne geotechnické riešenia a ich verifikácia, Bratislava 5-6 Jun 2017 (in Slovak)*

Ortuta, J. Bakoš, M. & Paločko, P. 2018. The Influence of Material Models for the Effective Design of the Primary Lining., *North American Tunneling*. Washington: 363–367

Tunnels and Underground Cities: Engineering and Innovation meet Archaeology,
Architecture and Art, Volume 12: Urban
Tunnels - Part 2 – Peila, Viggiani & Celestino (Eds)
© 2019 Taylor & Francis Group, London, ISBN 978-0-367-46900-9

Tunnel support design by geo-structure interaction approach

J. Pan & A. Kuras
WSP Australia Pty Ltd, Sydney, Australia

ABSTRACT: The North Strathfield Rail Underpass (NSRU) Project includes the construction of a 148 m rail underpass under live railway tracks at North Strathfield for freight trains. This paper presents design of the tunnel support structure concentrated on factors of geotechnical and structural aspects. The geotechnical aspect has been focused on opening stability, ground-support load share, tunnel deformation and surface settlement in consideration. Ground reaction curves are developed by applying material softening and load reduction methods to two dimensional models created using Phase^2. These curves were then used to estimate the likely amount of load share between the shotcrete lining and the surrounding ground. Simple 2D and 3D structural modelling of the first six meters of support immediately after excavation was conducted to assess the imposed structural actions on the temporary and permanent tunnel support. The structural actions were compared with the calculated Ultimate Limit State (ULS) structural capacity of the support system during construction and in the permanent conditions.

1 INTRODUCTION

The North Strathfield Rail Underpass (NSRU) project was jointly funded by the Australian and NSW governments. The project formed part of the Northern Sydney Freight Corridor Program, designed to improve capacity and reliability for freight and passenger services between North Strathfield and Newcastle.

NSRU project included the construction of a rail underpass for freight trains at North Strathfield that eliminates the flat crossing north of Concord West Station (Figure 1). A holding loop was constructed for freight trains diverging from the mainline, to enable a freight train to be held at a signal prior to entering the rail underpass dive structure. This avoids freight trains being stopped within the rail underpass and having to restart on an uphill gradient.

The rail underpass is 148 m in length and passes beneath operational rail tracks at a skew angle of around 80 degrees. The tunnel has a horseshoe profile in cross section was driven using roadheader that excavated Ashfield Shale of varying quality.

WSP, in joint venture with Jacobs, completed the detailed design and provided construction phase services for an alliance team, which comprised Bouygues Travaux Publics, John Holland, and Transport for NSW. As part of the design services, the design joint venture carried out proof engineering of the underpass design. This required a full and independent assessment, without the exchange of calculations or similar information with the design team, of all factors influencing the final integrity of the design, including all the associated temporary works. The proof engineering exercise included the following:

- undertaking of design calculations and modelling
- reviewing the safety, durability and functional requirements of the identified elements, the design documentation and construction methodology and
- performing an independent dimensional check of the design.

This paper describes the geotechnical and structural assessments that were undertaken for the proof engineering.

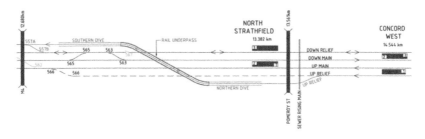

Figure 1.   North Strathfield Underpass Project.

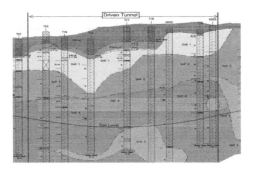

Figure 2.   Geotechnical longitudinal section.

2   SITE GEOLOGY

The ground investigation for the project included series of boreholes and test pits that targeted the tunnel alignment within the rail corridor. These were carried out during weekend track possessions (planned periods when Sydney Trains suspend rail services on a segment of the network to enable track maintenance). From the resulting borehole logs, a geological model along the tunnel alignment was prepared. Figure 2 presents this ground interpretation along the underpass alignment. The main features can be summarised as follows:

- fill material at surface to a depth 0.4 m to 2.2 m below ground level
- stiff to hard residual clays below the fill material that are up to 3.5 m in thickness and
- weathered siltstone ranging from 2.2 m to 6.2 m in thickness encountered at depths of 0.55m to 4.0 m below ground level.

3   UNDERPASS TUNNEL DESIGN AND CONSTRUCTION

The horse shape underpass tunnel profile (Figure 3) has a minimum excavated span of around 9 m and minimum excavated height of around 7 m. The ground cover above the tunnel crown varies from 2.5 m to 3.5 m.

Excavation of underpass was progressed under the support of an array of overlapping canopy tubes, pre-installed ahead of the advancing excavated face. Notably, the underpass was constructed while the rail corridor remained fully operational. The face of the tunnel was stabilised using an array of glass reinforced polymer (GRP) dowels.

Successive layers of shotcrete were sprayed onto the excavated profile and face of the tunnel as it was advanced to achieve a total nominal thickness of 250 mm. After which a smoothing layer, sprayed waterproofing membrane, final shotcrete layer and passive fire protection were applied to create a lining with a nominal total thickness of 395 mm.

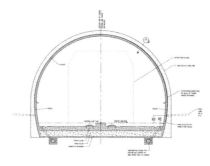

Figure 3. North Strathfield Rail Underpass - typical tunnel profile.

The prevailing site conditions and need to maintain an operational rail environment above necessitated a slow rate of tunnel face advancement. Typically, the face was advanced 1.0 to 1.5 m each day. The rate of advance was influenced by the quality of the encountered ground conditions and the outcome of continual assessment of retrieved monitoring data.

4 METHODOLOGY

Proof engineering of the tunnel structure design concentrated on the following:

- A geotechnical assessment conducted at the Serviceability Limit State (i.e. working stresses) and involved a quantitative investigation of opening stability, ground-support load share, tunnel deformation and surface settlement.
- A subsequent structural assessment of the tunnel support was then undertaken to determine the capacity of the lining and canopy tubes at the Ultimate Limit State (ULS) and the ability of the support system to resist structural actions as imposed during construction and once the underpass is in operation.

The ground and the lining was modelled and analysed using a range of methods and finite element software packages. The commercial available software packages Phase^2 (geotechnical) and Strand7 (structural) were used to predict the behaviour of the support system and surrounding ground under imposed loading.

5 GEOTECHNICAL APPROACH

5.1 Governing sections

To undertake the proof engineering the interpreted geotechnical model was interrogated to establish lengths of the alignment that could be defined as being within "Typical", "Favourable" or "Adverse" ground conditions. Geotechnical models were prepared to represent these ground conditions. The following geotechnical assessments were carried out using three models (Table 1):

- Ground reaction curves were developed by applying material softening and load reduction methods to numerical models that were created using Phase^2. These curves were used to estimate the likely amount of load sharing that would occur between the shotcrete tunnel lining and the surrounding ground.
- Based on the results from the load sharing assessment further Phase^2 modelling was undertaken to investigate the amount of tunnel deformation that could occur, taking into account the contribution from the tunnel support, namely the canopy tubes and the shotcrete lining. These predictions were compared the acceptance criteria that was nominated in the monitoring plan to ensure that the defined limits could be met.
- A sliding wedge analysis was conducted to prove the stability of the tunnel face during its advance.

Table 1. Governing sections.

	Ground Conditions
Favourable	1.6 m Fill material
	0.6 m Residual (Clay)
	0.8 m Rock Unit 1
	Approx. 3.8 m Rock Unit 2
	Then Rock Unit 4 & Cover to tunnel crown: 3 m
	The entire tunnel section is within Rock Unit 2 and Rock Unit 4
Typical	1.4 m Fill material
	Approx. 1.0 m Residual (Clay)
	1.7 m Rock Unit 1
	Approx. 2.6 m Rock Unit 2
	Then Rock Unit 4 & Cover to tunnel crown: 2.9 m
	Rock Unit 1 is 0.5 m above tunnel crown and 1.2 m at the upper portion of tunnel
Adverse	2.0 m Fill materials
	Approx. 0.7 m Residual (Clay)
	3.5 m Rock Unit 1
	0.6 – 1.2 m Rock Unit 2
	Then Rock Unit 4 & Cover to tunnel crown: 2.7 m
	Full thickness of Rock Unit 1 (3.5 m) within the tunnel section

❖ *Rock Unit 1 - Extremely low to very low strength, fractured to highly fractured siltstone.*
❖ *Rock Unit 2 - Low to medium strength, fractured siltstone.*
❖ *Rock Unit 3 - Medium to high strength, fractured to highly fractured siltstone.*
❖ *Rock Unit 4 - Medium to high strength, slightly to unfractured (RQD>70%) siltstone.*

Longitudinal modelling was undertaken by Phase^2 to assess ground movement ahead and behind the face to investigate face stability and tunnel convergence along the length of the tunnel.

5.2 *Geotechnical design parameters*

The main material types encountered during the site investigation were divided into the units for the purposed design assessment. These units are listed in Table 2 below. The material properties and parameters recommended for the engineering design of tunnel structure are provided in the table. As part of the proof engineering exercise these parameters were checked and then used to undertake the proof engineering geotechnical assessment.

5.3 *Ground reaction curves*

It was acknowledged at an early stage that the performance of the underground support system would need to be measured in terms of limiting ground movement and the induced surface settlement. A convergence-confinement method was used to the assess the theoretical performance of the proposed tunnel support system. Stresses and displacements in the rock

Table 2. Recommended design values for driven tunnel (for numerical modelling).

Unit	γ (kN/m^3)	c' (kPa)	φ' (degrees)	E (MPa)	ν
Filling	18	0	35	60	0.3
Residual	19	15	25	12	0.3
Rock Unit 1	20	60	25	100	0.25
Rock Unit 2	22	250	25	300	0.25
Rock Unit 3	23	670	27	900	0.2
Rock Unit 4	23	1000	33	2000	0.2

that surrounds the tunnel and in the lining would be influenced by the rock mass properties, in-situ stress fields, and the stiffness of the lining and the timing of its installation. The inter-dependence of these various factors for proof engineering was represented by ground reaction curves and reaction of support system.

To assess these impacts a typical cross section was modelled using Phase^2 based on the construction sequences presented in the detailed design. The proposed temporary tunnel support system comprised overlapping arrays of canopy tubes installed around the tunnel crown, ahead of the face. Each array of canopy tubes consisted of 19 tubes installed at an inclination of 5.5 degrees. The canopy tubes are 12 m long and 138 mm in diameter. The shotcrete support adopted for the analysis was assumed to be 150 mm thick with an elastic modulus of 12,000 MPa (i.e. representative of the shotcrete at 12 hours of age as defined in the shotcrete specification). Typical results are shown in Figure 4.

The proof engineering assessment of the canopy tubes followed an approach recommended by Dr Evert Hoek (2004). This involved developing basic two-dimensional equivalent numerical models, for which it was assumed that the canopy tubes increase the strength and stiffness of the ground immediately above the tunnel crown where they pass through. A weighted average approach was taken to estimate the increase in the relative strength/stiffness of the 'reinforced' ground.

Two commonly used approaches were using for proof engineering to generate the ground reaction curves, namely the 'material softening' and 'load reduction' methods (Rocscience 2004, Vermeer et al 2003). Figure 5 and Figure 6 below show, for the assumed "Adverse"

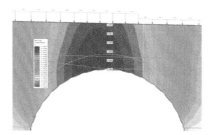

Figure 4. Vertical settlement: "Adverse" section with canopy tubes – 5.5mm at surface level.

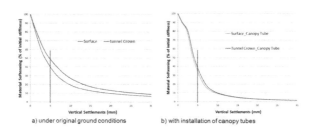

a) under original ground conditions   b) with installation of canopy tubes

Figure 5. Ground reaction curve – material softening approach.

ground conditions, the consequential ground reaction curves that were generated for the

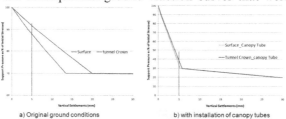

a) Original ground conditions   b) with installation of canopy tubes

Figure 6. Ground reaction curve – load reduction approach.

case of unsupported ground (a) and the case where the ground above the tunnel crown is reinforced with an 'umbrella arch' canopy tubes (b).

Appropriate levels of material softening and support pressure were selected based on an assumption of 5mm of maximum surface settlement. This amount of predicted settlement was based on the settlement acceptance limits (green trigger level) nominated in the monitoring plan, developed during detailed design stage. This plan imposed the following maximum settlement limits:

- Green: 5 mm
- Amber: 8 mm
- Red: 10 mm

The application of this deformation limit resulted in the following ground reaction curve assumptions:

- Material softening method – 40 per cent of original material modulus (Figure 5)
- Load reduction method - support pressure equivalent to 40 per cent of field stresses with canopy tube installation (Figure 6).

These assumptions were applied to predict the load sharing characteristics of the support system and assess its structural capacity, as discussed in further detail in this paper.

5.4 *Tunnel heading stability*

The stability of the tunnel heading was assessed by undertaking a sliding wedge analysis and longitudinal modelling using Phase^2 to assess ground movement ahead and behind the face to investigate face stability and tunnel convergence along the length of the tunnel.

5.4.1 *Sliding wedge analysis*
During excavation of the tunnel the stability of the tunnel head must be maintained. To undertake the proof engineering assessment the excavated tunnel section was simplified to a rectangle profile with an equivalent height and width of excavation. The forces that act on the wedge are separated into driving forces that produce failure and resisting forces that provide stability. The following proof engineering checks were completed to prove that face stability would not be an issue during construction:

- Driving forces including (a) the self-weight of the sliding wedge and (b) any additional load acting on the wedge
- Resisting forces – friction and cohesion on both the sliding plane and the side planes.

5.4.2 *Finite element analysis*
Additional finite element modelling was undertaken to further assess the stability of the tunnel using an approach proposed by Dr Evert Hoek (Hoek, 2004). This method involves creating a two-dimensional cross-section parallel to the tunnel axis which represents an infinitely long tabular excavation. The results of this analysis were used to obtain an indication of the behaviour of the rock mass surrounding a tunnel.

Following Dr Hoek's approach, a model was setup to represent the typical longitudinal section with the consideration of tunnel advance rate, material softening and shotcrete strength gain with times. Given the ground conditions, the modelling work was progressed on the assumption of one metre of tunnel advance each day. The following layers of shotcrete support were assumed for the analysis, in combination with the strength gain and stiffness assumptions listed in Table 3:

- 0 to 1.0 m from the tunnel face: 1st layer of 150 mm
- 1.0 to 2.0 m from the tunnel face: 2nd layer of 50 mm, total shotcrete thickness – 200 mm
- 2.0 to 3.0 m from the tunnel face: 3rd layer of 50 mm, total shotcrete thickness – 250 mm

Table 3. Shotcrete strength gain with time.

Times to strength gain	Strength (MPa)	Young's modulus (MPa)
3 Hours	1	1,000
12 Hours	6	12,000
1 Day	18	20,000
3 Days	26	24,000
7 Days	35	28,000
28 Days	40	30,000

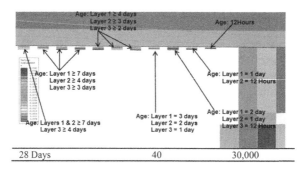

Figure 7. Shotcrete strength gain with tunnel excavated 7 m further.

Figure 7 presents the construction staging and shotcrete strength assumptions that have been adopted for the proof engineering over a tunnel length of 7 m from the face.

This tunnel advance model provides an indication of tunnel crown movement away from the tunnel face due to tunnel construction. Figure 8 below shows that at tunnel crown deformation profile for a tunnel face advance of 7 metres. Based on this profile it is possible to provide an interpretation of the deformation to determine ground loading to be applied for the structural assessment of the support system. It was judged that based on the results that full ground loading would be applied on the support system 5 m back from the advancing tunnel face (Figure 9).

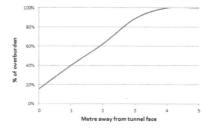

Figure 8. Total deformation at an excavation 7 m further (shotcrete strength gain model).

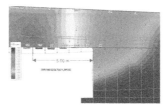

Figure 9. Interpreted loading 5 m away from the tunnel heading.

6 STRUCTURE ASSESSMENT

As part of proof engineering a structural assessment was carried out using the results from Phase[2] modelling to determine ground pressures that would need to be supported by the shotcrete lining both at the permanent and temporary stages. The structure analysis included the following:

- Assessment of the fibre reinforced shotcrete lining structural capacity in its final state and at an early stage during construction when the different shotcrete layers gain their strength.
- Two dimensional 'bedded beam' modelling of the shotcrete lining using Strand7 at the ULS to determine the imposed structural adequacy of the structural shell in its final state and during construction.
- Three-dimensional modelling of the first 6 m of support immediately after excavation of one metre full face advance to assess the capacity of the full array of canopy tubes and the contribution of the tubes to support imposed ground load, whilst the previous application of shotcrete gains strength.

6.1 Load cases

The following load cases were considered as part of the proof engineering assessment:

- ground loading: permanent condition – full overburden pressure and temporary stage during construction (Figure 10)
- live loading: train load (includes ballasted track structure) assumed to be 37 kPa at surface in compliance with the Australian Bridge Code (AS 5100.2). This load is assumed to distribute with depth to the level of the tunnel crown.

The following combinations of factored loading were found to be governing. The applied load factors were adopted in accordance with the requirements of the relevant Australian Standards:

- ULS: 1.2 x lining self-weight + 1.25 x overburden + 1.5 x train load (live).
- SLS: 1.0 x lining self-weight + 1.0 x overburden + 1.0 x train load (live).

6.2 Structural performance and capacity of tunnel support during construction

The following analyses were conducted to assess the criticality of the temporary support elements at construction:

- Loading on canopy tubes
- Loading share between canopy tubes and shotcrete lining
- Loading on shotcrete lining

Table 4 below gives the thickness and strength of shotcrete modelled and loads applied to Strand7 models.

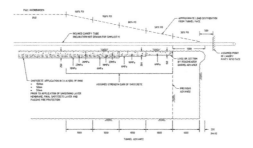

Figure 10. Load distribution and shotcrete strength gain assumptions.

Table 4. Strand7 model inputs.

Distance to tunnel face (m)	Thickness (mm)	Strength (MPa)	% of full overburden
1	150	6	50
2	200	18	80
3	250	23	95
4	250	26	100

The reaction curves (Figures 5 & 6) developed using the 'material softening' and 'load reduction' approaches were compared with the anticipated support reaction behaviour of the support system, namely the pre-installed canopy tubes in combination with shotcrete lining. This comparison enabled an estimate to be made of the loads that will need to be supported by each element, giving consideration to the anticipated strength gain at the various stages of construction.

The capacity of the grout filled canopy tube was calculated to have design moment capacity of 32 kNm and shear yield capacity of 380 kN approximately. The 3D structural modelling demonstrated that the canopy tubes would have adequate theoretical capacity to sustain the assumed loading patterns at the stage when the tunnel was advanced by one metre, before the first layer of shotcrete is applied (Figure 11).

The results from the proof engineering structure modelling indicated that the first layer of shotcrete with a minimum early strength value of 6MPa provided adequate capacity to support loading equivalent to 50% of full overburden, before advancing to the next excavation cycle. The 3 layers of shotcrete lining with a combined nominal thickness of 250 mm and strength gain as shown in Table 4 had adequate theoretical capacity to support the assumed combination of applied design loading (Figure 12).

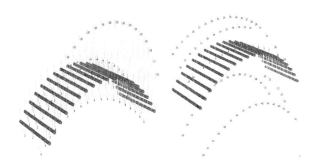

Figure 11. Calculated canopy tube forces.

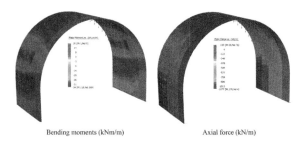

Bending moments (kNm/m)   Axial force (kN/m)

Figure 12. Calculated forces on the shotcrete lining due to the imposed loading (ULS).

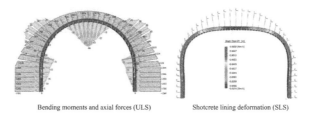

Bending moments and axial forces (ULS)   Shotcrete lining deformation (SLS)

Figure 13.   Results of shotcrete lining assessment under the permanent loading conditions.

6.3   *Structural performance and capacity of tunnel support during operation*

Two dimensional 'bedded beam' modelling was undertaken to assess the structural adequacy of the shotcrete in its final state, when the tunnel is in operation. The aim of this exercise was to calculate the structural actions that will be imposed on the lining based on combinations of assumed loading.

The results indicated that at the ULS the load cases described in section 6 above would imposed combinations of bending, shear and axial load along the "Adverse" section of the tunnels that could accommodated by the lining. Further the deformation of the lining would be limited to 6 mm at the SLS which was considered to be acceptable (Figure 13).

7   CONCLUSIONS

Proof engineering of the design NSRU tunnel design through the application of geotechnical and structural analysis demonstrated the adequacy of the designed support system. The combined geotechnical and structural approach provided a means of undertaking the design check without resorting to complex and time consuming 3D modelling. The combined approach allowed for a more complete understand the ground conditions/behaviour and response of the tunnel to construction. This case study has described the following procedures adopted for this approach:

- Development of ground reaction curves by applying material softening and load reduction methods to models created using Phase[2]. Then the use of these curves to estimate the likely amount of load share between the shotcrete lining and the surrounding ground.
- Tunnel heading stability check using a simplified approach and longitudinal modelling that incorporates construction sequencing.
- Applying geotechnical assessment to evaluate the performance of grout filled canopy tubes as a means of temporary support.
- The application of structural modelling to evaluate the capacity of the fibre reinforced shotcrete lining in its final state and at an early stage during construction when the different shotcrete layers gain their strength.

REFERENCES

Hoek, E. 2004. Numerical modelling for shallow tunnels in weak rock. Discussion paper # 3, Retrieved from http://www.rocscience.com.3pdns.korax.net/assets/files/uploads/7698.pdf.
Rocscience 2004. Developer's tip – 3D tunnel simulation using the material softening method in Phase[2]. Retrieved from http://www.rocscience.com.3pdns.korax.net/assets/files/uploads/8372.pdf.
Vermeer, P. A., Moller, S.C. & Ruse, N. 2003. On the application of numerical analysis in tunnelling. *12 Asian Regional Conference on Soil Mechanics and Geotechnical Engineering*, Singapore, Vol. 2.

Tunnels and Underground Cities: Engineering and Innovation meet Archaeology,
Architecture and Art, Volume 12: Urban
Tunnels - Part 2 – Peila, Viggiani & Celestino (Eds)
© 2019 Taylor & Francis Group, London, ISBN 978-0-367-46900-9

Numerical analysis on performance of EPB shield TBM by discrete element method

B. Park
University of Science and Technology, Goyang-si, Republic of Korea

S.H. Chang, S.W. Choi, C. Lee & T.H. Kang
Korea Institute of Civil Engineering and Building Technology, Goyang-si, Republic of Korea

ABSTRACT: The Discrete Element Method (DEM) has been widely used in civil engineering as well as various industrial fields to simulate granular materials. In this study, DEM was used to predict the performance of the earth pressure balance (EPB) shield TBM (Tunnel Boring Machine). For the analysis, EPB shield TBM with the diameter of 7.73 m was selected, having used previously for an excavation of the metro tunnel in Seoul. First, the ground modeling was conducted based on the result of the direct shear test simulation. Next, TBM 3D modeling composed of five parts (cutting tools, cutterhead, chamber, shield and screw auger) was carried out. Then, TBM excavation analysis will be performed. After that, the excavation performance indicators, such as cutter force at the cutterhead, thrust force, resistant torque will be estimated. Finally, we will compare the performance indicators obtained from numerical analysis with the maximum capacity of the real TBM design parameters selected for the numerical analysis.

1 INTRODUCTION

Tunnel excavation method can be generally divided into conventional tunneling method and mechanized tunneling method, represented by tunnel boring machine (TBM) (ITA, 2000). In recent days, the earth pressure balance (EPB) shield TBM has been widely used in urban tunnels where the rapid excavation and high safety are essentially required (KICT, 2015). To balance the pressure conditions at the tunnel face with the minimum surface settlement, the chamber of the EPB shield TBM is filled with the excavated materials, composed of soil pastes acted as the support medium at the tunnel face.

For an economical construction of tunnel by EPB shield TBM, it is critical to choose the optimized EPB shield TBM considering the various project conditions. Furthermore, it is also required to estimate the excavation performance of selected EPB shield TBM under the ground condition on the target. An application of the numerical analysis by discrete element method (DEM) would be a useful way to estimate the excavation performance of the EPB shield TBM. This DEM method was considerably verified by the previous studies (Wu & Liu 2014, Wu et al. 2013, Maynar & Rodríguez 2005) through PFC^{3D} commercial software and the author's recent studies (Lee et al. 2017a, Lee et al. 2017b) conducted by EDEM commercial software.

In this study, the excavation performance of EPB shield TBM was estimated by the DEM analysis. For the analysis, EPB shield TBM with the diameter of 7.81 m was selected, having used in 2012 for an excavation of the metro tunnel in Seoul (Yooshin Engineering Corporation, 2013). Through the analysis, several performance indicators including the force of cutting tools and cutterhead front face, resistant torque caused by the rotation of the cutterhead, and the relative wear generated from cutterhead and cutting tools, were mainly estimated and checked. Furthermore, these estimated values were compared with the maximum capacity of the practical TBM technical specifications.

2 GROUND MODELING

2.1 Site information

According to Yooshin Engineering Corporation (2013) as presented in Figure 1, the EPB shield TBM excavated through the weathered soils composed of compact silty soils, 15.2m below the surface for the metro tunnel construction. The related geotechnical design parameters of the ground were summarized in Table 1.

2.2 Direct shear test

Before the ground modeling work, a direct shear test was conducted first by EDEM software based on the 3D DEM as shown in Figure 2. Once the two large shear boxes were modeled with each size of 4m (width) × 4m (length) × 0.75 m (height), 25,789 globular shape particles with the radius of 50 mm were created and filled with both frictionless boxes. Some input parameters of the particle and contact properties between particles are summarized in Table 2. Lastly, five different normal stresses (49kPa, 98kPa, 147kPa, 196kPa, and 294kPa) were applied and the shear displacement rate was set to 0.01 m/s for the simulation.

Based on the Mohr-Coulomb theory, the result of the simulation and the geotechnical parameters (Table 1) were expressed as drawn in Figure 3. It refers to the difference between the result of the simulation and the calculated geotechnical design parameter under the five different normal stress conditions. Overall, shear strengths estimated from the simulation were about 4~5 % underestimated compared with those of the calculated geotechnical design.

2.3 Ground modeling

Once the mechanical properties of the DEM soil model were estimated by the simulation of the direct shear test, the ground model used for the TBM excavation were created. As presented in Figure 4, the size of the ground model is 16 m(length) × 21 m(height) and the thickness (length of excavation) is 1.5 m with the use of about 520,000 globular shape particles.

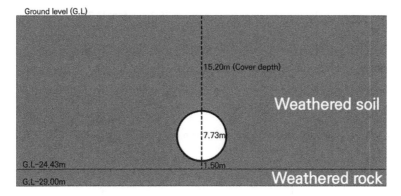

Figure 1.   Cross-sectional diagram of the site.

Table 1.   Geotechnical design parameters of the site.

Classification	Unit weight	Cohesion	Friction angle
	kN/m3	kPa	°
Weathered soil	19.5	26.3	27.3

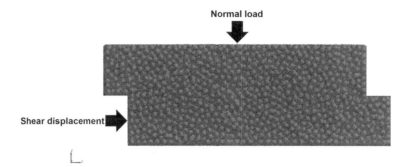

Figure 2.   DEM simulation of direct shear test.

Table 2.   DEM input parameters used for the simulation.

Number of particles	25,789
Particle radius (mm)	50
Poisson's ratio	0.3
Particle density (kg/m³)	3537
Coefficient of restitution (particle-particle)	0.1
Coefficient of static friction (particle-particle)	5.0
Coefficient of rolling friction (particle-particle)	0.2

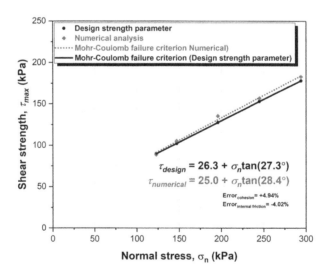

Figure 3.   Results of simulation and geotechnical design parameter expressed in Mohr-Coulomb line.

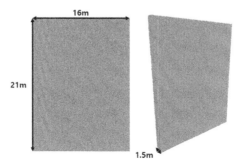

Figure 4.   Drawing of ground model.

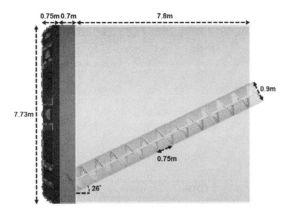

Figure 5. Drawing of 3D TBM model.

3 TBM MODELING

TBM 3D modeling works were performed by the 3D CAD software to be used in DEM analysis. As seen in Figure 5, 3D TBM model largely consists of five parts: cutting tools (red & pink), cutterhead (dark navy), chamber (green), screw auger (yellow) and TBM skin (dark grey).

Particularly, the cutting tools and cutterhead with the diameter of 7.73 m were modeled in detail based on the CAD drawings of the cutterhead as described in Figure 6. Both sides of eight main spokes are installed with a total number of 170 cutter bits (pink), and 52 disc cutters (red) are installed at the front face of the cutterhead. Also, the opening ratio of the cutterhead was calculated to be about 21 %.

4 TBM EXCAVATION

4.1 *Simulation setting*

Before the analysis, the ground and the TBM model were merged first as shown in Figure 6. After that, advance conditions, including translational and rotational motions depending on each part were given as summarized in Table 3. As shown in Figure 1, a cover depth above the TBM tunnel face is 15.2 m. However, because the TBM model was located at the center of the ground model, the cover depth in this simulation can be about 8 m, the same as the diameter

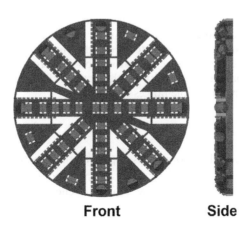

Front   **Side**

Figure 6. Drawing of 3D TBM cutterhead model.

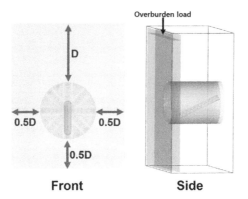

Front   **Side**

Figure 7.   Drawing of excavation setting.

Table 3.   Advance conditions of TBM during the simulation.

Component	Translational		Rotational	
	Direction	Condition	Direction	Condition
Disc cutters	+X axis	10mm/rpm	-	-
Cutter bits	+X axis	10mm/rpm	-	-
Cutterhead	+X axis	10mm/rpm	CCW	3rpm
Chamber	+X axis	10mm/rpm	-	-
Shield skin	+X axis	10mm/rpm	-	-
Screw auger	+X axis	10mm/rpm	CCW	12rpm

* rpm: revolution/minute
* +X: advance direction
* CCW: counterclockwise

of TBM. As a result, the overburden load can be different. To complement this problem, the normal load was also set on the ground model to apply the additional overburden load corresponded to the cover depth with 7.2 m.

5   CONCLUSION

The simulation of the TBM excavation has been still in progress. Once the simulation finishes, the excavation performance indicators of TBM, such as cutter force at the cutterhead, thrust force of the TBM, resistant torque generated from the cutterhead will be checked and estimated. Furthermore, these performance indicators obtained from the numerical analysis will be compared with the maximum capacity of the actual TBM design parameters to check whether they will be within reasonable ranges or not.

ACKNOWLEDGEMENT

This research was funded and performed as a part of the construction technology project, "Development of optimal cutterhead design and operating control systems for TBM" (project number: 17SCIP-B129646-01) provided by Korea Agency for Infrastructure Technology Advancement(KAIA).

REFERENCES

Ita, W. 2000. *Mechanized Tunnelling: Recommendations and Guidelines for Tunnel Boring Machines (TBMs). Publication No. MAI 2001 – ISSN 1267-8422.* Lausane: International Tunnelling Association.

KICT. 2015. *Development of optimized TBM cutterhead design method and high-performance disc cutter. Publication No. ISBN 979-11-954377-2-6.* Anyang-si: Korea Agency for Infrastructure Technology Advancement.

Lee, C., Chang, S., Choi, S., Park, B., Kang, T. & Sim, J. 2017a. Preliminary study on a spoke-type EPB shield TBM by discrete element method. *J. Korean Tunnelling and Underground Space Association.* Vol-19 (6):1029–1044 (in Korean).

Lee, C., Chang, S., Choi, S., Park, B., Kang, T. & Sim, J. 2017b. Numerical study of face plate-type EPB shield TBM by discrete element method. J. Korean Geosynthetics Society. Vol-16 (4):163–176 (in Korean).

Maynar, M. J. & Rodríguez, L. E. 2005. Discrete numerical model for analysis of earth pressure balance tunnel excavation. *Journal of Geotechnical and Geoenvironmental Engineering*, Vol-131 (10):1234–1242.

Wu, L. & Liu, C. 2014. Modeling of shield machine tunneling experiment by discrete element method. *Applied Mechanics and Materials*, Vol-556: 1200–1204.

Wu, L., Guan, T. & Lei, L. 2013. Discrete element model for performance analysis of cutterhead excavation system of EPB machine. *Tunnelling and Underground Space Technology*, Vol-37(August 2013): 37–44.

Yooshin Engineering Corporation. 2013. Passing Safety of Underground Leakage Section during Shield TBM Excavation/Construction Case. *Yooshin technical bulletin.* Seoul: Yooshin Engineering Corporation (in Korean).

Tunnels and Underground Cities: Engineering and Innovation meet Archaeology,
Architecture and Art, Volume 12: Urban
Tunnels - Part 2 – Peila, Viggiani & Celestino (Eds)
© 2019 Taylor & Francis Group, London, ISBN 978-0-367-46900-9

Sequential excavation method in soft ground: Monitoring and modelling

T.P. Perez
Technical University of Catalonia, Barcelona, Spain

ABSTRACT: The increasing urban transport demand in São Paulo, Brazil leads to the construction of new projects such as tunnels and roads. The main attention should be paid, particularly, to the range of induced soil deformations in depth and at ground level and potential effects on surrounding structures when the tunnels are excavated by conventional methods. The Rodoanel Norte Lot 5 is part of a 44 Km road project, which connects São Paulo, Guarulhos and Arujá. The one kilometer double tunnel was excavated by means of the Sequential Excavation Method, through a mixed geological formation. Ground surface displacements of approximated 150 mm in the initial 20 m of the east portal tunnel drive are measured. These displacements are been analyzed and correlated with excavation phases and ground topography. The approach presented on this article is a three dimensional numerical back-analysis of the section considered.

1 INTRODUCTION

A new 44 km road extension is currently being built to connect the cities of São Paulo, Guarulhos and Arujá in Brazil, which is called Rodoanel Norte. This extension, as it can be seen in Figure 1 (North Part) is part of the road ring around São Paulo, the Rodoanel Project. The project includes the excavation of large diameter tunnels at different depths using the Sequential Excavation Method.

In this paper, the first part will present an overview of the results obtained for the surface ground displacements measured in the first 20 m of the two tunnels excavation. The excavation sequence during this period corresponds to the four side drifts excavations and the two top headings advance. The second part is a three dimensional numerical back-analysis of the section with the objective to reproduce the displacements occurred.

2 EFFECT OF TUNNELING: MONITORING SECTION

The present section summarizes the results obtained by the instrumentation plan on the east tunnel portal. Numerical back-analysis is further proposed for the section to validate the ground parameters.

2.1 Description of the section

The east portal is located in the city of Guarulhos, approximately 30 km from São Paulo. The tunnels are located in a zone with a natural slope with an inclination of 15°, which provides a higher overburden for the right tunnel compared to the left tunnel.

The section analyzed has a superficial 5 m layer of colluvial soil, which is composed of silty clay with soft consistency. At lower depths, especially in the level of the tunnels, the soil is composed by silt with presence of rock blocks.

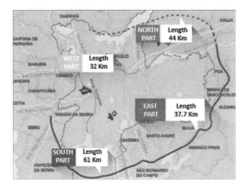

Figure 1. General overview of the Rodoanel map.

Figure 2. Tunnel face with presence of silty soil.

2.2 *Description of the tunnels and the instrumentation plan*

The tunnels are being constructed by means of the Sequential Excavation Method. Both of the tunnels are 20.45 m diameter maximum, with a first layer of shotcrete of about 30 cm thick for the top heading and 15 cm for the side drifts and lattice girders 16, 16, 20 mm. Through the soft soil section, which is analyzed in this paper, the tunnel advances are of about 0.8 m with a cross section been divided in five sections. First, the excavation of the side drifts, following the top heading, the bench and the invert for last. The results for last two sections are not presented in this paper.

The instrumentation plan can be seen in Figure 3, and it is composed by 42 superficial borehole extensometers, 9 deep borehole extensometers divided in four lines spaced at 10 m. The plan also includes 3 inclinometers for the measure of the horizontal movements.

The vertical displacements of the ground are essentially measured by precise levelling of the borehole extensometers heads. The results presented on this paper are those related with the

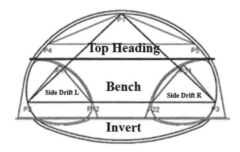

Figure 3. Cross section of the tunnels.

6037

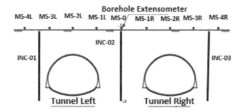

Figure 4. Instrumentation plan.

sections S0 and S1, corresponding to the initial 20 m of the excavation. The results obtained were measured during the excavation of the side drifts followed by the excavation of the top heading.

The Figure 5 and Figure 6 present the results of the section S0, while the Figure 7 and Figure 8 presents the section S1 results. It is important to note that the sequence of the extensometer, which the higher number (MS-30) is the last extensometers positioned at far right, while the lower number (MS-10) is the first extensometer, positioned at the far left.

The initial phase of the tunnel excavation as said before is the excavation of the four side drifts that were excavated almost simultaneously. The results presented above show that during this period, the ground experienced large surface displacements ranging between 60 and 70 mm. The field data showed that the tunnel right advanced faster than the tunnel left, so when the excavation of the top heading started, the ground experienced to different displacement steps. As the tunnel right excavation advanced the ground continues to settle and reach values between 85 and 120 mm. When the excavation of the tunnel left top heading started, is possible to see an increment of ground settlement ranging from 150 mm.

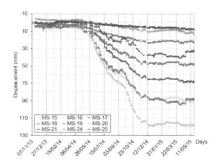

Figure 5. Vertical displacements of the section S0 – Tunnel Right.

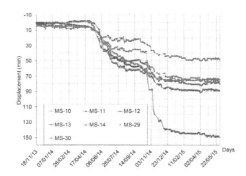

Figure 6. Vertical displacements of the section S0 – Tunnel Left.

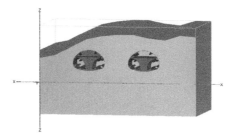

Figure 7. Overview of the numerical model.

The increase of ground displacement in the tunnel left side can be explained by three factors. First, according to Devriendt (2006) cited in Chapman, Metje and Stark (2010), the interference volume' effect occurs where the volume loss is commonly greater when a second tunnel is excavated adjacent to the completed tunnel (asymmetric effect in the final settlement trough will occur for two side-by-side tunnels of the same cross section and depth. Second, the accumulation of shear strain adjacent to the first tunnel results in a reduction of the ground stiffness and hence greater displacements where the left tunnel is constructed. Third, the ground topography, which combines low overburden and the natural slope.

After the excavation of the soft ground zone, the project was stopped. It is possible to note that during this time, the ground displacement increment is almost zero, and remains constant.

The displacements measured on the tunnel right side are in a range of magnitude acceptable according to the project. The tunnel left excavation induced larger displacements, which is considered as a danger alert by the project. According to Perez (2016), the magnitude of these displacements are related with the large volume loss due to the large cross section to be excavated, the low overburden, the natural slope without any type of retaining structure and the use of light-weight lattice girders with bar thicknesses of only 16, 16 and 20 mm.

3   THREE DIMENSIONAL NUMERICAL BACK-ANALYSIS

The results presented on the previous section have been used for the validation of a three dimensional numerical simulation procedure. The back-analysis has been performed with the software Plaxis 3D. The model includes an explicit description of the two tunnels excavation. Firstly the excavation and lining installation of the four side drifts and lastly the same procedure for the two top headings.

The soil behavior is described with the Mohr – Coulomb constitutive model. According to the horizontal displacements measured, the silty soil layer appears that the K_0 value is low, indicating a normally consolidated soil (0.5 – 0.8). The umbrella pipe injections were modelled by means of an improved soil volume as the excavation progresses. The soil parameters can be seen in the Table 1.

Table 1.   Soil parameters.

Soil	γ (KN/m³)	K_0	c´(KPa)	φ (°)	E (KPa)	ν
Colluvial	20	0.50	10	27	40000	0.3
Silt	20	0.50	20	27	50000	0.3

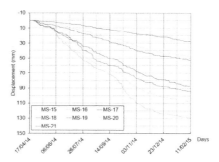

Figure 8. Vertical Displacements obtained by the numerical calculation - Tunnel Right.

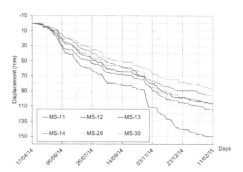

Figure 9. Vertical Displacements obtained by the numerical calculation - Tunnel Left.

The shotcrete + steel girders lining corresponding to each 80 cm advance step are composed of linear elastic plate elements. The top heading lining has an elastic modulus of 24 GPa and a Poisson´s ratio equal to 0.3.

In order to qualitatively understand the soil behavior and the interaction with the lining during the progression of the excavation, the model presented in Figure 7 considers some few simplifications of the excavation sequence. First, the model considers that the four side drifts are excavated simultaneously. The top heading excavation is modelled according to the field data, which the tunnel left started the excavation almost two months after the tunnel right. Following this assumption, the model considers that the tunnel right is excavated and supported first, and then, the tunnel left is excavated and supported.

3.1 *Numerical analysis results obtained*

Figures 8 and 9 presents the vertical displacements calculated by the model at ground surface in section S0.

Based on the results obtained, it is possible to see that the influence of the natural slope on the extensometers. The model overestimate the settlement and shows a tendency to continue the settlement.

4 CONCLUSIONS

This paper presented some results of ground settlement during the excavation of a double tunnel for the Rodoanel Norte Lote 5 project. The tunnels were excavated by means of the Sequential Excavation Method.

The observed movements at ground surface or within the soil show that, due to normally consolidated character of the ground and low overburden, the excavation induces large vertical displacement of the soil.

The proposed 3D numerical simulation procedure has proven, by an explicit representation of the different phases of progression of the tunnel face. The model used the Mohr-Coulomb constitutive model, and it is possible to see that it has the ability to reproduce the observed phenomenon but in some cases overestimating the settlement magnitude.

For further developments, the author recommends that the model expands considers a nonlinear constitutive model to improve the quality of the results. In addition, it is recommended the design of the cement injection at tunnel face during the advances.

REFERENCES

Chapman, D., Metje, N., Stark, A. (ed.3) 2010. *Introduction to Tunnel Construction*. London: Spoon Press.

Dasari, G. R. et al. 1996. Numerical Modelling of a NATM Tunnel Construction in London Clay. *Geotechnical Aspects of Underground Construction in Soft Ground*. Rotterdam, Balkema.

Divall, S., Goodey, R.J. 2015. Twin-tunneling-induced ground movements in clay. *Proceedings of the Institution of Civil Engineers: Geotechnical Engineering*, 168(3),247–256.

Karakus, M., Fowell, R. J. 2005. Back Analysis for Tunneling Induced Ground Movements and Stress Redistribution. *Tunneling and Underground Space Technology*, 20 (6): 514–524.

Perez, T. P. 2016. Análise de Deformações Ocorridas em um Maciço Classe V durante a Construção de Emboque em Solo Grampeado para Túnel Duplo. *Revista Fundações & Obras Geotécnicas*, 75 (1): 42–52.

Tunnels and Underground Cities: Engineering and Innovation meet Archaeology,
Architecture and Art, Volume 12: Urban
Tunnels - Part 2 – Peila, Viggiani & Celestino (Eds)
© 2019 Taylor & Francis Group, London, ISBN 978-0-367-46900-9

Northeast Boundary Tunnel: Applied lessons learned from the Anacostia River Tunnel Project, Washington, USA

M. Pescara & N. Della Valle
Tunnelconsult Engineering SL, Sant Cugat del Vallés, Barcelona, Spain

D. Nebbia
Salini Impregilo Lane, Washington, DC, USA

M. Gamal
Brierley Associates Corporation, Denver, CO, USA

ABSTRACT: The Northeast Boundary Tunnel (NEBT), the largest component of the Clean Rivers Project, is a deep, large sewer tunnel that will increase the capacity of the existing sewer system in the Washington DC area, significantly mitigating sewer flooding during large storm events and improving the water quality of the Anacostia River. The construction of this last portion of the project was awarded to Salini Impregilo Healy JV (Lane), already successfully leading the completion of the previous Anacostia River Tunnel (ART) project in 2018. The Salini Impregilo Healy JV is currently (Dec. 2018) excavating the new tunnel using the same ART machine after full refurbishment, recertification and size adjustment. This will operate in the same Potomac formation but under larger overburden and water head, thus introducing additional design and construction challenges. The NEBT project is benefiting from the lessons learned from ART in terms of fine tuning the TBM equipment, soil conditioning strategy, EPB pressure definition, segmental lining design and adits connections with shafts performed through soil improved areas.

1 INTRODUCTION

The DC Water and Sewer Authority's long term CSO control plan includes a system of tunnels to control combined sewer overflows.

The discharge of these tunnels will be into the Anacostia River, Rock Creek, and the Potomac River after the transport of those captured flows to the Blue Plains Advanced Wastewater Treatment Plant. The Anacostia River Tunnel is one component of this long-term control plan. The project was completed in 2018 and has contributed to an increased capacity of the District's sewer system, significantly mitigating the frequency, magnitude, and duration of sewer flooding. It is also helping to improve the water quality of the Anacostia River.

The Northeast Boundary Tunnel is the largest component of DC Water's Clean Rivers Project and it will be a large, deep sewer tunnel that will increase the capacity of the District's sewer system. The tunnel will be 16 to 53 feet below ground and run 8,9km from just south of Robert F. Kennedy Stadium to the intersection of Rhode Island Avenue NW and 6th Street NW. It will be aligned to intersect with the existing chronic flood areas along Rhode Island Avenue NW. With the completion of the NEBT Project in 2022—two years ahead of the Consent Decree schedule—the CSO Overflow Volume to Anacostia River will reduce by 98 percent from the 1996 Baseline.

Figure 1.   View of DC Water's Clean River Plan.

2   THE GEOLOGICAL CONTEXT

2.1   *Potomac formation*

The project area is located at the western edge of the Atlantic Coastal Plain Physiographic Province, with the Piedmont Physiographic Province lying to the west, separated by the Fall Line. Starting at the Fall Line and thickening eastward, a wedge of Coastal Plain sedimentary deposits overlie older Piedmont residual soils and crystalline bedrock.

Thus, the geological history of the site is characterized by successive periods of sedimentary deposition and erosion over millions of years. Figure 2 shows two typical stratigraphies located between ART and NEBT and at the intersection area between (Rhode Island Avenue/ 6th street).

A brief description of the soil is provided hereafter:

- Fill comprised of fine-grained to coarse-grained soils containing fragments of construction debris;
- Alluvium deposits consisting of soft clay, loose silt and fine sand, with varying amounts of organic material. Gravel deposits including cobbles and boulders are also present.
- The Cretaceous soils belonging to Potomac formation consist of hard and over-consolidated fine-grained cohesive soils and dense to very dense coarse-grained soils with variable amounts of fine-grained soils. The Potomac soil deposits can be divided into two sub formations: Patapsco/Arundel Formation (P/A) which has predominance of silt and clay and Patuxent Formation (PTX) which has predominance of sands and coarse material.
- Underlying the Potomac Group soils, the Pre-Cretaceous crystalline bedrock is present, formed by metamorphic rocks (predominantly amphibolite with schists and gneiss).

Table 1.   Description of the geotechnical units.

Groups	Description
G1	High plasticity, sticky clays with high swelling potential. Slickensided/fissured over-consolidated clays will behave in low-strength blocky manner; may slide or fall as slow to fast raveling ground. Fast raveling behavior when dewatered
G2	Lower plasticity, medium to high stickiness, with medium to high swell potential.
G3	Slow to fast raveling to flowing behavior when saturated depending on amount and plasticity of fines; running behavior when dewatered
G4	Below groundwater table, mixture of soil and water flows into the tunnel from all exposed surfaces; running behavior expected when dewatered

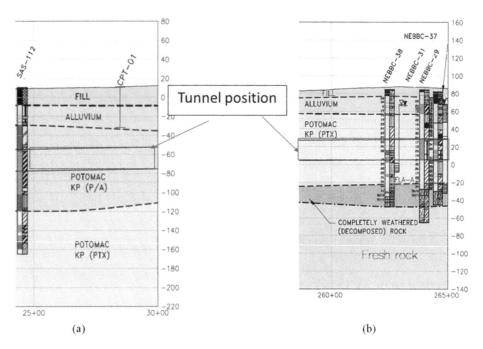

Figure 2.   Typical stratigraphy along Anacostia River System; (a) representative of ART and first portion of NEBT, (b) representative of second half of NEBT.

From the geotechnical point of view, the P/A sub-formation is characterized by G1 and G2 groups while PTX sub-formation by G3 and G4 groups which are described in Table 1.

2.2   *The alignment and profile for the two tunnels*

Although ART and NEBT are excavated in the same Potomac formation there are some pronounced differences mainly because of the distance of each alignment from Anacostia river (see Figure 1). The alignment of ART runs close to the river (under-passed once) with a relatively shallow overburden due to lower ground elevation in this area. The alignment of NEBT has a portion close to the river, but then it deviates away from the river, where ground elevation increases and thus increasing the thickness of the overburden. In this latter stretch, the P/A sub-formation (G1 and G2) becomes shallower and thinner. The P/A sub-formation is missing at locations, where the tunnel is excavated into the PTX sub-formation through the end of the alignment where the bedrock becomes closer to invert. At shallower overburden (almost all ART and southern portion of NEBT close to river) there is an alternation of the two sub-formations, while in the remaining portion of NEBT the presence of PTX is predominant.

3   FACE STABILITY CONTROL

Face stability is maintained by two main components while tunneling using an EPB machine: application of the correct face pressure and proper conditioning of the excavated material in the plenum.

3.1   *Face pressure calculation*

Among different methods to define the face pressure to be applied at the face to control the face stability while tunneling with and EPB machine, the method of Anagnostou & Kovári

(1996) was selected. Using this method, it was possible to better calculate the face stability in the two very different tunneling conditions where the alignment is always located under water table; the cohesive P/A sub-formation and the cohesionless PTX sub-formation.

The key-point has been to optimize as much as possible the applied face pressure to reduce cutterhead wear, to preserve the TBM components and to speed up the excavation process.

This criterion was successfully applied to the ART tunnel excavation and has been used again to define the face pressure for the NEBT tunnel underway between 2018 and 2020. The face pressure is calculated according to the following formulas

$$ST = s' + hf \tag{1}$$

$$s' = F_0\gamma'D - F_1c + F_2\gamma'\Delta h - F_3c'\Delta h/D \tag{2}$$

Where ST is the face pressure; s' is the effective component of ST that resists the ground load and filtration forces; F_0, F_1, F_2 and F_3 are dimensionless coefficients given by nomograms as a function of the friction angle and D, H and h_0 ratios; γ' is the submerged unit weight [kN/m3]; D is the diameter of the excavation [m]; H is the overburden height; c is the effective cohesion [kPa] and Δh is the piezometric difference between the fully hydrostatic head (h_0) and the piezometric height in the excavation chamber (h_F). Figure 3 shows the geometry and the main variables used in Equation (1).

In case of cohesionless material the second and fourth terms of the equation (2) are zero, while the third term become zero if no filtration of water through the face is allowed and the full water head is applied at the face.

On the other hand, in cohesive material the second term is not zero and the third and fourth are not zero as well if filtration is allowed.

Therefore, the key-point is thus the filtration, which is namely a function of the third term except in cohesive soil where a reduction is applied as reflected by the fourth term.

According to this principle, and with the aim to reduce as much as reasonable the pressure at the face to make more efficient and economical the excavation process, in soil groups G1 and G2 the calculated and required pressure at face could be set below the hydrostatic head, while in soil groups G3 and G4 the required and applied pressure should be closer or higher than the hydrostatic water head. This principle has been successfully implemented at ART tunnel as shown in Figure 4 where the applied pressure is plotted with the overburden and the water head.

The overall average face loss recorded in ART using topographic survey was 0.27% in the P/A and 0.18% in the PTX, which is the evidence of success of this kind of approach.

Salini Impregilo Healy is proposing the same Anagnostou & Kovári (1996) method for calculating face pressures for the NEBT Project. As of the submittal of this paper (Sept. 2018), the full EPB calculations and report are under reviewed by the Owner.

3.2 Soil conditioning

Given the geotechnical groups and behavior as described in Table 1, different strategies were implemented in ART to manage the different soil groups and proposed for NEBT:

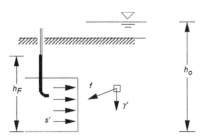

Figure 3. The geometry of the face stability problem from Anagnostou & Kovári (1996).

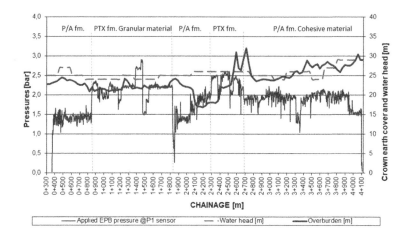

Figure 4.   Record of applied pressure for ART.

- Group G1 exhibits high plasticity and high potential for clogging
- Group G2 is similar to G1 but with less potential for clogging and thus results in an ideal material for EPB tunneling
- Group G3A has more sandy behavior with some fine which is good material for EPB tunneling
- Group G4 is a fine to coarse sand material which could create problems due to water inflow and EPB control

Each of the soil groups described above has different conditioning requirements.

Group G1 requires (1) significant amount of conditioning agent with a Foam Injection Ratio (FIR) of 60% or more obtained with a foam concentration (Cf) greater than 2%; (2) Foam Expansion Ratio (FER) of 1:6-1.7 and (3) Water Injection Ration (WIR) of 10-15% introduced directly in the excavation chamber in order flush out the clay from the excavation chamber and therefore reduce clogging.

G2 and G3A require (1) less FIR of 50-55% (Cf 1.6%); (2) FER of 1:6-1:7 and (3) almost no water due to the presence of groundwater.

G4 requires (1) FIR of 50-55% with Cf=1.6 but (2) a FER up to 1:10 to dry out the material with (3) adding of polymers and bentonite to control the EPB along the screw.

4   THE SEGMENTAL LINING

Both tunnels are designed using a fiber reinforced universal ring with a typical 6+1 configuration of bolted and gasketed 1828mm average length segments as shown in Figure 5.

The relevant difference is related to the thickness of the ring itself which changed from 30.48cm (12") of ART to 35.56cm (14") of NEBT.

The change is caused by the increase of the overburden, long-term maximum water head and finally the required TBM thrust related to this new condition.

The NEBT segmental lining is expected to support higher loads because of the reduced relaxation effect during excavation in the Potomac deposits which, in turns, leads to a higher load on the liner. In other words, the Potomac deposits do not exhibit as much arching compared to other formations thus increasing in the overburden loading on the liner in addition to the higher possible ground water load.

Moreover, most of the tunnel to shaft adit connections are in an urban area, far from the river where the PTX sub-formation is more present. The sandy deposits are characterized by a lower value of the at rest coefficient (ratio of horizontal to vertical effective in-situ stress)

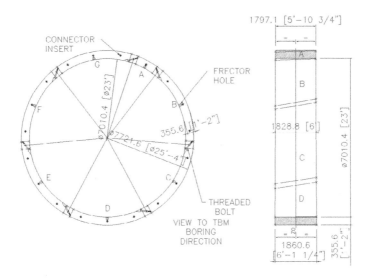

Figure 5. Segmental lining for NEBT.

resulting in higher differential pressure, thus inducing higher flexural loading on the liner. This effect becomes more pronounced at the openings where connections are created, thus requiring a special arrangement for the rings at and near the opening such as conventional rebar reinforcement.

5 UNDERPASS OF RELEVANT PRE-EXISTING STRUCTURES

To mitigate the impact from the TBM tunnel excavation underneath sensitive structures, a compensation grouting campaign was implemented in the ART Project. The crossing of the fragile and essential 108 inches force main located in Anacostia Park was of major concern for DC Water, as the force main is currently servicing a wide area of the District.

The TBM tunneling underneath the 108 inches force main was performed continuously within the area where the TBM area of influence could potentially impact the main. The ground underneath the utility was pre-grouted by means of Tube-A-Manchettes (TAM) ahead of the TBM arrival, followed up by the actual compensation grouting during the TBM crossing. Extensometers, Utility Monitoring Points and arrays of optical targets were continuously read and available to the Engineers and TBM crew during the TBM undercrossing.

Figure 7 shows in section the location of the TAM valves and grouting area, with reference to the ART tunnel and the force main: the targeted area is the ground immediately below the 108" foundation piles. The plan view of the same figure is presented in the same Figure 7 that shows the extension of the grouted area on each side of the TBM tunnel to the extent of the zone of influence.

A series of multi-base extensometers were installed in the same area, with 3 sensors at different elevations, the first one directly above the grouted area, the remaining 2 below the grouted zone and therefore between the grouting and the crown of the tunnel. Figure 7a shows Extensometer arrangement. The data from the instrument sensor located immediately above the new tunnel shows that the pre-grouting operation slightly heaved the ground at S1 sensor located above the grouted zone (Figure 7b), while downward displacement was observed at the other 2 sensors located within grouted zone.

The Utility Monitoring Point UMP-21 shown in Figure 8 was attached directly to the crown of the 108 inches force main, and optically surveyed in real time as the TBM was progressing. This point did not show any significant movement neither during the pre-grouting or during the TBM crossing.

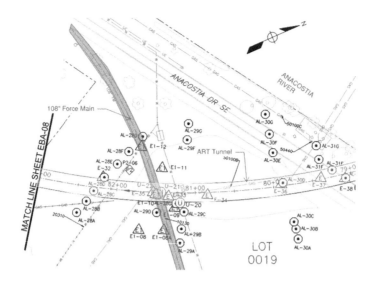

Figure 6. TBM Crossing 108 inches Force Main where (U) = Utility Monitoring Points (E) = Extensometers (o) = Ground Monitoring Points.

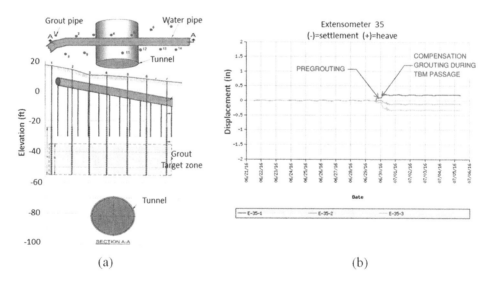

(a)   (b)

Figure 7. (a) Plan view and section of the TBM Crossing 108 inches Force Main with TAM locations. The red area represented the grouted zone (b) extensometers readings.

Overall, a total of 2437 liters of a cement-bentonite grout was pumped during the preconditioning phase, and a total of 6703 liters during the production phase. A total of 9140 liters were injected in the ground to prevent any settlement during the TBM passage.

Even though the compensation grouting or other means of ground improvement are effective methods to prevent or mitigate settlements, it should be recognized that the best way of controlling ground loss and possible impact on existing structures is through the optimization of the TBM excavation process.

Figures 9 to 11 show the TBM data recorded during the crossing of the 108 inches force main. The following main factors contributed to the success of the operation:

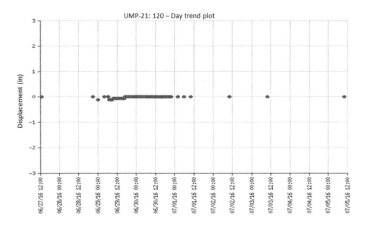

Figure 8. Vertical movement of the 108 inches Force Main at crown.

Table 2. Amount of grout injected.

Grout Pipe location	No. of Ports	Volume (gals)	No. of Ports	Volume (gals)	Total Program Volume (gals)
CMP 01	0	0	15	240	240
CMP 02	3	48	4	64	112
CMP 03	7	112	4	64	176
CMP 04	7	112	16	256	368
CMP 05	1	16	8	128	144
CMP 06	1	16	2	32	48
CMP 07	1	14	9	137	151
CMP 08	0	0	7	112	112
CMP 09	7	112	8	128	240
CMP 10	2	32	6	86	118
CMP 11	1	15	4	64	79
CMP 12	2	32	13	204	236
CMP 13	5	75	8	128	203
CMP 14	4	60	8	128	188
TOTALS	37	644	112	1771	2415

- Performing planned maintenance, add stoppage for precautionary maintenance before entering the zone of influence to the main (conveyor belt extension, cutterhead inspection, other).
- Making sure that all consumable materials are available at the construction site, including emergency stock, to secure continuity of the excavation operation.
- Ensuring that communication is well defined between all key personnel involved in the operation and defining the roles and responsibilities clearly for the crew and reviewing the procedures with each shift and making sure that each one of the key personnel has a backup.
- Review TBM parameters and fine-tune it based on feedback from surveying and instrumentation. This is not always possible as ground condition can change, or surface conditions may not allow. In this case, Impregilo Healy Parsons had a short tunnel length between the Anacostia shore and the 108 inches force main where the information from the optical points arrays was used to tweak TBM parameters such as EPB pressures, conditioning, apparent density and grout pressures.

- Focusing on few key indicators of **TBM** performance. Until Artificial Intelligence will be widely and commonly used with **TBM**, the flow of information and data can be overwhelming to inexperienced engineers and operators. Engineering judgement and experience are the key success factors for **TBM** excavation. Specifically, for this operation key parameters include: apparent density and EPB, grouting volumes and pressures, spoil balance and ground conditioning.
- Have a backup plan and detailed Emergency Response Plan in the event of failure of the 108" force main. These emergency procedures were identified well ahead of time in conjunction with the Owner of the utility DC Water.

The following figures shown the **TBM** recorded data during the crossing of the 108 inches force main. The figures are indicative of how the excavation proceeded uninterrupted and following rigorous control of key TBM parameters such as EPB, spoil weights, backfill grout volumes and pressures and others.

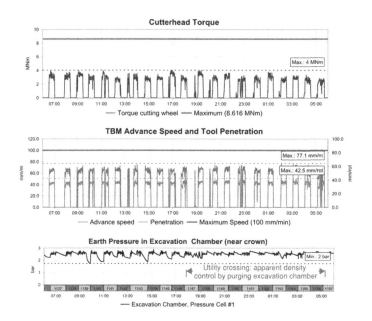

Figure 9.   Cutterhead and main parameters.

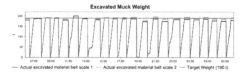

Figure 10.   Control of the amount of extracted material.

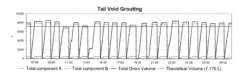

Figure 11.   Control of the tail void grouting volumes.

Under the above circumstances, with the TBM operations regularly performed, the grouting may not be deemed necessary but it was injected as an additional measure of protection to minimize the risk of settlement for fragile and essential 108 inches force main. This was a contractual requirement included in the Contract Specifications.

6 CONCLUSIONS

This paper presented the lesson learned by the construction of Anacostia River Tunnel in Washington DC. The design of the Northeast Boundary tunnel, which is the last portion of the Anacostia River Tunnel System and part of the DC Clean River Project, greatly benefitted from these lessons learned both in terms of segmental lining design and TBM operations.

This was possible as both Anacostia River Tunnel and NEBT are excavated within the Potomac formation despite some differences that were well identified in relation to overburden and ground water load.

The varying geological formation has been managed by applying a range of face pressures based on the methodology developed by Anagnostou & Kovári (1996), allowing filtration of water through the face whenever possible to reduce value of applied face pressure and by the smart application of soil conditioning technique with a wide range of FER-FIR and WIR. This guaranteed the success (based on settlements readings) of the ART excavation, where good productions were achieved with minimum impact on existing structures and utilities and is now applied for the NEBT Excavation.

Moreover, during the TBM tunnel excavation, Salini Impregilo Healy Joint Venture will have the opportunity to further take advantage of the knowledge of the ground and the already trained personnel, to further minimize the risk of potential impact on existing structures. In fact, as the experience of the crossing of the 108 inches force main has once again shown, ground improvement techniques associated with good knowledge of the ground, well-engineered TBM processes and controls are key factors for TBM excavation in urban areas.

REFERENCES

Anagnostou, G. and Kovári, R. 1996. Face stability conditions with Earth-Pressure-Balanced shields. *Tunnelling and Underground Space Technology, Vol.11, No 2, pp.165-173*. Amsterdam: Elsevier
EFNARC, 2005. *Specification and guidelines for the use of specialist products for mechanized tunnelling in soft ground and hard rock*. Farnham: Efnarc
Guglielmetti V., Grasso P., Mathab A., Xu S. 2007. *Mechanized Tunneling in Urban Area. Design methodology and construction control*. London: Taylor and Francis
Schanz T., Vermeer P.A., Bonnier P.G. 1999. *The Hardening Soil Model. Formulation and Verification. Beyond 2000 in Computational Geotechnics – 10 Years of PLAXIS*. Rotterdam: Balkema,.

*Tunnels and Underground Cities: Engineering and Innovation meet Archaeology,
Architecture and Art, Volume 12: Urban
Tunnels - Part 2 – Peila, Viggiani & Celestino (Eds)*
© 2019 Taylor & Francis Group, London, ISBN 978-0-367-46900-9

Theoretical analysis of ground deformation due to tunnelling through weak deposits of phosphatic chalk in area of the proposed Stonehenge tunnel

J. Pettit & C. Paraskevopoulou
University of Leeds, Leeds, UK

ABSTRACT: This study uses numerical analysis to preliminarily model the anticipated ground deformation and surface settlement that could occur during tunnelling though weak deposits of phosphatic chalk using the finite element program RS2. This paper focuses on the factors that lead to increased and severe surface settlement and proposes mitigation measures. According to this analysis, the main factors contributing to extreme settlement increases are: high ground water levels, weak tunnel crown lithology, stress interaction between the twin tunnels and discontinuities within the rock mass. It is suggested that further detailed site investigation work is required to refine the understanding of the ground conditions be-fore excavation can commence. It should be noted and highlighted that the results of this study should not be used for any design purposes as it was a theoretical exercise to highlight the factors influencing ground movements induced by tunneling through the phosphatic chalk.

1 INTRODUCTION

The A303 is a major trunk road that links the M3 motorway with the A30 forming the main artery between London and South West of England. Despite its importance, 35 miles of the A303 remains single carriageway and is over capacity. Potential plans to improve the road are complicated as the section between Amesbury and Berwick St James passes close to the prehistoric monument of Stonehenge, one of the UK's very first UNESCO heritage sites as shown in Figure 1.

In general, excavation of an underground opening causes the redistribution of in-situ stress within the ground which results in small amounts of deformation around the tunnel. At shallow depths, this deformation can propagate to the ground surface and form a depression directly above the tunnel, known as a settlement trough. Surface settlement has the potential to cause damage to buildings or structures above the tunnel.

The initial 2002 ground investigations discovered a number of unexpected ground conditions that create adverse conditions for tunnelling. These conditions consist of deposits of weak, friable phosphatic chalk up to 20m thick that extend across the intended tunnel route and high ground water levels that have a large seasonal variation (Mortimore et al., 2017).

This study is a theoretical assessment of ground risks to tunneling at Stonehenge using finite element analysis, the software program RS2, to model the ground deformation and surface settlement that could be caused during tunnelling through this kind of phosphatic chalk and weak geological environments such as the Stonehenge area. The objective is to investigate and optimise the tunnel design by performing a sensitivity analysis on various combinations of ground water levels and geology along the tunnel alignment. Stress interaction between the two tunnels and the effect that this interaction has on surface settlement is also investigated.

It should be noted that this paper is based on a thesis project submitted as part of an Engineering Geology MSc degree.

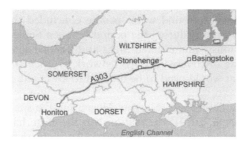

Figure 1. Location of Stonehenge and the A303 (BBC.com).

2 BACKGROUND

2.1 *Geological Setting and Ground Conditions*

The geology beneath the Stonehenge region is chalk from the upper Cretaceous period formed 100–65Ma. Chalk is soft white limestone that is formed of the fossils of sub microscopic calcareous marine algae. During the 2002 ground investigation drilling discovered deposits of phosphatic chalk greater than 20m thick; the thickest ever found in England (Mortimore, 2002).

Phosphatic chalk is white chalk that has been enriched with phosphatic material, probably due to an upwelling of deep nutrient rich water shortly after deposition (Mortimore et al., 2017).

Smaller deposits have been found in other locations in southern England and there are large economic deposits located in France and Belgium. The engineering importance of phosphatic chalk is that it forms very weakly cemented, friable sandy siltstone that has a much lower strength than normal white chalk. When lengths of phosphatic chalk core are lifted they fragment easily to silt, sand and gravel sized pieces. This lack of strength creates a number of issues including the likelihood of tunnel instability and greater amounts of settlement (Mortimore et al., 2017).

Figure 2 shows geological cross sections created with data from ground investigation (GI) boreholes (Mortimore et al., 2017). These three sections are used as geological models for this study as they contain the largest sections of phosphatic chalk and therefore provide the most conservative results (extreme cases).

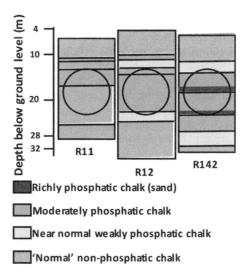

Figure 2. Geological sections used in modeling. (Modified after, Mortimore et al., 2017).

Initial assessment of the site ground water profile concluded that water levels were well below the tunnel invert. However, during the period between September and December 2002 ground water rose to previously unmeasured highs and flooded some of the trial pits (Mortimore, 2002). Pumping tests discovered a huge network of open fractures that are part of a large reservoir of water that would need to be drained to lower site ground water levels (Mortimore et al, 2017).

Due to the size of the underground reservoir dewatering the site by pumping is not financially viable (Mortimore et al, 2017). Excavation will therefore have to be carried out with groundwater at the level at which it is encountered. An increase in ground water level can affect tunnelling performance by decreasing the effective stress experienced by the ground and thus reduce the resistance to shearing (Terzaghi, 1923). This study uses the extreme cases of the highest and lowest ground water level values measured by the initial GI for modeling. High ground water levels and weak face conditions dictate the use of a closed face tunnel boring machine (TBM) for excavation; the small particle size of the phosphatic chalk may require that face support is provided by earth pressure balance rather than slurry as would typically be in case with chalk. A face pressure of 90% σ_v (vertical stress) at the crown of the tunnel is used for modeling, as this is the maximum face pressure that does not create a risk of tunnel blow out due to an over pressurised excavation face (Golder Associates, 2009).

During continuous tunnelling operations, the closed face TBM machine applies an earth (or slurry) pressure to support the face of the excavation. However, when maintenance of the cutter head is required the TBM has to be pulled back from the face to allow access, during this process face support may be provided by increasing the air pressure inside the tunnel. Head access is the activity that carries the greatest risk of ground deformation and surface settlement (Golder Associates, 2009).

2.2 *Inward tunnel wall deformation due to tunnelling*

Tunnel deformation happens in advance of the active tunnel face due to inwards deformation of this face and the changes in stress that result. Deformation is controlled by the ratio of p_o, the in situ stress, to p_i, the internal tunnel support. Prior to excavation $p_i=p_o$ and therefore no ground deformation occurs (Panet, 1993; Paraskevopoulou, 2016 etc.). Deformation starts to occur at a point around one and a half tunnel diameters ahead of the excavation face when deformation of the face starts to reduce the internal pressure provided by the rock mass. Radial displacement of the tunnel reaches about a third of its final value at the tunnel face and does not reach its final displacement until about one and a half diameters behind the excavation face when $p_i=0$ (Hoek, 2008; Paraskevopoulou and Diederichs, 2018). The rate and amount of radial displacement is dependent on the method of tunnel excavation and both can be decreased by supporting the excavation face.

Constructing the second tunnel within the zone of altered stress created by the first can result in large and irregular volume loss (Fang et al., 1994). On the Crossrail project the twin tunnels of drive C310, which was driven through chalk, were separated by a distance of between 2 to 5 tunnel diameters. Monitoring during construction established clear interference between the tunnels for sections where the separation was less than 3 tunnel diameters. Areas where separation was between 3 and 4.5 diameters showed occasional interference and areas where tunnel separation is greater than 4.5 diameters show no measurable interference (Cheng and Mikulski, 2014). This study analyses distances between the tunnels of 1, 3, 4 and 5 diameters to test if stress interaction occurs at the same distances as during drive C310.

2.3 *Tunnel alignment and depth*

This paper is intended to demonstrate that altering the depth of a tunnel can help to mitigate a number of ground risks. However, the final alignment and depth of a tunnel is often controlled by other factors. The residual risks (including ground risk) then have to be addressed during the remainder of the design and construction.

3 METHODOLOGY

Creating the numerical model requires knowledge of the ground conditions through which the tunnel will be excavated. This includes; the lithology, the groundwater level and the discontinuities in the rockmass. The required data for this preliminary study comes from on-site ground investigation, literature data and fieldwork at Seaford head, the type locality of Seaford chalk, the main lithology beneath Stonehenge. Seaford and Stonehenge have similar tectonic settings as they are both located in the Sussex-Wessex trough (Mortimore, 2009).

This study uses geological and ground water data taken from the 2002 ground investigation (Mortimore et al., 2017) and discontinuity data comes from observations at Seaford head. Three orthogonal joint sets were observed in the Seaford chalk; two vertical and one horizontal. As RS2 is a 2D program only one vertical joint set is included in the models. Joints are idealized as weathered from 0 to 10m due to observations made during the ground investigation, (Mortimore 2017, pers. comm., 19 June) and clean from 10 to 50m.

Below 50m, the chalk is idealised as unfractured as this is generally considered to be the case (Lord et al, 2002). Jointing at Seaford head is too closely spaced to allow the use of a Schmidt hammer therefore the joint compressive strength at depths of 0 to 10m is estimated to be 0.5xUCS of the rockmass based on the weathering of joint surfaces relative to the rock mass (Barton, 1973).

To estimate the coefficient of earth pressure, k, for the purposes of modeling (Sheorey, 1994) the following Equation (1) is used:

$$k = 0.25 + 7Eh\,(0.001 + 1/z) \tag{1}$$

where: z (m) is the depth below surface and E_h (GPa) is the average deformation modulus of the upper part of the earth's crust measured in a horizontal direction.

This equation gives a value of 0.8 which is the same as the value that was measured using hydraulic fracture tests at depths up to 90m during the construction of the North Downs tunnel on the channel tunnel rail link (Watson et al, 1999).

Tables 1 and 2 show the geotechnical parameters that are used in the numerical analysis.

Table 1. Discontinuity parameters used in numerical analysis.

Type of Chalk	JRC	JCS (MPa) 0–10m	Kn (GPa)	Ks (MPa)
Normal	11[2]	4[1]	10[2]	100[2]
Weakly Phos.	11	4	10	100
Mod. Phos.	11	4	10	100
Richly Phos.	11	4	10	100

1. (Barton, 1973). 2. (Barton and Steffanson, 1990).

Table 2. Intact rock parameters used in numerical analysis.

Type of Chalk	UCS (MPa)	c' (MPa)	θ (°)	E (GPa)	T (MPa)	ρ (Mgm⁻³)
Normal	8[1]	2.20[1]	32[1]	2[2]	0.8[1]	0.2[1]
Weakly Phos	4	1.10	32	2	0.4	0.2
Mod. Phos	3	0.83	32	2	0.3	0.2
Richly Phos.	2	0.55	32	2	0.2	0.2

1. (R Mortimore 2017, personal communication, 7 July). 2. (Lord et al, 2002. Note: v = 0.24)

The twin tunnel dimensions are taken from the Highways England design codes. Each of the tunnels has an internal diameter 12m with reinforced segmental lining. This gives an overall excavated diameter of 13m (Highways England, 2017).

3.1 Sensitivity analysis

The variables that are investigated in the parametric analysis are: the tunnel depth, the ground water level and the distance between the twin tunnels. The distance between the tunnel and Stonehenge is not taken into consideration and it is out of the scope of the work presented herein.

At the preliminary stages of the analysis it was noted the failure initiates in the tunnel crown. Examination of the geological cross sections shows that although the tunnels are located within phosphatic chalk their crowns are located in the most competent layer of the lithology that surrounds them. This is an ideal tunnel design but due to the uncertainties and heterogeneity in the ground model it is not realistic to expect that a design could achieve this throughout the tunnel's length. A second depth is also used for testing that places the tunnel profile at a less favourable point in the cross section to provide a range of settlement values that could be expected.

3.2 Staged modelling

In order to represent the continuous process of tunnel deformation using a discontinuous medium, RS2, it is necessary to create frozen snap shots of time using a staged model. Staged models allow the user to change the conditions that the model experiences from one stage to another to show how deformation evolves. As aforementioned tunnel deformation is the result of the decrease of internal tunnel support, pi, as the excavation face progresses. Deformation is a con-tinuous process that starts as soon as pi<po and con-tinues until pi=0. In the model pi is represented us-ing a distributed load inside the tunnel. The force ex-erted by this load is programed as a percentage of the principle stress exerted on the tunnel. This is incrementally reduced from one stage to another to repre-sent the decrease in pi. In the case of the lined mod-els some deformation is allowed before the lining is erected as this occurs during tunnel excavation (Hoek, 2008).

3.3 Lined and unlined models

Models are considered in both a lined and unlined state. The lined models are to reflect tun-nelling in the normal continuous process whereas the unlined models reflect a worse case scen-ario situation where the TBM has been withdrawn to allow head access and face support has not been adequately maintained.

Table 3 shows the different combinations of varia-bles, these give a total of 96 different models. Due to the number models, codes are used for sake of sim-plicity. The first part relates to the geological sec-tion, the next to the tunnel depth, either shallow or deep, then GWL, low or high, then tunnel spacing and lastly whether or not the tunnel is lined. For ex-ample: R142-D-L-L.

This study measures ground settlement using vol-ume loss, VL, the parameter typically used to assess settlement from tunnelling. It is defined as the vol-ume of ground lost as a

Table 3. Summary of the models performed in numerical analysis.

Tunnel Section	Tunnel Depth (m)		GWL (m)		Tunnel Spacing in Diameters			
R11	13	15	15	27	1	3	4	5
R12	13	18.5	13	23	1	3	4	5
R142	13	24	17	30	1	3	4	5

proportion of the final tunnel volume (Dimmock and Mair, 2007). To calculate VL ground loss was taken as the area of the calculated settlement trough expressed as a percentage of the ex-cavated tunnel cross-section. Based on recent experi-ences in central London, volume losses of less than 1% can typically be achieved in competent chalk us-ing modern TBMs, but are often over 2% for open face methods (Thames Tideway Tunnel, 2015).

4 NUMERICAL ANALYSIS

The results from section R11 show that surface settlement can be mitigated to below 0.5% volume loss along the majority of the tunnel's length by good tunnelling practice once more knowledge has been gained on the discontinuities within the rock mass. However, where ground conditions are adverse this is not the case. Figure 3 shows the average values of surface settlement of unlined models across all four separation distances for each of the three geo-logical sections. In the case of lined models face pressure reduced the amount of ground move-ment significantly but enough deformation still occurred as face pressures dropped from 100–90% of σ_v *crown* for surface settlement to occur. Results are grouped by ground conditions.

The single factor that has the greatest impact on surface settlement is the strength of the lithology. Section R11, which has the least phosphatic chalk, shows very small amounts of settlement, giving a volume loss of less than 0.05%, even when modeling uses the least favour-able combination of ground conditions. In contrast section R142, which has large amounts of phosphatic material, has a volume loss of 0.3% when modelled with ground conditions at their most favourable.

Modelling only results in large amounts of surface settlement when the phosphatic material is located in the crown of the tunnel. Figure 3 demonstrates that a tunnel design that positions the phosphatic chalk in the sides of the tunnel rather than the roof, +ve Geo, mitigates the excess settlement. A tunnel design that positions the phosphatic material in the crown of the tunnel, -ve geo, potentially results in large amounts surface settlement even when combined with low GWL.

Although the ground water level has a significant impact on settlement it is less than the effect of the lithology. As aforementioned an increase in groundwater level decreases the ground's resistance to shearing bringing it closer to failure. However, if the tunnelling lith-ology is strong and the joint sets positioned favourably relative to the crown of the tunnel then this decrease in strength will not be enough to cause large amounts of ground deform-ation and increased settlement. In combination weak geology and high groundwater have a

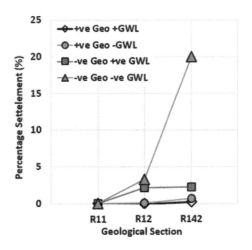

Figure 3.   Average settlement values for each unlined geological section showing the effect of ground conditions.

compounded effect; the sum of the increase in settlement is greater than the two halves, and in some models the result is complete tunnel collapse.

The increase in surface settlement caused by tunnel interaction is in some cases very large. A value is not given because a small number of extremely high values skew the overall trend. The models that show the greatest increase due to tunnel interaction were those that were strong enough to result in very little surface settlement after the first excavation but weak enough for this excavation to have brought the joints above the excavation very close to failure.

This study considers 24 different combinations of geology, tunnel depth and ground water level. Of these 24, interaction between the twin tunnels increases settlement when the tunnel separation is 1D on one occasion, up to 3D on seven occasions, up to 4D on three occasions and up to 5D on seven occasions. These distances are greater than those measured on the Crossrail project. The latter is not surprising as the ground conditions at Stonehenge are much less favourable for tunnelling.

As can be seen in Figure 4 modelled settlement troughs do not follow Gaussian distribution as is the case with tunnelling in soft ground (O'Reilly and New, 1982) and was observed on Crossrail drive C310. Modeled troughs are narrower, deeper, and somewhat irregular in shape

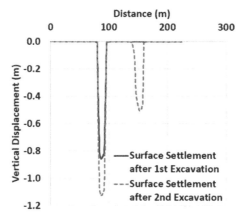

Figure 4.   Modelled settlement for R142DH4L after the excavation of each tunnel.

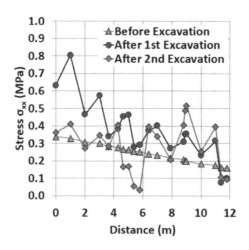

Figure 5.   Horizontal effective stress changes along the vertical joint in the roof of the 1st tunnel of model R142DH1UL due to tunnel excavation.

with smaller cross sectional areas than a trough of similar depth in soft ground. The shape of the settlement troughs are controlled by discontinuity spacing rather than particle interaction, as is the case with soft ground.

Another difference between these results and C310 is the location of the increased settlement caused by the second tunnel. In C310 increased settlement occurred above the second tunnel, however, Figure 4 demonstrates that this is not the case in this study where the increase happened above the first.

A different mechanism is in play in each instance due to the ground conditions. At Stonehenge the ground consists of jointed chalk. When the first tunnel is excavated large stress increases occur in the joints directly above the excavation. Excavation of the second tunnel allows the ground to relax sideways slightly reducing the shear strengths of the highly stressed joints and causing failure. During C310 soft ground above the chalk redistributed the stress more evenly through the ground. Joints surrounding the first tunnel did not fail and small stress changes surrounding the second tunnel caused a slight increase in settlement.

Figure 5 shows the release of horizontal stress along the main joint above the first excavation caused by the excavation of the second. This release of stress caused a large drop the shear strength mobilized along the joint.

6 CONCLUSIONS

Modelling shows that a combination of adverse ground conditions and stress interaction between the twin tunnels has the potential to result large amounts of surface settlement during tunnel excavation at this site. Adverse ground conditions at Stonehenge include; high ground water level, weak tunnel crown lithology and discontinuities within the rock mass. The effect on settlement of a combination of two or more of these factors is much greater than the sum of the individual parts.

It is the recommendation of this study that the following ground investigation work should be constructed to more accurately define the ground model: 1) rock face mapping via an exposed quarry to define the strength, aperture, location and orientation of discontinuities within the rock mass, 2) further investigation of drilling broeholes and mapping work to closely define the size and exact location of the phosphatic deposits.

Once the ground model has been more precisely defined it is also recommended that a tunnel design is used that places strong non-phosphatic lithology at the tunnels' crown and avoids the presence of discontinuities at the apex of the crown as far as is possible. It is also advised that the twin tunnels are excavated at a separation greater than five tunnel diameter apart as modelling shows that excavating twin tunnels spaced by a smaller distance carries a risk of increased surface settlement, interaction of tunnel stress at this site still requires further investigation to discover the absolute limit of interaction. This might show that the tunnels need to be located even further apart. Groundwater levels at the site cannot be decreased by dewatering, this could be mitigated in part by timing the excavation work so tunnelling through phosphatic chalk occurs during seasonal ground water lows.

ACKNOWLEDGEMENTS

The authors would like to acknowledge the help and information provided by Professor Rory Mortimore.

REFERENCES

Barton, N. 1973. Review of A New Shear-Strength Criterion For Rock Joints. *Engineering Geology*, Pp.287–332.

Cheng, M. And Mikulski, P. 2014. Ground Settlement Behaviour In *Chalk Due To TBM Excavations*. ICE Publishing.

Dimmock, P. And Mair, R. 2007. Volume Loss Experienced On Open-Face London Clay Tunnels. *Proceedings Of The Institution Of Civil Engineers - Geotechnical Engineering*, 160(1),Pp.3–11.

Fang, Y. S., Lin, L. S. And Su, C. S. 1994. An Estimation Of Ground Settlement Due To Shield Tunnelling By The Peck-Fujita Method. *Canadian Geotechnical Journal*, Vol. 31, No. 3, Pp. 431–443.

Highways England 2017. *A303 Stonehenge Amesbury to Berwick Down Technical Appraisal Report.* http://www.gov.uk/highways

Hoek, E., Carranza-Torres, C. and Diederichs, M. 2008. *The 2008 Kersten Lecture Integration Of Geotechnical And Structural Design In Tunnelling.*

Lord, J.A., Twine, D. & Yeoh, H. 1994. Foundations In Chalk. *Funders Report/Cp/13*. Ciria Project Report

Mortimore, R.N., 2002. *A303 Amesbury-Berwick Down Stonehenge (Incorporating 1320 The Winterbourne St Oke Bypass) Geological Report.* January 2002.

Mortimore, R. (2009). *An atlas of Stratigraphy Structure and Engineering Geology.* The London Basin forum.

Mortimore, R., Gallagher, L., Gelder, J., Moore, I., Brooks, R. And Farrant, A. 2017. Stonehenge— A Unique Late Cretaceous Phosphatic Chalk Geology: Implications For Sea-Level, Climate And Tectonics And Impact On Engineering And Archaeology. *Proceedings of The Geologists' Association.*

O'reilly, M. And New, B. 1982. 831153 Settlement Above Tunnels In The United Kingdom — Their Magnitude And Prediction. *International Journal Of Rock Mechanics And Mining Sciences & Geomechanics Abstracts*

Panet, M. 1993. Understanding deformations in tunnels. In: Hudson JA, Brown ET, Fairhurst C, Hoek E (eds) *Proc. of the Comprehensive Rock Engineering*, Vol. 1. Pergamon, London, 663–690.

Paraskevopoulou, C. 2016. *Time-Dependency Of Rock And Implications Associated With Tunnelling*, Phd Thesis. In: Queen's University Publications. Canada.

Paraskevopoulou, C., Diederichs, M., 2018. Analysis of time-dependent deformation in tunnels using the Convergence-Confinement Method. *Tunnelling and Underground Space Technology*, 17, 62–80.

Sheory, P.R. 1994. A Theory For In Situ Stresses In Isotropic And Transversely Isotropic Rock. *Int. J. Rock Mech. Min. Sci. & Geomech.*

Terzaghi, K. (1923) *Die berechnung der durchlassigkeitzifer des tones aus dem verlauf der hydrodynamischen spannungserscheinungen, Mathematish-naturwissenschaftliche*, Klasse. Akademie der Wissenschaften, Vienna, 125–138.

Thames Tideway Tunnel (2015). *Employer's design specification – Impact of ground movement.* p. Section 2: Works Information.

Watson, P.C., Warren, C.D., Eddie, C. & Jaegar, J. 1999. CTRL North Downs Tunnel. Tunnel Construction & Piling 99. Institution Of Mining & Metallurgy, *British Tunnelling Society & Federation British Piling Specialists*, 301–323.

Tunnels and Underground Cities: Engineering and Innovation meet Archaeology,
Architecture and Art, Volume 12: Urban
Tunnels - Part 2 – Peila, Viggiani & Celestino (Eds)
© 2019 Taylor & Francis Group, London, ISBN 978-0-367-46900-9

Excavation of connecting tunnels in Centrum Nauki Kopernik underground train station on line II in Warsaw

M.A. Piangatelli, A. Bellone & M. Bringiotti
CIPA S.P.A., Rome, Italy

ABSTRACT: The construction of the three connecting tunnels between the east side and the west side of the Centrum Nauki Kopernik underground train station, adjacent to the Vistula river, has a long and complex history, as they cross beneath one of Warsaw's main road arteries (Wislostrada). After the attempt to excavate the first tunnel, the face of the tunnel collapsed on the night of August 13, 2012, and a volume of about 10,000 cubic meters of liquid mass (water, silty soil and sand) entered and flooded the west side of the station, which was filled with more than seven meters of material. Fortunately, the workers managed to escape, but machines and equipment were lost. This paper deals with how the tunnels were completed, by adopting soil improvement, jet grouting, GRP bars, ground freezing and a careful choice of work steps.

1 INTRODUCTION

The Centrum Nauki Kopernik station on Line II of the Warsaw Underground is located alongside the Vistula river (Figure 1). This station hereinafter is referred to as station C13.

The station consists of three parts: the east and west part, and three connecting tunnels 40 meters long with a large cross section ($170 m^2$). It includes the rails along the sides and the platform along the centre. The reason for the underground connection of two shafts with a tunnel is the presence of a road tunnel (Wisłostrada, hereinafter called WS), which must be underpassed by the new metro line (Figure 2).

The existing tunnel (Wisłostrada) was built many years before on diaphragm walls and barrettes, so some foundation structures interfere with the underground tunnel. The challenge was to build the connectors without stopping the traffic of vehicles along the highway above. All works took place in very difficult geological and groundwater conditions due to the presence of Vistula alluvial deposits with poor mechanical characteristics. The soil profile consists of relatively loose sands on top, and silts overlying high plasticity Pliocene clays highly disturbed by the effects of the repeated over-sliding of glacial moraine deposits.

The station is located entirely on the west side bank of the Vistula River: the eastern boundary of the station shaft is less than 10 m from the river bank. The station has been constructed with top-down methodology, with diaphragm walls (1.40 m thick) supported by the slabs that have been constructed during the lowering excavation. After the construction of each slab supported by the soil in the bottom, the excavation has been carried out under the slab until the level of the next slab to be constructed. The maximum excavation depth reaches a level of 24 metres. The two side tunnels have an internal diameter of about 7 meters and their shape is perfectly circular to allow, after their completion, the TBM's passage.

The three tunnels were to be excavated in clayey soil using conventional methods. Instead, the soil turned out to be of sandy materials below groundwater. In essence, it is as if the tunnels were excavated in the river's sub bed, in the presence of flowing water. A further major

Figure 1. The Centrum Nauki Kopernik station and the Vistula river.

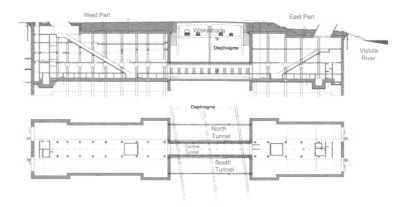

Figure 2. The two parts of the station and the connecting tunnels.

complication was due to the presence of the of WS tunnel foundations made by diaphragm walls, which would have been dismantled during tunnels excavation and in the presence of normal vehicle traffic. This paper focuses on the several excavation attempts that were implemented to complete the connector.

2 EXCAVATION ATTEMPTS

Several attempts to construct the three tunnels were carried out. The first attempt followed the original design that was based on a geological profile that turned out to be inaccurate. Only clay was expected along the tunnel path, but this was not the case.

2.1 The First Attempt –Steel pipe umbrella (Canopy Piles)

The original construction scheme entailed the excavation and lining of the two side tunnels (for the tracks), followed by the excavation and lining of the central tunnel (for the platform), connected to the two side tunnels by arches.

The excavation of the two side tunnels would have to be supported by steel pipe umbrella except for the sides bordering the central tunnels and Fiber Reinforced Plastic tubes (hereinafter called FRP tubes) at the face. On the sides of the two tunnels bordering the central tunnel, the design required FRP tubes instead of steel ones, to allow the subsequent excavation of the central tunnel, whose final lining was to have been supported by the final lining of the two side tunnels (Figure 3).

After the first core drilling on the diaphragm walls of the entrance of the north tunnel on the west side, on February 2, 2012, a massive flow of sand and water occurred from the hole into the shaft (Figure 4).

There was no clay soil behind the wall, as expected, but something else. The need to proceed with geological surveys along the tunnel path became clear. Beforehand the surveys could not be executed in order not to close the high flow road WS. The hole was plugged and a new investigation campaign begun.

2.2 The Geological Survey of the Path

Once the east and west parts of the station were excavated, it was possible to perform a geological survey of the tunnel path, with horizontal and sub-horizontal borings, starting from the station diaphragm walls and without having to close the busy WS tunnel above.

As a result of this survey, it appeared that the level of the clayey soil was lower than expected and that the top of the tunnel sections were in sandy material below ground water.

The decision was made to perform a soil improvement, but since the interventions could not be implemented from the WS tunnel in order not to close it, a new design including columns of jet grouting on top of the tunnel was drawn up (Figure 5).

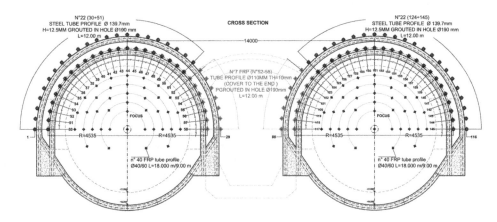

Figure 3. Original design of the two side tunnels to be excavated.

Figure 4. Water inflow.

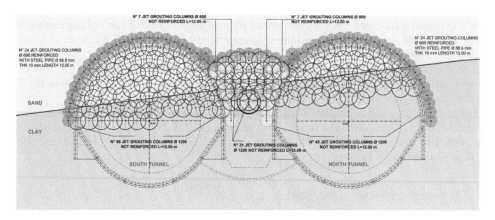

Figure 5. Updated design with columns of jet grouting.

2.3 The Second Attempt – Jet Grouting

The new work schedule planned the north tunnel excavation first, followed by the south tunnel. The central tunnel would have been the last one. Excavation would have been done from both the east and west heading of each tunnel, up to the central diaphragm walls supporting the WS tunnel.

Jet grouting and the steel pipe umbrella started on the diaphragm walls of the north tunnel's west side with a big machine, then the big machine moved to the east side of the station, to the other face of the north tunnel, and a smaller one arrived in the west side. Afterwards, the east face of the other two tunnels was dealt with. The remaining face of the south tunnel in the east side would have been the last.

During the soil improvements carried out at the east side, the excavation of the north tunnel started from the west side, following the phases shown in Figure 6.

After the demolition of the station's diaphragm walls, the first excavation step was finished. The diaphragm walls supporting the WS tunnel were also demolished and the primary lining was applied, consisting of steel ribs, wire mesh, struts on the invert and shotcrete.

The second excavation step began and continued without problems. Every 75 cm, a layer of shotcrete was applied to the face of the tunnel during the primary lining of the previously excavated portion of tunnel. On the evening of August 13, 2012, the excavation arrived at the fifth steel rib of the second excavation step (Figure 6). The workers applied the shotcrete to the face of the tunnels in order to stabilize them, and went to dinner at 10:30 PM.

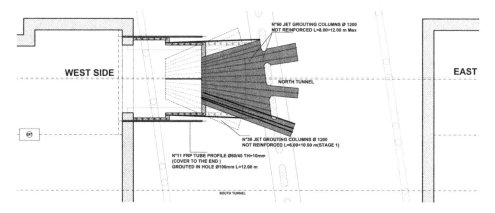

Figure 6. Plan view of an excavation phase of North tunnel from west side.

2.4 The Collapse of the Tunnel Face

At 10:45 PM, a foreman saw that something strange was happening. Some water was inflowing from the top and from the face of the tunnel. He immediately sounded the alarm, and a team of workers ran down into the station, at the tunnel face. At 11:00 PM they began to spray another layer of shotcrete on the face of the tunnel, but because the sprayed shotcrete was swelling and cracking in some areas on the face, the action taken was interrupted and all the workers attempted to place material against the face to contain it (Figure 7).

At the same time, because the face was continuing to move despite the containment action, the workers also managed to save equipment and machines using the gantry crane, but were unable to finish this operation because the immediate evacuation of the station was ordered.

To avoid the feared collapse of the west side of the station, the flooding in progress was intentionally accelerated by pumping water from the Vistula river in order to balance the external pressure of the soil on the walls (Figure 9). Access to the WS tunnel was forbidden for safety reasons.

Many machines were lost in the station (Figure 10). Fortunately, no one was injured.

2.5 The Concrete Filling Under the WS Tunnel

After non-destructive fact-finding surveys, it was determined that there was a deposit of solid material, mostly sand, of approx. 6,500 cubic meters inside the station, and a loss of the equivalent amount of material beneath the road slab of the Wisłostrada tunnel. The cavity underneath the WS tunnel was filled with water up to approximately 2 meters from the bottom of the slab. The road slab thus was not supported by natural subsoil, and was suspended in risky equilibrium between the lateral diaphragm walls.

Figure 7.   The collapse of the tunnel face.

Figure 8.   Water, sand and clay inflow from the tunnel.

Figure 9.   The station before the event and after the intentional flooding.

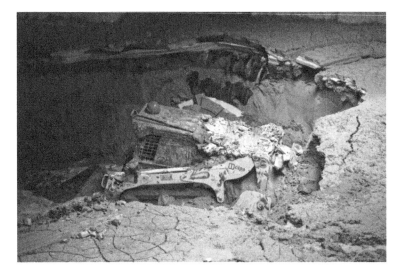

Figure 10.   One of the buried machines.

In order to fill the cavity with concrete, there was an initial campaign with a massive injection of concrete, self-levelling, with mixes designed to cure underwater. The second campaign for deep consolidation was obtained with a low-pressure injection of cement mixes followed by chemical mixes, using manchette tubes.

The second campaign stopped and was not completed because the grouting lifted the road slab up to 20 cm in some points.

2.6   *The Third and Final Attempt – Soil Freezing*

Soil freezing was the final solution adopted to improve the soil (see Capata et al. 2015, Balossi Restelli et al. 2016).

Divers inspected the bottom of the station and found a large pile of sand which obstructed the tunnel entrance, so the station was carefully emptied of inflow material. The pile of sand was carefully removed, layer by layer, and a reinforced concrete wall was built by underpinning methodology. Horizontal and vertical freezing probes were installed. The excavation phases were quite complex.

The Figure 11 shows some design phases for the north tunnel (similar to those for the south and the central tunnels).

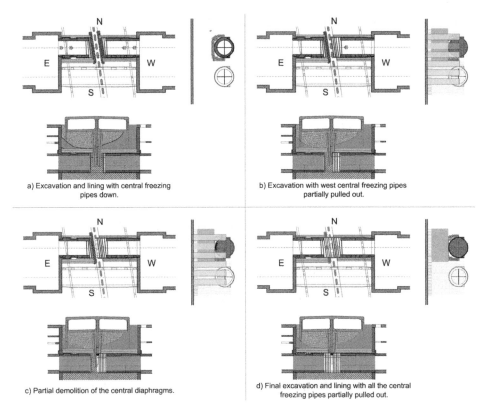

a) Excavation and lining with central freezing pipes down.

b) Excavation with west central freezing pipes partially pulled out.

c) Partial demolition of the central diaphragms.

d) Final excavation and lining with all the central freezing pipes partially pulled out.

Figure 11.   Excavation and lining phases.

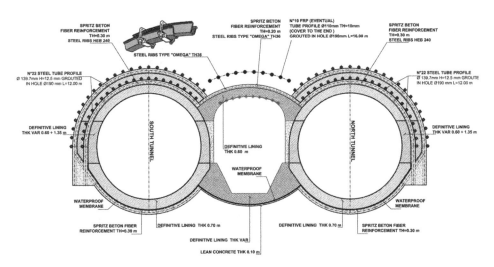

Figure 12.   Central tunnel with adjustable Omega Type steel ribs.

The horizontal probes were put in place all around the tunnel section, and vertical probes were placed at the sides of the central diaphragm walls supporting the WS tunnel. The excavations were carried out from the two headings and stopped near the diaphragm walls. The two sections were lined and after the vertical probes were partially pulled up, the central section was excavated and lined.

The two side tunnels were completed and then the central tunnel was excavated adopting, in the primary lining, steel ribs bearing on the steel ribs of the side tunnels, uncovered during the excavation. The central ribs were welded on the side tunnels' ribs and, at the end of the excavation cycle, the side tunnels' piers were cut. Omega-type steel ribs (Figure 12) were chosen to compensate for the variability of the several lengths of arc.

3 CONCLUSIONS

The first attempt failed due to the presence of sand and water along the excavation path, the second attempt probably failed because the water was in motion and washed away the injected grout. In both cases, the failures are attributable to the water that negatively affected the consistency of the soil and the consolidation attempt.

The third attempt, the freezing of the soil, worked well because the same water, whose presence was harmful in the two previous attempts, was successfully used as a support system in an ice form.

4 EPILOGUE

As shown in Figure 13, it took 33 months to complete the excavation and structural works of the station:

- 9 months for flooding,
- 5 months for freezing,
- 18 months of effective works

Six of the 18 months were used for the tunnel during the 3rd attempt (Figure 14) and the work was completed 15 days in advance of the last milestone.

It took more than two and a half years to complete the tunnels, but, thanks to the efforts of the people who worked to solve the serious issues of this work in order to safeguard people, structures and equipment as far as possible, the tunnels were finished. Figure 15 and Figure 16 show some pictures of the tunnels during the works and how it appears today.

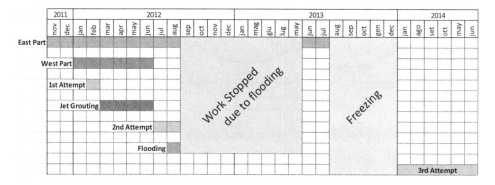

Figure 13.   Chronology of events.

Nome attività	Inizio	Fine	gennaio 2014	febbraio 2014	marzo 2014	aprile 2014	maggio 2014	giugno 2014
North West Excavation & lining	mer 08/01/14	gio 13/02/14						
North East Excavation & lining	mer 29/01/14	mar 08/04/14						
South West Excavation & lining	gio 16/01/14	mar 25/03/14						
South East Excavation & lining	mar 18/02/14	mer 07/05/14						
Central West Excavation & lining	mer 26/03/14	mer 04/06/14						
Central East Excavation & lining	sab 10/05/14	ven 13/06/14						
North D-walls crossing and lining	lun 17/03/14	lun 14/04/14						
South D-walls crossing and lining	gio 27/03/14	sab 10/05/14						
Central D-Walls crossing and lining	ven 23/05/14	mar 10/06/14						
Central Invert	ven 13/06/14	gio 26/06/14						

Figure 14. The chronological of the 3rd attempt in detail.

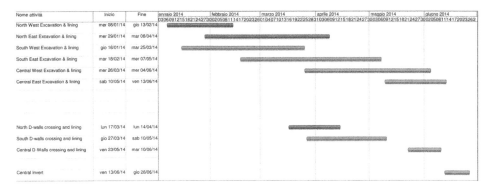

Figure 15. At top, a side tunnel; at centre, the central tunnel; at bottom, a panoramic photo of the three tunnels.

Figure 16. At left, a side tunnel; at centre, at right, the central tunnel.

REFERENCES

Capata, V., Lombardi, A., & Bizzi, F. 2015. Line II of Warsaw Underground: the excavation of tunnel connector (Centrum Nauki Kopernik station). *Infrastructure Conference Proceedings – ICE Publishing*

Balossi Restelli, A., Rovetto, E. & Pettinaroli, A. 2016. Combined Ground Freezing Application for the Excavation of Connection Tunnels for Centrum Nauki Kopernik Station—Warsaw Underground Line II. *ITA-AITES World Tunnel Congress 2016 (WTC 2016) - Proceedings of a meeting held 22–28 April 2016*, San Francisco, California, USA

*Tunnels and Underground Cities: Engineering and Innovation meet Archaeology,
Architecture and Art, Volume 12: Urban
Tunnels - Part 2 – Peila, Viggiani & Celestino (Eds)*
© 2019 Taylor & Francis Group, London, ISBN 978-0-367-46900-9

Case study of a shallow tunnel construction underneath an operating metro and railway in fine grained soil environment

S.M. Pourhashemi & M. Sadeghi
Municipality of Tehran, Engineering & Development Organization, Tehran, Iran

H. Maghami
CVR Consulting Engineers, Tehran, Iran

ABSTRACT: Constructing a 105m length traffic tunnel with a depth of 3m underneath a two-way railway and a two-way metro line is discussed in this paper. With due attention to fine grained soil terms 'ML' in tunnel excavation environment and low overburden, settlement and face stability control of excavation in return of dynamic rail loads were this project main challenges. For excavation stabilization, some integration methods for pre-consolidation according to sensitiveness and project restrictions were used, which are: Excavation of five parallel micro tunnels in the top of the tunnel section and convert them to reinforced concrete longitudinal beams, and implementation of umbrella grouting system on top of the tunnel section. Finally, this project was inaugurated with a maximum ground surface vertical deformation of 31mm without any disturbance on the trend of railway and metro utilization. Results of numerical modeling and analysis were compared and discussed with monitoring results.

1 INTRODUCTION

Considering many existing constraints, construction of civil engineering projects in metropolises always has its own sensibilities and difficulties. These include the implementation of projects related to reducing traffic load of intersections in urban communication arteries that, there are always several options to study in this regard that, depending on the intersection analysis conditions, these are divided into two general categories of intersections and grade separations.

Ultimately, the summing up of technical, executive, economic, and environmental studies which determines preferred choice to the implementation of overpass and underpass options at a grade separation. This issue is more important in a metropolitan area like Tehran, which also has a dense and old texture area in its southern parts.

Figure 1. Location of the Project.

2   ZARBALIZADEH UNDERGROUND PROJECT

The purpose of the project is to construct Zarbalizadeh underground in the 16th District of Tehran is the East-West connection between two urban areas and reducing traffic and travel time in this dense and populated demographic region. Project has been constructed by going through the subway line of Tehran Metro Line 1 and the North-South Railway. This project is located between the South Passenger Terminal and two Tehran's subway stations. Figure 1 shows the project location.

Zarbalizadeh underpass project is an urban tunnel located at south of Tehran with an average over-burden of about 3.5m, excavation width of 14m and height of 11.5m. This tunnel with a total length of 105m consists of a multi-arc section and was excavated by the NATM method. The initial liner has been designed with the shotcrete and lattice girder system and the final liner has been designed with reinforced concrete with a thickness of 50cm. Figure 2 shows the plan and cross sections of the route.

In order to obtain the required design data, three machine boreholes and one test pit on the track have been excavated. Laboratory tests including triaxial and uniaxial tests, direct shear test, consolidation test, grading and Atterberg limits and chemical tests on samples taken from test pit and boreholes, were performed. In situ tests have also been performed on plate loading and SPT tests in boreholes and pit. Figures 3 and 4 show the longitudinal geological profile and the proposed geotechnical parameters, respectively. Types of soil of the project area range from ML to CL-ML according to the USCS classification.

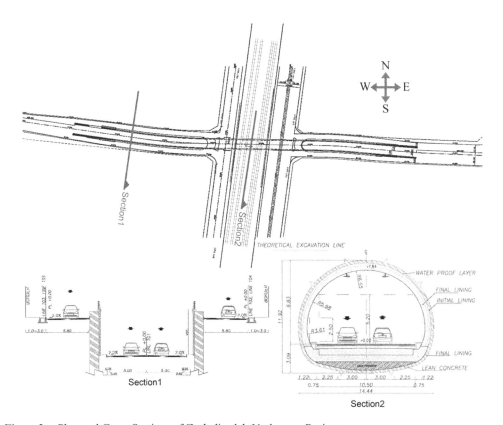

Figure 2.   Plan and Cross-Sections of Zarbalizadeh Underpass Project.

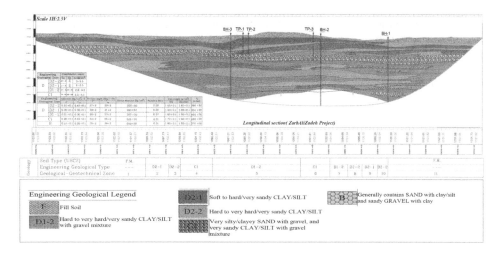

Figure 3. Longitudinal Geological Profile in the Project Axis.

Engineering Geological Type		Cohesion (kg/cm²)	Friction angle (deg.)	Elastic Modulus (kg/cm²)	Poisson's Ratio	Unit weight (g/cm3)		
						Dry	Sat	Nat
D	D2-2	0.32±0.1	27±2	260±50	0.30	1.65±0.1	1.85±0.1	1.8±0.1
	D2-1	0.30±0.1	28±2	280±50	0.30	1.60±0.1	1.85±0.1	1.8±0.1
	D1-2	0.31±0.1	30±2	300±50	0.32	1.65±0.1	1.85±0.1	1.8±0.1
C1		0.20±0.1	34±2	520±50	0.31	1.70±0.1	1.90±0.1	1.85±0.1
B		0.12±0.1	35±2	600±50	0.31	1.80±0.1	1.95±0.1	1.9±0.1

Figure 4. Soil Geotechnical Parameters of the Project Area.

2.1 Study Limitations of the Project

There were some limitations for this project including:

- Maintaining the transit traffic of national railway and metro lines during operation
- Observing geometric design standards
- Obtaining the consent of the railway and metro authorities to cross the railway area and observing the relevant restrictions
- Maintaining the safety and security of the railway and metro route
- Choosing reliable and efficient construction method
- Construction restrictions
- Urban view and landscape

2.2 Project Construction Options

Considering the project constraints, various executive variants were predicted and studied:

- Bridge and overpass option
- Stage deviation option for railway lines and open cuts
- Pipe ramming and underground construction
- Using dense forepoling pre-consolidation system with micro pile network
- Using a hybrid pre-consolidation system including the implementation of leading beams (Micro tunnel) and fore-poling

One of the important criteria considered during the design and construction of urban shallow tunnels, especially in inappropriate areas is controlling ground displacement and keeping it at an allowed level. To doing this, usually in underground ways, the inevitable option is the use of pre-consolidated systems. In this project, considering the importance of operation of rail structures which is passing about 3.5m above the tunnel, keeping continuous and safe operation of rail structures was a must. Furthermore, the stability of the excavation section, reducing surface settlement as much as possible and controlling the deformation of the railway facilities at the permitted level were other regulations for this project, in order to reduce the risk of any disruption in the Tehran Metropolitan Railways.

Among the various methods used to construct the Zarbalizadeh tunnel with respect to parameters such as:

- Low overburden (average 3.5m)
- Low-level geo-mechanical factors of the soil such as fine-grained soil and the presence of about 1m of filling soil and some fiber optic and electrical installations on top of the excavation area
- Presence of high dynamic loading of subway and rail trains and
- Congested and old texture of peripheral buildings and facilities, and after reviewing different options
- Finally, combination of the implementation of the micro tunnel with fore-poling as a pre-consolidation method and then the SEM (Sequential Excavation Method) with the installation of the lining was chosen for final and accurate analysis.

3 STUDIES AND PROJECT DESIGN

Generally, the deformations along the tunnel and during the excavation operations follow a certain trend and curves. According to studies, in general terms, at least 30% of the section deformations occur behind the excavation front (Eberhardt 2001). Therefore, even if a rigid maintenance system is used, a part of the deformations in the tunnel would be occurred. Accordingly, the use of pre-consolidation methods as an effective option for reducing surface settlements along with implementation of lining with proper rigidity should be considered in order to control deformations (Oke et al. 2014).

3.1 Numerical Modeling and Analysis

Among the various behavioral models used in the analysis of geotechnical projects, the Hardening Soil Model (HS), which considers the history of stress and it also differentiates between loading and unloading modes, has better consistency with actual results in analyzing issues such as excavation of tunnels.

Since the excavation and installation of the supporting system of different parts are performed at different times proportional to the progress of the excavation front, the amount of stress release and subsequent loading on the support system at different installation times will change as a function of the progress of the working front. One of the common methods for considering stress release in the two-dimensional modeling is to use the ground response curve (GRC). The ground response curve can be obtained by measuring the closure of the tunnel wall for the release of stress at different levels. On the other hand, based on the relationship provided by Panet (Panet et al. 1982), the amount of displacement of excavation space can obtain as a function of the distance from the excavation front as defined by equation (1) as follows:

$$\frac{U_r}{U_r^M} = 0.25 + 0.75 \left[1 - \left(\frac{0.75}{0.75 + x/R} \right)^2 \right] \tag{1}$$

U_r: Radial displacement; U_r^M: Maximum displacement
R: Tunnel radius; X: Distance from tunnel face

Therefore, at each step, using the GRC, one can determine the percentage of stress release per calculated displacement and apply it in two-dimensional modeling. The support system is installed after obtaining the stress release rate applied in each stage (Vlachopoulos et al. 2009).

Numerical modeling was performed in two-dimensional and three-dimensional format using PLAXIS 2D Ver.8.6 and PLAXIS 3D Ver.1.2 software. The hypotheses considered in the numerical modeling are as follows:

- Dimensions of the model are considered to be at a minimum acceptable level for the effects of boundaries.
- The lateral boundaries of the model are restrained along the x-axis and the lower boundary has been fixed along the y axis.
- A distributed load of 93.05 kPa as the equivalent load due to the passing of the railway and metro lines above the tunnel is considered in accordance with the loading regulations.

Two-dimensional modeling details have been shown in Figure 5.

The properties of the forepoling zone estimate based on the weighted average, (Hoek 2004) as shown in Figure 6 and defined by equation (2).

$$\bar{X} = \frac{\sum(X_i \times A_i)}{A_{\text{effect zone}}} \quad X = \{E, \gamma\} \ i = \{\text{soil, Grout, Steel}\} \tag{2}$$

$$E : \text{Young Modulus}; \ \gamma : \text{Unit weight}; \ A : \text{Area}$$

The tunnel construction method is a sequential excavation in seven stages with approximately parallel excavation of the side drifts (I, II) and the excavation of the middle drift (III) at a distance of 4m from the excavation front of the side drifts (Figure 7).

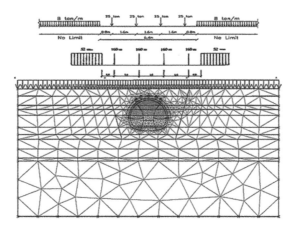

Figure 5. Details of two-dimensional modeling and the equivalent loading of the rail and metro railway fleet.

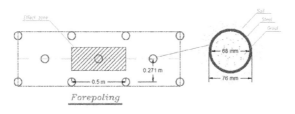

Figure 6. The schematic equivalent properties of the fore-poling zone.

6075

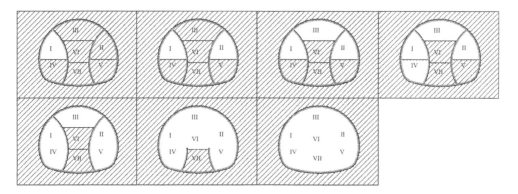

Figure 7.   Excavation steps for the Zarbalizadeh Tunnel Project.

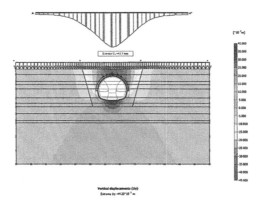

Figure 8.   Vertical displacement contours and ground displacement curve of excavated tunnel model after initial lining installation in the 2D model.

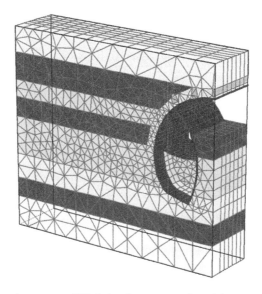

Figure 9.   One of construction stages of 3D finite element tunnel model.

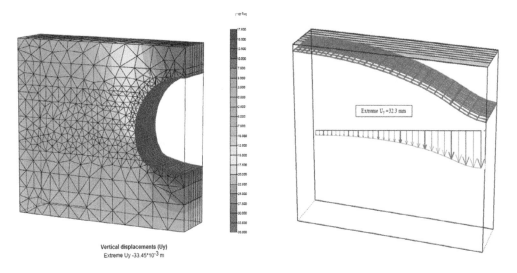

Vertical displacements (Uy)
Extreme Uy -33.45*10⁻³ m

Figure 10. Vertical displacement contours and ground displacement curve of excavated tunnel model after initial lining installation in a 2D model.

In Figure 8, vertical displacement of the pre-consolidated tunnel section in a 2D modeling is shown.

Considering the great importance of the project and the relatively low accuracy of the results of the 2D modeling of micro-tunnel and forepoling as pre-consolidation systems, a simplified 3D modeling was also carried-out, the model and its results of which have been presented in Figures 9 and 10.

4 COMPARISON OF GROUND SETTLEMENT WITH NUMERICAL MODELING

In the Zarbalizadeh tunnel project, to accurately monitor and control the deformations, in addition to installing leveling point and target prism stations (LPS & TPS) on the ground, convergence monitoring and underground survey point stations (CMS & USPS) were used at different stages of the tunnel section (Figures 11 and 12). Figures 13 and 14 show some monitoring results.

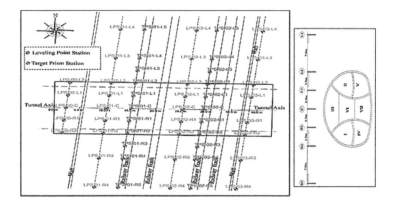

Figure 11. Position of the surface monitoring stations in the range of the rail corridor.

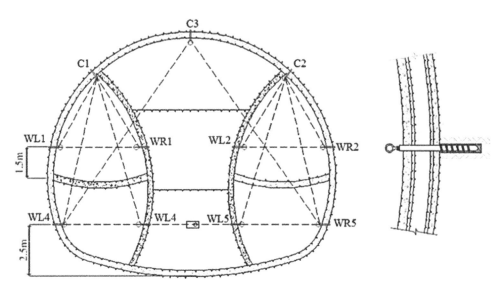

Figure 12. The location of the installation of convergence pins in the walls and tunnel roof.

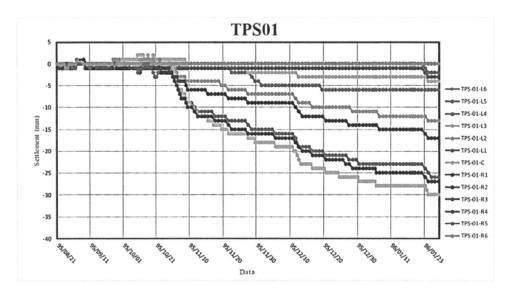

Figure 13. Target prism station monitoring results (TPS01, adjacent to the railway).

According to the results of the 3D model, the deformational behavior of tunnel in the case of using pre-consolidation system such as micro tunnels and forepoling will have more logical results than the 2D model, because determination of the percentage of exact release of stress in the 2D modeling steps with the predicted consolidation systems is simply not possible. Figure 15 compares the ground displacement of 2D and 3D pre-consolidated modeling with site monitoring data.

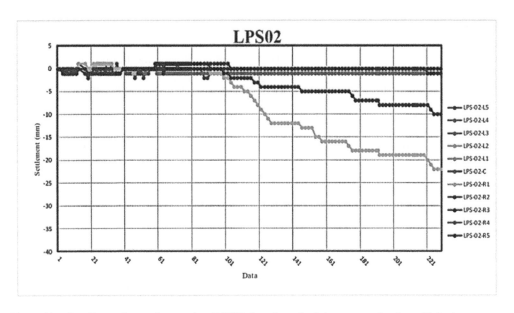

Figure 14. Leveling point station results, (LPS02, location of minimum over-burden of 2.5 m).

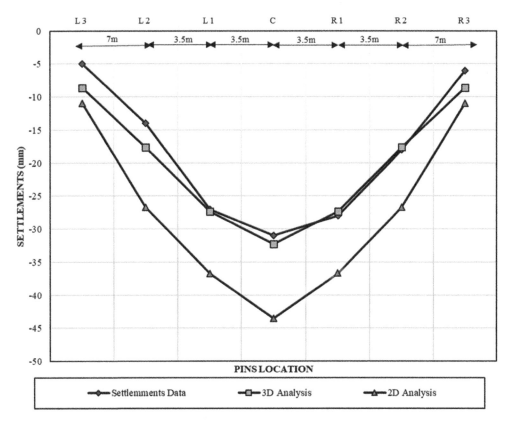

Figure 15. Comparison of ground settlement with the 2D and 3D models results.

5 CONCLUSION

Considering high sensitivity of the rail roads to ground settlement and to prevent interrupting their operation, Zarbalizadeh tunnel project was constructed by sequential excavation method (SEM) together with pre-consolidated system. In this paper, the design and construction details in order to control the settlements, have as low as possible interaction effect on the rail road passing above the project and have face stability during construction were explained. Through comparing the numerical modeling (2D and 3D) with monitoring results, it was shown that the deformational behavior of tunnel in the case of using pre-consolidation system such as micro tunnels and forepoling have more logical results than the 2D model, because determination of the percentage of exact release of stress in the 2D modeling steps with the predicted consolidation systems is simply not possible. Finally, this project was inaugurated in 2017 with a maximum ground surface vertical deformation of 31mm without any disturbance on the trend of railway and metro utilization.

REFERENCES

Aghamalian, M. 2012. Inverse analysis of monitoring data resulted from multi- stage construction of twin tunnels of Resalat highway in Tehran. Journal of Civil Engineering Infrastructures, 45th years, No. 5.

Eberhardt, E. 2001. Numerical modelling of three-dimension stress rotation a head of an advancing tunnel face. Rock mechanics and mining science, No. 38:499–518.

Oke, J.M., Velachopoulos, N., & Diederich, M. S. 2014. Numerical analysis in the design of umbrella arch systems. Journal of Rock Mechanics and Geotechnical Engineering, 6:546–564.

Panet, M., & Guenot, A. 1982. Analysis of convergence behind the face of a tunnel. Proceedings, International Symposium Tunelling'82, IMM, London:197–204.

Vlachopoulos, N., & Diederichs, M.S., 2009. Improved longitudinal displacement profiles for convergence confinement analysis of deep tunnel. Rock Mech. Rock Eng. 42;131–146.

Tunnels and Underground Cities: Engineering and Innovation meet Archaeology,
Architecture and Art, Volume 12: Urban
Tunnels - Part 2 – Peila, Viggiani & Celestino (Eds)
© 2019 Taylor & Francis Group, London, ISBN 978-0-367-46900-9

The response of underground structures to seismic loading

J. Pruška, V. Pavelcová, T. Poklopová, T. Janda & M. Šejnoha
Faculty of Civil Engineering, Czech Technical University in Prague, Prague, Czech Republic

ABSTRACT: The paper focuses on the evaluation of the effect of earthquake on underground structures using the finite element method. The dynamic effects can be addressed either by a fully dynamic approach or using a pseudo-static calculation. Fully dynamic calculations are based on a direct application of recorded accelerations to the actual computational model of the tunnel. In the pseudo-static analysis the effect of earthquake is transformed into an equivalent shear strain constant within individual soil layers. These are introduced in the form of displacements prescribed along the vertical boundaries of the FEM model in a usual static manner. This also implies the basic assumption of the geometry and material parameters of the geological profile not varying in the horizontal direction. In both cases it is generally accepted that vertical propagation of the pressure and shear waves can be described as a time-dependent 1D analysis of so called free-field column. Example comparing both approaches together with popular analytical methods is presented to highlight their advantages as well as drawbacks.

1 INTRODUCTION

The response of underground structures subjected to seismic load can be solved in several ways. The most accurate predictions of the effect of earthquake on the underground structure can be obtained by a fully dynamic finite element analysis. This approach requires numerical model of the solved area and complex definition of boundary conditions. Engineering practice then often calls for more simplified pseudo-static methods where the actual time dependent seismic load is converted to the equivalent static load applied either directly to the underground structure or to a simplified geometrical model accounting, at least to some extent, for mutual interaction of the underground structure and the surrounding soil. Such approaches, assuming that the deformation of the structure should conform to the deformation of the soil in the free field under the design earthquakes, can be classified as:

– An analytical solution (Wang (1993), Power et al. (1996), Hashash et al. (2001) providing close form equations of internal forces developed on a circular lining being located in the homogeneous linear elastic rock/soil layer. Although this is the most conservative method having only a limited application, it provides a first-order estimate of the liming stresses to potentially serve for the accuracy check of more complex calculations.

– Pseudo-static finite element calculation provides solution for any shape of underground structure in a generally nonhomogeneous layered rock/soil mass with potentially nonlinear material response. The basic idea of this approach is that the maximum shear deformation in free-field condition of the layer represents the maximum dynamic earthquake stress in this layer and can be input to the numerical model as a boundary condition. The shear strain equivalent to the actual dynamic loading can be calculated from the particle velocity estimated either in the simplified manner based on limited information (Hashash et al. (2001) or from the solution of one dimensional simulation of free-field response to actual earthquake (Poklopová (2018a,b).

The above mentioned three approaches will be theoretically addressed in Sections 2 and 3. A simple numerical example will be presented in Section 4 to compare prediction of the lining forces provided by these methods.

2 ANALYTICAL METHOD

The methodology of the analytical method for the seismic loading design is in that the static condition has to incorporate the additional loading imposed by ground shaking and deformation. In general, the seismic loads for underground structure are characterized in terms of strains imposed on the structure by the surrounding ground or their interaction. So we can distinguish two approaches:

– Soil-structure interaction approach (it will not be pursued further).
– Free-field deformation approach.

In free-field deformation approach, the ground deformation caused by seismic waves is assumed to occur in the absence of structure or excavation. The advantages and disadvantages of this method have been reported by Wang (1993). If seismic waves are propagating perpendicular to the tunnel longitudinal axis the circular tunnel lining undergoes the ovaling deformation. The solution for this mode of deformation is well described in Hashash et al. (2001). The ovaling deformation is determined by the assumed shear strain (Figure 1 a) to give the following relations

$$\Delta d_{ff} = \gamma_{mean} \frac{d}{2}, \Delta d_{lin} = R^n \Delta d_{ff}, R^n = \frac{4(1 - v_m)}{\alpha^n + 1}, \alpha^n = \frac{12 E_l (5 - 6)}{d^3 G_m (1 - v_l^2)} \tag{1}$$

where γ_{mean} is the average shear strain of the model, d is the tunnel radius, R_n is the lining – massive racking ratio under normal loading, a_n is the ratio between the stiffness of the lining material and the massive, v_m is the Poisson number of the massive, v_l is the Poisson number of the lining, G_m is the elasticity modulus of the massive and E_l is the Young modulus of the lining.

It is evident that parameters R^n and α^n take into account the rigidity of the interface tunnel lining-surrounding massive. If we express the curvature of the elliptical excavation in relation to Δd_{lin}, then the moment in the lining can be written using the polar coordinate φ (Figure 1 (b)) as

$$M(\varphi) = \frac{6 E_l \Delta d_{lim}}{d^2 (1 - v_l^2)} \cos\left(2\left(\varphi + \frac{\pi}{4}\right)\right) \tag{2}$$

where I is the moment of inertia of the cross-section and d is the tunnel diameter. Referring to Figure 1(c), the normal force on the curved element can be expressed in the form:

$$N(\varphi) = -\frac{1}{R} \frac{dM(\varphi)^2}{\varphi} + p_n(\varphi) R. \tag{3}$$

where p_n is the contact normal pressure and R is the lining diameter. After rewriting Eq. (3) using Eq. (2) we get

$$N(\varphi) = -\frac{48 E_l \Delta d_{lin}}{d^3 (1 - v_l^2)} \cos\left(2\left(\varphi + \frac{\pi}{4}\right)\right) + R p_n(\varphi) \tag{4}$$

Notice that expression (4) contains also the pressure dependent term due to rock/soil – structure interaction not taken into account in calculations according to Hashash et al. (2001), which may prove highly inaccurate particularly for shallow tunnels, see Kučera (2017) for more details.

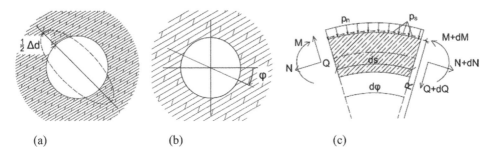

Figure 1.   a) Parameter $\Delta d(\Delta d_{ff}, \Delta d_{lin})$, b) Orientation of polar coordinates φ, c) Internal forces and contact stress on the differential element.

Recall finally that the above formulation assumes a homogeneous material surrounding the tunnel. In case of layered subsoil profile some sort of homogenization is needed adopting for example the Voigt assumption of constant deformation at each point of the solid, see e.g. Kučera (2017), Susanti et al (2017). For a particular example of two layered massive we obtain the effective shear modulus G_V^{hom} and the average shear strain of the model γ_{mean} needed in the above equations as the weighted average of the relevant values of the particular layers of the soil/rock mass

$$G_V^{\text{hom}} = \frac{h_1}{h} G_1 + \frac{h_2}{h} G_2, \gamma_{mean} = \frac{h_1}{h} \gamma_1 + \frac{h_2}{h} \gamma_2 \tag{5}$$

where h_i is the thickness of the layer i, h is the thickness of the soil deposit, G_i is the shear modulus of the layer i and γ_i is the shear strain introduced in the layer i.

3   FULL DYNAMIC CALCULATIONS USING FEM

The numerical solution of dynamic load effects is based on the D'Alembert principle, which states that the sum of all forces acting on the body is equal to zero. Applying this principle to a problem with many degrees of freedom with domain loaded by vertically propagating waves we obtain upon discretization and with reference to Figure 2 the system of second order differential equations in the form

$$\mathbf{M}\ddot{\mathbf{u}}_R(t) + \mathbf{C}\dot{\mathbf{u}}_R(t) + \mathbf{K}\mathbf{u}_R(t) = -\mathbf{M}\ddot{\mathbf{u}}_0(t) + \mathbf{C}\dot{\mathbf{u}}_{I0}(t)|_{y=0} \tag{6}$$

where $\mathbf{M}$, $\mathbf{C}$ and $\mathbf{K}$ are the mass, damping and stiffness matrices, respectively. Equation (6) assumes that the actual displacement vector $\mathbf{u}(\mathbf{x},t)$ can be split as

$$\mathbf{u}(x, t) = \mathbf{u}_0(t) + \mathbf{u}_R(x, t) \tag{7}$$

where $u_0(t)$ is the part of the total displacement $u(x,t)$ and is constant within the whole domain, thus independent of the position x. This particular format of Eq. (6) assumes the so called absorbing or quiet boundary conditions applied at the bottom boundary of the domain. This condition allows for absorbing the downward travelling wave u_d, so it is not reflected back to the domain. Such boundary is typically introduced in cases where the geometrical domain has to be truncated in the vertical direction within a given layer. The load is then introduced in terms of the incoming wave u_0 $(t)=u_{I0}(t)$, recall Figure 2.

Providing the domain can be truncated at the interface between compliant and infinitely rigid layers, the absorbing boundary condition can be replaced by the so called fixed

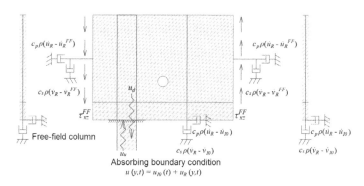

Figure 2. Loading and boundary conditions introduced in 2D fully dynamic analysis assuming absorbing boundary introduced within an "infinitely" long limestone layer.

boundary. Because $u_0(t)$ then corresponds to the total motion we have $u_R(x,y=0,t)=0$ and the second term on the right hand side of Eq. (6) drops out. This boundary condition should, however, be used with caution as it may greatly overestimate the strains developed in the domain. This is also the case of the model in Figure 2 to be considered in the comparative numerical analysis in Section 4. Therein, the clayed top layer is supported by the layer of limestone of considerably higher stiffness, but still far from being infinitely rigid. While for static analysis the geometrical domain could be safely truncated at the soil-rock interface, the fixed boundary would not be an adequate boundary condition for dynamic analysis, see Poklopová (2018a,b) for more details. Given this, only the model with absorbing boundary will be considered henceforth. This will apply also to the calculation of free-field strains adopted in the pseudo-static analysis.

As seen in Figure 2 the absorbing boundary is imagined in the form of viscous dampers where the pressure and shear wave velocities c_s and c_p are provided by

$$c_p = \sqrt{E_{oed}/\rho}, \ c_s = \sqrt{G/\rho} \tag{8}$$

where E_{oed}, G, ρ are the oedometric modulus, shear modulus and density of the soil, respectively. For particular derivation of these boundary conditions we refer the interested reader to Zienkiewicz et al. (1986).

Apart from the bottom boundary we should also accord our attention to the boundary conditions applied along the lateral edges. As pointed out in Zienkiewicz et al. (1986) such boundary conditions should account for the disturbance from the free-field conditions caused by the presence of structure, e.g. tunnel. This is schematically shown in Figure 2, suggesting that only the difference $(u_R - u_R^{FF})$ between the actual u_R and free-field motion u_R^{FF} should be absorbed. The free-field motion in particular is derived from an independent one-dimensional free-field column analysis assuming the same loading conditions as in case of the two-dimensional analysis (Kučera (2017). To ensure that the solution of the 1D free-field column analysis is properly reproduced by the 2D analysis in case of no structure, it is necessary to supplement these boundary conditions by the prescribed shear stress provided by the 1D column analysis, see Poklopová (2018a,b) for additional details. Thus in case of no material damping Eq. (6) becomes

$$\mathbf{M\ddot{u}}_R + \mathbf{Ku}_R + \mathbf{C}_p\dot{\mathbf{u}}_R\big|_{y=0} + \mathbf{C}_s\dot{\mathbf{u}}_R\big|_{x=0,L} = -\mathbf{M\ddot{u}}_0(t) + \mathbf{C\dot{u}}_{J0}(t)\big|_{y=0} +$$
$$+ \mathbf{C}_s\dot{\mathbf{u}}_\mathbf{R}^{\mathbf{FF}}\big|_{x=0,L} - \mathbf{R}_\tau\big|_{x=0} + \mathbf{R}_\tau\big|_{x=L} \tag{9}$$

where R_τ represents the vector o nodal forces associated with the prescribed boundary shear stress. The element damping matrices $\mathbf{C}_p^e$ and $\mathbf{C}_s^e$ take the form

$$\mathbf{C}_p^e = \begin{bmatrix} c_s\rho & 0 \\ 0 & c_p\rho \end{bmatrix}, \ldots\ldots \mathbf{C}_s^e = \begin{bmatrix} c_p\rho & 0 \\ 0 & c_s\rho \end{bmatrix} \tag{10}$$

4 CASE STUDY

The analyzed structure is a circular track tunnel in Baku, the capital city of Azerbaijan. The section is 1 500 m long with two directional curves of 1 200 m and 1 600 m radius. The track tunnel crosses the height difference of 25.65 m and has a slope of 1.90%. Tunnel excavation was carried out using a TBM with a diameter of boring head of 6 m. The tunnel lining consists of reinforced concrete segments with a thickness of 0.3 m. The geological conditions in this area are: under the thin layer of sediments with sand and sandstones deposits there is subsoil formed by clay layers (medium and low plasticity) and a rock mass of limestone.

As seen in Figure 2, the computational model was simplified considering only clay (top) and limestone (bottom) layers. In numerical analysis the limestone was assumed to behave linearly elastic while the response of the clayey layer is governed by the Mohr-Coulomb (MC) model. This is because the GEO5 FEM program used in the present study does not allow for using in this case more suitable critical state models in dynamic analysis. To that end, the parameters of the MC model were tuned to provide response close to that derived by the Cam clay model in the excavation step performed in Section 4.1 (Poklopová (2018a). The adopted material parameters are listed in Tables 1 and 2. As evident from Table 2 the MC model considered the modulus of elasticity varying with depth. Upon unloading during excavation step the material stiffness is governed by the unloading/reloading modulus E_{ur}. In case of dynamic analysis, the value of E_{ur}, set equal to the dynamic moduslus E_{dyn}, was assigned to the whole clayey layer.

Table 1. Material parameters of limestone – linear elastic material model.

Parameter	symbol	value	unit
Unit weight	γ	21.70	kN/m^3
Poisson umber	v	0.30	–
Elastic modulus	E	630.00	MPa
Coefficient of earth pressure at rest	K_0	0.60	–
Unit weight of saturated soil	γ_{sat}	22.00	kN/m^3

Table 2. Material parameters of clay – Mohr-Coulomb model.

Parameter	symbol	value	unit
Unit weight	γ	19.60	kN/m^3
Poisson number	v	0.40	–
Coefficient of earth pressure at rest	K_0	0.60	–
Elasticity modulus on the surface	E	5.00	MPa
Elasticity modulus on the interface	E	40.00	MPa
Unloading/reloading modulus	E_{ur}	80.00	MPa
Dynamic modulus	E_{dyn}	80.00	MPa
Angle of internal friction	φ_{ef}	25	°
Cohesion	c_{ef}	1.00	kPa
Unit weight of saturated soil	γ_{sat}	22.00	kN/m^3
Angle of dilatancy	ψ	0.00	°

The following calculations were carried out, see Pavelcová (2018a) and Poklopová (2018a) for additional details,

- Static 2D FEM analysis of the excavation step to generate the initial stress prior to dynamic analysis.
- 1D free-field column analysis to acquire the maximum shear strain applied both in the analytical and pseudo-static analysis.
- Pseudo-static 2D FEM analysis.
- Fully dynamic 2D FEM analysis.

All calculations were carried out with the help of GEO5 FEM program (Fine (2018)). Details are presented next in Sections 4.1–4.4.

4.1 Static 2D FEM analysis of excavation step

To generate initial stresses before introducing earthquake we performed the excavation analysis. Focusing on the comparative study on various computational approaches we limited our attention to the east track tube (excavated first) and one particular section where the tunnel is located in the clayey soil. As mentioned earlier the interest in application of absorbing boundary conditions required the introduction of a layer of limestone as already depicted in Figure 2. In order not to unnecessarily increase the computational burden this layer was assumed 10 m thick only. Given the axis of the tunnel tube at a depth of 40 m thus resulted in a reasonably deep model with the bottom boundary at a depth of 59 m. The vertical boundaries of the model were set at a distance of 40 m from the center of the tunnel. The interface of the geological layers is located at a depth of 49 m. This model geometry together with the finite element mesh is plotted in Figure 3(a).

The actual construction of the tunnel was conducted using the TBM. Modeling such a complex problem in 2D is very problematic. For example, setting the grouting mixture between the segments and the soil or the soil force on the shield in the case of inadequate grouting cannot be taken into account in the 2D FEM model. We therefore settled for a more traditional convergence confinement method to approximate the 3D behavior of the structure. The percentage of soil excavation prior to introduction of lining was set such as to arrive at the maximum terrain settlement of 2 mm at the end of the excavation step. The lining was modeled with one-dimensional beam elements. Because the longitudinal gap between the segments is not continuous, a fixed link between segments along the perimeter was adopted. This first calculation phase thus proceeded in three computational stages – the geostatic stress was estimated in the first stage using the K_0 procedure, this was followed by the first excavation step deactivating 30% of the geostatic pressure. The remainder of the supported pressure was applied in the last step while activating the lining.

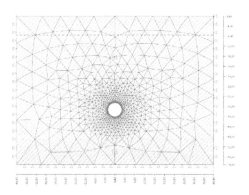

Figure 3.   Geometry and finite element mesh.

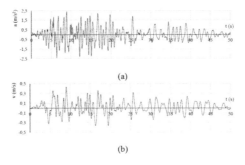

Figure 4.   Recorded rock outcrop motion: a) acceleration, b) velocity.

For illustration we plot the distribution of the resulting bending moment in Figures 7(a,c) to compare the location of the corresponding maxima before and after imposing earthquake.

4.2   *1D free-field column analysis*

This section reports on the results from the solution of 1D free-field column analysis. The computational model is the one displayed in Figure 2 assuming one-dimensional elements in plane-strain setting. In case of absorbing boundary we obtain the governing system of equation of this problem, compare with Eq. (9), in the form

$$\mathbf{M\ddot{u}}_R + \mathbf{Ku} + \mathbf{C}_p\mathbf{\dot{u}}_R|_{y=0} = -\mathbf{M\ddot{u}}_{I0} + \mathbf{C}_p\mathbf{\dot{u}}_{I0}|_{y=0} \qquad (11)$$

Figure 4 plots the acceleration and velocity profiles generated by earthquake and adopted as loads on the right hand side of Eq. (11). Assuming the recorded acceleration corresponds to the rock outcrop motion only one half of the velocity magnitude is applied as the incoming part of the total motion (Poklopová 2018b).

The FEM analysis provides the time variation of shear strain the maxima of which are displayed in Figure 5. As often used in analytical approach to calculate the shear strain as the ratio of particle and shear wave velocities we also plot, for illustration, the variation of the maximum particle velocity. The average maximum shear strain then serves as loading in both analytical and pseudo-static FEM analysis, see the next section.

4.3   *2D pseudo-static analysis*

It has been advocated in Kučera (2017) that in layered structures a piecewise constant shear strain corresponding to individual layers should be considered to properly represent the free-

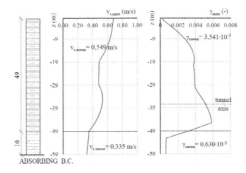

Figure 5.   Vertical variation of maximum velocity and shear strain.

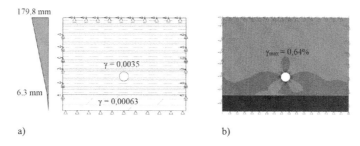

a)   b)

Figure 6.   a) Kinematic boundary conditions, b) Distribution of shear strain.

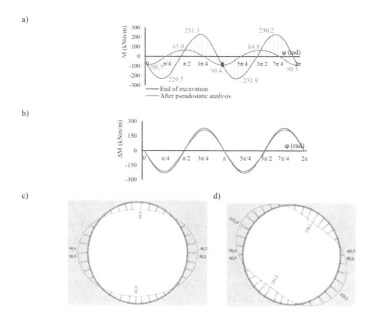

Figure 7.   Variation of bending moment: a) as a function of the cylindrical coordinate φ, b) comparison of 2D pseudo-static analysis and analytical solution, c) at the end of excavation (FEM), d) at the end of pseudo-static analysis (FEM).

field conditions. In most FEM programs such strains must be converted to corresponding displacements introduced as kinematic boundary conditions as shown in Figure 6(a). Notice that the resulting shear strain seen in Figure 6(b) is more or less constant within a given layer except for the vicinity of tunnel. Point out that the elastic behavior of the clayey layer was considered in order to avoid unrealistic plastic strains developed along the terrain surface where in reality the shear strain is equal to zero. Otherwise, the same material properties as stored in Tables 1 and 2 were used with constant dynamic moduli E_{dyn} in both layers.

Figure 7(a) shows variation of the bending moment around the perimeter of the lining comparing the results at the end of excavation (gray line) and with superposed values of the bending moment caused by earthquake effect (red line). Clearly, due to ovaling not only the actual value but also the location of the maximum moment changed corresponding well to anaytical solution given by Eq. (2). This is further demonstrated in Figures 7 (c,d) for illustration.

Figure 7(b) compares the variation of bending moment caused solely by the prescribed strain and derived from the 2D pseudo-static FEM analysis (red) line and by the analytical method (grey line) using Eq. (2). In the latter case the strain γ_{mean} was set equal to that of the

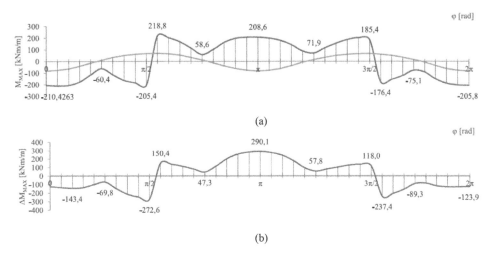

(a)

(b)

Figure 8. a) Variation of the maximum bending moment as a function of the cylindrical coordinate φ, b) Variation of bending moment attributed to dynamic loading only.

clayey layer. In this case, the results are close confirming a good accuracy of analytical solution for such simple geological conditions.

4.4 2D fully dynamic analysis

The fully dynamic 2D analysis was carried in the light of Eq. (9) considering the same acceleration and velocity records as used in the previous section, recall Figure 4. Although not limited to, the elastic behavior was again considered to keep mutual similarity with the pseudo-static analysis outlined in the previous section.

Graphical representation of the moment variation at the end of dynamic analysis appears in Figure 8a (red line). The moments pertinent to the end of excavation are again plotted in grey for the sake of comparison. Figure 8b finally offers, similar to Figure 7b, the variation of the maximum of moment caused by the dynamic loading only. These results can be compared with the results in Figures 7(a,b) to suggest a reasonable agreement between all methods, at least in the light of bending moments, and as such to promote more simple methods at the preliminary stage of design.

5 CONCLUSIONS

The present paper compares various approaches available to engineering community to address the effect of seismic loading on underground structures. A simple example of a circular tunnel located in the homogeneous layer has demonstrated the applicability of analytical methods to provide reasonably accurate estimates of the lining internal forces at a fraction of computational time (Hashash et al. (2001) compare to FEM analyses. However, the latter computational strategy overcomes some of the shortcomings of the analytical methods since being free of any limitations associated with subsoil profile and tunnel geometry. Even the pseudo-static method, the FEM extension of the analytical approach, can properly address the soil structure interaction (Kučera (2017) and to some extent also the soil nonlinearity if limited to the vicinity of the tunnel. Because of the basic assumption of layer-wise constant shear strain, the subsoil close to the terrain surface, where this assumption violates the actual boundary condition of zero shear strain, must be assumed elastic (Poklopová (2018a,b). On the other hand, no such obstacle arises if adopting a fully dynamic analysis allowing for an arbitrary constitutive model to be assigned to individual soils/rocks in the profile. Although not in general limited to this requirement, the height of lateral boundaries of the computational model should be approximately the same and the soil layers should be, similar to the pseudo-static analysis, more or less horizontal. This is because of the free-field conditions being assumed prior to introduction of any disturbance, e.g. embankments, construction ditch, tunnel, etc., to the original domain. The present work, see also (Pavelcová (2018a,b), promoted the application of so called absorbing boundary conditions at the bottom boundary of the computational model as providing more realistic loading caused by propagating waves. In this regard, the so called fixed boundary, causing the downward travelling wave to be fully reflected back to the domain, should be used with caution.

ACKNOWLEDGEMENTS

The article was processed under financial support by Competence Centres program of Technology Agency of the Czech Republic (TA CR), project no. TE01020168 Centre for Effective and Sustainable Transport Infrastructure (CESTI).

REFERENCES

Zienkiewicz O.C., Bicanic N. & Shen F. Q. 1986. Generalized smith boundary – a transmitting boundary for dynamic computation, *Institute for Numerical Methods in Engineering*, University College of Swansea, Vol. 207, pp. 109–138.

Wang J.N. 1993. *Seismic design of tunnel – A simple state of the art design approach*. Parson Brinckerhoff Quade & Douglas, Inc., New York, Monograph 7.

Power M.S., Rosid D. & Kaneshiro J. 1996. *Strawman: screening, evaluation, and retrofit design of tunnels. Report Draft. Vol. III*, National Center for Earthquake Engineering Research, Buffalo, New York.

Hashash Y.M.A., Hook J.J., Schmidt B. & Yao J.I.C. 2001. Seismic design and Analysis of underground structure. *Tunnelling and Underground space Technology*, Vol. 16, No.4, pp. 147–293.

Kučera D. 2017. *Analysis of geotechnical structures subjected to earthquake*. Diploma thesis, CTU in Prague, Faculty of Civil Engineering, in Czech.

Susanti, E. 2017. *Numerical evaluation of the bearing capacity of the All Saints Church walls in Broumov*. Diploma thesis, CTU in Prague, Faculty of Civil Engineering.

Šejnoha M., Janda T., Pruška J. & Brouček M. 2015. *Metoda konečných prvků v geomechanice: Teoretické základy a Inženýrské aplikace*. ČVUT v Praze, Prague, in Czech.

Pavelcová V. 2018a. *Evaluation of real underground structure subjected to earthquake – fully dynamic analysis*, Bachalor's thesis, CTU in Prague, Faculty of Civil Engineering, in Czech.

Pavelcová V., Poklopová T., Janda T. & Šejnoha M. 2018b. The influence of boundary conditions on the response of underground structures subjected to earthquake, *Acta Polytechnica*, Accepted.

Poklopová T. 2018a. *Evaluation of real underground structure subjected to earthquake – pseudostatic analysis*, Bachalor's thesis, CTU in Prague, Faculty of Civil Engineering, in Czech.

Poklopová T., Pavelcová V., Janda T. & Šejnoha M. 2018b. Evaluation of real underground structure subjected to earthquake – pseudostatic analysis, *Acta Polytechnica*, Accepted.

Fine 2018. *GEO5 FEM* https://www.finesoftware.eu/geotechnical-software/fem//.

Tunnels and Underground Cities: Engineering and Innovation meet Archaeology,
Architecture and Art, Volume 12: Urban
Tunnels - Part 2 – Peila, Viggiani & Celestino (Eds)
© 2019 Taylor & Francis Group, London, ISBN 978-0-367-46900-9

Metro tunnelling in Singapore – Thomson Line tunnelling in close proximity to existing railway lines

A. Raedle
Arup, Singapore

S.S. Marican & C.N. Ow
Land Transport Authority, Singapore

ABSTRACT: Singapore's 30km-long all-underground Thomson Line (TSL) involves the operation of approximately thirty Tunnel Boring Machines (TBMs) to complete the twin tunnels and the construction of twenty-two underground stations, including five interchange stations. This involves challenges such as tunnelling in urbanized areas and poor geological conditions such as reclaimed land. Bored tunnels of 5.8m internal diameter are aligned in side-by-side or stacked configuration to accommodate these constraints. In addition to challenges posed by densely-urbanised landscape, the project runs through a number of interfaces, which necessitates tunnelling in close proximity to existing railway lines and stations, such as interface at Outram Park and Robinson Road. Another challenging construction contracts, T226, includes the construction of a portion of Marina Bay Station, a TBM launch shaft, two working shafts to facilitate the underpinning of live operational rail tunnels and station, mining of a pedestrian link directly underneath with clearance of 1.0m, as well as mined mainline tunnels.

1 PROJECT INTRODUCTION AND GENERAL OVERVIEW OF THOMSON LINE

The Thomson East Coast Line (TEL) is the sixth Mass Rapid Transit (MRT) line in Singapore and will be fully automated and driverless. The line will be opened in 2019 onwards over five stages. It is expected to eventually service one million commuters daily when fully operational. The entire TEL stretches over 43km and the entirety of construction will be subsurface, with thirty-one stations and seven interchange stations. The alignment runs from the north to south, cutting through the Singapore's CBD and to the east. The alignment will connect with the existing MRT lines at several locations.

The Thomson Line (TSL) is a portion of the TEL line which runs from the north to the south of Singapore between Woodlands North and Gardens by the Bay. The project owner and developer, Singapore's Land Transport Authority (LTA) has divided the TSL line into four contract packages for engineering consultancy services. Arup was successfully appointed to two out of the four packages, TSL Package A (TSL-A) at the north, and TSL Package D (TSL-D) at the south. TSL-A begins at Woodlands North Station and consists of three stations and associated tunnels. TSL-D is located at the southern tip of the TSL, and interfaces with the Eastern Regional Line (ERL) at Garden by the Bay Marina East shaft along the east coast of Singapore to form the overall TEL. This paper focuses on TSL-D. Figure 1 below illustrates the overall alignment of TEL in relation to the existing MRT lines in Singapore.

TSL-D consists of approximately two nos. of 6km running tunnels and six stations, beginning at the southern end of Havelock Station and ending at the intervention shaft near Gardens by the Bay Marina East shaft for future connection with the ERL. To enhance

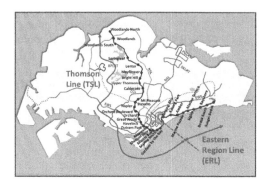

Figure 1.   Overall TEL Alignment in relation to Existing MRT Lines.

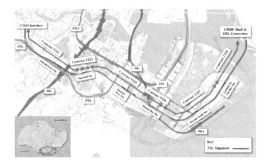

Figure 2.   Overall Alignment of TSL-D and Construction Contract Packages.

connectivity, TSL-D will interface with four existing MRT lines at two existing stations namely Outram Park Station and Marina Bay Station. Since the alignment of TSL-D runs through the CBD, the twin bored tunnels of 5.8m internal diameter are aligned in stacked configuration at some locations to accommodate the constraints of the densely-built urban environment. Figure 2 below illustrates the overall alignment of TSL-D, including the different contract packages for station and tunnels construction.

2   GEOLOGICAL CONDITIONS OF SINGAPORE

Singapore's geological formations can be broadly categorized into four major classifications: Bukit Timah Granite from the Mesozoic era is commonly encountered in the central region of Singapore; Jurong Formation, which is a sedimentary rock formation from the Mesozoic era,

Figure 3.   TSL-D Alignment Overlying Singapore's Geological Map.

and is distributed in the western regions of Singapore; Old Alluvium (OA) from the Pleistocene age, and commonly encountered in the east; and the Kallang Formation, which consists of interbedded marine and terrestrial sediment layers, deposited in the Holocene period, most commonly found in the coastal areas and low-lying regions of Singapore.

TSL-D encounters Jurong Formation, Kallang Formation, Old Alluvium, and Fort Canning Boulder Bed (FCBB), which is a colluvial deposit of Jurong Formation material from seismic activity in the past. Towards the southern portion of the TSL-D alignment, construction works encounter reclaimed land, which comprises of consolidating marine clay, fluvial sand and clay of Kallang Formation. Figure 3 below illustrates TSL-D alignment overlying an extract of geological map of Singapore.

3 TUNNELLING NEAR EXISTING RAILWAY LINES AT OUTRAM PARK

One of the main engineering challenges in TSL-D is the running of TSL tunnels close to existing MRT tunnels. In addition, these existing railway lines need to remain in operation throughout the TSL constructions works, which renders the project even more challenging. There were three main instances where the close proximity of tunnels required further study and detailed impact assessment, close monitoring during the project and implementation of additional mitigation measures specified in construction contract, as elaborated in sections below.

3.1 West of Outram Park (Undercrossing EWL)

One of the instances is at the western side of Outram Park Station where the TSL tunnel alignment is running underneath the existing EWL tunnels for approximately 410m long stretch. The bored tunnels are in the sedimentary Jurong Formation and have a minimum clearance of 6m to the existing tunnels. Figure 4 below illustrates the location of the undercrossing and minimum clearance between the MRT lines.

Undercrossing of existing tunnels with such minimal clearance could result in settlement and, consequently, potential opening of circumferential joint and water ingress into the tunnel due to gasket relaxation. Ultimately, excessive settlement could impair the railway operation arising from failure of rail fasteners, track sag and twist. Therefore, the impact of TSL tunnelling on the existing EWL needs to be checked against the criteria set-out in the LTA DBC Code of Practice for Railway Protection (CoPRP). The CoPRP allows either compliance with the deemed-to-satisfy provisions or alternative solution achieving performance-based requirements. Arup in-house software Xdisp has been used to calculate the ground movement due to TSL construction works, including TSL bored tunnels and a mined cross passage. The analysis in Xdisp is based on the 3-dimensional Gaussian settlement trough

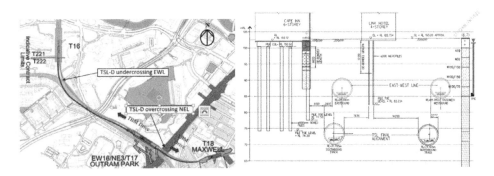

Figure 4.   TSL-D, EWL and North East Line (NEL) Interfacing Near Outram Park Station Layout and Section.

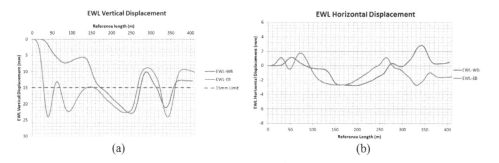

Figure 5. (a) Estimated maximum vertical displacement and (b) maximum horizontal movement of EWL tunnels along interface section.

Table 1. Volume loss values (in %) for 10 limit state assessment.

Model	Full Face Jurong Formation Rock (Grade III or better)	Mixed Face of Jurong Formation Rock/Soil	SCL Construction of Cross Passage 1
Case 1	0.5	1.0	0.5
Case 2	0.5	1.5	0.5
Case 3	1.0	1.5	1.0
Case 4	Mixed face conditions along both TSL bored tunnels and mined CP1 construction, 1.5%		

suggested by Mair et al. (1993). Figure 5 shows the cumulative vertical and horizontal movement of the EWL tunnels estimated using Xdisp. In this analysis the ground movement is taken at the EWL tunnel invert without considering the EWL tunnel structural stiffness, yielding to conservative results.

Nevertheless, the estimated movement of the EWL tunnels exceeds the limit set out by the deemed-to-satisfy provision for EWL tunnels which is 15mm movement in any direction. Hence, an alternative solution involving detailed assessment of the TSL construction impact according to the 10 limit states proposed by Shirlaw et al. (2000) was undertaken. The 10 limit states govern the serviceability and operability of railway tunnels, namely tunnel ovalisation, squat, step, opening of circle joints, additional water make, structural gauge clearance, failure of rail fasteners, track dip/peak, track horizontal versine and track twist. The results of the detailed assessment show that the impacts on the EWL are well within the critical levels for all the 10 limit states. Table 1 shows the volume loss values adopted in the above-mentioned assessment of the EWL and for all cases, the assessment is satisfactory. Case 2 was set as the target performance for construction.

3.2 East of Outram Park (Parallel to EWL)

At the east side of the Outram Park station, the TSL runs in parallel to the existing EWL bored tunnels with closest separation of 6m. The related construction works include a launch shaft, a temporary shaft, TSL twin bored tunneling and a mined pedestrian link, as shown in Figure 6. The movement of existing EWL tunnels were estimated using the software Xdisp. Figure 7 (a) shows the cumulative displacements along EWL-EB invert due to various relevant construction works as described above calculated using Xdisp. As shown in Figure 7 (a), the predicted cumulative movement, which is approximately 58mm, exceeds the the limit set out by the deemed-to-satisfy provision for EWL tunnels. Detailed assessment of the construction impact according to the 10 limit states were hence carried out. The impacts on the EWL tunnels are shown to be well within the critical levels for the 10 limit states. Actual settlement at

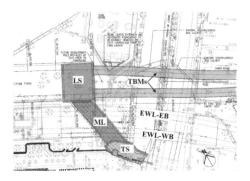

Figure 6.   Location of assessed EWL tunnels and related TSL works.

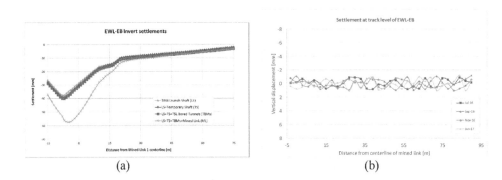

(a)   (b)

Figure 7.   (a) Cumulative vertical displacement along EWL-EB Invert and (b) Monitoring results of the settlement at track level of the EWL-EB tunnel along its alignment across the construction period.

the invert of the EWL tunnels estimated from the settlement data recorded at the track level were much smaller than predicted, which maximum is shown to be within 2.0mm (Figure 7 (b)).

3.3   East of Outram Park (Overcrossing NEL)

The TSL tunnels interface with the NEL line to the east of Outram Park Station, as shown in Figure 4. The TSL-D line over-crosses the NEL tunnels with a minimum clearance of 2.4m in Jurong Formation. Finite element simulation was conducted to predict the movement of the NEL caused by overcrossing of the TSL tunnels. The simulation results show that the NEL tunnel movements and track distortions are within the CoPRP Deemed-to-Satisfy provisions and hence, no further assessment is required. Similarly, NEL tunnels were monitored by ATMS (Automated Track Monitoring System). Monitoring data shows that the movement of the NEL tunnels is below 2mm, which is within the construction alert level.

4   TUNNELLING UNDER-CROSSING EXISTING RAILWAY LINES AT ROBINSON ROAD

The third notable interface is the undercrossing of the EWL tunnels at Robinson Road with a minimum clearance of 3.7m in a mixed face condition of Fort Canning Boulder Bed (FCBB) and Kallang Formation (KF). At this location TSL soil investigation reveals abrupt transition of geological formations from Fort Canning Boulder Bed (FCBB) at the North-West, to Kallang Formation overlying Old Alluvium, at the intersection of existing EWL tunnels and proposed TSL tunnels at the junction of McCallum Street and Robinson Road, as shown in

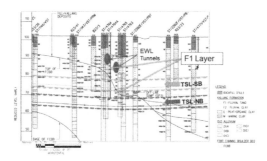

Figure 8. Longitudinal Geological Profile along TSL undercrossing EWL at Robinson road.

Figure 8. Tunneling through drastic geological change, i.e. mixed-face condition poses risk of high ground loss as it is relatively more difficult to balance the penetration rate against the excavation across the tunnel face. The excessive ground loss may potentially jeopardise the stability of the existing EWL tunnels. At this location, the twin TSL tunnels are in stacked configuration.

To mitigate the engineering risks and hazards associated with the close proximity of the tunnels, a variety of measures such as pre-tunnelling protection measures involving ground improvement and micro-tunnelling of canopy pipe piles and TBM control measures at the upper TSL tunnels were proposed.

Fluvial sand layers were also detected along the KF-FCBB interface. Fluvial sand layers act as an aquifer channel which may cause high water ingress into the TBM and groundwater drawdown and consequently, ground settlement at a distance away from the TBM location. In addition, Fluvial sand layers are not easy to treat and treatment may not be effective. TAM grouting was proposed to lower the permeability of the Fluvial sand layers around the TSL upper tunnel and hence, acting as water cut-off.

4.1 Protective hood of canopy pipe piles

Arup proposed two temporary shafts, North West (NW) and South East (SE) shafts, to be constructed to facilitate installation of canopy pipe piles (see Figure 9). Steel pipe piles are then jacked from one shaft on one side of existing EWL, steered using slurry Micro TBM underneath the EWL bored tunnels towards the shaft on other side of EWL to retrieve the Micro TBM. The canopy pipe piles are interlocked to each other and fix-ended to the shaft walls. In this manner, the canopy pipe piles serves as a suspended roof which is able to minimise potential localised settlement of the EWL tunnels in the event that large ground loss is induced during TSL upper South Bound (SB) TBM drive.

However, in view of the challenges associated with to divert the massive network of utilities underneath Robinson Road in order to construct the retrieval shaft, another pipe-jacking scheme has been adopted by the contractor. Instead of two shafts, only the SE launch shaft is constructed and the canopy pipe piles are socketed into the FCBB which practically creates fix-end condition similar to the function provided by shaft wall as the Micro TBM proposed has been designed with a retractable cutterhead which omits the need of a 2nd NW shaft designed for retrieval of the Micro TBMs.

4.2 EPB TBM equipped with AFC

Experience with EPB TBM in mixed face condition in Singapore indicates potential risk of loss of face pressure and risk of potential sinkhole formation due to inability to anticipate exact location of abrupt geological change at the face and potential sudden drop of face pressure. To mitigate this risk, the EPB TBM is specified with built-in AFC (Automatic Active Face Support System), which will automatically regulate the pressure inside the

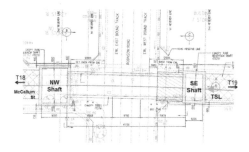

Figure 9.   Location of the proposed temporary NW and SE shafts.

plenum in real-time to minimise potential ground loss and hence, movement of the existing EWL tunnels.

4.3   *Settlement analysis considering various construction works and assessment of EWL*

Movements of the EWL tunnels have been calculated considering the effects of various TSL construction activities and mitigation measures. These include NW and SE shafts construction, ground improvement works, canopy pipe piles micro-tunnelling, construction of TSL-NB (lower) and TSL-SB (upper) bored tunnels as well as consolidation settlement.

Canopy pipe piles are prescribed to reduce potential settlement of the existing EWL tunnels in case of ground loss when TSL TBM encounter mixed soil condition underneath the existing EWL tunnels. When the face pressure is at 80% of overburden with the presence of canopy pipe piles, maximum settlement is predicted as 55mm. And this has been assessed against the 10 limit states.

Figure 10 shows the cumulative settlement of the EWL-WB tunnel induced by various construction works with mitigation measures considered. As shown, the movements of both EWL tunnels (maximum 55mm) exceed the CoPRP Deemed-to-Satisfy provisions. Thus, alternative approach was undertaken by detailed assessment of the effects against the 10 Limit States. The results of the alternative assessment shows that the impacts on the EWL are within the critical levels for all the 10 limit states when mitigation measures are adopted.

Figure 11 shows the ATMS monitoring readings of the total movement (vector sum of horizontal and vertical displacement) of the EWL-WB tunnels at different construction stages. Monitoring prisms "225WC19" and "225WR19" are located at the interface with TSL bored tunnels and are hence expected to produce the highest displacement readings (see Figure 11). Figure 12 plots the total movement recorded by different monitoring prisms installed along the EWL-WB alignment. The curves of the monitoring results generally exhibit the shape of a Gaussian settlement trough (Figure 12), suggesting the validity of the data.

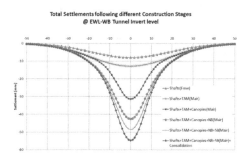

Figure 10.   Predicted settlements at EWL-WB invert.

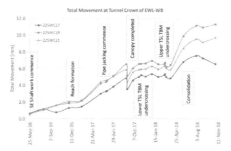

Figure 11. Total movement readings of EWL-WB tunnel at crown across construction period.

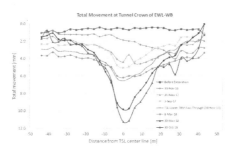

Figure 12. Total movement readings of EWL-WB tunnel at crown along the alignment.

The ATMS-recorded data indicates that the maximum movement of the EWL-WB Tunnel is within 12mm (as shown in Figure 12), which is well below the predicted value (Figure 10). The smaller movement of the EWL tunnel can be attributed to the deviation in protective measures, i.e. constructing only one shaft, lower actual volume loss induced during canopy pipe pile construction. In addition, the estimated ground movement by Xdisp is conservative in nature and was expected to be higher than the actual measurements. The conservativeness of the analysis comes from the fact that the model does not consider the stiffness of the EWL tunnels, mobilised ground arching or soil-structure interaction, which hold significant impact. Nevertheless, the monitoring results indicate the successful implementation of the notional design of the TSL undercrossing EWL.

5 DESIGN AND CONSTRUCTION OF MINED TUNNELS BELOW AN OPERATIONAL THREE-WAY INTERCHANGE STATION AT MARINA BAY

One of the most challenging construction contracts of the TSL-D is contract T226, which included the construction of the TSL Marina Bay Station, a TBM launch shaft, two working shafts to facilitate the underpinning of North South Line (NSL) tunnels, mining of a pedestrian linkway directly underneath the NSL tunnels and station platforms, as well as mined mainline tunnels in a stacked arrangement underneath the pedestrian linkway. It was also stipulated that the design and construction works are to be strictly compliant with the allowable movement of existing tunnels with maximum movement of 15mm, and that no existing MRT services are permitted to be closed or degraded in any way during the construction of works. The consolidating Marine Clay of the Kallang Formation was expected to extend to a depth of 35m. Figure 13 (a) below illustrates the layout of Marina Bay Station and the proposed construction works under Contract T226.

The pedestrian link is 40m in length, 10.7m wide, and 4.1m in height. It was constructed directly underneath the base slabs of the NSL tunnels. The first challenge is to carry out mining safely in the very soft Marine Clay (typical SPT value ranges from 0 to 2). Vertical jet grouting from a limited work site space directly between the NSL and CCL tunnels as well as horizontal jet grouting from working shafts were utilized to improve the ground condition for mining construction. The required minimum undrained shear strength ("S_u" or "C_u") of the treated soil is 300 kPa, as stipulated in Arup's notional design. Figure 13 (b) shows the 3D model of the ground improvement layout. The pedestrian linkway is constructed under a few stages namely:

1. Mining of the two adits using a rectangular open-face shield with compressed air support while cutting NSL piles along the alignment and supporting the adits using temporary steel segments;
2. Mined enlargement of the adits and construction of the transfer beam to underpin the NSL tunnels; and
3. Mining the ground in between the two adits, cut the remaining NSL piles within the linkway space and construct the linkway permanent structure.

Another challenge is the removal of the existing foundation piles of the NSL tunnels during shield mining of the pedestrian linkway. Stitch coring method was adopted to cut the piles as cutting work needs to be carried out in a confined space in the front of the shield machine with limited impact to the NSL tunnels above (Figure 14). A total of 18 piles were cut to construct the pedestrian link. Ground freezing is proposed as groundwater cut-off during mining works for the new TSL tunnels as ground improvement is known to be ineffective in Old Alluvium formation. Transfer beams parallel to the pedestrian linkway have been designed to support the loads of the NSL tunnels since part of the existing pile foundations need to be removed. Flat jacks were provided as a provision to jack up the NSL tunnels in the event of settlement. 4 nos. of flat jacks were provided below the NSL tunnel base slab, as shown in Figure 15.

Real-time monitoring of the settlement of the NSL tunnel soffit was conducted during the shield driving to excavate the adits for the pedestrian linkway, which involves pile cutting work. Monitoring data shows that the actual settlement of the NSL tunnels is within 1 mm for both Adit 1 and Adit 2. This is likely due to the high JGP strength installed below the NSL tunnels which supports the tunnels and prevents excessive settlement. The tests on core samples from the treated soil shows that the achieved undrained shear strength was between 600 kPa to 1000 kPa, which exceeds the minimum design requirement by a large extent. The existing foundation piles were safely removed without adverse effect on the existing NSL tunnels. These indicate the effectiveness of the solutions offered to the project challenges and highlights the success of the concept design.

Minimal displacement of the NSL tunnel soffit was captured by the instrumentation system. This could be a combined consequence of two possible factors. First, the flexural stiffness of the NSL tunnel boxes which might be conservatively estimated during design. Second, the undrained shear strength of the JGP-improved ground which is much higher than the design-required value. As a result, the provision of flat jacks becomes redundant.

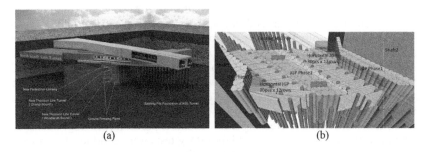

Figure 13. (a) Section Illustrating the Construction Works at Marina Bay Station and (b) 3D view of the ground improvement configuration (from Hashida et al. (2018)).

Figure 14. Excavation face with existing foundation pile during shield driving (from Hashida et al. (2018)).

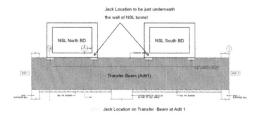

Figure 15. Arrangement of flat jacks between transfer beam and NSL tunnels (from Hashida et al. (2018)).

Figure 16. 3D model of the frozen soil based on temperature monitoring data (from Chua et al. (2018)).

Mining of the 5.8m internal diameter stacked mainline tunnels underneath the constructed pedestrian linkway with excavator and temporary ground support was proposed due to the challenging ground condition and the short tunnel span of 40m. This also leads to the adoption of the sprayed concrete lining (SCL) method, which requires dry condition for construction. The lower mainline tunnel is in mixed soil condition of Marine Clay and Old Alluvium. However, since ground improvement is deemed to be ineffective to treat Old Alluvium, there is a risk of water ingress during mined excavation. As such, ground freezing technique is proposed to prevent water ingress through the Old Alluvium layer during mined excavation. Vertical ground freezing tubes were installed from within the completed pedestrian link to the depth of Old Alluvium layer to prevent excessive water ingress and to ensure stability of the mining faces (Figure 16). The permanent lining of the mainline tunnels is constructed by cast-in-situ concrete lining.

The main concern with ground freezing includes soil expansion (heaving) during soil freezing and settlement during thawing. Special laboratory tests were conducted to investigate the behaviour of the frozen local soil, and to determine its engineering properties as well as design parameters. 3D finite element modelling was then carried out to simulate the mining of the

Table 2. Comparison between predicted and actual movement of existing tunnels.

Location of interface	Predicted movement (mm)	Actual movement (mm)
East of Outram Park (Parallel to EWL)	58 *(cumulative displacement by several construction activities)*	2
East of Outram Park (Overcrossing NEL)	4.3	< 2
Undercrossing EWL at Robinson road	55 *(with mitigation measures)*	< 12

mainline tunnels in conjunction with ground freezing. The impact of the construction on the existing MRT tunnels was then investigated and controlled.

The finite element analysis predicted a maximum heave of NSL tunnels to be 7mm and the overall settlement to be 7 mm, considering all construction phases. The vertical displacement of the NSL tunnels was continuously recorded during the construction of the mined tunnels using the ATMS. The actual deformation of the NSL tunnels was significantly smaller than the predicted 7 mm. This is attributed to the fact that the structural rigidity and restraints of the NSL tunnels were not fully incorporated in the finite element model.

In addition, the heaving of the frozen soil was not observed according to the monitoring data. This suggests that the test results of the frozen soil sample may not represent the actual behaviour of the frozen in-situ soil mass.

6 CONCLUSION

The TEL project in its entirety can make a good argument to be one of the most challenging MRT design and construction projects in Singapore and South-East Asia. The alignment encounters all of Singapore's major geological formation and drives into the heart of its highly developed CBD. TSL-D embodies the most complex of engineering challenges, with contrasting geological conditions and close proximity tunnelling to existing railway and stations.

Table 2 summarises the predicted movement of the existing tunnels at interfaces with TSL tunnel alignment and compared against the actual data recorded via the monitoring system. From the table, it can be seen that the design analysis is conservative as the achieved movement with closed face TBM tunnelling control measures and additional considered mitigation measures implemented is much lower.

REFERENCES

D.S.T.A. 2008. Geological Map of Singapore. Singapore: Defense Science and Technology Agency.
Hashida K., Tada H., Watanabe T., Chua T.S., Lew M. and Nair R.S., 2018. Linkway Construction Underneath Existing MRT Tunnels by Open Faced Rectangular Shield for Marina Bay Station. Underground Singapore 2018.
Chua T.S., Lew M., Hashida K., Tada H., Takeda S., Marican S. and Nair R.S., 2018. Ground Freezing for the Design and Construction of SCL Tunnels underneath Marina Bay Station, Singapore. Underground Singapore 2018.
Mair R. J., Taylor R. N. and A. Bracegirdle, 1993 Subsurface settlement profiles above tunnels in clays Géotechnique 43. No. 2, 315–320
Shirlaw J. N., Tham-Lee S. K., Wong F. K., Ang-Wong L. P., Chen D. C., Osborne N. and Tan C. G., 2000. Planning the monitoring required to confirm the movement and rotation of tunnels and track-work due to excavation and tunnelling. Proceedings of International Conference of Tunnelling and Underground Structures, Singapore.

Tunnels and Underground Cities: Engineering and Innovation meet Archaeology,
Architecture and Art, Volume 12: Urban
Tunnels - Part 2 – Peila, Viggiani & Celestino (Eds)
© 2020 Taylor & Francis Group, London, ISBN 978-0-367-46900-9

Crossing of tunnels between the piles of operational metro station

A. Raj & A. Kumar

Delhi Metro Rail Corporation Limited, New Delhi, India

ABSTRACT: This paper is a case study of crossing of Delhi Metro's Line-7 tunnels between the piles of operational Line-6 Lajpat Nagar Metro Station. The pile foundations of existing station which were falling within the zone of influence of the tunnels were likely to get affected by the tunnelling and construction activities. To keep lateral deflection of pile and ground settlement under acceptable limit, it was necessary to reduce the effect of vibrations generated during TBM operation and movement of piles induced due to tunnelling. To reduce the mentioned effects, contiguous RCC piles were casted in front of existing piles of Lajpat Nagar metro station. The casted curtain of RCC piles reduced the lateral deflection and ground settlement under acceptable limit. Also to reduce vibrations, TBM speed was reduced while passing existing station. An extensive instrumentation monitoring system was placed adjacent to piles, piers and station structure to ensure safe working. This monitoring was covered on a 24 hour, 7 days basis by two shifts of monitoring team.

1 INTRODUCTION

The total length covered by metro after the completion of phase-III of project assigned to Delhi Metro Rail Corporation is 329 km. A total of 140 km of the total length is added to the existing network of 189 km in phase-III. In this 140 km, underground twin tunnels of length 53 km (106 km tubes) are constructed in phase-III. The current work discusses about the challenges faced and process used in constructing these underground tunnels.

The network from Mukundpur-shiv vihar also known as line-7, constituted of 58.596 km. Line-7 has 39.479 km of elevated network and 19.117 is km underground network. There are total 38 stations out of which 26 are elevated and 12 are underground. The tunnels between South Extension and Lajpat Nagar metro stations of line-7 were crossing between the piles of operational line-6 Lajpat Nagar Metro Station. As shown in Figure 1, the depth of line-7 tunnels from the platform of existing Lajpat Nagar station was 10.5m and the minimum clearance from the existing piles was 2.4m. The pile foundations of existing Lajpat Nagar Metro station which were falling within the zone of influence of the tunnels were likely to get affected by the tunnelling and construction activities.

1.1 *Ground Condition*

The urban area soil in Delhi consists of a layer of alluvium, silt with contents of sand and clay including pockets of coarse material, on a rock bed of quartzite with varying weathered state according to depth. The construction of tunnels of line-7 below existing Lajpat Nagar station was carried above the water table. The ground condition at this location is below in Table 1.

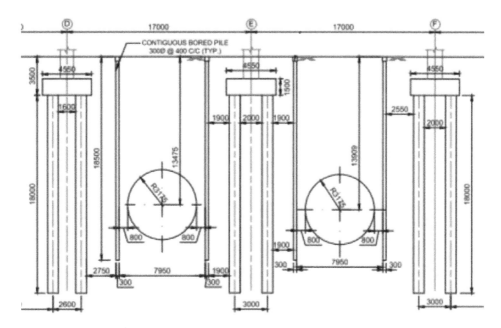

Figure 1.　Section showing distance between piles and tunnels

1.2　*Assessment of Existing piles*

A typical pile-tunnel configuration used for the assessment is shown in Figure 2. Depending on the location of the tunnel centre line with respect to the pile toe level, the pile can be assessed as either short pile (Lp/H < 1) or long pile (Lp/H > 1).

The stress-strain development mechanism induced by tunnel excavation is different for buildings founded on piles. It is not appropriate to consider green field settlements for building risk assessment for buildings founded on piles because the building will not settle along the ground settlement trough. Building settlement is the result of pile settlement as sown in Figure 3.

Tunnelling-induced effects on existing pile foundations was assessed under long pile category since Lp/H > 1. To overcome this care has been taken to restrict the ground settlement with in 5mm.

1.3　*Potential Risk to Existing piles*

Due to tunnelling induced ground movements and vibrations generated during TBM operation, existing pile foundation of existing Lajapat Nagar Station falling within the

Table 1.　Soil Parameters

Layer #	Depth (m)	N60	CN	(N1) 60	Ko = (1-sin Φ') OCR0.5	Permeability (x10-6) cm/s	Design Parameters				
							Bulk Density	Strength Parameters		Stiffness Parameters	
							γb	c'	Φ'*	E'	v'
							kN/m3	kPa	deg	MPa	
Layer -1	0–5	14	1.22	17	0.53	3	18	-	28	10	0.30
Layer -2	5–15	46	0.85	39	0.46	8	19	-	33	32	0.30
Layer -3	15–28	78	0.55	43	0.44	24	20	-	34	55	0.30

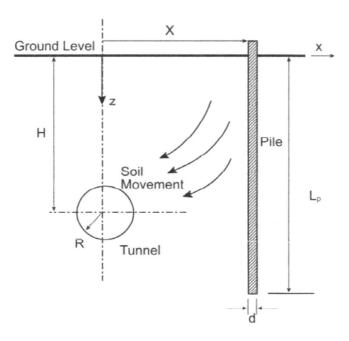

Figure 2.   Typical pile-tunnel configuration

zone of influence of tunnels was subjected to high lateral deflection during twin tunnel excavation. By analyzing the behavior of piles during tunnel excavation in PLAXIS-2D, it was observed that Lateral deflection in few of the piles was more than 5mm, which was not acceptable. The vertical settlement of ground observed at above the twin tunnel locations was in the range from 23mm to 95mm, which was too high. Ground settlement adjacent to the pile also induced negative skin friction on pile. This negative friction added extra load on the pile.

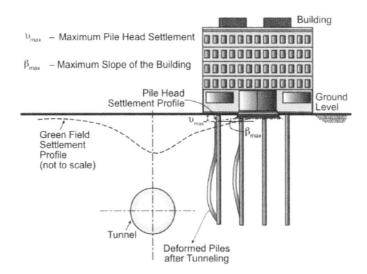

Figure 3.   Schematic diagram of pile settlement profile and ground settlement

Also, the vibrations generated during tunneling operation was to be minimized as the same would have resulted more lateral deflection and ground settlement in the proximity of twin tunnels.

Considering the safety of commuters and importance of the existing Lajpat Nagar Station, protection measures were required for the structure to reduce the impact of tunnelling on the pile foundations and on the station structure.

2 PROTECTION MEASURES

Considering the safety of commuters and importance of the existing Lajpat Nagar Station, protection measures were required for the structure to reduce the impact of tunnelling on the pile foundations and on the station structure.

2.1 *Contiguous piles*

To control the lateral deflection of existing piles and vertical settlement of ground, it was considered to drive the contiguous piles in between existing piles and tunnels. In order to reduce the lateral deflection of existing piles and vertical settlement of ground as shown in Figure 4, 300mm diameter of contiguous piles with 400mm c/c spacing was driven to reduce the volume loss of soil. The distance between contiguous piles and tunnel was kept 800mm.

The actual site photographs showing the machinery used for carrying piling work and the progress of work is shown in Figure 5.

The insertion of contiguous bored pile formed a curtain in front of existing piles (as shown in Figure 6) and from PLAXIS analysis, it was observed that there was drastic reduction in vertical ground settlement and lateral deflection of pile after driving the contiguous pile at and around the twin tunnel surface and deflections of

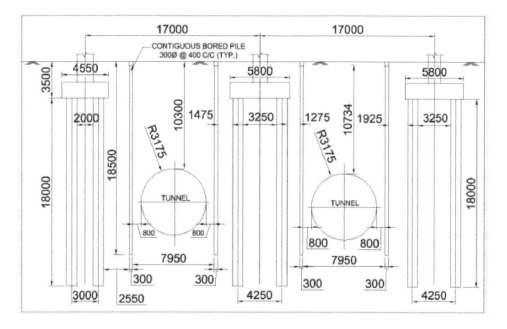

Figure 4.   Insertion of contiguous bored pile

Figure 5. Machinery used for carrying piling work and the progress of work

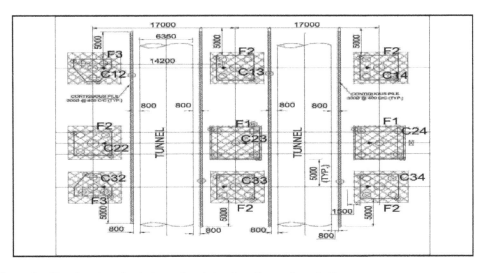

Figure 6. Plan showing pile cluster and curtain of contiguous bored pile

existing piles analyzed after insertion of contiguous piles were within the permissible limit.

The movement of ground before and after insertion of contiguous piles is shown in Figure 7.

The comparison of lateral deflection of existing piles and vertical settlement of ground before and after the drive of contiguous piles for the most severely affected piles has been shown in Table 2 and Table 3 respectively. The ground settlement at most critical location at the time of excavation of twin tunnels has been shown in Table 4.

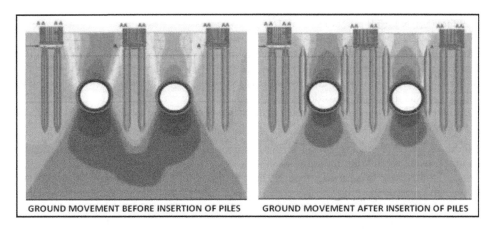

GROUND MOVEMENT BEFORE INSERTION OF PILES | GROUND MOVEMENT AFTER INSERTION OF PILES

Figure 7.   Comparison of ground movement before and after contiguous piles

Table 2.   Lateral deflection of existing piles

Column/Footing	Pile	Length of Pile	Lateral deflection (mm)	
			Before contiguous piles	After contiguous piles
C12/F3	P3	18	10.664	2.683
	P4	18	10.669	2.681
C13/F2	P5	18	6.774	3.022
	P6	18	6.018	3.026
C14/F2	P7	18	7.477	3.082
	P8	18	3.917	2.267

Table 3.   Vertical deflection of existing piles

Column/Footing	Pile	Length of Pile	Vertical deflection (mm)	
			Before contiguous piles	After contiguous piles
C12/F3	P3	18	57.338	3.473
	P4	18	57.049	3.146
C13/F2	P5	18	51.505	4.237
	P6	18	51.867	4.568
C14/F2	P7	18	48.799	2.041
	P8	18	48.658	1.688

Table 4.   Ground settlement during tunnel excavation

Stage	Ground Settlement (mm)	
	Before contiguous piles	After contiguous piles
Excavation of Twin Tunnel	93.62	19.95

3 INSTRUMENTATION MONITORING

Instrumentation monitoring by precise leveling was undertaken to measure actual ground movements. This monitoring was carried out to verify that the actual ground/building settlements were as predicted, which ensured structural movements and damages to existing piles and buildings was under the considered limits.

Monitoring of buildings and structures within the zone of influence of tunnel was carried by using following instruments:

a. Building settlement marker - Monitoring by precise leveling was carried to measure actual ground movement on all buildings
b. Optical survey targets - On the piers, at height optical targets were placed and 3-D monitoring was carried
c. Ground Settlement marker - To monitor the settlement of ground level, ground settlement markers was used and precise leveling was carried to monitor the movement of ground
d. Tilt Meters - To monitor the movements of piles, tilt meter was placed on piers and pile caps.

To ensure safety of commuters using the operational line, it was necessary mitigate any damage to the existing metro line. For the same at track level of existing metro line, 24 hours real time monitoring was carried with the help of beam sensors.

3.1 Frequency of monitoring

Based on the instrumentation type and location, the construction progress and criticality of activity, frequency of monitoring adopted is mentioned in Table 5.

3.2 Review levels

Instrumentation monitoring review levels was defined in accordance with the contract and acceptable parameters. Trigger Values was be established as approximately 70% of the Design

Table 5. Frequency of monitoring

Instrument	Monitoring frequency		
	Prior to tunnelling	During tunnelling and within 10m boundary region of station location	After tunnelling
Building Settlement Markers	Weekly	Every 6 hours	Weekly
Optical Survey targets	Weekly	Daily	Weekly
Ground Settlement Markers	Weekly	Every 6 hours	Weekly
Tilt meters	Weekly	Daily	Weekly
Beam Sensors	24 hours real time monitoring		

Table 6. Review levels

Instrument	Trigger level	Allowable level	Design level
Building Settlement Markers	7mm	9mm	10mm
Optical Survey targets	3.5mm	4.5mm	5mm
Ground Settlement Markers	17.5mm	22.5mm	25mm
Tilt meters	1/4000	1/2550	1/2000

Values (limiting values) and allowable levels as 90% of design values. The Limiting Values will be set as stated in Volume 4 of tender document. The design value for the review levels were selected based on the actual calculation done on the PLAXIS model. The review levels for different instruments are shown in Table 6.

4 CONCLUSION

The project was challenging and allowed us to learn quite a lot of things. Firstly, it was observed that vibration can significantly increase in case the TBM speed was high. Hence, a bare minimum speed was kept. The reduction in the speed of TBM also helped in controlling the deflection and settlement. Further, isolating structures were found more suitable than going for the ground stabilization. Also, instrument monitoring is a rigorous process which was religiously followed for identifying any movement of ground structure. After the insertion of contiguous piles, the deflection of most critical pile was also brought under acceptable limit i.e. below 5mm. The ground settlement also reduced to below 20mm.

The exhaustive instrumentation monitoring explained above and by taking protection measures, the tunnels were successfully constructed below the operational Lajpat Nagar metro station. No damage has been observed on the structures coming in the zone of influence of tunnels and no change has been observed in the track level of existing metro line.

Tunnels and Underground Cities: Engineering and Innovation meet Archaeology,
Architecture and Art, Volume 12: Urban
Tunnels - Part 2 – Peila, Viggiani & Celestino (Eds)
© 2019 Taylor & Francis Group, London, ISBN 978-0-367-46900-9

The Fereggiano River Diversion in Genoa

P. Redaelli
PAC S.p.A., Bolzano, Italy

M. Bringiotti
GeoTunnel S.r.l., Genoa, Italy

ABSTRACT: Genoa is frequently hit by disastrous flooding events; therefore, the City has recently implemented part of the works included in the Final Project of the Bisagno river diversion tunnel. Wednesday the 17.02.2016 the Italian Minister of the Ambient stated: "In Genoa we are building one of the most important hydrogeological restructures ever done in Italy". In this document the project related to the first allotment contract, called the Fereggiano River Diversion, is described: a 3.7 km long, 6 m wide tunnel that runs from the intake on the river, in the north of the city, to the east seaside, under some of the most densely populated neighborhood of the city. The excavation under the city with low overburden, the two principal construction sites on the richest seaside of the city and in the narrow riverbed of the Fereggiano, the secondary construction sites in the city tissue are the main issue of the contract.

1 INTRODUCTION

The construction of the Fereggiano River Diversion started in the late eighties, as a service tunnel for the excavation of the more complex project of the Bisagno River Diversion. The work stopped in early nineties because of administrative issues and was left undone for more than twenty years.

After the flood of the Fereggiano River in 2011 that caused the death of 6 people, the Municipality of Genoa, with a noticeable economical effort, financed the first part of the first allotment contract consisting in: the Fereggiano Tunnel (lining inner diameter: 5.2 m; length: 3700 m), the intake on the Fereggiano riverbed with a vortex shaft, the sea outfall and four ventilation shafts. This was the minimal configuration to prevent river flood in the city neighbourhood lying in the Fereggiano Valley. In 2015 the contract between PAC and the Genoa's Municipality was signed.

2 THE MAIN TUNNEL

The Fereggiano Tunnel is a horse shoe shaped 5.2 m wide and 3.723 m long tunnel, with a 0.3% longitudinal slope (Figure 1).

The first 909 m were excavated via an open TBM in the early nineties and left unfinished because of financial problems, after being safeguarded with a cast in situ concrete lining. The section of this first stretch (Figure 1) is narrower than the new Fereggiano Tunnel and with a 2.75 m free span for vehicles is one of the logistic main issue of the work.

The Fereggiano Tunnel (Figure 2) starts with the sea outfall on the "Corso Italia Seaside", one of the most beautiful promenades of the region, then mainly passes under the rich neighbourhood called "Albaro", with an overburden variable between 35 m and 160 m, and ends under the Fereggiano riverbed with a 22 m deep intake shaft.

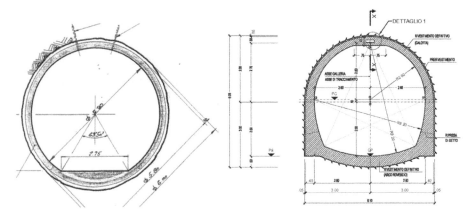

Figure 1.   Cross section of: existing tunnel (left) and new Fereggiano Tunnel (right).

Figure 2.   Fereggiano Tunnel general plan.

2.1   Geology

Almost all the tunnel length pass through the mixed carbonate-siliciclastic flysch of the Monte Antola Formation (Figure 3), having compressive resistance varying from 20 MPa in the pelites layers to 160 MPa in the sound sandstone layers.

This very ancient formation suffered strong tectonic deformation, resulting in a chaotic geometry of the original layers, that sometimes could not be recognised (Figure 4).

Only in a 100 m long stretch the excavation passed through the Ortovero Clays formation, dealing with over consolidated clays and argillites.

2.2   Advancing methods

The four typological sections envisaged in The final Design are:

– A0/A0x – with PPFR shotcrete and rock bolts (Swellex type);
– A1 – with lattice girders and PPFR shotcrete;
– B0 – with fore poling, 2 x IPE 160 steel ribs and PPFR shotcrete;
– B1/B1a – with steel pipes, 2 x IPE 160 steel ribs and PPFR shotcrete.

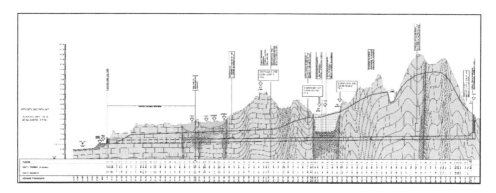

Figure 3.   Geological section along Fereggiano Tunnel.

Figure 4.   Left: excavation intersect a tectonic fold; because of layers bending, it was hard to keep the correct tunnel shape. Middle: a fully tectonized rock mass, with no sign of the original layering; thanks to re-cementation, the front face was stable. Right: accurate Robodrill robotized drilling protocol in the tunnel.

The B0 section with forepoling was widely applied in poor rock since it allows to advance with the two booms Robodrill jumbo, thus efficiently coping with the fast-changing rock mass condition, instead, the B1 steel pipes section (and consequently the tunnelling rig) was only applied for the most critical stretch.

Given the high resistance of some portion of the rock mass, it was impossible to excavate only by mechanical means (that could minimize annoyance to the urban surface); consequently all the tunnel was excavated choosing day by day between mechanical equipment (either hydraulic breaker or roadheader) and Drill & Blast, depending on rock mass condition.

The use of explosives in a highly populated urban area has been a noticeable issue. Blasting was only allowed in daytime; moreover, the shallow depth of the tunnel required an in-depth study of blasting schemes, day by day updated thanks to the results of the monitoring system. In fact, more than 40 vibrometers were installed along the tunnel alignment, sending data in real in time via gsm to a remote server.

The principles of the blasting design were:

– The use of millisecond delay blasting cap combined with ordinary delayed action cap, in order to minimize the single-phase loading;
– The drill of small diameter holes loaded with small diameter cartridges, in order to minimize linear loading.

The result was a medium incidence of 2.5 m of mining holes each rock mass cubic meter.

All the drilling activities were carried out by the French partner Robodrill, using a two booms fully computerized jumbo, equipped with feeds and clamps for forepoling with self-drilling anchors.

2.3 The Genoa-La Spezia Railway tunnel Underpass

Between the chainage km1+560 and km1+596 the tunnel underpass the twin tube railway tunnel "S. Martino", one of the oldest railway tunnel in Italy (half XIX century), with a 3 m vertical diaphragm (Figure 5).

The agreement with RFI (Italian Railway managing authority) established:

– The installation of a monitoring system within the railway tunnels, in order to check in real time strains and displacements in both the rails and the tunnel lining (Figure 6);
– The adoption of a procedure for the risk management, linked to an automatic system of monitoring data analysis and alarm;
– The ban on the use of explosives closer than 20 m to the historical tunnels.

The expected deformations were evaluated by the designer Lombardi Ingegneria via a 3D FEM parametric model and were in real time compared to monitoring results.

After having checked the rock mass quality in the underpass stretch with a 45m long horizontal core drill, a tailor-made typological section called B1a was applied, with the following features:

– n. 33 steel pipes at the crown, L=12 m, step 6 m;
– 2 x IPE 160 steel ribs, step 1 m;
– Double phase PPFR shotcrete.

Given the ban on explosives, the 60 m long stretch was entirely excavated with a roadheader, that experienced poor advancing rates, due to the high resistance of the rock mass; consequently, the excavation of the underpass took about 60 days, with an advancing rate of 1 m/day. Thanks to the very good rock mass quality, monitored deformations were lower than expected and almost nonexistent.

2.4 Rovare and Noce Rivers diversion chambers

Along its track the Fereggiano Tunnel will collect waters of two rivers called rio Noce and rio Rovare, by means of two intake-tunnel systems that aren't included in this first allotment. The

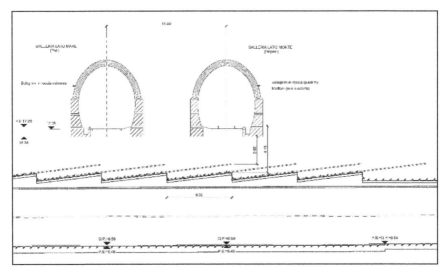

Figure 5. The twin tube railway tunnel "S. Martino" crossing the Fereggiano Tunnel in construction.

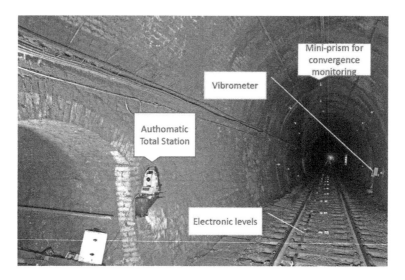

Figure 6. Monitoring system within the historical railway tunnel.

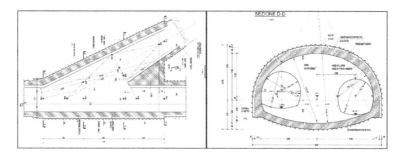

Figure 7. Rovare e Noce Rivers diversion chambers.

only works included in the present project are the two twin diversion chambers, that increase the tunnel span to an internal width of more than 17 m (Figure 7).

The huge sections (as compared to the 6 m span of the line tunnel) were excavated in two phases, with a pilot tunnel on Fereggiano alignment, later widened to the final shape.

The split in two phases of the excavation was compulsory for two reasons: limiting explosive quantity to that allowed in the permit given by the authority and keeping low vibration induced to surface.

The blocky nature of the rock mass, together with the generous span of the excavation (Figure 8), brought to a systematic installation of variable span steel ribs immediately after each advancement phase.

In the widening phase an average advancing rate of 1.5 m/day was experienced, with excavation width varying between 6 m and 19 m.

3 THE INTAKE

The intake system (Figure 9) consists in:

- a concrete structure to be built within the riverbed of Fereggiano River, with a spiral wall that creates a vortex in the water;
- an intake shaft, with 5.9 m internal diameter and 2 2m depth;
- a Venturi chamber excavated at the tunnel level.

Figure 8. Rio Noce diversion chamber and, on the right, the Fereggiano Tunnel.

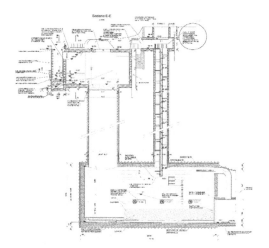

Figure 9. Lay-out of the intake system.

The underground (Venturi) chamber is connected to the surface by means of an access shaft, with 2.5m internal diameter and 35 m depth, that will house a spiral staircase.

The Venturi chamber is a 12 m wide, 26 m long tunnel, excavated via mining methods at the end of the Fereggiano Tunnel, beyond the chainage km 3+723. The excavation will be executed in two phases, with a first phase 8m x 8m pilot tunnel, widened to the final shape after the excavation of the shafts.

The access shaft was excavated via Raise Boring machine (3.03 m nominal head diameter), after having improved a shallow landfill layer 8 m thick with VTR reinforced concrete micropiles and cement moisture injections. Shaft walls have been safeguarded by rock bolts and hexagonal rock net, installed top down by cragsmen.

The concrete final lining of the shaft will be cast down-top with a raising formwork.

The intake shaft has 6.3m excavation diameter; it could not be carried out via Raise Boring because it would require a too large construction site as compared to the available area.

Figure 10. Left - Raise boring pilot hole hits the crown of Fereggiano Tunnel, surveyors breathed a sigh of relief. Right - Beginning of intake shaft widening.

Furthermore, such a large unlined cavity, as it would be if excavated via Raise Boring, could experience stability problems in relation to the blocky nature of the rock mass.

Therefore, a solution combining down-top raise boring and top down enlargement excavation with contemporary lining and support installation has been carried out.

In the first phase, a 2.3 m wide pilot hole has been bored via Raise Boring, in order to have a comfortable way to demuck. In the second phase, the hole has been widened with a hydraulic hammer; each a 2 m step, the excavation was stopped and the shaft wall was safeguarded with wire mesh reinforced shotcrete and rock bolts (Figure 10).

4 THE OUTFALL

The Fereggiano River Diversion will share the sea outfall with the biggest Bisagno River Diversion, that will be soon contracted. Since the Fereggiano Tunnel must operate during the construction of Bisagno Tunnel, the outfall will be completed in this first allotment.

The last hundred meters of the Fereggiano Tunnel lay for a half on the seaside and for the other beyond the coastline; this second section will be hosted in a temporary Landfill, safeguarded by a 3.5m high cliff.

All this stretch will be excavated via cut & cover method, with bulkheads providing support and waterproofing. The choice of bulkheads technology was influenced by two conditions:

– the proximity to the sea, involving the risk of water pollution;
– the presence of a shallow bedrock, unfavorable for diaphragm walls or large diameter pile walls.

Finally, a mix of two technologies was adopted:

Figure 11. Left: 1280mm secant piles on the landfill. Right: excavation under the sea level.

- on the seaside, the wall is composed by bi-fluid jet grouting columns Ø1000mm and micro-piles Ø350m i.d. reinforced by HEB200 steel beams;
- in the off-shore landfill, a pile wall with 1280mm diameter secant piles is envisaged.

5 VENTILATION SHAFTS

Along the track of the tunnel four ventilation shafts 600mm i.d. had to be built, varying between 35m and 90m in depth.

Depending on ground conditions, two technologies were applied:

- In stable rock, shafts were bored via R.B.M. machine,
- In soft soil or mixed ground conditions, shafts were bored by a light drilling Rig.

At the end, two shaft was excavated by R.B.M. and the other two via Drilling Rig.

One shaft, intersecting the tunnel near to the middle, was over-bored to an internal diameter of 1000mm and used to provide fresh air to the tunnel advance. The main issue of the ventilation shaft sites, all located within densely populated neighborhood, were the lack of space and the difficulties to access by truck and other heavy vehicles.

The R.B.M. technology proved to be the most versatile, being possible to divide the site in a very small area at the head of the shaft (Figure 12) and in a second area, linked via pipes and electric cables (housing electric generator, air compressor and other facilities), that could be hundreds of meters far away. Founding the area for the Drilling Rig was more difficult:

- in one case the closure of a carriageway of via Ricci, a street in the city center, was needed;
- the other site was located within a flowerbed in S. Martino Hospital area, taking up not more than 200square meters.

6 LOGISTIC

The most critical issue of the contract was undoubtedly the location of the main site on the seaside, within a unique landscape and in the center of a prestigious neighborhood.

In the tender phase competitors were asked to find the way to mitigate the impact of the site on this sensible context.

PAC submitted to the Commission the idea of constructing a large hangar, lined with phono-absorbent panels and painted in order to minimize sight impact on the environment, that would house all the facilities needed by the site; the idea was very well considered and brought PAC to win the tender.

Figure 12. Left: R.B.M drilling site. Right: Drilling Rig within S. Martino Hospital area. All very narrow areas.

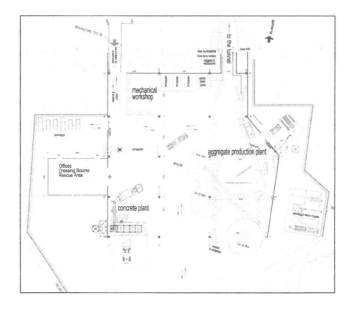

Figure 13. Main Site lay-out.

Figure 14. The Hangar from the promenade (left) and inside – quite different!.

Figure 15. Barge loading in front of the hangar.

The hangar houses site offices and plant for aggregate production, concrete plant with aggregate storage area, mechanical workshop, Electric power plants, fresh air supply plant and water treatment plant.

As prescribed by the contract, al the muck was carted away by the sea, brought to Genoa harbor and then carried by track to final destinations. The loading of the muck (Figure 15) was carried out by a crane equipped with a clamshell, working from a barge.

7 CONCLUSION

The primary task of this job was probably not only to excavate the tunnel, a complicate long one with quite a restricted section and different other engineering tasks, but mainly to cope with the delicate urban environment.

Company PAC S.p.A. designed a complete on shore plant installation, comprise of ventilation, dedusting, dewatering, crushing, belt conveying, ship muck loading plants, everything protected and hidden to the human habitants in a very delicate habit.

Only well-organized Planning, sensitive Technicians and Miners, good selected Partners and an intelligent Client could lead to the success such a job. Lesson to be learnt and followed!

Tunnels and Underground Cities: Engineering and Innovation meet Archaeology, Architecture and Art, Volume 12: Urban Tunnels - Part 2 – Peila, Viggiani & Celestino (Eds)
© 2020 Taylor & Francis Group, London, ISBN 978-0-367-46900-9

The T3 stretch of Line C in Rome: TBM excavation

E. Romani, M. D'Angelo & V. Foti
Metro C, Rome, Italy

A. Magliocchetti
Rocksoil, Milan, Italy

ABSTRACT: Line C is the third line of Rome underground system. Two TBMs are being used to excavate the tunnels at a depth ranging from 30 to 60 m below ground level. The job sites are located in the centre of Rome, in an area characterized by several historical buildings: as a consequence, technical and operative choices have been imposed, to minimize the job sites dimension. This approach required *Metro C* to use unconventional methods to start the TBM excavation from shaft 3.3, located in the "Via Sannio gardens" very close to the Aurelian Walls at Porta Asinaria. The two TBMs were lowered into this shaft to build the two tunnels which will run as far as Amba Aradam station, over a length of about 400 m each. In order to complete the shaft itself and the tunnels to San Giovanni station, as soon as the TBMs arrive at Amba Aradam station, the TBM site equipment will be moved into this station area to complete the excavation planned for the remaining stretch. Tunneling started last March and an accurate structural and geotechnical monitoring system was installed in the soil and on the existing buildings adjacent to the line, in order to check in real time the level of interaction with the excavations.

1 INTRODUCTION

Line C is the third line of Rome underground. Once completed, it will run under the city from the South-eastern to the North-western area, for a total length of 25.6 km and 30 stations.

Metro C, the General Contractor led by Astaldi, manages the construction of Line C in all its phases: design, archaeological surveys, tunnel drilling, excavation and construction of stations, building of trains and start-up. Line C is a driverless fully automated rail system, with automatic platform doors. The automation of the system will ensure a greater frequency for the vehicles (90 seconds, theoretically) providing greater transport capacity (24,000 passengers per hour per direction). Now 22 stations and 19 km of line are open to the public.

The T3 stretch is presently under construction: it runs over a length of about 2.8 km right underneath the historic centre of Rome. The new line was divided into four different sections. The first section will link San Giovanni Station, currently in operation, to Amba Aradam Station. San Giovanni Station is an interchange between the existing line A and the new line C. At present, it is linked only to the T4 stretch of Line C, which is already in operation.

Metro C started TBM excavation from shaft 3.3, located near san Giovanni station, in the "Via Sannio gardens", very close to the Aurelian Walls at Porta Asinaria The second section of the T3 stretch will connect Amba Aradam Station to Shaft 3.2. The third section will connect this shaft to Fori Imperiali Station. The fourth section is functional to the extension to the T2 stretch, where the TBMs will stop under Via dei Fori Imperiali, waiting for the continuation of the excavations. The T3 stretch includes 2 underground stations, Amba Aradam and Fori Imperiali, and 2 ventilation shafts: shaft 3.3 and shaft 3.2. The two tunnels (Odd and Even Rail) are being bored by two Earth Pressure Balance TBMs: the tunnels have internal

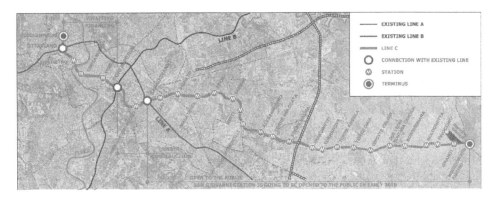

Figure 1. Line C lay-out.

Figure 2. T3 first track between shaft 3.3 and Amba Aradam Station.

diameters of 5.8 m and an external one of 6.4 m, whereas the excavation diameter is 6.71 m. The tunnels will run at a depth ranging from 30 to 60 m below ground level and they will be excavated in difficult soils. The job sites are located in the centre of Rome, in an area characterized by several historical buildings: as a consequence, technical and operative choices were imposed, to minimise the job site dimensions.

Tunnelling in the first section started in March 2018 and an accurate structural and geotechnical monitoring system was installed in the soil and on the existing buildings adjacent to the line, in order to monitor in real time the level of interaction with the excavations.

2   THE FIRST SECTION UNDER CONSTRUCTION

Design and construction were carried out applying the principles of the ADECO-RS approach (Lunardi, 2008), whose main stages are reported hereafter.

The survey phase regards the acquisition of information on the pre-existing equilibriums (characteristics of the buildings affected by the excavation, the geology and geotechnical characteristics of the soil, the natural stress-strain states, etc.).

In the diagnosis phase numerical analyses are performed, to assess the soil behaviour category depending on its response to the excavation and the level of risk associated with each building.

The therapy phase identifies the appropriate interventions (TBM parameters) which are to be implemented in order to get a category A ("stable core-face") soil response to the excavation.

The operational phase and monitoring during construction allow to continuously verify the design. In addition, appropriate adjustments are introduced, based on the monitoring of results and on the subsequent back-analysis, if necessary.

3 GROUND CONDITION

The area of the city of Rome is located on the Tyrrhenian side of the Central Apennines, where the Meso–Cenozoic carbonate rocks are covered by a succession of marine and continental Plio–Pleistocene sediments and products of the Sabatino Volcanic Apparatus and the Volcanic Apparatus of the Alban Hills. Based on the results of the surveys carried out and the general geology of the Roman area, along the route under examination the following lithological complexes have been identified, which will be excavated in the first track between Shaft 3.3 and the Amba-Aradam Station:

- Anthropogenic land complex (R)
- Complex of the Holocene deposits of the Tiber and its tributaries:
- Clayey- sandy and sandy grey-blackish silts, with variable organic content (LSO)
- Complex of the Pleistocene alluvial deposits
- Unit B of the Paleo-Tiber 2 (St)
- Unit A of the Paleo-Tiber 2 (Ar, Sg)
- Complex of the Pliocene deposits (Apl)

As regards the following tracks, the soils to be excavated are:

- Complex of the Pleistocene Volcanic deposits:
- Valle Giulia Unit (Tb)
- Pyroclastics of the volcanic system of the Alban Hills (Tl)
- Units of pyroclastic soils (Ta)

The landfills are loose, heterogeneous soils with a sandy clayey matrix and consist mainly of more or less altered pyroclastic materials reworked with fragments of stone tuffs and bricks of various sizes. The recent floods of the valley bottom consist essentially of clayey and sandy silts and medium to fine-grained, silty sands, with some gravel intercalation. The pyroclastic deposits consist of an alternation of more or less cineritic layers and of more or less cemented scoriaceous lapilli layers (Granular Tuffs). In the Pleistocene fluvial-lacustrine and fluvial-marshy sediment complex it is possible to distinguish three main horizons:

- the summit (St), consisting essentially of silty sands which fade into coarse calcareous sand with a low degree of cementation;
- the intermediate one (Ar), consisting of weakly clayey silty sands;
- the basal one (Sg) consisting of a of heterometric coarse bank gravel with a sandy matrix.
- Finally, the Pleistocene fluvial-lacustrine and fluvial-marshy sediment complex, consisting of very thick clays with frequent very thick sand intercalations.

The following tables show the geotechnical parameters adopted in the numerical analyses for each lithological unit in the first track between Shaft 3.3 and Amba Aradam Station:

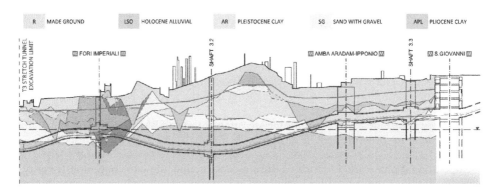

Figure 3.   Geological profile of the stratch.

4 TBM EXCAVATION

The tunnels for the T3 stretch of Rome Underground will all be constructed using EPB (Earth Pressure Balance) TBMs. The final lining of the tunnel will be made of precast segments placed by the machine immediately after excavation at a small distance behind the face. The ring for the 5.8 m internal diameter tunnels consists of six 30 cm thick segments (6 + 1 keystone). In fact, the pressure exerted in the excavation chamber allows to exert the confinement pressure of the front necessary to guarantee the conditions of stability of the face in the short term and to limit the volume loss in progress, on whose extent depend the settlements generated on the surface.

The T3 section presents high ground coverings and a soil of poor geomechanical characteristics, therefore concrete segments with a characteristic strength of 50 MPa were needed. In the project the behavior of the front during the phase of excavation has been examined, identifying the range of variability of the pressure value to be maintained in the excavation chamber along the route of the tunnels, depending on the cover, the geotechnical characteristics of the land and the groundwater level.

Pressure values have been calculated by analytical methods. It is expected, throughout the phase of transit of the machine, a continuous control of the affected area, through a network **monitoring**, consisting of various types of measurement stations, located both transversally and longitudinally to the tunnel line, in order to detect movements at ground level following excursions of the pressure value at the front. Along the route some maintenance interventions will be scheduled in order to guarantee the full operation of the machine front (tool replacement, pressure sensor calibration, etc.).

Anyhow, before launching the TBM under the monuments, the machine is set up so as to reduce the time related to all operations to the minimum necessary for the excavation phases (brush maintenance, instrument calibration, extension of cables and tapes, etc.). The final lining, besides performing and ensuring the normal function of support in both the short and the long term, must also provide the required hydraulic seal.

For this reason, the segments are fitted with watertight neoprene seals along all surfaces in contact with other segments, and arranged in corresponding housings on the sides of the segment, to ensure the required water-tightness under hydrostatic pressures with the planned clamping forces. In order to ensure the water-tightness between the segments of adjacent rings, as well as for safety reasons during the transitory phases of handling and laying the segments themselves, the connection is provided by means of longitudinal mechanical dowels (Biblock System or equivalent type) arranged at regular intervals around the circumference. The TBM is equipped with 19 thrust plates arranged in groups of three for each segment plus one for the keystone, or a total of 38 jacks acting in pairs on each plate. The dimensions of the

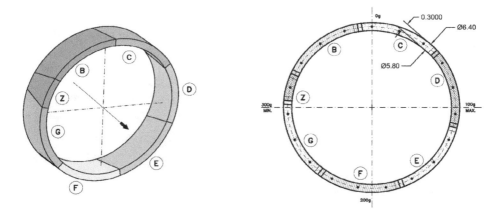

Figure 4. TBM section type: Universal ring – Particular of precast.

plate are 26 x 75 cm. The maximum thrust that the machine can exert is equal to 56,000 kN or 2,650 kN per thrust group.

5 TBM JOBSITE AND CONSTRUCTION PHASES

The peculiar urban environment in which the jobsites of the T3 stretch are located required the reduction of their sizes as much as possible, with significant consequences on the design and operational choices, in particular those related to the TBM excavation of the tunnels.

The launching shaft of the TBMs, called Shaft 3.3, is located in the Via Sannio gardens. Starting from that shaft, the gallery stretch up to the Amba Aradam – Ipponio station was constructed, for a length of about 385 each.

Once completed this gallery track, the TBM jobsite will be relocated to the Amba Aradam station to complete the mechanised excavation as far as the end of the stretch, for a total length of about 2.4 km for each direction, whereas the shaft and the galleries which extend up to the San Giovanni station will be completed by a traditional excavation.

The Via Sannio shaft is 62 m long and 22 m large: it is totally undersized as compared to those generally needed to enable the operation of these TBMs, which are equipped with a 100 m back-up. Moreover, at the same time as the excavation of the 2 TBM galleries, consolidation work of two 150 m long microtunnels was carried out, heading for the line stretch already in operation, starting from the same shaft.

The 2 TBMs (of about 450 tons each) were entirely assembled on the surface and lowered through a temporary slot down to the bottom of the shaft by a superlift Liebherr LG1750 750-ton lattice boom crane. Once laid on purpose-built steel structures, placed at the bottom of the shaft, the TBMs were moved as far as the head diaphragm wall to start the excavation.

Two false tunnels for the launch of the TBMs were constructed in shaft 3.3. These structures must be able to create an adequate confinement for the TBM head in order to reach the designed support pressure in the first meters of excavation outside the station, until the first ring is placed and backfilled in correspondence to the station wall. Before the TBM arrival, the lower part of the cradle is cast and anchored to the bottom slab and the reinforcement in the area between the wall and the TBM. When the TBM arrives, all the reinforcement bars can be mounted around the TBM and then the false tunnel is cast, protecting the station wall and the TBM shield with polystyrene pads. After the passage of the TBM back-up, the false tunnel structure will be demolished and the slab completed with an integrative casting until the final level.

In order to enble the operation of the machine, (350 m^3) slurry tanks were constructed at ground level. On the shaft, a 35 +38 tons Goliath crane was installed, dedicated both to the feed of the equipment needed for the excavation (precast segments, rails, pipelines, lubrication and sealing means, etc.) and to the extraction of the spoil.

The dimensions of the shaft also required to:

* start the excavation with only 4 out of 7 muck cars installed at the rear of the TBM;

Figure 5.   TBM job site in shaft 3.3.

Figure 6.   Archaeological remains in Amba Aradam Station.

- up to the (15th thrust, the excavation was performed by unloading the conveyor belt through a transverse conveyor on a car powered by a locomotive at the bottom of the shaft on a track adjacent to the TBM. The spoil car was equipped with bits, needed to extract it (through the central slot of the shaft) and unload it in the slurry tank through one of the two winches of the Goliath crane;
- up to the 35th thrust, once cleared the bottom of the central slot with the fourth car, it was possible to place the spoil car and add another car, at the rear of the TBM.
- starting from the 66th thrust, it was possible to complete the back-up of the TBM with the last 3 cars. By assembling also an elevated single switch at the tunnel mouth, it was possible to complete the final configuration for the excavation up to Amba Aradam station placing the back-up with a locomotive and 4 spoil cars, as well as another power unit car, precast segments and spoil cars.

Once completed the excavation of the first gallery, the second was started, with the same methodology and operative timing.

In order to respect the construction timing of the launching station (Amba Aradam), which was involved in important archaeological findings, both the TBMs accessed the area before the excavation had reached the foundation slab.

6   INTERACTION EVALUATIONS WITH EXISTING BUILDINGS

The hazards associated with the tunnel construction in urban areas include poor ground conditions, presence of a water table above the tunnel, shallow overburden (20m) and ground settlements induced by tunneling, with potential damage to the existing structures and utilities above the tunnel. In the case of the T3 stretch of Rome Underground a study to assess the interference between underground excavations and the existing buildings was carried out. The excavation is performed by a mechanized system, by means of an EPB type TBM. The subsidence trough related to the excavations have been evaluated, to assess the expected damage class for each building. Three stage of analyses should be considered:

- Empirical evaluation of settlement, which defines the subsidence basin with Dependent Gaussian on the Lost volume functions (Peck, 1969; O'Reilly and New 1982), has been carried out ("green field" analyses). As a result of the evaluation of the subsidence basin and of the surface and foundation level settlements, the effects induced on the buildings were evaluated by calculating the damage categories as reported in the literature (Mair et al., 1996). The envelope of the area affected by the deformations is therefore the function of the distance from the vertical axis of the tunnels and depends on the size of the tunnels, their dimension, the lost volume and the resistance and deformation parameters of the excavated ground.
- Settlement analysis by FDM analyses were performed only for the most significant cases (historical buildings), where the interference problem is more severe; so, a more reliable

prediction of the ground deformation response is achieved, taking into account the interaction with the building.

- A 3D Model was implemented only for the critical cases (historical buildings), taking into account the real geometry and the structural-construction system of the interference.

7 DIAGNOSIS PHASE – BUILDING RISK ASSESSMENT IN THE CURRENT SECTION

In the analysis of the excavation, the shape parameters of the Gaussian curve k=0.4 were assumed; the maximum volume loss is 0.5%. The following table shows the results of the analysis carried out in the current stretch. The damage criterion used for Line C provides two main categories (depending on the information deduced from technical datasheets) for the buildings:

- Category A: ordinary buildings with no previous structural damages
- Category B: ordinary buildings with previous structural damages and/or sensitive sites (school, hospital, offices)

Since settlement profile was produced at the foundations level, building risk assessment at this stage was carried out, in compliance with Burland & Wroth's theory (equivalent simple beam), calculating the maximum tensile strains induced by tunnelling. It is possible to define the following block diagram (Unified criteria for tunnelling induced damage assessment):

- Category A (ordinary buildings with no previous structural damages)

  Settlement prediction in free-field condition with empirical methods
  Damage assessment (Burland & Wroth, 1974)

- Category of damage equal/lower than 1 → no protective measures
- Category of damage higher than 1 → protective measures
- Category B (ordinary buildings with previous structural damages and/or sensitive sites)

  Settlement prediction in free-field condition with empirical methods:

- $W_{max} \leq 5mm$ OR $5mm < W_{max} < 10mm$ and $\beta \leq 1/1000$ → no protective measures
- $W_{max} \leq 10mm$ OR $5mm < W_{max} < 10mm$ and $\beta \leq 1/1000$ → no protective measures

8 MONITORING SYSTEM

A fundamental aspect of tunnel construction in urban areas is the control of subsidence induced by excavation, which is directly proportional to the volume loss values. The subsidence that occurs at the surface is, in fact, due to the deformation behavior of the core-face and the convergence values at the face, along the shield and in the area in which the precast segments are installed and grouted. If applied properly, the EPB technology can minimize the subsidence induced at ground level, in a manner which is compatible with the urban environment and the shallow overburden in some parts of the route. Controlling the pressure at the face, as well as the grouting pressures, allows for a careful construction; besides, the volume of grout mix injected behind the segments are of primary importance in limiting the volume loss. Correctly controlled tunnelling thus allows to keep under control the volume loss during excavation, and hence the predicted subsidence on the surface. Surface monitoring is primarily concerned with the buildings and their foundations, monuments and their foundations, the land on which these buildings are situated, and their variations in terms of geotechnical and hydro-geological characteristics.

- **External monitoring:** several ground settlement measurements are considered, through levelling pins, inclinometers (horizontal displacement), extensometer (vertical displacement) and vibrating-wire piezometer (water table movement).

- **Internal monitoring:** it checks correspondence between design analysis and in situ measurements, in order to ensure precast segmental lining to work properly. Stress state of tunnel lining had to be monitored by vibrating-wire strain gauges BE/BC, lining convergences and diametrical distortion by optical levelling MP (targets and prisms). TBM parameters (face pressure, volume of the excavated material, advancing rate) had to be monitored.

The control system considers several sections, perpendicular to the tunnel axis, provided by surface and internal instrumentation.

- The topographic monitoring sections (MON-02 ÷ MON-07) provide for the monitoring of the ground level (topographic monitoring). It consists in the installation of no 9–11 levelling pins installed at ground level, aligned perpendicular to the tunnel axis.
  - The geotechnical monitoring sections (MON-01) provide for the monitoring of the ground level (topographic monitoring), the stress-strain behavior of the ground (geotechnical monitoring) and the stress-deformation state of the lining (structural monitoring). It consists in the installation of the following:
    i.   no 10 levelling pins installed at ground level, aligned perpendicular to the tunnel axis;
    ii. no 3 inclinometers, 2 of which external to the excavation area and 1 along the tunnel axis;
    iii. no 2 boreholes, external to the excavation area, each with 2 vibrating-wire piezometers (the one for the surface water table and the other for the possible deep water table);
    iv. no 2 extensometers, installed along tunnel axes, whose depth depends on the tunnel cover;
    v.   no 10 mini-prisms, anchored to the tunnel lining: 5 within ring 116-Odd Track line and 5 within ring 12-Even Track line (pentagonal arrangement);
    vi. no 84 VW strain-gauges, for monitoring the stress state in the lining, within ring 116-Odd Track line and ring 12-Even Track line.

Mini-prisms, data-loggers and VW strain gauges identify the position of the instrumented rings (ring 116 for the Odd Track line and ring 12 for the Even Track line). In order to better define the stress-state in the ground around the tunnel in free-field condition, a further monitoring section was suggested:

- The special monitoring section (MON-05 "Campo Romulea") provides for the monitoring of the ground level (topographic monitoring) and the stress-strain behavior of the ground (geotechnical monitoring). It consists in the installation of following:

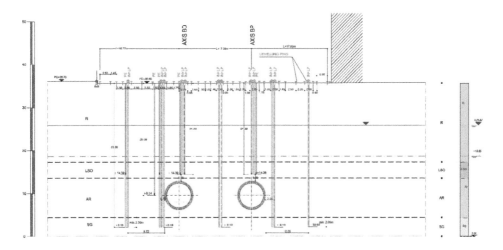

Figure 7.   MON 05: Monitoring section.

i. no 33 levelling pins installed at ground level, aligned perpendicular to the tunnel axis;
ii. no 5 inclinometers, with mini-prisms at the top;
iii. no 5 boreholes, external to the excavation area, with 2 vibrating-wire piezometers each (the one for the surface water table and the other for the possible deep water table);
iv. no 7 extensometers, with mini-prisms at the top.

The subsidence analysis is intended to provide a prediction of the subsidence values, which will be determined in compliance with threshold values. In particular, "warning" (80% of expected settlement, that is 9.2 mm) and "alarm" (100% of expected settlement, that is 11.5 mm) thresholds are identified, defined along tunnel axis in free field conditions. It is provided that also the parameters of the TBM excavations must be continuously monitored and compared to the design values.

9 BACK ANALYSIS – ODD TRACK SECTION ALREADY EXCAVATED

In order to calibrate the design parameter in the design range of variability and to monitor the excavation, the data are automatically recorded by the TBMs and processed in real time. In this case, the TBM face design pressure had provided limited values in terms of settlements and volume loss, in accordance with the threshold limits suggested. The following pictures show a comparison between the monitored settlement profile and the design settlement profile along the Odd Track line monitoring section.

As shown in previous figures, the design parameters had been correctly defined. The Volume loss can be assessed as 0.3–0.4%, while k=0.4–0.45 for most of sections; these values are close to the design parameters (V_L=0.5%, k=0.4). So, the design methods and assumptions were confirmed and so they were used to contain settlements and volume loss within the thresholds for the Even Track as well.

10 TBM ADVANCING DESIGN PARAMETERS – ODD TRACK LINE

The main parameters defined in the design and monitored during construction were those related to the pressure stabilization in the excavation chamber and the backfill pressure of the grout injections behind the concrete segments. The results of the monitoring were used to carry out numerical back analyses continuously, which made it possible to check the design machine parameters in real time and promptly take appropriate corrective action in order to conform to the safety requirements.

Pressure at the face: the pressure at the face and at tunnel contour were analysed and defined, in order to keep the loss volume within the design limit. In this case the design and

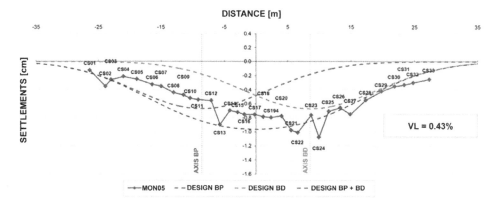

Figure 8.   MON 05: Transverse profile of settlements.

back analyses confirmed that this pressure is between the condition of active pressure (Rankin formulation) and at-rest pressure.

Grout injection pressure at the back of the TBM shield, to fill the voids between the precast segment lining and the excavation, was estimated according to front pressure, increasing it by 0.5-1bar. This value must be controlled, in order to prevent high pressure. Back-filling pressure must not overcome the average in-situ lithostatic stress.

Grouting injection volume: In order to control the balance between the tunnel volume, the volume of excavated material and the grouting volume, the weight of the excavated material was assessed. The grouting volume was assessed as the difference between the excavation volume (related to Ø=6.71m) and the extrados ring volume (related to Ø=6.40m). For each advancement (1.4m), equal to a single ring length, the design volume to be injected is 4.47m³

Weight of the excavated material: In order to control the balance between the tunnel volume, the volume of the excavated material and the grouting volume, the weight of the excavated material was assessed. The weight of the excavated soil corresponds to the volume of each advancement (1.4m) multiplied by the weight per unit volume of the excavated material (expected 20 kN/m³).

11 THE TBM ADVANCE DATA – ODD TRACK LINE

The analysis concerning the TBM data consisted in the registration of all TBM parameters and in their subsequent summarization. The data described above were summarized for each thrust and compared to the design value in term of: earth pressure, grouting volume, grouting pressure, volume of excavated material. Furthermore, in order to keep the thrust under control, other parameters were checked: speed and torque, force main thrust, advance speed. The following figures summarize the face support pressure values recorded by the sensors of the TBM, during the track tunnel excavation. The volume of the excavated material was controlled through the weighting of the muck extracted during the excavation. According to the TBM data recorded, the following can be established:

- The average face pressure are consistent with the design range within the defined track, with only a distortion between central sensors
- The back-filling injection pressures are oscillating, with an average value in accordance with design parameters

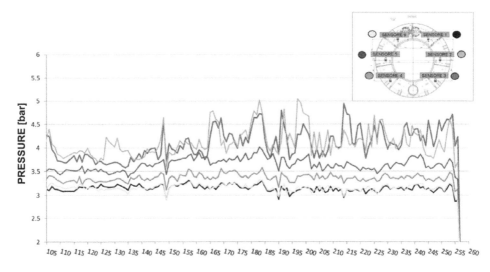

Figure 9.   Earth pressure balance.

- The weight of the extracted material appears to be constant and in accordance with the design value
- The back-filling injection volume appears to be slightly higher than the design value (about 20%). This value must absolutely be expected, due to the continuous maintenance

The other TBM parameters (torque and main thrust force) are consistent with the expected values.

REFERENCES

Boscardin, M. D., et al. 1989. *Building response to excavation induced settlement*; Journal of Geotechnical Engineer ing, ASCE, 1.

Lunardi, P., 2008. *Design and Construction of Tunnels. Analysis of Controlled Deformation in Rock and Soils (ADECO-RS)*. Springer, Berlin, Germany.

Mair, R. J., et al. 1996. *Prediction of ground movements and assessment of risk of building damage due to bored tunneling*.

O'Really, M. P., et al., 1982. *Settlement above ls in the United Kingdom – their magnitude and prediction*; *Proceedings of Tunneling Symposium 1982*

Peck, R. B., 1969. *Deep excavations and tunneling in soft ground*; *Proceedings of 7th International Conference on Soil Mechanics and Foundation Engineering*. Mexico City

Tunnels and Underground Cities: Engineering and Innovation meet Archaeology,
Architecture and Art, Volume 12: Urban
Tunnels - Part 2 – Peila, Viggiani & Celestino (Eds)
© 2020 Taylor & Francis Group, London, ISBN 978-0-367-46900-9

Estimated settlements on buildings and facilities. An application to Grand Paris metro

M. Russo & S. Bonaccorsi
Icaruss, Paris, France

ABSTRACT: A more than 30 billion € project of 200 km of new metro lines is planned in the Paris region in the Grand Paris Express scheme. In particular the stretch of Line 15 South – Lot T3C from Fort d'Issy-Vanves Clamart to Villejuif Louis-Aragon, including 8 km tunnel, 5 stations, 8 shafts has been awarded in March 2017 for 926 M€ to a JV lead by Vinci Construction Grands Projets, and composed by Spie Batignolles TPCI, Dodin Campenon Bernard, Vinci Construction France, Spie Fondations et Botte Fondations. ICARUSS is the JV subcontractor in charge for assessing the vulnerability of the buildings and structures on this stretch. In this paper is given an overview of the application of a systematic method for assessing the vulnerability of the structures affected by the construction of the structures of this stretch.

1 INTRODUCTION

A huge underground metro project is going on around Paris. Southern portion of the project is currently under construction. The stretch named T3C links the station of Villejuif Luis Aragon to the station of Fort d'Issy-Vanves-Clamart. It includes a 8 km twin track tunnel excavated by a TBM and 5 stations:

- VLA – Villejuif Luis Aragon
- IGR – Villejuif Institut Gustave Roussy
- ARC – Arcueil-Cachan
- BAG – Bagneux M4
- CHM – Châtillon-Montrouge

The analysis focuses on the 5 stations of the lot.

This paper focuses on first returns yield by the excavation of Arcueil-Cachan station, in particular the settlement of the adjacent ground and the lateral deformation of diaphragm walls.

2 THE EFFECTS OF THE WORKS ON STRUCTURES

The Client approach is to provide absolute thresholds for settlement, rotation and elongations. This approach differs from other in literature where a combined threshold set of values is given and has the advantage of being of more simple use for practitioners although more limiting thresholds are specified.

The Client ranked three classes of structural sensibility on the alignment, in terms of intrinsic sensitivity to the settlements from low to high sensitivity. Buildings and utilities have been ranked by location, actual condition, foundation, function, type of occupants, etc. For each sensitivity class à contractual threshold not to be overcome have been defined. The classification system is summarized in Figure 2. The system defines acceptable deformation thresholds to contain the damage in 3/4 categories: architectural, functional (moderate and important), structural.

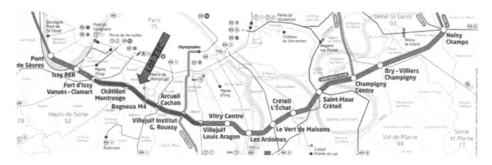

Figure 1.   Line 15 south alignment and lot T3C localization.

VULNERABILITY		Sensibility		
		Low	Medium	High
Category of damage	CD1-architectural damages	Low vulnerability	Low vulnerability	Vulnerable
	CD2.1-moderate functional damages	Low vulnerability	Vulnerable	Vulnerable
	CD2.2-important functional damages	Vulnerable	High vulnerability	High vulnerability
	CD3-structural damages	High vulnerability	High vulnerability	High vulnerability

Figure 2.   Vulnerability assessment principles basing on structures sensibility.

For each class of sensibility, limit values of deformations (settlements, horizontal strain et angular distortion damage) are defined. Low values of sensibility mean higher capacity of the building to resist to deformation.

The matrix shows the correlation between the thresholds (for each sensibility) and risk categories through vulnerability index evaluation. Since the contractual threshold is respected, the damage is repaired by the Client. If a contractual threshold is overgone, the Contractor has to repair the induced damage.

Only green cases are accepted by the Client, and the Contractor design has to, at least, induce a compatible maximum displacement. If the vulnerability case is orange or red, the building requires a specific measure (monitoring, consolidation, ...) before or during construction.

In terms of monitoring system, most vulnerable building's façades are continuously monitored by robotic total stations. Ground deformations are controlled along the tunnel and close to stations and shafts, at every critical phase by inclinometers, and groundwater pressure has been verified by piezometers.

The vulnerability study aimed to determine TBM pressures and soil improvement treatments to be done before the TBM excavation through critical areas.

3   THE EFFECTS OF THE TUNNEL ON STRUCTURES

The effects of tunnel excavation on structures have been evaluated in two steps; one empirical, and the other basing on FEM 2 and 3D analyses.

Following Peck (1969) and Schmidt (1969), the surface vertical settlements (Sv) have first been estimated using empirical equation presented in Eq. (1).

$$S_v = S_{max} \exp\left(\frac{x^2}{2i^2}\right) = \frac{vl}{i\sqrt{2\pi}} \exp\left(\frac{x^2}{2i^2}\right) \qquad (1)$$

Where: S is surface displacement, Smax is maximum surface settlement at the point above tunnel centerline, x is distance from tunnel centerline in transverse direction, and i is horizontal distance from tunnel centerline to point of inflection of settlement trough, and it's defines by the Eq. (2).

$$i = K \cdot (z - z_0) \qquad (2)$$

Where z is tunnel axis depth and z0 is the depth beneath ground surface of the considered settlement trough. K is a dimensionless parameter depending from soil type.

Vl value changes with the position of the TBM and it is dependent on the tunnel construction method and the type of soil. The final Vl value is defined as the ratio between volume of the surface settlement trough over the area of the tunnel face and it usually ranges from 1% to 2% for tunnels excavated with the conventional method.

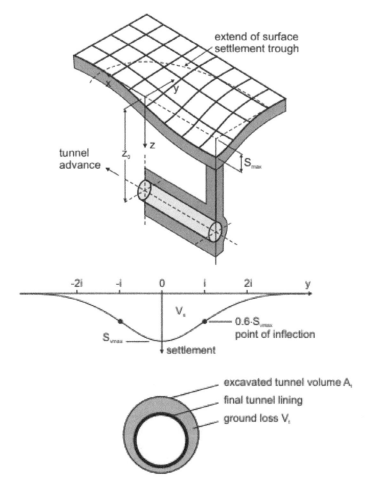

Figure 3. Settlements principles.

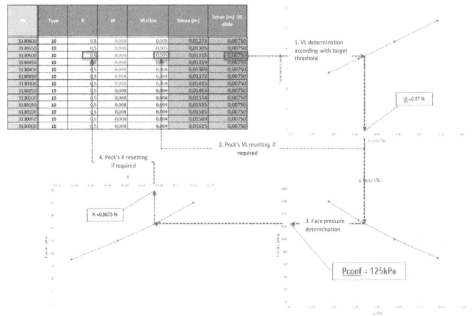

Figure 4. Relationships among Vl-Pface and K.

Besides vertical displacements, the threshold included rotation and elongation displacements. The rotation is the derivative of the vertical settlement, the elongation is the derivative of the horizontal displacement obtained by considering total displacement directed towards tunnel axis by equation (3).

$$S_h = S_v \frac{x}{(z - z_0)} \tag{3}$$

Thirty finite element 2D cross sections analyses were realized to characterize all the track by specific ground properties and specific pressures. The 2D models allowed to determine, for any face pressure P, the Volume Loss (VL) and K to be considered. For each section, VL and K were estimated as a function of pressure by analyzing the settlements gaussian at surface.

This analysis allowed to determine the extension of the Zone of Influence. The Zone of Influence is conventionally stopped at an induced settlement of 2 mm. At this threshold the risk of damage to structures is considered negligible.

Indeed, K values are significantly constant, while Smax varies with the face pressure.

3.1 The buildings

For all the buildings within the ZOI (zone of influence), settlements in free field have been predicted. The other control parameters (horizontal strain et angular distortion) are calculated from the predicted settlements in green field. This is on conservative side because it tends to overestimate settlements by neglecting structure stiffness.

As shown in Eq.(1) and (2), Peck established a correlation between the relative depth of tunnel and settlements. The greenfield displacements are applied to the foundation level of the building: the effects are thus more important for deep founded buildings close to tunnel axis.

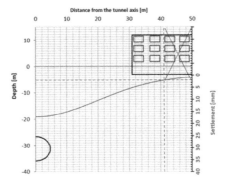

Figure 5. Example of calculated building-tunnel settlements.

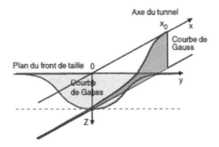

Figure 6. Deformations ahead of tunnel face.

3.2 *The utilities*

The excavation of tunnels may also affect other surface or subsurface structures, such as railway's lines, sewers, installations for electric cables, aqueducts and pipelines.

The utilities constitute 1D structure, not punctual as for buildings. Deformations are thus estimated by green-field analysis. Utilities can be considered as elements with L≫H. Deformations induced by tunneling are dependent not also on distance to the tunnel centerline but also on relative orientation angle Tunnel-utilities.

For this reason, settlements in longitudinal direction must be also considered. Indeed an utility whose alignment is parallel to the tunnel doesn't experience long term induced deformations (rotation or elongation) and no differential settlements are experienced once the TBM has passed.

Therefore, the development of settlements ahead of the face has to be considered in parallel and subparallel utilities. The development of deformations ahead of the face are thus estimated by Panet formulas:

$$S_x = 0 \text{ for } x > x_0 \tag{4}$$

$$S_x = S_0 \cdot [1 - \exp(-A \cdot X^2)/(1 + X^2)] \text{ for } x < x_0 \tag{5}$$

With:

$$A = \frac{a \cdot H}{(R + H)^2} \tag{6}$$

$$X^2 = \frac{(x - x_0)^2}{H^2} \qquad (7)$$

4 SITUATION CLOSE TO STATIONS AND SHAFTS

Several empirical relationships are available to predict green field displacements based on wall deformations. These methods are often based on the equivalence between the volume of the ground after the assessment of the bulkhead, and the volume delivered by the ground for the vertical yields.

Observed magnitudes and patterns of settlements adjacent to excavations supported by sheet pile wall were illustrated by Peck (1969) in which measurements were generalized by "envelopes" of maximum displacement for different ground types. Goldberg et al. (1976) separated deformation behavior by both soil and wall types, they showed that the settlement model do not only depend on the soil type but also on the wall lateral deformations, and degree of constraints of the wall. Clough and O'Rourke (1990) summarized methods to estimate maximum ground surface deformations associated with construction of excavations, they explained the lateral walls deformations according to the method of construction in two modes: cantilever mode, and bulging mode. Boone et Westland (2006) implement this method to estimate both lateral and vertical displacements induced by deep excavation, providing a set of closed form equations.

In particular, final surface settlement is a sum of a cantilever beam type induced deformation and a deformation linked to a beam that is blocked on top.

Boone and Westland proposed some ratios of vertical and lateral displacement areas = Ratio of vertical and horizontal displacement areas for differently constrained wall:

- Cantilever walls
  - Avs/Ahs=Avc/Ahc=Avt/Aht =1

- Supports remain in place
  - Avs/Ahs=Avc/Ahc=Avt/Aht =0.85

- Supports removed
  - Avs/Ahs=Avc/Ahc=Av/Ah =1.1 (no dilation)

Spandrel portion of settlement trough:

$$\delta_{vc\ max} = 3\,A_{vs}/D_s$$
$$Ds \approx 1.2H\ to\ 1.5H$$

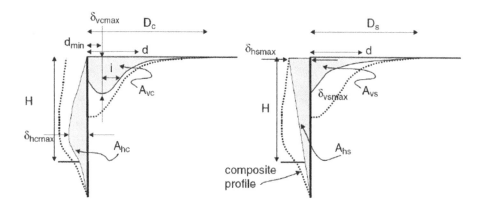

Figure 7.   Principles of deformations induced by deep excavations.

Concave settlement portion of settlement trough:

$$\delta_{vc\,max} = \frac{A_{vc}}{[1 - \Phi(0, d_{\min,i})]\sqrt{2\pi i}}$$

Complete settlement profile:

$$\delta_v = \delta_{vc} + \delta_{vs}$$

However, this approach showed its limits in line 15 application as:

- It should be used only for excavations in specific/homogeneous soils and conditions, and for specific supports;
- It does not allow to estimate deformation at different depths;
- It focuses on the maximum deflection and settlement behind the wall, but it does not give the shape of the settlement through it.

5 THE FEM METHOD AT WALLS AND ITS OUTPUTS

Since the correlation of the structural deformation of the wall and the settlements induced showed its limits, a FEM model was developed to check the effects of wall displacements on settlements, by the finite element code Plaxis 2D (Plaxis, 2018). Coherently to French application of Eurocodes, a coefficient reaction model has to be used to evaluate wall internal forces. Mooving from these analyses, a FEM model has been set out in which the horizontal deformations are introduced on the wall, neglecting soil-structure friction. The greenfield deformations are thus evaluated without having to define soil-structure interaction laws that are already accounted for in MISS K model.

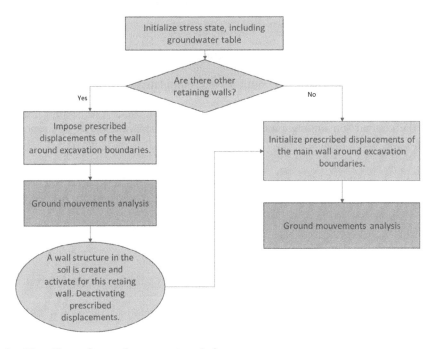

Figure 8. Flow Chart of ground movement analysis.

Once the geometry of model been created, the material properties were assigned with fine mesh generation. Model analysis was performed considering groundwater level, geometrical configurations and initial state of stress.

After the initialization of the stress field in the soil, wall displacements were imposed directly on the model, ignoring the presence of the wall structure.

The deformation induced in the wall construction phase was considered. In some cases, additional retaining structures (temporary or final) displacements have been included in the models accounting for settlements induced by previous excavation.

Wall displacements introduced in the model have been calculated with soil reaction models.

Some 3D models have also been developed. A parametric study of 3D FEM model showed that the geometry of the excavation zone and the mesh set-up has great influence over the response of the wall. If the excavation section was longer than 50m (25 m wide), the effects can be considered as plain, and the results in 2D and 3D FEM models were similar.

The settlement assessment accounts for both tunnel and station settlement both in terms of settlements and rotations and elongations. The output is a section line from which is possible to calculate the deflection ratio (both hogging and sagging) and the horizontal strain (tension and compression) for any given building and utility geometry along the section line.

Settlements are calculated along sections perpendicular to the tunnel or to walls.

6 FIRST RESULTS AT ARCUEIL CACHAN STATION

At the moment, works are on-going thus the results are not yet available to make a comparison among the predicted-experienced displacements.

Arcueil-Cachan station has been treated as exposed above. The station was critical for its distance to the surrounding buildings and rail tracks of Paris Regional Network RER B. The station is a semi-rectangular box, 109m long, 17–30m wide, and 30m deep. The structure is excavated in three sections separated by temporary walls. This solution is required to allow a TBM launch before the final completion of the station. The three sections are:

- An eastern section used as launch shaft for the TBM;
- A western part under the existing railway of suburban train RER B. This part must support the charges of train platform;
- A central part linking the two.

Comparison between pre-construction prediction suggest that empirical method is very conservative compared to finite element method.

Total settlements induced by the station were cumulated with tunneling settlements to define the zone of influence.

Monitoring results are available only for eastern portion. Since the bottom has not been reached, the comparison among predicted-experienced displacements is not representative.

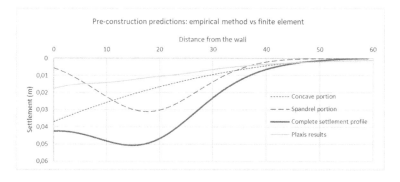

Figure 9. FEM VS Analytical settlement estimation.

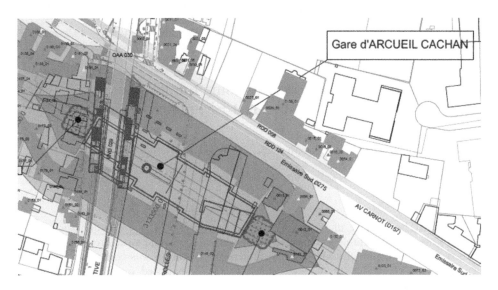

Figure 10.   Example of induced settlements.

Only can be said that the recorded movements are about 1/3 of the predicted on the external buildings at the moment this article is sent.

7   CONCLUSIONS

The outcome of the analyses shows the advantage, compared to simpler models, of using finite element methods. The results show that the advanced method is conservative compared to monitoring results although easy to use.

The work's advancements are not yet representatives of the final excavation phases calculated, nonetheless it can be noticed that the displacements experienced are lower than predicted.

The authors thank the entire VINCI-SPIE team for the collaboration during the development of the mission.

REFERENCES

AFTES, GT32R2F1. 2012. Caractérisation Des Incertitudes et Des Risques Géologiques, Hydrogéologiques et Géotechniques. *Recommandation de l'AFTES.*

AFTES, GT25R3F1. 2015. Maîtrise économique & contractualisation. *Revue TOS N.249.*

Boone, S.J., Westland, J. 2006. Estimating Displacements Associated with Deep Excavations.

Boscardin, M.D. & Cording, E.J. 1989. Building Response to Excavation-Induced

Settlement. *Journal of Geotechnical Engineering, ASCE, Vol. 115*, N. 1: 1–21.

Clough, G.W. & O'Rourke, T.D. 1990. Construction Induced Movements of In situ Walls. *Geotechnical special publication N.25*: 390–470.

Kleivan, E. 1989. NoTCoS: The Norwegian tunneling contract system. *Tunn Undergr Sp Technol* 4 (1):43–45.

Mair, R. 2011. Tunneling in urban areas and effects on infrastructure. *Muir Wood Lecture 2011.*

Peck, R.B. 1969. Advantages and Limitations of the Observational Method in Applied Soil Mechanics. *Géotechnique* 19(2):171–187.

Goldberg, D.T., Jaworski, W.E. & Gordon, M.D. 1976. Lateral support systems and underpinning. *Volume 3: Construction methods.*

Zhou, T. 2015. 3D FEM analysis for sequential excavation.

*Tunnels and Underground Cities: Engineering and Innovation meet Archaeology,
Architecture and Art, Volume 12: Urban
Tunnels - Part 2 – Peila, Viggiani & Celestino (Eds)*
© 2020 Taylor & Francis Group, London, ISBN 978-0-367-46900-9

Study on the influence of closely crossing tunnel on existing underground subway station by 3D-FEM models

M.H. Sadaghiani & M.R. Asgarpanah
Sharif University of Technology, Tehran, Iran

ABSTRACT: The increase of underground space applications, as a solution for congested urban areas makes it inevitable to cross closely spaced underground structures. The analysis of new TBM EPB-bored tunnel under existing underground station was conducted by 3D finite element method. The influence of new tunneling was investigated on the deformation and moments of final lining of existing underground station. The effectiveness of interaction analysis depends on factors such as station overburden, existing station supporting stiffness and traffic loads. In this paper, a parametric study was conducted to study ground movements and lining moments. It is shown that moment increase is significant when the tunneling face is in the closest distance to existing underground station. The increase of circumferential moment was appeared in sides; therefore, more attention should be paid on station side walls. The comparison of controlling parameters shows that the deeper structures intensify interaction effects on station crown deformation and side walls moments. The effect of station liner stiffness was considered. The results showed that stiffer (i.e., thicker) final lining does not have notable influence on station deformations but it causes significant increase in side walls moment. Lastly, the influence of surface traffic loads has been investigated (i.e. traffic loads of 0, 20 kPa and 40 kPa). The results show that regional stress increase by traffic load should be considered. In conclusion, less overburden loads and traffic live loads reduce interaction unfavorable effects and stiffer station supporting system is not a practical solution to improve response of existing structures.

1 INTRODUCTION

Increasing traffic congestion in urban areas and growing land values makes underground structures increasingly attractive for highways and transit compared to other options. Underground transportation systems expansion (roadways and subway networks) causes not only tunneling under existing surface structures, but often also under existing tunnels (Asano et al., 2003), (Liu et al. 2008), (Liu et al. 2009), (Afifipour et al., 2011), (Cooper, 2002). Ground movement is an inevitable risk to nearby structures which must be carefully assessed, both at the planning stage and operation phase. This, in addition to the potential negative effect on the safety of construction and the project cost, means that the ability to make these predictions accurately is crucial. So, the analysis of new tunneling effect on existing underground station, not only extends our knowledge about interaction between new and old underground structures, but it also reduces the risk of unexpected instability events in operation of existing underground station.

The problem of soil-structure interaction is relatively complex. The patterns of settlement, rotation and distortion, when considered together, provide a graphic description of a complex soil–structure interaction response to stress changes in the ground. The data can be used to predict responses in other soils and for other geometries by extrapolation (Cooper, 2002). Field observations, analytical methods and numerical analysis are three main approaches to investigate new tunneling effects on existing structures.

Field observations remain the key to understanding the interaction behavior between adjacent and crossing tunnels. The construction of the Heathrow Express tunnels as they were excavated under the Piccadilly Line at Heathrow 3 is one of the recorded recent histories (The monitoring data allow a detailed picture of the way in which existing segmental concrete-lined tunnels behave as a result of tunneling works below). Long-term settlement records indicate that the maximum settlement increased by 27% for a period of three years after tunnel completion. These results together represent the first comprehensive study of tunnel and lining response to adjacent tunneling published in the literature (Cooper, 2002). Li and Yuan monitored effects of shield tunneling on the double-decked tunneling (Li and Yuan, 2012).

A series of physical model tests on closely-spaced tunnels in kaolin clay samples was performed by Kim (Kim, 1996). In his tests, three tunnels were constructed in which the two new tunnels were either parallel to the existing tunnel or perpendicular to it. For perpendicular tunnels, settlements would be induced in the upper tunnel if a new excavation was constructed beneath an existing tunnel. Finally, he concluded that the interaction mechanisms between adjacent tunnels were extremely complex and further studies were needed.

Twin tunnels of Shenzhen metro Shekou Line excavated by EPB shields. Under the guidance of monitoring, the under-crossing project was smoothly completed without interruption of metro traffic. Migliazza, et al. (Migliazza et al., 2009) compared the results of analytical and 3-D numerical modeling with experimental subsidence measured in extension of Milan metro and found a suitable correlation between the results.

Meguid et al. (2008) and Li et al. (Li and Wang, 2009) presented a good summery of analytical and empirical methods to determine surface settlement. Application of the analytical solutions has limitation. Therefore, numerical method can be used to simulate support-ground interactions more accurately. Numerical modeling, such as finite element method is a useful tool for analysis of the stability of underground space in sequential construction and determination of the influence of effective parameters (Galli et al., 2004), (Karakus and Fowell, 2003). As the behavior of the tunnel heading is essentially three-dimensional (3D), it is not possible to reproduce such behavior accurately in a 2D plane strain analysis. Thus, there has been considerable interest in recent years in developing 3D numerical models to investigate the tunnel interaction problem. Liu et al (Liu et al. 2008), (Liu et al. 2009) carried out a series of 3D finite element analyses to investigate the interactions between perpendicularly crossing tunnels in the Sydney region are investigated. It was found that construction of the new tunnel causes the shotcrete lining of the existing tunnel to be in tension in the side facing towards the tunnel opening and in compression at the crown and invert. When a new tunnel is excavated perpendicularly beneath an existing tunnel, significant increases are induced in the bending moments in the shotcrete lining at the lateral sides of the existing tunnel and in the axial forces at its crown and invert. It is concluded that in order to ensure the stability of the existing tunnel, local thickening is needed at the sides of the existing shotcrete lining if the shallow tunnel is installed first.

A few researches have been reported for investigating the impact of new tunneling on existing underground station. It is clear that the station excavation procedure and supporting system are more complex than a tunnel and this intensifies the complexity of new tunneling under existing underground station.

The present research is intended to study the effects of new tunneling of line 7 of Tehran Metro on the support system of an existing Tohid underground metro station of line 4 in Tehran, Iran, using 3D numerical finite element method by ABAQUS (Dassault, 2010). The main objectives are: (1) to simulate the construction process of the existing underground station (Liu et al. 2008); to investigate the response of the existing support system to the construction of new crossing tunnel; and to study the effect of station overburden, existing station supporting stiffness and traffic loads on interaction mechanism (Liu et al. 2009).

2 THREE-DIMENSIONAL FINITE ELEMENT ANALYSIS

2.1 *Finite element model*

The 3D view of perpendicularly crossing tunnel of Line 7 and underground Tohid station is showed in Figure 1. The station dimensions are 17 m in width, 9m in height of crown with 6 m

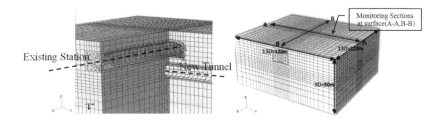

Figure 1.   3D Finite element model for perpendicularly crossing tunnel and station.

of overburden, Figure 2. The deep tunnel is 9 m in diameter excavated by the EPB TBM at a depth of 24 m, Figure 2. The center-to-center distance between two structures is 13.5 m. The construction of station are modeled in a step-by-step CAPS construction procedure, i.e. first pre-supporting system is constructed, the main inner part is excavated and then supported using a shotcrete and final lining.

In order to elimination the boundary conditions effects and minimize the required computational time, the finite element mesh is 120 m long, 50 m high and 120 m wide. Two monitoring sections (A-A and B-B) are located on the model surface at the middle of the mesh. No slippage between the soil and underground structures is permitted in the analysis. By roller supports, the boundary sides were restrained in perpendicular movements and pin supports were applied to the base boundary of the mesh so the movements in x, y, and z directions for these three boundaries were restrained.

2.2   Excavation procedure modeling

Concrete Arch Pre-supporting System (CAPS) was introduced by Sadaghiani and Dadizadeh (Sadaghiani, Dadizadeh, 2010). After a central adit excavation, Concrete Arch Pre-supporting System (CAPS) is constructed. Excavation and supporting of main underground space for station is carried out in several stages. Every excavation step has a minimum of 8m length. Each subsequent excavation stage is executed at 8 m distance (lag) behind the previous stage. Final lining of the invert and wall section is executed at 8 m lag and final lining of the crown is

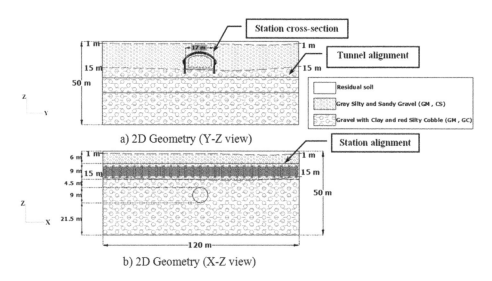

a) 2D Geometry (Y-Z view)

b) 2D Geometry (X-Z view)

Figure 2.   Geological profile of Underground Tohid Metro Station area, Line 4 and Line 7 tunnel.

executed at 16 m behind the invert. Figure 3 illustrates the sequential excavation of CAPS in underground station.

The numerical model simulates the sequential construction procedure of an EPB-TBM by deactivating the soil elements within excavated area and activating the lining elements in the finite element mesh. To simulate the tunneling rate, an optimum construction lag distance of 2.5 m was assumed in model. The thickness of final segmental lining is 0.35m.

To model pressure in the excavation chamber of tunnel, the face is supported by a distributed load which has notable influence on the stability of the face itself and the extent of the surface displacement. The pressure to simulate the face pressure, (i.e., EPB value) is assumed to be 150 kPa,

3 CONSTITUTIVE MODELS AND SOIL PROPERTIES

Tohid Station is an intersecting station of Line 4 and Line 7 of Tehran Metro located at the west Tehran downtown. An elasto-plastic model using the Mohr-Coulomb failure criterion is adopted. A thorough geotechnical site investigation is carried out by Zamin Fanavaran (Zamin Fanavaran, 2000). The ground consisted of silty and sandy gravel and coarse silty sand, and the ground water level was observed approximately at the depth of 36 m, Figure 2. Table 1 illustrates the soil parameters around the site. Lateral earth pressure coefficient, k is about 0.5 in the area.

4 SELECTED ELEMENTS IN THE MONITORING SECTION

In this study, the interaction mechanism is evaluated based on ground movement, deformations and moment of existing station. These quantities are monitored from the start of excavation to the end. Ground surface settlement monitoring paths were illustrated in Figure 1. The behavior of the final lining in the crossing area is more significantly affected by the

Table 1. soil parameters around site.

Soil layer	γ (KN/m^3)	c (kPa)	Φ (°)	E (MPa)	v	Depth (m)
Residual soil	17.5	15	27	20	0.22	0 to 1
Silty and sandy gravel	18	40	30	75	0.25	1 to 15
Coarse silty sand	19	55	35	150	0.35	15 to 34

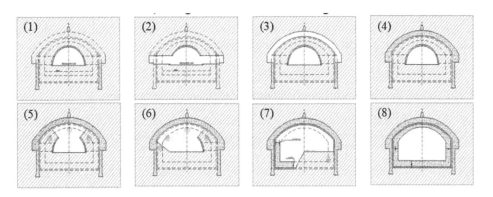

Figure 3. CAPS construction stages.

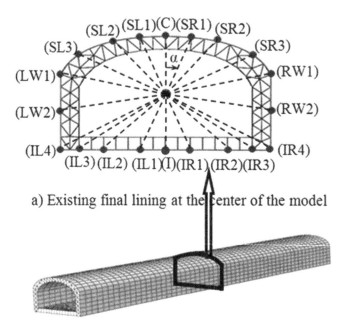

a) Existing final lining at the center of the model

b) Final supporting system of existing station

Figure 4.   the location and arrangement of calculation monitoring points on station final lining.

excavation of the new tunnel than at other places. Therefore, this location was selected to monitor interaction behavior. The location of section and the arrangement of monitoring points which is chosen to monitor station final support deformations and moments are shown in Figure 4.

In order to compare the results at different locations around the underground station, monitoring point deformation are selected as: at the Crown (C), Right and Left side of crown (SR1, SR2, SR3, SL1, SL2 and SL3), Right and Left Walls (RW1, RW2, LW1 and LW2) and Invert (I, IR1, IR2, IR3, IR4, IL1, IL2, IL3 and IL4) are monitored throughout the analyses, Figure 4b.

5   PARAMETRIC STUDY

The main purpose of a parametric study is to investigate performance of any existing station under the changes of loads and stiffness. The changes of loads include overburden load, traffic loads and the change of stiffness supporting system includes station final lining thickness. Details of considered analysis modes are summarized in Table 2.

Table 2.   Summary of parametric cases considered in the finite element analyses.

	Parameters		
	Station Overburden	Station Final Lining Thickness	Traffic Load
Cases (1)	6 m[*]	1 m	-
Cases (2)	12 m	1.5 m	20 kPa
Cases (3)	18 m	2 m	40 kPa

* Parameters in shaded box relates to the base model.

6 ANALYSIS RESULTS

6.1 Station Overburden

Three different overburden depths are analyzed. The EPB was adapted to the depth of tunnel face according to the increase of the earth pressure with depth.

6.1.1 Ground surface settlement and final lining deformation

The most important design criterion for underground excavations in urban areas is surface settlement. It can cause unbalanced settlements in buildings. So it should be evaluated, whether the settlement induced by tunneling is in allowable range or not.

Three different values according to Table 2 are used for the parametric study. The increase of overburden depth causes a considerable deviatoric stress increase in the vicinity of side walls and the invert. These stress changes in the soil are accompanied by plastic deformations around side walls and the invert and they become significantly larger for deeper overburden depth. As a consequence, larger surface settlement is predicted by numerical analyses. As overburden increases, the settlement trough becomes deeper in the station center line (Figure 5a) and tunnel center line (Figure 5b). The settlement trough is in accordance with plane strain condition along new tunnel axis (path B-B, Figure 5b) for the overburden depth of 6m. But the plane strain assumption is not true for other overburden depths. As a result, increase in overburden depth makes interaction mechanism more complex, Figure 5. For more accurate analysis of interaction between two structures and to find out how station cross-section will deform, displacements of station cross-section due to passing new tunnel is shown with exaggeration in Figure 6.

At the start of new deep tunnel excavation, the deformation caused by existing station excavation is reset to zero in the analysis. Thus, the calculated deformation of the existing support

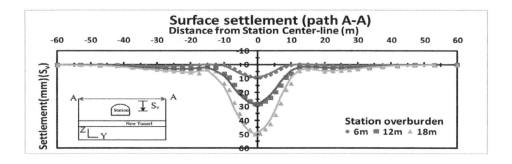

a) Surface settlement along the new tunnel(path **A-A**) after new tunneling

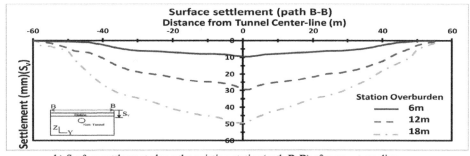

b) Surface settlement along the existing station(path **B-B**) after new tunneling

Figure 5. Developments of surface settlement due to new tunneling (paths A-A and B-B).

system around the station during new tunneling is actually the incremental deformation induced only by new tunneling. The results of the station final lining deformation indicate an increasing of inward horizontal displacement in side walls with increase of soil lateral pressure caused by overburden depth increase. The station final lining deformations due to new tunneling are highly depended on the overburden depth of both structures. In deeper structures, the settlement of the whole station and inward station deformation increased due to higher burden load.

The model predicts downward movement of the tunnel lining which is also observed in physical model by Kim (Kim, 1996). The results show that crown and invert are deformed inward. The arc-shaped crown causes smaller deformation in contrast to side walls and invert. Side walls and invert are flat, thus causes larger deformation and larger bending moments. The bending moment and deformation are not notable compared to side walls and invert. The interaction mechanism is complex for side walls. The station moved to right side when the new tunnel approached the station from left side due to ground stress redistribution. It is the case for all three overburdens. The curvature of side walls for 6 m and 12 m overburdens are outward and it is inward for the overburden of 18 m due to larger lateral pressure. Lateral pressure increase caused by larger overburden depth dominate interaction effects. Significant influence of overburden depth on downward movement of the existing station lining is predicted by the numerical modeling.

6.1.2 Circumferential bending moments of existing supporting system

If the bending moment tends to put the outer face of side wall into compression and the inner face into tension, it is regarded as positive. Otherwise, it is negative. A polar coordinate system is used to demonstrate values, Figure 7. The other coordinates are the magnitudes of the bending moments in kN.m/m. Moreover, the modeled bending moments in final lining of the station before and after the excavation of the new tunnel are compared. The initial distributions of bending moments in the circumferential and longitudinal directions indicate that the inner face of final lining of side wall is in compression (negative values). But in contrast, the inner face of final lining of crown and invert are in tension (positive values). The model predicts the same circumferential bending moment pattern which is also observed in physical model by Kim (Kim, 1996).

The circumferential bending moment changes are more notable in lower-half of section (side walls and invert). A relatively large increase of bending moment can be observed in side walls and invert by new tunneling.

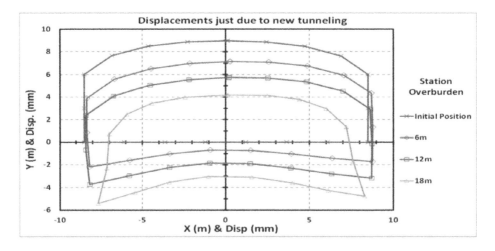

Figure 6.   Displacements due to new tunneling (Length in m & Displacement in mm).

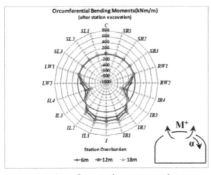 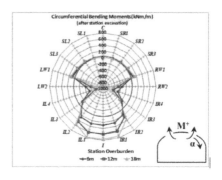

a) after station excavation   b) after new tunnel excavation

Figure 7. Variation of circumferential bending moments of station final lining.

6.2 Traffic Load

When the underground station is located under an existing road, the traffic live load is applied on it. Therefore, support system of excavation area must be designed to handle the imposed traffic live loads. Two traffic loads of 20kPa and 40kPa are considered to investigate the effect of this parameter. The traffic load is applied alongside of station alignment.

6.2.1 Ground surface settlement and final lining deformation

Increase in traffic loads causes larger plastic deformations of the soil in the vicinity of the excavated area and therefore leads to an increase of surface settlements. But there is a difference between traffic load induced settlement trough and overburden induced settlement trough. The vehicular traffic live load is concentrated to a limited band width and is applied to roof of underground structures that are constructed under roadway; therefore, this load is directly imposed on the roof of underground structures. Consequently, settlement increase is not affected at location about 20 m away from station centerline. As it was expected, the plane strain condition is not valid by imposing traffic load (Figure 8b). Figure 9 shows that traffic load has no significant effect on two structures interaction.

6.2.2 Circumferential bending moments of existing supporting system

The variations of circumferential bending moments monitored at the left (leading) side (LW2), the crown (C), the right (far) side (RW2), and the invert (I, IL1 and IR1) of the final lining of the station, at central section plane perpendicular to the axis of the station during the excavation of the crossing tunnel are illustrated in Figure 10. It is observed that overburden and traffic load had the least effect on the crown deformation and bending moments. It confirms that the soil arching has significant effect on minimizing bending moment in the lining. Therefore, polar presentation is ignored. Before the step 32, final lining does not exist and moment is zero, thus at step 32, the calculation monitoring section is activated, so there is a quick increase in moments after activation. After some excavation and calculation steps advancement, the moment values are nearly constant because they are at far excavation face from calculation monitoring point and stresses and moment levels in supporting system are uniform.

On the other hand, before the excavation of the deep tunnel, i.e. at Step 38, the bending moments at corresponding locations (i.e. left, crown, right and invert) have the same values, which reveal that plane-strain conditions are approximately achieved in each sectional plane and the station has been constructed over a sufficient length. Moreover, the bending moments at the sides (i.e. left and right) of the final lining are equal since the model is symmetrical about the axis of the tunnel.

At the left and right sides of the station as the tunnel face of the new tunnel approaches the monitored points, the negative bending moment of the final lining gradually increases, after

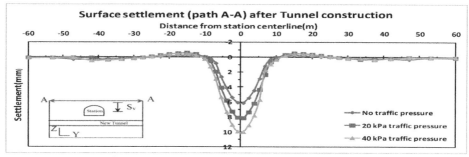

a) Surface settlement along the new tunnel (Path A-A) after new tunneling

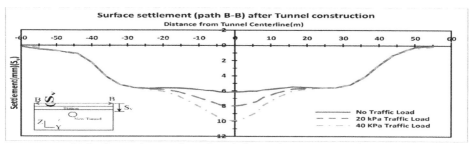

b) Surface settlement along the existing station (Path B-B) after new tunneling

Figure 8.   Developments of settlements at surface (A-A and B-B) due to new tunneling.

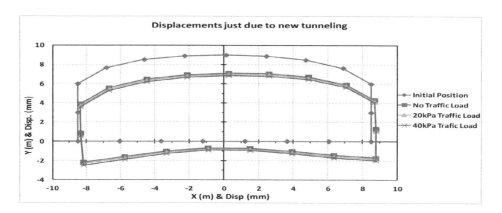

Figure 9.   Displacements due to new tunneling (Length in m & Displacement in mm).

step 38. As tunneling face passes the monitored points, the negative moments gradually increase to their maximum values, step 44. After tunneling face has passed the monitored points, the negative moments in the final lining decrease from their maximum values to stable values larger than their initial values, and finally an equilibrium moment condition is developed in the final lining.

As it is shown in computed station final lining deformations, vehicular live load has no significant effect on two structures interaction. This condition is also the same in circumferential bending moments of existing supporting system. As a result, traffic live load does not have a large influence in interaction mechanism of two underground structures.

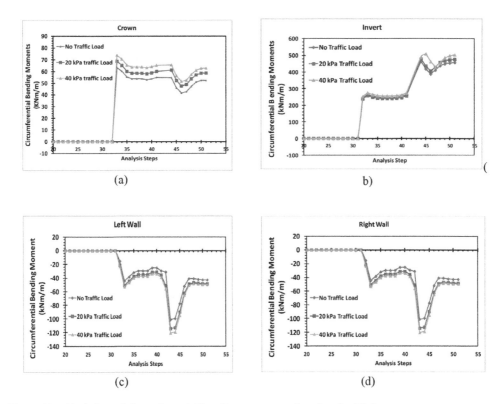

Figure 10. Variation of circumferential bending moments of station final lining.

7 CONCLUSION

The ground movement effects on existing underground structure are an important issue for future tunneling, as an increasing number of urban tunnels and underground structures are constructed with more crossings. A good understanding of the behavior of these crossings is therefore essential to manage the risk of future construction. The main purpose of this study is to investigate the existing underground station performance under the changes of loads and stiffness when a crossing new tunnel is constructed. The increase in overburden depth makes interaction mechanism more complex. Although, side wall is deformed outward in 6 m and 12 m overburden depths, it is deformed inward in 18 m overburden depth. The computed deformations of station final lining are almost identical for variable traffic loads. It shows that vehicular traffic load has no significant effect on interaction of two structures.

ACKNOWLEDGMENTS

Authors would like to thank Sharif University of Technology in providing facilities for this research and Tehran Urban and Suburban Railway Company for cooperation in allowing the access to the required information in completing this study.

REFERENCES

Asano T., Ishihara M., Yasuaki K., Kurosawa H., Ebisu S. An observational excavation control method for adjacent mountain tunnels, Tunneling and Underground Space Technology, 2003; 18(1),291–301.
Liu H.Y., Small J.C., Carter J.P. Full 3D modelling for effects of tunnelling on exiting support systems in the Sydney region, Tunnelling and Underground Space Technology, 2008; 23 (4), 399–420.

Liu H.Y., Small J.C., Carter J.P., Williams D.J. Effects of tunnelling on existing support systems of per-pendicularly crossing tunnels, Computers and Geotechnics, 2009; 36 (5), 880–894.

Afifipour M., Sharifzadeh M., Shahriar K., Jamshidi H. Interaction of twin tunnels and shallow at Zand underpass, Shiraz metro, Iran, Tunneling and Underground Space Technology, 2011; 26(1),356–363.

Cooper M. L., Chapman D. N., Rogers C. D. F., Chan, A. H. C., Movements in the Piccadilly Line tunnels due to the Heathrow Express construction, Géotechnique, 2002; 52(4), 243–2257.

Li X.G., Yuan D.J. Response of a double-decked metro tunnel to shield driving of twin closely under-crossing tunnels, Tunneling and Underground Space Technology, 2012; 28(1),18–30.

Kim S.H. Model testing and analysis of interactions between tunnels in clay, PhD thesis, Department of Engineering Science, University of Oxford (1996).

Migliazza M., Chiorboli M., Giani G.P., Franzius J. N., Potts D. M. Comparison of analytical method, 3D finite element model with experimental subsidence measurements resulting from the extension of the Milan underground, Computers and Geotechnics, 2009; 36(1),113–124

Meguid M.A., Saada O., Nunes M.A., Mattar J. Physical modeling of tunnels in soft ground: A review, Tunnelling and Underground Space Technology, 2008; 23(1),185–198.

Li Y., Wang Z. Study of Influence of Subway Station Excavation on Existing Metro Deformation, Crit-ical Issues in Transportation Systems Planning, Development, and Management (ICCTP), 2009; 3339–3345.

Galli G., Grimaldi A., Leonardi A. Tree-Dimensional Modeling of Tunnel Excavation and lining, Com-puters and Geotechnics, 2004; 31(1),171–183.

Karakus M., Fowell R. J. Effects of different tunnel face advance excavation on the settlement by FEM. Tunneling and Undergroun Space Technology, 2003; 18(2),513–523.

Dassault Simulia Corp., Abaqus Analysis User's Manual, Ver. 6.9-1 (2010).

Sadaghiani M.H., Dadizadeh S. Study on the Effect of a New Construction Method for a large Span Metro Underground Station in Tabriz-Iran, Tunneling and Underground Space Technology, 2010; 25 (1),63–69

Zamin Fanavaran, Consult. Engr. Co. Geological and Geotechnical Investigation Report, Lines 4 and 7. Tehran Urban and Suburban Railway Co., 347–405, (2000)

*Tunnels and Underground Cities: Engineering and Innovation meet Archaeology,
Architecture and Art, Volume 12: Urban
Tunnels - Part 2 – Peila, Viggiani & Celestino (Eds)
© 2019 Taylor & Francis Group, London, ISBN 978-0-367-46900-9*

Use of DInSAR technique for the integrated monitoring of displacement induced by urban tunneling

V. Santangelo, D. Di Martire, E. Bilotta, M. Ramondini & G. Russo
University of Napoli Federico II, Naples, Italy

A. Di Luccio & G. Molisso
Ansaldo S.T.S., Naples, Italy

ABSTRACT: The Line 6 of Naples Underground represents an important system of transport for the city of Naples., that will contribute to develop an integrated transport system implemented on highly interconnected and structured networks, in order to provide a balanced division of mobility between various transport means. The tunnel excavation was realized in urban environment, caracterized by very complex geological and geotechnical ground conditions. The ground deformations induced by the excavation tunnel, were controlled by an integrated monitoring system coupling conventional (optical leveling) and remote sensing technique between Mergellina and Chiaia of metro line during the work from March 2010 to December 2011. A good agreement with the measurements of displacements detected with the two different techniques can be observed; the satellite monitoring also allowed, in some sections, as to detect a continuous evolution of ground deformations even after the excavation phase.

1 INTRODUCTION

The excavation of tunnel produces the deformation of the ground surface which can be detected by using satellite data (Strozzi et al., 2011; Parcharidis et al., 2011, Mark et al., 2012). In urban areas it may induce a degree of damage to the structures and buildings next to the excavation site (Burland et al., 1977; Boscardin and Cording, 1989). Therefore the design must be oriented to the control of ground deformations, both to limit subsidence and to optimize preventative interventions in the various excavation and construction phases of the tunnel. The monitoring system has a fundamental role for the quickly management of any unexpected, potentially harmful situations. In this study, attention was paid to monitoring the subsidence induced by the excavation of a tunnel in an urban environment (Line 6 of Naples Underground), and on the applicability of the Differential Interferometry SAR (DInSAR) technique of integrated monitoring (Perissin et al., 2012; Barla et al., 2015; Bayer et al., 2017; Milillo et al., 2018), by making a comparison between displacement detected with optical levelling and DInSAR obtained by radar images of the COSMO-SkyMED constellation. A good agreement between the vertical displacements measured with the two different techniques was obtained. The results are discussed in the paper.

2 STUDY AREA AND GEOLOGICAL SETTING

The study interested the Line 6 of the Naples Underground (Russo et al., 2012; Aversa et al., 2013; Russo et al., 2015; Bilotta et al., 2017) along the stretch Mergellina-Chiaia (Figure 1). Line 6 of the Naples Underground is part of the network of public rail provided by the Municipal Plan of Transport for the Metropolitan Area of Naples (Italy), which aims to develop an

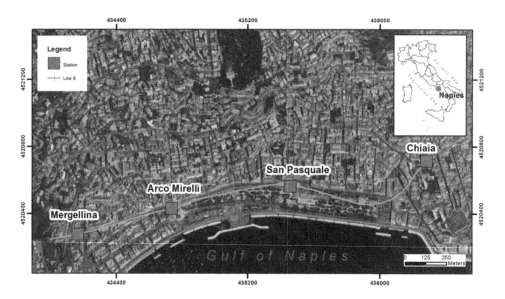

Figure 1.   View of Line 6 subway path.

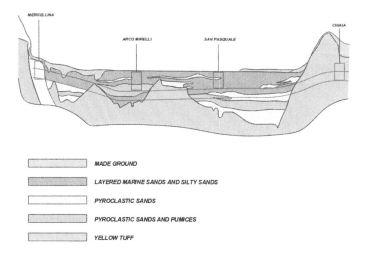

MADE GROUND

LAYERED MARINE SANDS AND SILTY SANDS

PYROCLASTIC SANDS

PYROCLASTIC SANDS AND PUMICES

YELLOW TUFF

Figure 2.   Synthetic geological section along the Line 6 path.

integrated transport system articulated on highly interconnected and structured networks in order to achieve a balanced distribution of mobility between the different modes of transport. The urban and territorial layout of Naples, the rich status of infrastructures, the intensity of transport request, the traffic congestion levels, show that the main element of system transports is an integrated and enhanced rail network.

The stretch in question is introduced in an area of volcanic origin that takes its name of "Distretto Vulcanico Flegreo-Napoletano". The ground conditions are characterized mainly of soils and rocks of pyroclastics origin and marine sands, both have a state of medium-high density with a medium or high permeability (Figure 2).

The stretch from Mergellina to Arco Mirelli is mainly interested by loose soils of pyroclastic origin. The most superficial layers are processed by the waters and sediments in the marine environment or backshore. On the surface are found landfills of varies constitution, with thicknesses that come up to ten meters. Both the pyroclastic soils and landfills are characterized by a state of medium-high compaction and a medium or high permeability.

In the stretch between the stations of Arco Mirelli and San Pasquale, the roof of the tuff reaches up to an altitude of - 6.00 m above sea level, and therefore is at a depth of about 9.00 m below the ground surface. In this area, the tunnel affects the tuff layer for a certain length. The made ground is flat and is located at an altitude of a few meters above mean sea level. Therefore, the water table is found at shallow depth, at an altitude of 1.00 to 1.50 m above sea level. The line tunnel and stations are then immersed in water.

In the area from San Pasquale to Chiaia, the stretch is excavated partly in soil and partly in the tuff formation.

3 DESCRIPTION OF WORKS AND OPTICAL LEVELLING

Depending to the ground conditions and the environment characteristics, the stretch was made by using an Earth Pressure Balance Tunnel Boring Machine (EPB-TBM), this machine turn the excavated material into a soil paste that is used as pliable, plastic support medium. This makes it possible to balance the pressure conditions at the tunnel face, avoids uncontrolled inflow of soil into the machine and creates the conditions for rapid tunneling with minimum settlement. The EPB-TBM used is produced by WIRTH, model TB816H / GS T with a maximum diameter D = 8.15 m. In the EPB-TBMs the excavated material is used to support the tunnel face during the excavation process. The excavated material enters the plenum in a semi-fluid state after having been mixed with a conditioning agent.

Particular attention was paid to the monitoring system, which was designed in a way to provide all the elements necessary to carry out an analysis of the situation during the construction phases, and its possible evolution in the most complete and rapid manner, aimed at defining possible corrective actions (intensifications of the measurements, installation of additional instrumentations, interventions on the executive phases, etc.) aimed at avoiding the emergence of dangerous situations. Monitoring with topographic instrumentation was performed by installing leveling benchmark along the stretch, thus identifying the orthogonal sections (Figure 3) to the excavation axis along its length. The data obtained by optical levels are referred to the period from March 2010 to December 2011.

The acquisitions data by optical levelling was obtained every day for in function of the excavation progress. The instrumented sections along the stretch was monitored from one to two months after the excavation phase. The displacements obtained by optical leveling technique showed the millimetric values.

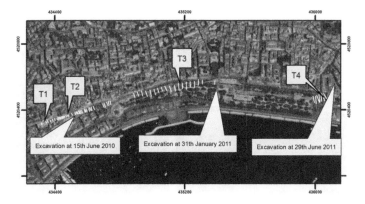

Figure 3.   Installed levelling benchmark along the stretch (in white); excavation phases (in red).

4 DIFFERENTIAL INTERFEROMETRY SAR TECNIQUE AND DATA

For the processing of SAR data, the DInSAR (Gabriel et al., 1989) technique was applied, which is based on the comparison of a stack of radar images, acquired by the satellite traveling in the same orbit and on the same portion of territory, performing a difference of sensor-target distances measured in two different moment of acquisition, in order to highlight any surface movements (Di Martire et al., 2017; Tessitore et al., 2017; Infante et al., 2018). Radar images were obtained by means of an agreement between the Department of Earth Sciences, Environment and Resources of the University of Naples Federico II and the Italian Spatial Agency (ASI). For the case study the images were selected in accordance of the study area, an observation period of interest and a type of SAR sensor, so a study area relating to Line 6 of Naples Underground (section Mergellina-Chiaia) has been set up, an observation period from March 2010 to December 2011 corresponding to the tunnel excavation phases, and finally, a SAR sensor, COSMO-SkyMED was chosen with an 8 day of revisit-time, thus obtaining 97 images in ascending orbit.

The algorithm used for the interferometric analysis is the Coherent Pixel Technique (Coherent Pixels Tecnique - CPT – Mora et al., 2003, Iglesias et al., 2015), implemented in the SUBSIDENCE software developed at the Remote Sensing (Remote Sensing Laboratory – Rlab) of the University Polytechnic of Catalunya (UPC). This technique allows to extract, from a series of differential interferograms, the evolutions of deformations on large areas and for long time intervals. We processed and interpreted the images provided by COSMO-SkyMED in ascending orbit. In the Figure 4 is shown an example of a high resolution image (0.98 m x 1.30 m in SAR coordinates) of the study area compared with optical image.

The images were processed, obtaining the mean displacement rate map and the time-series of displacements along the Line of Sight (LoS) for each selected pixel (Persistent Scatterers – PS) as reported in Figure 5. In order to carry out a comparison in terms of the displacements, the vertical component of the displacements was calculated, in particular, for the acquisition geometry in ascending orbit, the module of the detected displacement is indicated with D_{LoS}, and is given by:

$$D_{LoS} = D_v \cdot \cos\theta \qquad (1)$$

where:

- D_{LOS}: displacement along the LoS;
- D_v: vertical component of displacement;
- θ: incidence angle between the LoS and vertical direction.

Therefore, it is possible to evaluate the vertical component of displacement D_v by using the following equations:

$$D_v = D_{LOS} / \cos\theta \qquad (2)$$

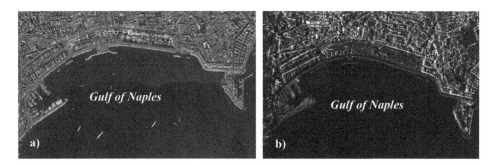

Figure 4.   Optical image from GoogleEarth (a); Single Look Complex radar image (b).

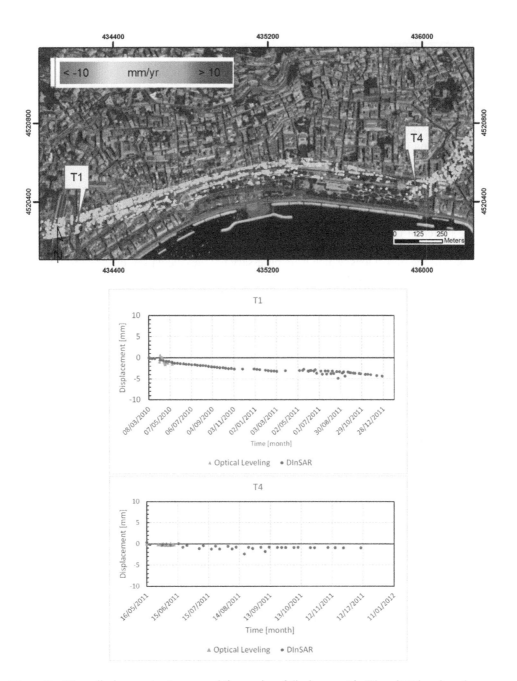

Figure 5.   Mean displacement rate map and time-series of displacement in T1 and T4 benchmark.

For the different instrumented sections subsidence profiles were elaborated, obtained through a fitting according to the available SAR and optical levelling data. Applying the expression of the gaussian distribution that well describes the progress of the displacements induced by the excavation of tunnel (Peck, 1969):

$$S = S_{max} \cdot exp\left(\frac{-x^2}{2i^2}\right) \tag{3}$$

where:

- S: vertical settlement at the surface;
- S_{max}: maximum vertical settlement over the axis of the tunnel;
- i: transverse distance to the point inflexion of the curve;
- x: transverse distance from the tunnel axis.

Furthermore, the maximum settlement S_{max}, i parameter and subsequently the volume loss during excavation (V_s) were calculated.

5 RESULTS

Following the elaboration and analysis of the data carried out for each section instrumented along the stretch of Line 6 of the Naples Underground a local reference system has been established, in such a way to measure the distance of both the topographic benchmark and the PS with respect to the tunnel axis, which were necessary for the calculation of the subsidence profiles. The comparison of the experimental data was carried out through the elaboration of a time-series from March 2010 to December 2011 (Figure 6). The zero of the satellite measurements were made to coincide with those of the leveling, in such a way to have the same correspondence of the measurements.

The frequency of acquisition of the two techniques is very different, as COSMO-SkyMED has a revisit time of 8 days, while the topographic levels are daily.

From the elaboration of time-series, it is immediate to observe that the satellite measurements exhibit the same tendency with respect to those of the optical levels. In some instrumented sections, even if the trend seems to be the same, a noise level is observed for the optical measurement, which fall within the 8 days of revisit time of the satellite measurements. In the other instrumented sections, a rapid evolution of the displacement from topographic surveys is observed in a time of about 8 days. In fact, because of ground conditions (tunneling occurred mainly in sandy and silty-sandy layers) the subsidence due to tunnel excavation developed rather quickly, as shown by optical levelling. However satellite provided data for several months after the last leveling registered optics, showing clearly that the trend is null (Figure 5).

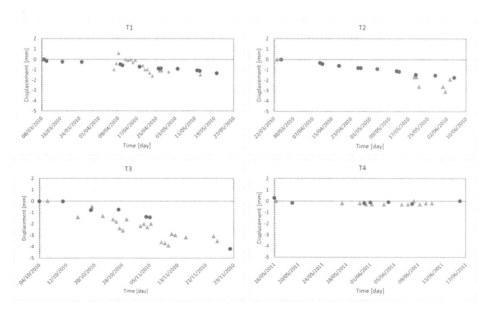

Figure 6. Comparison between optical leveling (green triangles) and DInSAR measurements (blue points) time-series.

6156

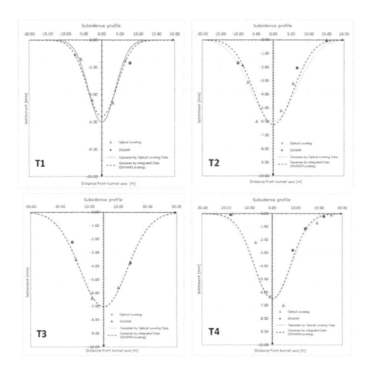

Figure 7. Comparison between optical leveling and DInSAR measurements, subsidence transverse profiles.

The measurements from optical leveling were characterized by a different frequency of acquisition in comparison to those from the satellite. This indicates what is currently the main limitation of the interferometric technique that is unable to measure displacements with higher frequency, as it would be required to monitor the construction stages at least on a daily basis. Nevertheless, both techniques provide similar trends of measurements, with a good correspondence in coincident days of acquisition.

According to the elaborated time-series, the trend is very similar, and this confirms the reliability of the measures obtained by the satellites in terms of displacement.

A comparison of the subsidence profiles in the transverse section along the stretch was also carried out (Figure 7), elaborating two curves with respect to the available data, a first curve requiring only the measurements obtained from leveling (green curve), and a second "integrated curve" obtained by adding the SAR measurements to the first curve (red dashed line).

A comparison was also made the volume loss V_s during the excavation with TBM, and the maximum settlement w_{max} above the tunnel axis.

The comparison of the gaussian curves fitted on the two different sets of data, shows that they are almost coincident in most of the sections.

6 CONCLUSION

The case study concerned the applicability of the DInSAR technique to the monitoring of subsidence induced by the excavation of a tunnel in urban environment, following the entire process of processing the SAR images to obtain the final data needed for the development of time-series and subsidence profiles. This technique has allowed to calculate the time evolution of ground settlements, through the superimposition of a stack of 97 images in orbit ascending and the application of an algorithm for interferometric processing.

From the comparisons carried out, good results of overlapping measurements have been obtained, with millimetric precision, both with respect the time-series and subsidence profiles.

The results show that this technique allows monitoring of very extensive areas and the high density of measurement points and offers the possibility to measure surface displacements to supplement information where in-situ monitoring instrumentation is not installed. It is therefore complementary to more conventional and real-time alert systems, which guarantee the safety of underground works. The main advantage of archive satellite images is that they make historical studies possible, covering long periods of time in order to identify unstable areas, define the perimeters affected, and better understand their causes.

Furthermore the temporal and spatial coverage offered by DInSAR provides both historical data on the natural movements of buildings and surrounding ground, as well as an ongoing check of the stability of instrumentation reference points, which is fundamentally important information when millimetric accuracy measurements are required, and the database of measures is continuously updated for long periods of time.

ACNOWLEDGEMENTS

Thanks are due to Dares Technology for providing support to differential interferometric analysis. Project carried out using COSMO-SkyMed® PRODUCTS, © ASI (Italian Space Agency) – PROVIDED UNDER LICENSE OF ASI.

REFERENCES

Aversa, S., Bilotta, E., Russo, G., Di Luccio, A. 2015. Ground movements induced by TBM excavation under an historic church in Napoli. In: *Proceedings of the XVI ECSMGEGeotechnical Engineering for Infrastructure and Development. ICE Publishing*, pp. 425–430. ISBN 978-0-7277-6067-8.

Bayer, B., Simoni, A., Schmidt, D., & Bertello, L. 2017. Using advanced InSAR techniques to monitor landslide deformations induced by tunneling in the Northern Apennines, Italy. *Engineering Geology* 226, 20–32.

Barla, G., Tamburini, A., Del Conte, S., & Giannico, C. 2016. InSAR monitoring of tunnel induced ground movements. *Geomechanics and Tunnelling* 9(1),15–22.

Bilotta, E., Paolillo, A., Russo, G., & Aversa, S. 2017. Displacements induced by tunnelling under a historical building. *Tunnelling and Underground Space Technology*, 61, 221–232.

Boscardin, M.D., Cording, E.G., 1989. Building response to excavation induced settlement. *J. ASCE J. - Geotech. Eng.* 115 (1), 1–21.

Burland, J.B., Broms, B.B., de Mello, V.F.B., 1977. Behavior of foundations and structures. *Proceedings of the 9th International Conference on Soil Mechanics and Foundation Engineering. IS-Tokyo*, vol. 2, pp. 495–546.

Di Martire, D., Paci, M., Confuorto, P., Costabile, S., Guastaferro, F., Verta, A., Calcaterra, D. 2017. A nation-wide system for landslide mapping and risk management in Italy: The second Not-ordinary Plan of Environmental Remote Sensing, *International Journal of Applied Earth Observation and Geoinformation, Volume* 63, Pages 143–157, ISSN 0303-2434, http://dx.doi.org/10.1016/j.jag.2017.07.018.

Gabriel, A. K., Goldstein, R. M., & Zebker, H. A. 1989. Mapping small elevation changes over large areas: differential radar interferometry. *Journal of Geophysical Research: Solid Earth*, 94(B7), 9183–9191.

Iglesias, R., Monells, D., López-Martínez, C., Mallorqui, J.J., Fabregas, X., Aguasca, A., 2015. Polarimetric optimization of Temporal Sublook Coherence for DInSAR applications. *IEEE Geosci. Remote Sens. Lett.* 12 (1), 87–91. http://dx.doi.org/10.1109/LGRS.2014.2326684 (2015).

Infante, D., Di Martire, D., Confuorto, P., Tessitore, S., Ramondini, M., Calcaterra, D. 2018. Differential SAR interferometry technique for control of linear infrastructures affected by ground instability phenomena. *International Archives of the Photogrammetry, Remote Sensing and Spatial Information Sciences - ISPRS Archives*, 42 (3W4), pp. 251–258.

Mark, P.P., Niemeier, W.W., Schindler, S.S., Blome, A.A., Heek, P.P., Krivenko, A.A., Ziem, E.E., 2012. Radar interferometry for monitoring of settlements in tunnelling. *Bautechnik* 89 (11), 764–776 Nov.

Milillo, P., Giardina, G., DeJong, M. J., Perissin, D., & Milillo, G. 2018. Multi-Temporal InSAR Structural Damage Assessment: The London Crossrail Case Study. *Remote Sensing*, 10(2), 287.

Mora, O., Mallorquí, J.J., Broquetas, A., 2003. Linear and nonlinear terrain deformation maps from a reduced set of interferometric SAR images. *IEEE Trans. Geosci. Remote Sens.* 41, 2243–2253. http://dx.doi.org/10.1109/TGRS.2003.814657.

Parcharidis, I.I., Lagios, E.E., Sakkas, V.V., Raucoules, D.D., Feurer, D.D., Le Mouelic, S.S., King, C. C., Carnec, C.C., Novali, F.F., Ferretti, A.A., Capes, R.R., Cooksley, G.G., 2006. Subsidence monitoring within the Athens Basin (Greece) using space radar interferometric techniques. *Earth Planets and Space* 58 (5), 505–513.

Peck, R.B. 1969. Deep excavations and tunnelling in soft ground. In: *Proc. of 7th ICSMFE, Mexico City. State of the Art*, pp. 225–290.

Perissin, D.; Wang, Z.; Lin, H. 2012. Shanghai subway tunnels and highways monitoring through Cosmo-SkyMed Persistent Scatterers. *ISPRS J. Photogramm. Remote Sens.*, 73, 58–67.

Russo, G., Viggiani, C., Viggiani, G.M.B., 2012. Geotechnical design and construction issues for lines 1 and 6 of the Naples underground. *Geomechanik Tunnelbau* 5 (3), 300–311.

Russo, G., Autuori, S., Corbo, A., & Cavuoto, F., 2015. Artificial Ground Freezing to excavate a tunnel in sandy soil. Measurements and back analysis. *Tunn. Undergr. Space Technol.* (50), 226–238.

Strozzi, T.; Delaloye, R.; Poffet, D.; Hansmann, J.; Loew, S. 2011. Surface subsidence and uplift above a headrace tunnel in metamorphic basement rocks of the Swiss Alps as detected by satellite SAR interferometry. *Remote Sens. Environ.*, 115, 1353–1360.

Tessitore, S., Di Martire, D., Calcaterra, D., Infante, D., Ramondini, M., Russo, G. 2017. Multitemporal synthetic aperture radar for bridges monitoring. *Proceedings of SPIE - The International Society for Optical Engineering*, 10431, art. no. 104310C, DOI: 10.1117/12.2278459.

Tunnels and Underground Cities: Engineering and Innovation meet Archaeology, Architecture and Art, Volume 12: Urban Tunnels - Part 2 – Peila, Viggiani & Celestino (Eds)
© 2019 Taylor & Francis Group, London, ISBN 978-0-367-46900-9

Innovative methods for excavation and ground improvement in Oslo

S. Santarelli & P. Ricci
Società Italiana per Condotte d'Acqua, Rome, Italy

ABSTRACT: The D&B contract realized by Società Italiana per Condotte d'Acqua is part of the largest infrastructure project in Norway, named Follo Line. It includes the construction of several tunnels in hard rock (gneisses) in an urban site and it has required a large use of special excavation methods, like the Drill & Split; furthermore, the great attention to minimize the excavation effects on the water table, demanded a great use of high pressure cement injection named pre-grouting. Other challenges were the excavation of a large cavern of over 400 sqm with variable geometry, named Three Track Tunnel as well as many important interferences with existing underground infrastructures like the Oil Depot (17 big caverns in which petrol and others oil products are stocked), the Alna river tunnel and the E6 Motorways Tunnels. These interferences required other special works like water curtain and the installation of temporary big steel pipe in the Alna river tunnel.

1 INTRODUCTION

The Follo Line Project is currently the largest infrastructure project in Norway and will include the longest railway tunnel in the Nordic Countries. The new double track rail line forms the core part of the InterCity development southward from the Oslo capital.

Due to the complexity of the works, the whole Project is split into 5 different main contracts and Condotte was the Contractor of the Drill & Blast and the Civil Oslo S, both EPC Contracts.

The Drill & Blast contract consists in different tunnels to be realized in the Ekeberg Hill (see Figure 1, next page General View). The numbers in the figure show the different tunnels, as follow:

- "1" Three track tunnel with a max excavation section of about 500 m^2, that cross a fault zone and has a reduced coverage in the northern section.
- "2" Crossing of the three single truck tunnels under the E6 Motorway tunnels, only 3 meters under the bottom of these road tunnels.
- "3" Twin tubes of the new Follo Line excavated at a minimum distance of only 5 meters from the oil depot big caverns.
- "4" Single truck tunnel for the connection to the existing Østfold Line excavated at a short distance from the oil depot caverns and the access tunnels of the oil depot.
- "5" (in Yellow) access tunnels, from the existing Sjursøy tunnel (number **8**), to the Østfold tunnel (the only accesses available) and access to Alna and Pumping station.
- "6" Alna a hydraulic tunnel, in which a river was collected, that is crossed by the three railways tunnels, with a distance of only 1 meter under the bed of the river.
- "7" Pumping station a cavern in which all the water drained from the tunnels is collected and removed by electric pumps.

The tunnels were excavated in hard rock (Precambrian gneisses formed of tonalitic, granite, quartz-feldspathe, biotite minerals) in an urban environment which meant important

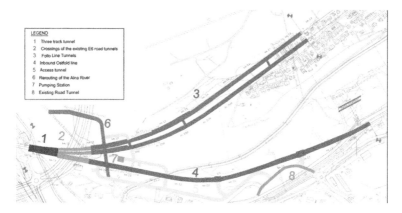

Figure 1.   General View of the Drill & Blast contract.

interferences with existing underground structures so that non-conventional solutions for the construction were required.

These solutions, having an EPC contract, were developed by the Condotte's technical and tunneling departments. In this paper, chapter 2) describes the main interferences, chapter 3) the other challenges and chapter 4) illustrates the innovative solutions adopted.

2   URBAN TUNNEL INTERFERENCE

Referring to the list of the chapter 1, we have the following main interferences:

- "1" The Three Track Tunnel, see fig 2 below, was excavated close to Grönlia road tunnel and only 5 meters under the Mosseveien public road, both open to the traffic.

In addition, in the north portal area, there is an important railway line open to traffic and there are several civil buildings; therefore, the excavation of this big cavern was realized only starting from the Østfold Line single track tunnel.

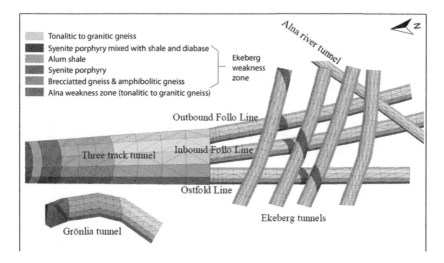

Figure 2.   Three Track Tunnel and E6 tunnels, general view showing interferences and geology.

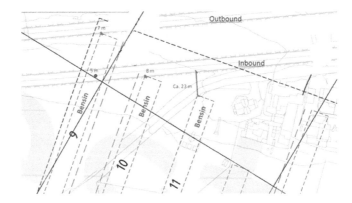

Figure 3.   General view of the crossing of new railway tunnels over the oil storage cavern.

- "2" The twin tube of Follo Line and the single Østfold Line tunnel pass under the four Motorway tunnels, see Figure 2 in the previous page, with a minimum distance of only 3 meters between the vault and the bottom of the road tunnels.
- "3" The twin tubes of Inbound and Outbound Follo Line, see fig 3 below, cross, with the minimum distance of few meters, the big unlined caverns of oil deposit in which petrol is stocked, having important risks of damage for the underground storage plant and risks of explosive gas inlet (benzene) in the new railway tunnels, both under the construction time and during the exercise of railway.
- "4" Also the new Østfold Line single truck tunnel is excavated at short distance from the caverns of oil deposit. It crosses over several service tunnels utilized for oil pipes, for access to the oil depot plant and as maintenance workshops. Particularly, the new Østfold Line tunnel requires the construction of a big niche for a transformer just in the zone of crossing over the existing Hallen Tunnel, a service tunnel of the oil depot.
- "6" The Alna river tunnel, a hydraulic tunnel in which a river was collected, is under passed by the three new single truck tunnels. In this case, the distance between the existing tunnel and the new railway tunnels is so reduced that isn't possible to assure the roof stability and the hydraulic seal during the construction; therefore, the original project foresaw the re-routing of the hydraulic tunnel. An alternative solution, described in chapter 4, was applied to minimize the risk under construction and to maintain the hydraulic tunnel in place.

In the chapter 4, all the details about this interference are given.

- The Østfold Line tunnel has important interferences with the E18 motorway too. In fact, the new tunnel, close to the south portal, is excavated under this important road only three meters below the road platform, without traffic restrictions.
- In addition, for the construction of the portal, it's necessary to cut a pier of the viaduct and transfer the load over the new portal.
- Finally, the presence of several buildings in the Ekeberg Hill imposes limitation in the blasting: the vibration speeds are limited for any receptor and the blasting is not possible during the night and the week-end.

3   OTHER CHALLENGES

The Urban environment imposes important limitations in the job site; the excavation was possible only starting from an existing road tunnel, closed to the traffic, and only a restricted rig area was available outside, close to the portal of existing tunnel, no temporary muck deposit was possible outside and was forbidden to install the fans outside.

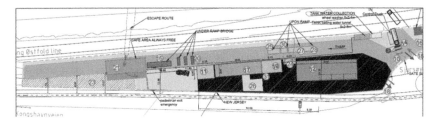

Figure 4.   Installation outside the existing Sjursøy tunnel.

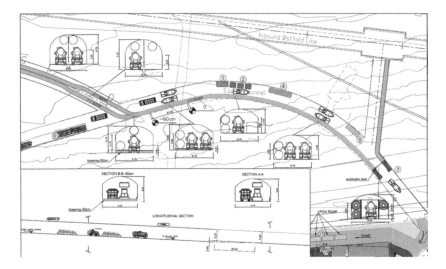

Figure 5.   Installation inside the existing Sjursøy tunnel.

In the Figure 4, we can see the installations in the restricted area outside of existing tunnel: in orange the workshop, in bleu the water treatment plant, in red the electric cabin and the emergency generators, in green an area to maintain completely free at service of an emergency exit of the Oil Depot.

In the Figure 5, we can see the existing road tunnel and the two access tunnels to reach the Østfold Line tunnel. In this sketch is shown the solution for the ventilation, using a door not to mix the return air, the zone of temporary mucking deposit, realized digging a deep trench in the final part of the existing tunnel and the reduced space for the transit of the trucks.

Other important aspects and challenges of this project came from the conservative approach of the design. In fact, as usual in Norway, the contractual technical requirements impose severe limits to minimize the excavation effects on the underground water level.

So, the permeability of the rock mass will be reduced, where necessary, by using the pre-grouting method in order to maintain the maximum inflow of mountain water into the tunnel less than 6 liters for minute every 100 meter of tunnel. The pre-grouting concepts will be described in the following chapter 4.

4   INNOVATIVE METHODS FOR EXCAVATION AND GROUND IMPROVEMENT

4.1   *Control of water inflow in tunnel and pre-grouting*

As anticipated in the third chapter, the design requirements imposed severe limits in water leakages (inflow of mountain water in tunnel less than 6 liters for minute every 100 meters).

To comply with these limits, a complete procedure was executed. See the steps below.

- Systematic probe holes at faces: every 15–16 meters of excavation, 4 probe holes, 24 meters length, were drilled with the jumbo, provided with an automatic rod charger; during drilling the parameters were recorded (MWD System) to detect quality of rock and discontinuities; the records were collected and stored in a data base (BIM);
- Measurement of water flow arriving from the holes; if the flow exceeds 1 liter/minute a pre-grouting fan will be put in place; in case of doubt, additional Lugeon tests can be realized. If we haven't any flow, excavation can restart for the next 15–16-meters.
- The pre-grouting fan, see fig 6, was realized with several holes, 24 meters length, drilled all around the tunnel, starting at a distance of about 1 meter from the excavation profile and with a divergence angle to reach about five meters outside the excavation profile at the end of holes. Other holes were drilled in the tunnel face.
- All the holes have a diameter of 64 mm to allow the installation of packers.

- The packers were of the mechanical type, placed at a distance of about 3 meters from the head of holes. They were connected with strong chains to the bolts installed in the face, to assure the safety in case of expulsion of packers during injections. First of all, the probe holes were grouted, after the contour holes, starting from the bottom to the top and having more contemporary injection lines (up to 4). The injections were performed with high pressure (up to 80 bar) and all the parameters (flow rate, pressure, volume) were automatically recorded; all the data were stored in a data base (BIM).
- At the end of grouting, other 4 probe drill were realized in the face, to test the effect of injections; if the total flow didn't exceed the limit, the excavation could restart, otherwise a supplementary fan of injection would have put in place.

The pre-grouting was applied in this project with success in several parts of the tunnels to minimize the water inflow; it was applied also under the Alna river to reduce the permeability of the temporary rock arch of the pilot tunnel and in the three truck tunnels to improve the rock mechanical characteristic, assuring the rock arch stability. See the following paragraphs.

4.2 The use of pre-grouting in the Three Track Tunnel

In the north terminal path of Follo line, the twin tubes flank the Ostfold Line connection tunnel, entering in a very big cavern called Three Track Tunnel (TTT), element number 1 in Figure 1.

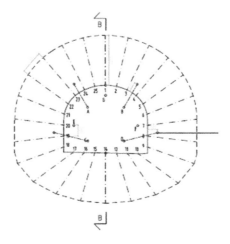

Figure 6.   Typical pre-grouting schema.

This big cavern has a variable geometry with a max span of about 35 m, a max height of about 18 m and a max excavation section of about 500 m^2. The dimensions of this big chamber decrease following the curvilinear curse of tracks, arriving to a width of 25.7m, a height of 14.7m and a section of excavation of approximately 300 m^2.

The geological conditions of rock, see Figure 2, are good in the first stretch (Tonalitic Gneiss of fair to good quality, Q> 4) but change rapidly in the following stretch in Brecciated Gneiss and especially in the following fault zone. This is a very important fault with a displacement of several hundred meters with a very important reduction in the rock quality and the presence of Alum Shale and Clay Shale. The Alum Shale is the oldest sediment in the Oslofeltet sedimentary deposit. It is an organic black shale that can swell when in contact with air and water. It contains more or less amount of radioactive material. In the Alum Shale the Q-value is typically in the range of 0.03 to 0.2 with the lowest value of 0.01 and an average value of 0.09, that corresponds to a rock class F (extremely poor). The situation is complicated by the interference with the Grönlia tunnel, a road tunnel open to the traffic at a minimum distance of about 13 meters from the TTT cavern, and by the crossing, under very little coverage, of an urban road open to the traffic, close to the northern portal. Furthermore the presence of a railway close to the north portal and the urban environment imposed the excavation of TTT only from the south side.

The big dimensions of TTT impose the excavation in five steps, three steps in the top heading and two in the bench. The excavation started from the Inbound Østfold Line at the roof level in the west sector; after the first 15–20 meters of advancement in this sector, the excavation continued perpendicular to this, up to reach the east side of the cavern. This work was performed with controlled D&B to limit the vibration in the near E6 motorway tunnels, see Figure 2.

The interference with the Grönlia Tunnel was tested by a complete 3 D model whit a big extension of 830 x 365 x 190 meters, including the Grönlia Tunnel and the E6 tunnels.

According to this model, to minimize the deformativa effects in the Grönlia tunnel a reinforced support was installed in west and central sector of Three Track Tunnels, for the thirty meters of interference zone. The excavation, in this stretch, was performed with very accurate controlled blasting that allowed the excavation of all this part without any need of Drill & Split.

In the fault zone the quality of rock was "Very Poor" (0.1<Q<1.0) or "Extremely Poor" (0.01<Q<0.1) and was decided to improve the rock quality using the pre grouting. The effect of pre grouting was to improve the quality of the rock from "Very Poor" to "Fair" (4.0<Q*<10) and improve the quality of the rock from "Extremely Poor" to "Poor" (1.0<Q*<4).

The injections were performed from the surface in the last zone with low coverture and from the tunnel faces starting from the sector west and east, interesting the whole section, see Figure 7.

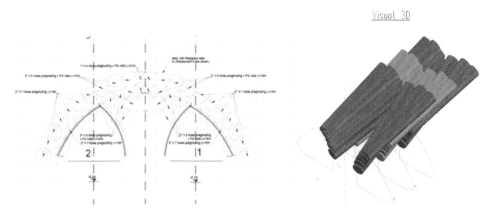

Figure 7.   Pre grouting treatment in advance from lateral sector, cross section and 3D vision.

The injections performed with high pressure, according to the pre grouting concept, gave good results allowing the excavation with the support designed without any problems.

4.3 Crossing under the Alna river hydraulic tunnel

The new Ostfold line tunnel and the twin tubes of Follo Line cross under the existing Alna river tunnel with a distance of about 1 meter between the bottom of the hydraulic tunnel and the vault of the new railway tunnels; this situation required the special construction sequence and the provisional installations described below:

- The crossing below the Alna was realized, in a first step, with a reduced section (pilot tunnel) that allowed maintaining a rock diaphragm of 5 meter between the tunnels.
- Before the excavation of the pilot, the rock diaphragm of 5 meter was injected by a fan of holes with water cement mixture at a pressure of 20–30 bar starting from the holes closest to the Alna River Tunnel. (upper holes).
- The excavation of the pilot tunnel was performed by controlled blasting to avoid the opening of new fractures in the rock diaphragm between the tunnels.
- The water flowing in the hydraulic tunnel was temporary collected, in the zone of interference, in two parallel steel pipes of 1.8 meters of diameter; the pipes were assembled in open air and transported in the tunnel by floating (the extremity was temporary closed). The tubes were placed over iron supports installed previosly in the Alna and sealed to the tunnel bottom and to the tunnel walls. So, the water flowed only in the iron pipe and this section of tunnels was dried by dewatering pumps.
- After dewatering the bottom of Alna was cleaned and a thin concrete slab, reinforced with wire mesh, was casted in place.
- At this point the pilot tunnel was enlarged to the final profile by a very soft blasting; during the excavation a special support with lattice girder was installed; this was designed to resist to the maximum water pressure foreseen. (10 meters of water).
- The final lining in this zone was designed with a special waterproofing system: twin layer of membrane with an external drain layer completely filled with water-cement- bentonite mixture. This was to avoid any leakage of water, during tunnel life (100 years).

The external filling is necessary to avoid any possibility of water flow between the waterproof system and the rock that could transfer the water from this section to the adjoining sections in the sides.

4.4 Gas hazard and Water curtain

As seen in the second chapter, the twin tubes of Follo Line are excavated at very short distance from the existing unlined big caverns of oil depot in which petrol is stocked.

It is well known, that the petrol includes very volatile elements like benzene that mixed to air can produce an explosive atmosphere (a mix with a percent between 1.2 % and 7.8 % can explode). The petrol tanks are maintained completely full, when the petrol is taken away, the free space is filled with water, these to avoid a massive presence of dangerous gas generated from the hydrocarbons at the top of the tank; however, a limited presence of these gases in the upper part of the caverns can't be excluded.

These gases are very light and they can go up along the rock fractures. Before the excavation of Follo Line the danger is totally eliminated because the rock mass is very good and has few fractures completely filled by water; in fact, the water table is about 90–100 meters above the caverns, while the caverns of Oil Depot have a pressure close to the atmosphere one. This means that all the fractures connecting this water table with the caverns are full of water that circulates from the surface to the oil depot tunnels. This situation acts a complete safety condition against any possibility of escape of gas from the tank caverns.

The construction of the Follo Line Tunnels, close to the oil depot caverns, can change this situation. In fact, the tunnels can intercept the existing fractures and can produce new ones;

the main risk arrives if the water that fill the fractures is intercepted from the tunnel and the fractures under this one remain empty of water; so they can be available for the ascent of gas that can arrive to the tunnel. To restore the safety condition, it is necessary to ensure that all the fractures between the new tunnel and the existing oil depot caverns remain full of water at any time. This is obtained by the water curtain; this is an array of drill holes filled with water under pressure. The water will percolate into fissures and joints that the borehole intercepts. In our case the water is supplied via a system of boreholes situated between the tunnels and the caverns.

Each hole shall be equipped with an injection pipe and packer; the pipes shall be connected with a manifold to water pump. The water feeding pipe shall be equipped with a closing valve, throttle valve, manometer, and a flow meter.

The position of the fractures is not known; the geological survey that will be realized during the excavation can give additional information but not complete.

To adopt a safety approach, we have to consider fractures with an angle on the horizontal of 30°, consequently the possible zone of interference has a big extension. As show in the Figure 8, the cavern 9 is overpassed from the Inbound Follo Line with a distance of only 6,5 m; therefore, the drillings to realize the water curtain need an accurate geometry that can't be reached with long holes realized at a safety distance from the dangerous zone. Therefore, the solution adopted includes the following steps:

- Before the excavation reached a distance of about 50 m from the cavern, the cavern was empty of petrol and filled with water;
- In this safety condition, the excavation can continue and the water curtain is drilled, put in service and tested;
- After the positive test of the water curtain, the cavern can be refilled with petrol.

The excavation close to the deposit was certainly realized with accurate controlled blasting to assure a max vibration speed of 25 mm/sec in the wall of cavern; where this condition was not possible, the excavation was performed by Drill & Split, see chapter 4.4.

The water curtain was maintained in service during all the construction works; in this period, the holes are constantly filled with water under pressure with a dedicated water pump and the daily water flow for each hole is measured recorded and checked to establish the correct running; the potential presence of gas is constantly monitored.

This procedure assures the safety under the construction period, but what about the safety in a 100 years life design?

This important problem was solved with a special waterproof system; the concept of the system is that of maintaining the protection of the water around the tunnel. For this purpose, see Figure 9 in the next page, is installed a double layer full round waterproof membrane; the two layers are divided in sectors of 50–60 sqm; each sector is connected with a control and

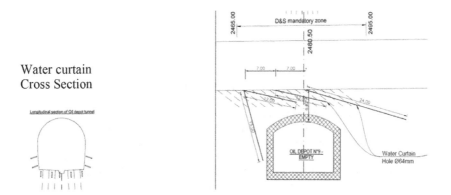

Figure 8.   Cross section and longitudinal profile of the water curtain for the crossing of IFL over cavern 9.

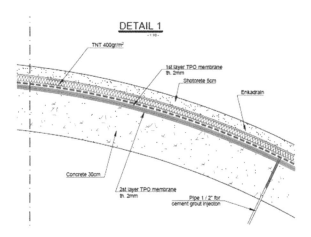

Figure 9. Waterproof system with double TPO layer with injectable compartments and external open layer realized with high performance drain.

injection pipe bringing up to a little niche in the wall of the tunnel. If any leakage is detected in one of the pipe, this can be injected with a mix of water, cement and bentonite to assure the perfect waterproofing of the sector.

Outside of this double layer of waterproofing membrane is installed a drain system that maintains a high permeability also under the pressure of concrete, during casting or long terms rock mass charges. The purpose of this layer is to allow the free circulation of water around the waterproof membrane; this condition assures that the water arriving from the rock over the tunnel (in this zone the water table is 90 meters over the oil depot caverns) fills the drain around the

tunnel and feeds all the fractures between the tunnel and the caverns of oil depot. In this way the safety situation existing before the tunnel construction is re-established and the tunnel is protected by a natural water curtain that prevent the possibility that benzene gas can enter in contact with any element of the tunnel; particularly, the waterproofing membrane is permanently in contact with water and not with gas, that can damage the membrane. To assure a correct running of this system it's necessary to realize, at the start and at the end of this zone, two plugs in which water can't go through. This is necessary because the standard lining profile of tunnel is drained and the water is taken away, instead, we want to maintain the drain layer all around the tunnel in the zone of interference with the oil depot full of water. An upper pipe crossing the plug, assures that the pressure of the water around the lining can't reach dangerous value for the static.

4.5 Vibration control and Drill & Split excavation

Due to the several sensitive interferences upon described, the excavation in the hard rock of Ekeberg Hill was conducted, for extensive sections, by drill & blast using special blasting path that allows the firing of one charging hole at a time and an accurate vibration control, to check and improve, step by step, the blasting path adopted.

Nevertheless, in several sections, the severe vibrational limits (10 mm/sec in the most sensitive infrastructures) and the precautionary approach of the third parts, imposed the not allowance of the drill & blast excavation and the use of the drill & split method.

This last method, used for the first time in Norway, has been chosen for the Follo Line from the Client in the tender phase to solve the most severe interferences. In fact, this method minimizes the excavation effects in terms of vibrations and induced displacements.

The drill & split was studied by Condotte during the tender phase and applied in the project for long sections. By using this method, it is necessary to drill a large number of holes (about

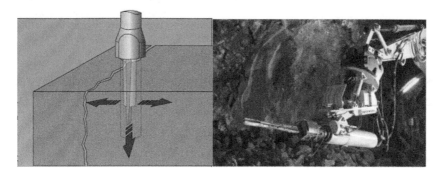

Figure 10. Operation mode of splitting wedge and the equipment installed on an excavator.

650 in a face of 85 m^2); the system requires big holes (76 mm), spaced of 40 cm and with a depth of 1.6 meter; the step of advance obtained was about 1 meter.

The experience has showed the necessity of following a very precise geometrical pattern and this was possible using a fully automatic Jumbo for the drilling. Another important element was the needs of a free surface for the starting of excavation. This was obtained by several large holes, drilled tangent to each other, to have a cut 1 meter large in the face.

The split starts from the holes close to the cut and continues one hole at a time up to the all tunnel profile; the split is made by a hydraulic cylinder pressing a wedge into the drill hole and thereby breaking the rock into smaller pieces (see Figure 10). The split unit is installed on a hydraulic excavator, it's very important to select and to train the operator for this particularly task. The excavation is very slow with this method and, for a tunnel of 85 m^2 is expected to dig between 0.5 and 0.8 meters every 24 hours of work. The D & S experience acquired in this project shows that the productivity of this system is strongly influenced by the presence of fractures in the rock and by the filling of the fractures. The orientation of rock joints is very important and the presence of rock joints parallel to the face makes the excavation favourable.

5 CONCLUSION

The D&B lot of Follo Line proposed a lot of challenges in several design and constructive aspects and required the use of special and innovative excavation methods, like the Drill & Split utilized for the first time in the very hard rock of Norway and with little examples in others countries. Other special works, like pre-grouting and water curtain, was applied to solve delicate interferences with oil depot and water table. The successful experiences acquired in this work have increased the possibility of applying these innovative methods in the next projects to overcome the future challenges.

REFERENCES

Volden, J.A.B. 2015. Engineering geological evaluation of the applicability of Drill & Split in tunnels at the Follo Line project. Trondheim: Norwegian University of Science and Technology.
Zhongkui, Li. 2016. Design and operation problems related to water curtain system for underground water sealed oil storage caverns. Journal of Rock Mechanics and Geotechnical Engineering 8-2016.
Butron, C. 2012. Pre-Excavation Grouting in Sjökullen Tunnels: Design, Evaluation and Experiences. Gothenburg: Chalmers University of Technology.
Abersten, L. 1985. Vibration control during blasting through the Italian Dolomites. Journal of Rock Mechanics and Geotechnical Engineering Volume 23, issue 3, June 1986. Page 100.

Tunnels and Underground Cities: Engineering and Innovation meet Archaeology,
Architecture and Art, Volume 12: Urban
Tunnels - Part 2 – Peila, Viggiani & Celestino (Eds)
© 2019 Taylor & Francis Group, London, ISBN 978-0-367-46900-9

Forrestfield airport link project in Perth, Western Australia – precast concrete segmental lining

S.C. Scaffidi
Salini Impregilo S.p.A., Milan, Italy

A. Anders, S. Porto & E. Torres
SI-NRW JV, Redcliffe, Perth, Western Australia

ABSTRACT: The purpose of this article is to provide an overview of the the production activities regards to the precast segmental lining for Forrestfield airport link project. The segmental lining was designed to meet the project's concrete requirements of 120 year service life with a minimum water/binder ratio of 0.35. The precast yard has a surface area of 40,000 square meters and is located within the jobsite. It is equipped with six overhead cranes with a 10 tons capacity, automatized concrete batching plant (60 m^3/hr capacity), a carousel system with 54 moulds (nine sets per six segments), controlled steam curing room, blanket curing zone, epoxy painting zone and an accredited laboratory for quality control.

This article draws its attention to the specific concrete mix design from the design up to the construction phases of the precast segmental linings.

1 PROJECT DESCRIPTION

1.1 *Location*

The Forrestfield airport link is a "design and construct" type of project; it will deliver an 8.5 km extension of the existing PTA urban rail network in Perth, Western Australia, connecting the Midland Line, just past Bayswater Station, to Forrestfield. The twin-bored tunnels will travel underneath the Swan River, Tonkin Highway and Perth Airport. The project will include three new stations; Redcliffe Station (located underground in Redcliffe), Airport Central Station (located underground at Perth Airport to service both domestic and international terminals) and Forrestfield Station.

Figure 1. Project location map (▬▬ tunnel alignment).

In august 2018, both TBMs had reached their first milestone – arrival at airport central station – and had re-launched to continue their mining towards the second station at Redcliffe

1.2 Base facts

The tunnel excavation diameter is 7,100 mm; the segmental lining has an inside diameter of 6,170 mm, and thickness of segment is 300 mm with an average length of 1,600 mm; the ring is made by universal type of five segments plus a key, namely respectively A, B, C,D, E and K. The segments have a double gasket which makes them state of the art and still quite unique. The configuration of rings provides a minimum tunnel radius of 300m.

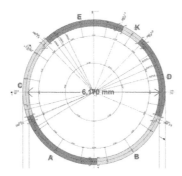

Figure 2.   General geometry of segmental lining.

The segments for the tunnels are made in a warehouse at 11 Carolyn Way, Forrestfield, near Forrestfield Station. The facility has a dedicated concrete batching plant, carousel system, vaporized curing room, blanket curing zone, painting zone in order to facilitate the production of the concrete segments.

2   CONCRETE MIX DESIGN STUDIES

2.1 Concrete durability

The concrete mix design studies started in June 2016 immediately following the award of the Design and Construct contract to SI-NRW JV.

Requirements for concrete durability of 120 years were studied in relation to the ambient, with the below considerations:

- Rising of the sea level will not affect the structures, as the groundwater chemistry is unlikely to became significantly aggressive, and ground structures will not become exposed to airborne chloride salt;
- Carbonation is expected to be the most likely corrosion mechanism for above ground and atmospherically exposed concrete elements.

Carbonation rate modelling has been conducted using the TR/61 Carbuff model that primarily input depends on chemical buffering of C_3A (Aluminate Tricalcium) in General Purpose cement but also on relative humidity; this model includes a probabilistic approach providing a "typical" case and a 95% confidence "design" case. Concrete life is calculated to initiation of corrosion and to onset of cracking in both cases. Based on the above assumptions and the model's results, several documents have been prepared such as a durability plan and durability assessment to defined the parameters regarding the cementitious materials on concrete mix design, fixed the concrete grade and the parameters to be obtained on typical tests (as an example, water permeability, residual flexural strength,.etc).

2.2 Mix design segmental lining

Various trial tests in different stages were performed in order to follow and satisfy the project's technical specifications, specifically on concrete constituents and the concrete mix. A very important aspect in the development of concrete mix was related to the production process and early strength requirements for the de-molding, and lifting of the segments in order to meet the required production targets. The concrete mix design was concluded in accordance with the projects' Durability Plan and the requirements of the Technical Specification.

The mix S60/20/85 GB utilized in the project has a water:binder ratio of 0.37 and the Mix Code as represented below:

- S60 – Special Class Concrete grade 60 MPa (28 days)
- 20 – Maximum Aggregate size, mm
- 85 – Target Slump, mm
- GB Blended Cementitious materials – Type GP (General Purpose) cement + Slag + Silica Fume

Table 1.   Concrete mix design.

Material constituents	Mix S60/20/50 GB			Factory	Type
	Kg/m³	%			
Cement	230	54		Cockburn	Kwinana Grey GP
GGBS	170	40	100	Cockburn	Munster Milled Slag
Silica Fume	25	6		Simcoa	Silica Fume
Water	157			Municipal water	
Granite 20mm size	272	15		Orange Grove Q.	Granite
Granite 14mm size	400	22		Orange Grove Q.	Granite
Granite 10mm size	400	22	100	Orange Grove Q.	Granite
Washed coarse sand	418	23		Gaskell Q.	Quartz
Fine sand	327	18		Gin Gin site	Quartz
Retarder	2*			Sika	Viscocrete PCHRF2
Polypropilene Fibers	2			Sika	Sikafibre 12-34
Steel Fibers	35			Bekaert	Dramix 4D 80/60 BG

* litres/m³ of concrete

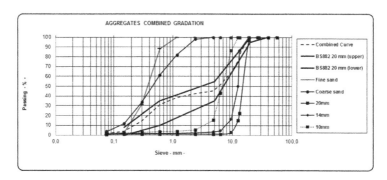

Figure 3.   Combined aggregates gradation.

The initial laboratory design approach was to fine tune the best aggregate-combine gradation and the precise water/binder ratio in order to satisfy the ability of the concrete to flow via a conveyor belt system to cast concrete in the mould without affecting its durability requirements whilst maintaining the mix stability by ensuring steel fibres are properly distributed all throughout the segment to meet its structural requirements e.g. Residual Flexural strength. The importance of having the proper aggregate-combine gradation in this specific concrete mix dictates the behaviour and final quality of the precast segmental lining in combination with the fines content, cementitious components and its very low water binder ratio. Strict quality control in terms of maintaining workability of concrete within a slump range between 70 and 100 mm paves its way to ease of placement from batching plant to mould via conveyor belt. Due to its high fines content (sand and binders), the concrete mix behaves as very dense and self-consolidated concrete. Design of segmental lining considered special type of steel fibres 4D with strength of 1800 N/mm². These fibres were designed for durable and liquid-tight structures and can perfectly combine with traditional steel reinforcement. The production of concrete segments started with the mix design that had steel fibre content of about 40 kg/m³; however, during the early stages of production, it was observed that the residual strength of concrete with 40kg/m³ of steel fibre provided much higher residual strength than that required by design and specification. On this basis, a reduction of 5 kg/m³ was implemented during the succeeding production. Several trials were performed with cast concrete beams according standard BS EN 14651 (single trial was made with #9 concrete beams) with 35 kg/m³ and the results are shown in graph 2. The characteristic values of these results were more than the (see the green triangle in the graph) required serviceability state 5.08 MPa at CMOD1 and 5.28 MPa at CMOD3. The results have confirmed the assumptions for the quantity reduction and the mix with 35 kg/m³ of steel fibres content were adopted in project works.

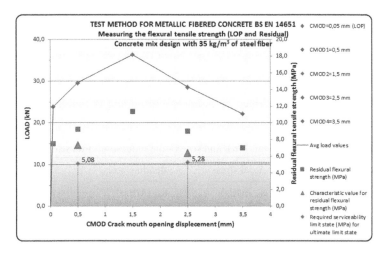

Figure 4. Residual flexural tensile strength.

The equation for to calculate the residual flexural strength is:

$$f_{RJ} = \frac{3F_J l}{2bh_{sp}^2}$$

where: $f_{R,j}$ = residual flexural tensile strength corresponding with CMOD = CMODj or $\delta = \delta j$ (j = 1.2.3.4), in MPa; Fj = load corresponding with CMOD = CMOD = CMODj or $\delta = \delta j$ (j = 1.2.3.4), in N; l = span length, in mm; b = width of the specimen, in mm; hsp = distance between the tip of the notch and the top of the specimen, in mm; δ is equal to 0,85 CMOD + 0,04; CMOD = crack mouth opening displacement.

2.3 Fire testing segmental lining

Another important project requirement was related to extensive fire testing of the segmental lining. The fire resistance tests were done on full scale segments under load and on small scale using square beams (panels) of 200x200x30 cm without load applied.

The prototype segments and beams were prepared in Perth in November 2016 and were sent to laboratories situated in Germany and Melbourne respectively. At that time, the moulds etc. were not ready and therefore the yard of the concrete supplier was used for sampling and casting. The wooden formworks were supplied by a local supplier and the trial was organized for the casting and curing operations as experimental.

The Scope of Works and Technical Criteria requires fire testing to be conducted on two separate fire temperature Vs time curves – the standard curve according to standard AS 1530.4 for 240 minutes and the RABT ZTV curve for 60 minutes (see Graph 3). The Contractor's intention was to perform the tests with two different polypropylene fiber content (1.75 and 2.00 kg/m^3) and was prepared No.8 for each scale prototypes and panels.

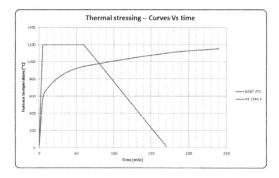

Figure 5.   Heating curves versus time.

The full scale testing was performed on stress condition and the table 2 below summarizes the stress induced to prototype segments.

Table 2.   Bending moment, force and stress state induced on Full scale.

Side of segment	Axial stress σ_N (MPa)	Bending stress σ_M (MPa)	Total stress σ_T (MPa)
Inside	−6.86	+3.07	−3.79
Outside	−6.86	−3.07	−9.93

The stress was applied in the middle and in the upper quarters of prototypes as per graph below:

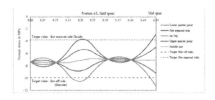

Figure 6.   Stress applied.

The acceptance criteria for fire testing was:

- maximum values for steel reinforcement temperature in compression and tension sides < 605°C;
- maximum depth of spalling < 25 mm.

The results obtained on prototype segments and panels shown the performance of the FAL project concrete to withstand under fire conditions.

Based on the above results, the concrete mix adopted in production contains 2.00 kg/m^3 of polypropylene fibers. It must be noted that the requirements for fire testing of segmental

Table 3. Full and Small scale test results.

Specimen	Heating curve	Full Scale (prototype segment)		Small Scale (panel)	
		T* (°C)	Spalling** (mm)	T* (°C)	Spalling** (mm)
1.75-1	RABT ZTV	378	24	266	19
1.75-2	RABT ZTV	319	18	463	21
1.75-3	AS 1530.4	558	26	245	23
1.75-4	AS 1530.4	617	25	338	0
2.00-1	RABT ZTV	292	21	623	15
2.00-2	RABT ZTV	295	19	531	13
2.00-3	AS 1530.4	614	21	529	0
2.00-4	AS 1530.4	592	23	502	0

* *Characteristic values obtained by interpolation in the corresponding measurement points.*
** *95% Quantile spalling depth measurements*

lining for the project are one of the most restrictive compared to other international projects, where the depth of spalling typically is limited to 40mm or 50mm. For this project maximum depth of spalling is limited to 25mm.

Figure 7. Casting operations of prototype segments with thermocouple incorporate.

Figure 8. Full scale Fire testing on prototype.

Figure 9. Spalling measurements.'segment in MFPA laboratory.

2.4 *Quality control laboratory*

An accredited on-site laboratory is set-up in the precast area in order to control the process of the concrete mixing, casting and curing. Accurately performed tests on fresh (slump, temperature, density, air content, fiber content, etc.) and hardened (compressive strength, flexural strength, …etc) concrete are carried out as per the frequency stated in the specification to verify the conformity of the concrete mix. The working time of laboratory personnel follows the precast production in two shifts per 24 hours. An example of laboratory tests results is shown in the graph below that regarding the compressive strength of concrete segmental lining at 28 days age based on the requirements of AS 1379 standard ($f_{cm} \geq f'_c + k_c*sd$). For 15 set of samples, the value of k_c is 1.25. It is observed that the results fluctuate in a range between 75-85 MPa. The represented period covered the precast production from July 2017 up to June 2018.

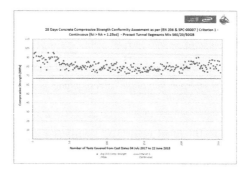

Figure 10.   Compressive strength segmental lining.

3   SEGMENTAL LINING

3.1   *Segment and reinforcement types*

The solution utilized for the manufacture of segmental linings for the project is based on a hybrid system using steel fibre-reinforced concrete and light steel reinforcement rebar cage. In addition, Polypropylene fibers are also incorporated within the concrete mix to comply with the fire resistance requirements of the tunnel lining. Mix-design was developed based on triple-blend cementitious materials, an accurately graded fine and coarse aggregates with 20 mm max size as specified. This is to ensure that the segmental linings are of highest quality with very low permeability and porosity, can withstand any deterioration caused by external conditions and meet its design durability life span. The table 4 below summarizes the reinforcement types:

Table 4.   Segment and reinforcement types.

Segment type	Concrete Quantity (m³)	Reinforcement type 1 (kg)	Reinforcement type 2 (kg)
Segment A	1.84	61.82	243.53
Segment B	1.82	61.72	233.47
Segment C	1.85	61.78	233.79
Segment D	1.81	61.58	233.40
Segment E	1.83	61.71	233.64
Segment K	0.61	32.63	96.64
No. 1 Ring	9.76	341.24	1,247.47

4   CONCRETE PRECAST

4.1   *General precast description*

The concrete precast is situated in Carolyne Way near the Forrestfield Station and has a total surface of 40.000sqm. A concrete mix plant was installed in order to facilitate the segments production. The carousel has 54 moulds, which means nine sets of segments. The steam curing room has capacity to contain No.42 moulds in three lines, at 55°C in 100% of humidity. The segments exit the steam chamber after 6.40 hours, with a compressive strength results on concrete >15 MPa necessary for demoulding.. The blanket curing zone has a capacity to contain almost 60 rings; the segment exits from blanket curing after 48 hours for type 1 and 72 hours for type 2. When the compressive strength results on the concrete reach >40 MPa, segments are ready for transport. The precast facility has 6 overhead cranes of 10 tonne-capacity for handling operations. A fork-lift is used to move the segments to the different work zones. The outer layer of the segments is coated with a corrosion protection system and an epoxy painting zone is dedicated. The segments are ready to use in 28 days, within a compressive strength results >60 MPa and the residual flexural strength > 5.08 and 5.28 MPa at CMOD1 and CMOD3 respectively (see par. 2.2). The people involved are almost 100 of wich 93 workforces and the remain professional staffs (precast manager, precast engineers, quality engineers and supervisors).

4.2 Concrete batching plant

The concrete mix plant dedicated for precast segment production was purchased from Marcantonini italian factory. It is equipped by 2 planetary mixers with a 3 cubic meters capacity, 3 silos for cementitious materials, an automatic weighting system, a conveyor belt for handling the aggregates from hoppers to mixers, and dedicated aggregates stockpiles covered and compartment sets for each granulometric size. The total capacity for this kind of concrete batching plant is 60 cubic metres per hour; obviously this capacity in precast is not demand as well as is working just one of the planetary mixers and the other is in stand-by. With these expedient, the batching plant is able to provide the concrete continually as per precast requirements. The concrete is full automatically batched, this means that all constituents are automatically weighed as well as the water demand is controlled by humidity/moisture sensors installed in the aggregates hopper. Also the polypropylene and steel fibers have dedicated hoppers with full automatic weighting systems. The concrete is delivered to a hopper in the casting room by conveyor belt and discharged in the mould. The operation and control is carried out through by a trained operator automatically from the console cabin.

Figure 11. Plant view.

Figure 12. Aggregates stockpiles and hoppers.

Figure 13. Dosage system of PP and steel fibers.

Figure 14. Concrete conveyor belt.

4.3 Precast work zones

The precast facility has different work zones necessary to prepare the mould, casting, steam curing, removing the mould, blanket curing, painting and storage of segments before sending to the Forrestfield construction site. The carousel is equipped with different working stations for each activity required – demoulding, cleaning, gasket and embedded items installation, reinforcement placement. The process is fully automatic with regular intervals foe each activity.

Figure 15.   Mould preparation.

Figure 16.   Mould ready for cast.

Figure 17.   Casting room.

Figure 18.   Finishing operations.

Figure 19.   Steam curing entrance.

Figure 20.   Demoulding.

After the casting and steam curing process, which typically takes up to six hours, there are other two main work zones dedicated to blanket curing and painting of the outer segment with epoy paint. Scaffolding is used to insert and remove the blanket (see figure 18). Humidity is full controlled under the blanket during the curing process and if required, is adjusted manually by spraying water inside.

The outer layer of the segments is painted with an epoxy coat, with three Mapei products. The first application is done with Mapei coat Primer G. Usage is based on 0.2 kg/m^2 + 10%, this that means in one ring the total product applied is 2.5 kg in half hour comprehensive of maturation. The second application is done by Mapei coat W SP type A and usage is based on achieving a film thickness of 400μ + 15%, that means in one ring the total product applied is 11.08 kg in three hours comprehensive of maturation. The last application is done by Mapei coat W SP type B and usage is based on achieving a film thickness of 400μ + 15%, which means in one ring the total product applied is 28.44 kg in six hours comprehensive of maturation.

Figure 21.   Blanket curing and painting zone.

Figure 22.   Blanket curing zone.

Figure 23.   Painting zone.

Figure 24.   Painting operations.

Figure 25.   Segment's tilt zone.

Figure 26.   Cargo zone.

The remaining work zones in the facility are tilt and cargo. The segments are rotated before handling the cargo for transport to a dedicated storage yard at the Forrestfield construction site (tunnel entrance).

5   CONCLUSION

The article draws an overview of the production activities in precast concrete segmental lining for the Forrestfield-Airport Link project, and in particular presents the concrete mix which was establish to ensure the durability, fire resistance and production requirements. All operations and processed mentioned are the results of studies that were performed during the start-up phase and applied, verified and improved at the beginning of the production line. It is planned that production of linings will be conducted by December 2018, once more than 9000 rings are casted.

REFERENCES

Drawing No. 26-S-320-0001 – 2016 - General geometry of tunnel segmental ring.
Fal-Sinrw-Msf-0034-A – 2017 - Concrete mix design report for precast tunnel segmental lining.
Fal-Sinrw-Cm-Rpt-00024.B.Fr – 2017 - Concrete Mix Design with 35 kg/m3 of steel fibre.
Fal-Sinrw-Cm-Rpt-Bor-00036.A.Fr – 2017 - Concrete Mix Design with increased slump.
Sinrw – 2017 - Interpretative report of fire testing segmental lining.
Fal-Sinrw-Cm-Mst-00018 – 2017 - Segment production and quality control.
Fal-Sinrw-Pm-Pln-00004 – 2017 - Durability Plan Vol. 0;
Fal-Sinrw-Pm-Pln-00011 – 2017 - Durability Plan Vol. 2.

Tunnels and Underground Cities: Engineering and Innovation meet Archaeology,
Architecture and Art, Volume 12: Urban
Tunnels - Part 2 – Peila, Viggiani & Celestino (Eds)
© 2019 Taylor & Francis Group, London, ISBN 978-0-367-46900-9

The need for development of station's design for reducing risks and disturbance during construction

H. Schwarz & L. Banlin
Egis, Lyon, France

ABSTRACT: Urban transport systems using underground space are necessary for attending the needs of increasing urban population. Modern standards, technology and experience are key factors for reliable, fast and safe construction. TBM technology allows successful construction of running tunnels. Station's and other structure's construction causes disturbance to the urban environment, limits accessibility and requires traffic or utility diversions. Station's design shall attend architectural and functionality features, taking in account social-economic impacts during construction. Risk assessment requires a reliable investigation of the subsurface conditions. The design of underground structures should consider all constraints. Including constructability and safety criteria in early design stages reduces risks. Riyadh Metro project includes the design and construction of large diameter tunnels and underground stations. The time schedule required specific solutions in sequencing tunnelling and deep excavations. Stations include construction with pre-cast elements. Some technological challenges for designers and constructors in future projects of underground infrastructures are proposed.

1 INTRODUCTION

The reference project for this paper are the design and build contracts for Riyadh Metro implemented by Ar Riyadh Development Authority (RDA, previously ADA) according to its mission as published in (ADA 2015). The High Commission for the Development of Ar Riyadh was established in 1974 as the organizational, planning, executive and coordinating body responsible for the development of metropolitan Riyadh. The ADA, established in 1983, is the High Commission's executive, technical and administrative arm and is responsible for implementing the High Commission's decisions and missions.

The Metro network is to be the backbone of the public transport system developed as part of the King Abdulaziz Project for Riyadh Public Transport in Riyadh. With six lines and a total length of 176.5 km (Table 1) and 85 stations, the Metro network will cover most of the densely populated areas, public facilities, and the educational, commercial and medical institutions. The public transport plan includes a bus network with 22 lines on 1200 km and 6765 stops that will be the main feeder for the Riyadh Metro network; it will also be the main mode of transportation within and around Riyadh districts (ADA 2015). The design and build contracts for the Metro project were granted in three independent packages.

The underground sections include bored and mined tunnels as well as cut and cover structures on 60.7 km (34%) of the system. The running tunnels in the underground sections of Line 1, Line 3 and Line 5 were constructed by seven EPB-TBMs. Only the tunnel section of Line 2 and tunnel adits between running tunnel and shafts were constructed using mining methods with shotcrete lining and reinforced concrete for the permanent structures.

These deep underground sections are developed mostly in layered limestone and brecciated limestone as described in (Lord 1981) or (Abdeltawab 2013). As expected from geological and geophysical investigations, locally karstic voids were found (SGS 2007). Water of perched groundwater tables circulates through voids and vugs, mostly in relation to a

Table 1. Lengths of Riyadh Metro Lines.

LINE	L1 Blue	L2 Red	L3 Orange	L4 Yellow	L5 Green	L6 Violet	Total
Total Length [km]	38.0	25.3	40.7	29.6	12.9	30.0	176.5
Bored Tunnel [km]	17.3	2.9	7.4	-.-	12.9	-.-	40.5
Cut & Cover [km]	-.-	-.-	3.2	7.0	-.-	-.-	10.2
At Grade [km]	4.8	17.0	5.2	6.7	-.-	-.-	33.7
Elevated [km]	15.9	5.4	24.9	15.9	-.-	30.0	92.1
Package	P-1	P-1	P-2	P-3	P-3	P-3	

system of Wadis or dry valleys (Al-Othman 2011) that are partially covered by the fast developing city.

The deep underground stations were built as cut and cover boxes with bottom-up methodology. The excavations were protected by retaining systems according to the individual geotechnical conditions and available space. Where necessary, temporary structures and provisional decking systems were used for maintaining the road and pedestrian traffic during construction. Different methodologies were used for building emergency exit and tunnel ventilation shafts, depending on the local ground conditions, available space for construction and logistics, depth and geometry, etc.

In the shallow underground sections running tunnels and stations were built as cut and cover structures using mostly the bottom-up methodology. In these sections, the ground conditions are characterized by weathered limestone. Quaternary deposits of alluvial and aeolic sands are found in increasing depth towards the east of lines 2 and 3 of Riyadh Metro project.

2 TBM TUNNELLING

TBM technology with pre-cast segmental lining allows successful construction of running tunnels in reliability of production, limiting risks and achieving appropriate quality of the works. Therefore, TBM tunnelling was selected as the preferred technology for most of the deep underground sections of Riyadh Metro project. The real production rate achieved was better than the one foreseen in the time schedules during the early design stages (Schwarz 2016). The critical paths, which initially included the TBM drives, changed and converted some of the tasks for construction of the deep underground stations into elements on the critical path.

For the production of the pre-cast lining segments, two design and build contractors built up their own pre-cast yards and facilities, meanwhile one contractor subcontracted a local pre-cast factory for production and transport of the tunnel lining segments.

The design and shop drawings of the segments had to take in account that most reinforcement steel available in Saudi Arabia is considered as not weldable. Due to this constraint, overlapping of re-bars and anchor lengths had to be added to the structural effective length of steel re-bars. Manual steel fixing was employed. Observing the details and steel fixer's operations in assembling the re-bar cages some optimizations to the initial shop drawings were introduced, improving especially the corners and borders, as well as the stiffness of the cage. The improved design allowed also for reducing time for steel cage's assembly.

Following ACI 305.1-06 Specification for Hot Weather Concreting, project specific considerations for concreting in hot and dry weather had to be taken in account during the heat period in summer. During this period with air temperature reaching 50°C, the temperature of the fresh concrete is controlled by substituting partially or totally water by ice; and taking aggregates (gravel and sand) from shaded stockpiles. For maintaining humidity in the concrete elements during the curing process, these are covered with impermeable sheets and computer controlled steam curing was applied.

For reducing the effects of differential temperature, hot weather and intense sunlight on the segments, these were initially stored inside the production hangars. Once the segmental lining

Figure 1. Segmental lining on stockpile.

rings were at outside stockpile, it was necessary to provide humidity to the concrete during the first days and weeks.

3 UNDERGROUND STRUCTURE CONNECTED TO SURFACE

Underground structures that connect to surface are the cut and cover sections of running tunnel, Metro stations, as well as emergency exit and ventilation shafts. Station's design has to attend functionality and architectural features for the passengers, whereas the design of the below surface part of shafts is mainly based on functional criteria. All stations include ventilation and air conditioning for passenger areas and technical rooms.

The architectonic design of the elements above surface needs to be integrated in the urban environment and improving it. As these structures interfaces urban surface and underground utilities, the design constraints should not disregard the social-economic impacts during construction. The construction activities cause disturbance to the urban environment, limit accessibility to neighbours and required traffic diversions or utility relocations.

The main risks identified for station's design and construction are:

- Ground and groundwater conditions;
- Stakeholders (utilities, traffic, etc.);
- Uncharted utilities;
- Internal coordination with TBM-drive, MEP and systems.
- Safety at work

Risk management followed recognized risk standards (e.g. ISO 31000) and the Code of Practise (ITIG 2006) for the underground sections. Geological, geotechnical and geophysical investigations were analysed bi-dimensional and in three dimensions. With structural design in 3D and using BIM techniques, for structures and combined service drawings, construction risks were designed out. Systematic Instrumentation and Monitoring included all deep underground constructions, as bored and mined tunnels, cut and cover sections, stations and shafts.

3.1 Ground and groundwater conditions;

The geotechnical design of temporary works and permanent structures is based on geological, geotechnical and geophysical investigation and its interpretations. Groundwater circulating through interconnected voids and vugs, as well as between layers of the karstic limestone was identified as the main uncertainty. Deep vertical wells for lowering the groundwater level only could be successful where they connected to the voids and vugs filled with water.

The groundwater appearing during excavation was clear, not transporting any fine materials and the limestones in Riyadh are quite stable. Locally chemical and biological load of the groundwater found close to surface indicated its origin from broken sewer lines or uncontrolled infiltration.

Considering these conditions, observational methods were established for controlling groundwater inflow to construction sites with horizontal drainage systems when humidity on excavation walls appeared or water flow was observed during excavation. All underground construction included topographical, geotechnical and structural monitoring.

3.2 Stakeholders (utilities, traffic, etc.);

An effective stakeholder management is necessary for reducing the risks associated to utilities and traffic diversions. These risks are substantially time related, as for example electricity and water supplies cannot be interrupted during heat period. For specific tasks, only subcontractors authorized by the utility owners could be employed.

First contacts to the neighbours of the underground sections were established while performing the pre-construction dilapidation surveys. RDA maintains an active communication policy using all types media in specific informative campaigns. A call centre is enabled for responding concerns of the neighbours and a large visitor centre brings the vision of the new transportation systems to the population.

3.3 Uncharted utilities;

In a fast growing city like Riyadh, uncharted utilities had to be expected as not all drawings are precise or complete. Subsurface Utility Engineering (SUE) was not implemented in Riyadh when Metro construction started.

Recently several countries published standards or specifications for SUE/SUM, as there are for example ASCE 38-02 in the USA or PAS 128:2014 in the United Kingdom. Based on these documents other countries continue with the evolution and innovation in this field. The implementation and development of these methodologies could be in benefit of the utility owners and for underground construction for Metro and other infrastructures.

3.4 Internal coordination with TBM-drive, MEP and systems.

In a fully integrated project like Riyadh Metro, all internal interfaces are handled by the design and build contractor. An integrated management allows reducing the risks of interfaces and can improve coordination between the areas of construction and installation.

Concurrent design and construction allows for projects on the fast track. The procurement, especially of long lead items, needs to be well aware of the time constraints of construction and installation of equipment.

3.5 Safety at work

Working in confined space and simultaneous TBM operation and station's excavation require special attention on all Health and Safety related risks. Taking constructability criteria as part of the design process, many risks can be designed out.

Exclusively dedicated teams for each package and stretches managed and controlled the deep underground construction. Appropriate method statements and protection measures

Figure 2. Pre-cast elements on stockpile (R-beams, boundary wall, TT-slabs, track slabs).

were put in place. During the erection of the permanent structure of cast in situ concrete and pre-casted elements the main risks of accidents were identified from lifting equipment and for working at height. With specific method statements and training the frequency of accidents and lost-time injuries could be minimized below the average of construction industry in the country.

4 USAGE OF PRE-CAST ELEMENTS

Pre-cast elements are widely and successfully used in Riyadh Metro construction. Systematic usage of pre-cast elements is appropriate for projects with tight time schedule as it allows for simultaneous production on site and in the pre-casting yard. The design and methodologies have to consider constructability, logistics, transport to site and erection.

Lift plans shall identify exactly the required lifting equipment and its locations on site. These locations have to be accessible and free before erection of pre-cast elements starts. The accuracy of in situ concrete structures needs to be according to the accuracy required for positioning the pre-cast elements correctly.

A résumé of pre-cast elements used in Riyadh Metro is given in Table 2. Depending on the level where beams are installed in the underground stations, they are designed for vertical loads and a strutting function. The type of beams employed have rectangular shape or I-beam shape (pre-stressed), also TT-beams/slabs are used. For slabs either hollow-core elements or solid slabs as stay-in-place structural formwork are used. Boundary walls for the train depots consist in double-sided sandblasted elements including the Riyadh Metro logo on the outside.

Furthermore, in Riyadh Metro project the viaducts and elevated stations are mostly built with large pre-casted elements, either as post-tensioned short elements or pre-stressed long beams. For all pre-cast concrete elements specific methods according to ACI 305.1-06 for concreting in hot and dry weather had to be taken in account during the heat period in summer. During this period, the temperature of the fresh concrete is controlled by substituting partly or totally water by ice; and taking aggregates (gravel and sand) from shaded stockpiles. For maintaining an appropriate humidity during the concrete curing of pre-cast elements, these

Table 2. Usage of structural precast elements in underground sections.

Tunnel	Underground Station	Others
Segmental lining	Track slab	Retaining walls
Base infill	Pillars	Pipes, duct banks, man holes
Duct banks, man holes, lids	Slabs (solid, hollow core)	

usually are covered with impermeable sheets and water is sprinkled on the stockpile where required.

5 CASE STUDIES

The tight time schedule required specific solutions in sequencing tunnelling and deep excavations. Station's design includes pre-cast construction methods. Taking in account that the geotechnical conditions are mostly favourable, the major risk for time and safety of deep excavations arouse from subsurface utilities.

5.1 *Case study 1 - Interface with utilities in trenchless construction methods*

Trenchless construction methods like micro-tunnelling and HDD (horizontal directional drilling) reduces surface affection and minimizes construction time. Specialized contractors are working successfully in the Kingdom of Saudi Arabia and its capital city Riyadh since decades. Several utilities relocations in Riyadh Metro project were built using these technologies.

Nevertheless, applying these methods may induce new risks and uncertainties to following underground constructions. Detection and diversion of utilities built by applying trenchless methods can be difficult due to the depth of the infrastructure. It shall be mandatory to include all drilling or driving data in as built documentation of utilities built with trenchless methods. As built drawings should reflect the real position of the underground infrastructure, providing three-dimensional information.

Excavation of a cut and cover section was started after having obtained "utility clearance" following the method statement for utility detection and diversion. In the drawings provided by the telecom operator a main fibre optics cable was indicated at a depth of about 2.00 to 2.50 m below surface and crossing the axis of the Metro project. The known utility did not appear in over 3.00 m deep trial trenches in the zone of the expected cable crossing. No manholes were found at the locations indicated in the as built drawing of the cable. A request for clarification to the utility owner did not provide further information.

Temporary retaining system and shoring were only needed in the upper part or the trench for a layer of quaternary sandy deposits overlaying brecciated limestone. The controlled excavation proceeded without any hint on utility trench in this area; the rock was undisturbed not showing any previous excavations. Finally the cable was found at the bottom of the excavation of the cut and cover section at over 8.00 m of depth. The excavator only caused slight damage to the outer shell of the cable, and after being checked by the utility owner, and due to its position below foundation level, it was decided and agreed to protect the cable without diverting it.

In benefit of the utility owners and for reducing risks for underground construction, as built documentation of subsurface utilities should be reliable and mapped in 3D using GIS and BIM technologies. Subsurface Utility Engineering and Subsurface Utility Mapping (SUE/SUM) have methodologies developed in ASCE 38-02 and PAS 128:2014. The practical applications are still under development for this specific branch of underground engineering and are supported by various software packages. Utility owners should be interested in protecting their assets and provide reliable information, especially for utilities built with trenchless construction methods.

5.2 *Case study 2 - Underground station design with pre-cast elements*

The central sections of Line 1, 3 and 5 of Riyadh Metro are deep underground sections with a single double-track tunnel excavated by EPB-TBMs of approximately 10 m diameter. The deep underground stations have lateral platforms in straight and horizontal alignment. A redundant system of escalators, elevators and fixed stairs provides vertical passenger transport. Staircases for emergency evacuation of station and adjacent tunnel are located at both extremes of each platform. From top to bottom three levels for passengers (concourse, mezzanine and platform level) and a technical under platform level are available (Gómez 2017).

In the initial 30% design stage, the design and build contractors designed the structures of the underground stations with cast in situ concrete structural elements and secondary structures of blockwork and cladding elements in public areas. Before proceeding with further design stages, the project management requested a detailed time schedule for these works including the times for preparing scaffolds and formwork, concrete stripping and re-propping wherever necessary.

During the following stages of 60% and 90% design delivery, the design and build contractor proposed changing main structural elements as roofs and pillars to be constructed with pre-cast concrete elements. The project management, requesting at the same time fully designed stations including all major MEP systems (escalators, elevators, ventilation, etc.), rail systems and preparing combined service drawings (CSD) using BIM technologies, approved this decision. It was suggested involving the technical offices of the pre-cast specialists in early design stages.

From an overall point of view, using pre-cast concrete elements in structural works of deep underground stations can be considered appropriate in terms of civil works construction and time schedule. Nevertheless, difficulties arose where definitions in CSD were incomplete or where due to changes in procurement MEP and rail systems equipment had different dimensions or points of connection than foreseen in the approved design.

Pre-cast production, transport and erection is a specific work cycle with own conditions. Structural designers should understand the constraints of pre-cast production (type of moulds, introduction of reinforcement steel, pre-stressing, quality control etc.), transport

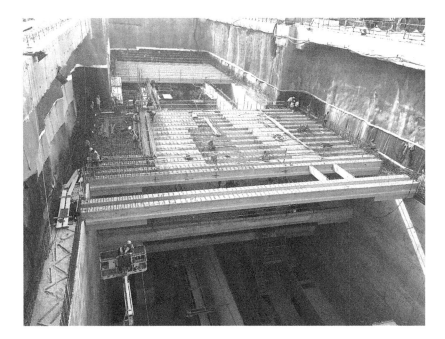

Figure 3.   Erection of pre-cast elements in deep underground station.

(dimensions and load limitations) and erection (supply, crane capacity, etc.). A close coordination between the different designers and construction managers of the design and build contractor and the supplier of the pre-cast elements can minimize interfacing risks.

As produced in factories, pre-cast elements can be prepared with high standards of quality. Nevertheless, the contracting party (in this case the design and build contractor) shall have clear specifications for production, quality control and testing. Additionally to typical quality control of concrete, load tests should be mandatory and close quality control of the supplier in all production steps is highly recommendable. Detailed design with BIM and taking the real dimensions of all elements including MEP and rail systems equipment can give response to interfaces. Major changes during procurement of equipment should be avoided, suppliers of MEP and rail system equipment shall be aware of available openings for cables and ducts.

6 CONCLUSIONS

Design, construction and operation of urban mass transport systems using underground space is necessary for attending the needs of increasing urban population. The implementation of a new Metro system in the city of Riyadh was granted to three design and build contractors. These projects are integrated projects including civil works, architectural work, MEP installation, rail systems, rolling stock, streetscaping, etc. The integration of the projects allows for an internal interface management and concurrent design, procurement and construction.

Modern standards, technology and experience are key factors for a reliable, fast and safe underground construction. Risk management forms part of the design and construction process, for designing out as many potential risks as reasonable possible or providing mitigation and contingency measures were deemed necessary. In regards of the tight time schedule civil construction of deep underground structures included pre-cast elements. A design in three dimensions and BIM technologies including the coordination between civil works and installations is required as pre-cast construction is limiting options for field changes.

Utilities diversions and the associated risks of uncharted utilities were handled in a traditional way. In future projects, utility related risks could be minimized with Subsurface Utility Engineering and three-dimensional Subsurface Utility Mapping. The utility owners should be the first interested party in having a precise documentation of their own assets.

Technological challenges for designers and constructors in future developments of underground construction can be attended by using modern tools and appropriate interface management including constructability as well as Health and Safety in construction works. Concurrent design and construction obliges for a close coordination between designers, procurement, main contractor and relevant sub-contractors along the whole duration of a project.

REFERENCES

Abdeltawab, S. 2013. Karst limestone foundation geotechnical problems, detection and treatment: Case studies from Egypt and Saudi Arabia; *International Journal of Scientific & Engineering Research*, Volume 4, Issue 5: 376

ADA. 2015. *Arriyadh Development Authority website*; www.ada.gov.sa

Al-Othman, A. 2011. Implication of gravity drainage plan on shallow rising groundwater conditions in parts of ArRiyadh City, Saudi Arabia. *International Journal of the Physical Sciences* Vol. 6 1611–1619

ASCE 38-02; *Standard Guideline for the Collection and Depiction of Existing Subsurface Utility Data* (2002)

EN1997 - *Eurocode 7: Geotechnical design* (2006)

Gómez, F.J., Martín A., Zarrabeitia, G. (2017) Estaciones enterradas de la línea 3 del Metro de Riad. Estaciones profundas con contrabóveda - Underground stations of Riyadh Metro project line 3. Deep stations with counter-vault, *Hormigón y Acero* 68(283) 209–220

ITIG (The International Tunnelling Insurance Group) 2006. *A Code of Practice for Risk Management of Tunnel Works*

Lord J.A., Marcetteau, A.R. 1981. Problems of Deep Basement Construction in Riyadh; *Proc. Sym. Geotech. Problems in S.A.*: 129–167

PAS 128:2014, *Specification for underground utility detection, verification and location* (2014)

Saudi Geological Survey (SGS), 2007. *Saudi Cave Unitmaps of Caves Surveyed by Saudi Geological Survey*, Kingdom of Saudi Arabia

Schwarz, H., Almousa, A., Al Arifi, F. 2016. Riyadh Metro Design and Construction - Design and Construction on the Fast Track; *Proc. WTC San Francisco*

*Tunnels and Underground Cities: Engineering and Innovation meet Archaeology,
Architecture and Art, Volume 12: Urban
Tunnels - Part 2 – Peila, Viggiani & Celestino (Eds)*
© 2019 Taylor & Francis Group, London, ISBN 978-0-367-46900-9

Design and construction of artificial tunnels in urban context

S. Sdoga, S. Giovenco & R. de Falco
Italferr S.p.A., Rome, Italy

ABSTRACT: In urban context, the improvement of rail infrastructures leads to develop underground railways by means of bored and artificial tunnels, aimed at respecting the environment and giving back urban spaces to citizens, with a particular attention to city centers and historical sites. Different design solutions can be adopted depending on depth of excavation, overburden, ground geotechnical characteristics, water table level, logistic and site layouts, potential issues with construction sequences and line disruptions, available techniques and equipment proposed by the contractors and other several variables. The aim of the present paper is to provide an overview of different design solutions and construction approaches for artificial tunnels, with reference to the ones designed and built in Italy, as the majority of the railway lines and hubs have been strongly renovated during the last years. Then, by highlighting advantages and disadvantages of each alternative, the best practices for artificial railway tunnels will be defined.

1 INTRODUCTION

Starting from the 1960's, the unexpected rapid anthropization and urban growth throughout Italy, have brought many collateral issues, including traffic congestion, air pollution, loss of available space for new infrastructures and major traffic disruption during their construction or maintenance.

In Italy, the transport policy is focused mainly on road transport and the major cities have nearly two-thirds of their central districts occupied by road infrastructures (e.g. freeways, streets, and parking facilities), while the remaining is left for productive or recreational services.

During the last decades, a new awareness that cities can develop underground without other surface consumption has been addressed, and new sustainable urban planning is increasingly relocating many facilities underground, such as rapid transit, parking, utilities, sewage and water-treatment plants, warehouses, light manufacturing and so on.

The new strategy carried also new technical solutions for underground construction, with reference to material, temporary and permanent retention system, provisional works, boring machines and so on. This is ascertained in the case of rail infrastructures, due to its large amount of constraints, related both on the line geometry requirements (e.g. tracks, cant, horizontal radius and slope) and on the dense geographical and anthropogenic context. Therefore, tunneling for rail infrastructure projects has been widely adopted to overpass limitation and obstacles, leading to challenging solutions and improvement of new techniques.

The aim of this paper is to investigate the relationship between urban context and artificial railway tunnels, through an overview of different design solutions depending on the different ground and site condition, depth of excavation, water table and availability of techniques.

Examples of artificial tunnels on the Italian railway system will be highlighted to define advantages and disadvantages of each alternative and to provide the state-of-the-art.

2 TYPICAL ARTIFICIAL TUNNEL SOLUTIONS

More than 70% of Italians lives in the cites and the urbanization process has become increasingly rapid and aggressive. During the last decades, local administrations started to be aware on the necessity of defining new technical solutions to deliver a widespread infrastructure system, that could link urban areas without affecting the surface, usually filled with a huge concentration of historical sites and art heritage cities. This is particularly true in the case of the railways, which are more constrained from a planimetric and altimetric perspective and usually are element of division within the cities.

Regarding the complexity of the design of a new railway and the high cost of expropriation as well as longer time for authorization, the tendency is to realize underground railways and stations. When realizing a tunnel, many variables must be considered as they affect considerably the design and the construction of the work. Therefore, a large amount of technical solution has been defined and adopted, depending on depth of excavation, ground conditions, availability of equipment, potential disruption of the existing lines, water table, presence of buildings in the surrounding area and so on. Typically, in the case of the artificial tunnels, "cut & cover" is widely used as construction technique; the variables above-mentioned may influence whether to pursue a Bottom-Up or a Top-Down approach.

In urban areas, due to the limited available space, the tunnel is frequently constructed within a neat trench line using braced or tied back excavation supporting walls. Wherever construction space allows, in open areas beyond urban development, it may be more economical to employ open cut construction.

In the following paragraphs an overview of technical solutions designed and supervised by *Italferr SpA* will be shown.

3 ARCISATE – STABIO RAILWAY LINE: TUNNEL GA02

This paragraph is concerning the design of the "Completion works for the new rail connection Arcisate - Stabio", along the section between the Olona River Bridge and the Swiss-Italian border.

The tunnel under examination is the GA02 tunnel, and it is located between the chainage 4 +051.55 and 5+003.65 on the new Varese Railway, for a total length of 952 m.

The new railway line is double track and overlaps planimetrically the existing line (previously single track), while the altimetric profile of the new alignment is lowered by roughly 10 - 15 m; hence, the need to include the GA02 tunnel to the track.

Between chainage 4+051.55 and 5+003.65, the existing railway line, develops in open cut trench and fully urban context, with the presence of numerous buildings located near the railway line (see Figure 1).

Figure 1.   Single Track Layout.

The new layout, larger than the previous one because of the widening of the cross section, shows many interferences with adjacent buildings and makes necessary an expropriation plan. The lowering of the line is due to reduce as much as possible the impact of the project on the community. The Figure 2 shows the interfering buildings with the new rail alignment, and the progressive numbering assigned to each of them during the census activities. This activity has been carried out through specific surveying activities (technical report, census forms, photographic survey, etc.). Also, the interference of the new railway line with the close residential area has been considered with regards to noise, vibration, subsidence and settlements on the ground level that could have arisen during the installation of temporary supports for the excavation.

The structure of the tunnel is characterized by two secant pile walls, which are acting as support of the excavation and permanent structural walls. Being a temporary works, secant pile walls require an additional internal reinforced concrete wall as final lining.

The considerable depth of excavation, between approximately 13.00 m and 20.00 m, together with the need to mitigate the impact of operational noise and vibration on the buildings nearby, has made it necessary the adoption of double-chamber sections, almost all along the tunnel development. Therefore, the tunnel section includes three slabs at different levels.

The construction is carried out using a top-down method: the structural cast-in-situ slabs will act as internal bracing to support the excavation thus reducing the amount of temporary propping required.

To speed up the construction activities, it has been proposed to divide the excavation in two stages. The excavation of the first phase is carried out up to the intrados of the mezzanine slab, before the installation of the precast top slab. Then, the surface is reinstated before the

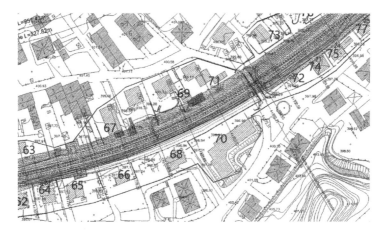

Figure 2.   Double track layout.

Figure 3.   Construction sequence.

completion of the construction once the precast top slab is installed, allowing early restoration of the ground surface above the tunnel.

The peculiarity of the top slab is the adoption of a Dywidag post-tensioning system, that allows to tension tendons at the same way. To ensure a safe work condition and verify that no damage is occurring to the neighboring buildings, a displacement monitoring system has been installed.

The feasibility study has identified the application of the above-mentioned solution only for homogeneous sections of the tunnel, where the net distance between top and mezzanine slab is equal to 3.00 m.

The presented solution is not only economically efficient, as it results in lower cost for the structure, but also beneficial for the construction program since it enables the maximizing of space and the overlapping of activities during the construction stage. Therefore, the design solution proves to be the most suitable and reliable alternative in relation to the depth of excavation and crossed sections.

4   GENOVA- VENTIMIGLIA RAILWAY LINE: GENOVA CASTELLO TUNNEL GA05

The "Genova Castello" tunnel is one of the artificial tunnel on the "San Lorenzo al Mare - Andora" segment, which is part of the widening project of the "Genova - Ventimiglia" railway line.

The design requirements allow to increase the speed on the railway line up to 200 km/h; due to the peculiar geomorphological context of the region, many tunnels along the track alignment have been included.

The "Castello tunnel" is in Diano (Savona) and its total length is about 700 m, as it is composed by a 485 m-long natural tunnel double track, and two artificial tunnels at the entrances. "Genova Castello" is one of the two artificial tunnels, between chainage 38+450.0 and 38+540.00, with a global extension of 90 m.

The tunnel is needed not to expropriate the surrounding buildings and to provide access to them through an overpass on the railway line (Via San Pietro).

The "Genova Castello" is right next to the Diano Station, therefore the first section of the tunnel has a wider cross section than the standard adopted on the line, to accommodate the 3 m wide platforms of the Station itself.

The tunnel, lies entirely within a very stiff silty clay; the first wider section (total length of 57.00 m), is defined by two circumferences: the crown is encircled by a 6.75 m radius circumference, whose center is located at 1.60 m above Top of Rail, while the invert is defined by a 10.60 m radius circumference, whose center is 8.7 m above Top of Rail (Figure 5).

Then, the second section of the artificial tunnel narrows; the crown is defined by a 5.25 m radius circumference, the invert by a 1.16 m radius circumference (Figure 6).

Due to the constant radius and thickness of the lining of the crown, the solution of a precast segmental lining has been implemented, thus the installation must be performed after the excavation as usually done in a bored tunnel; the invert is generally cast in situ. the monolithic static scheme is ensured by the execution of rigid joints at the crown and at the toes of the vault.

Figure 4.   Genova Castello tunnel – Cartography of railway line and Plan view.

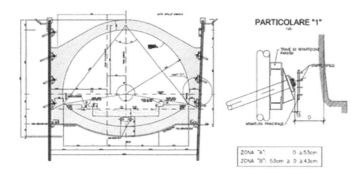

Figure 5. Genova Castello tunnel with platform - section and detail of the anchorage.

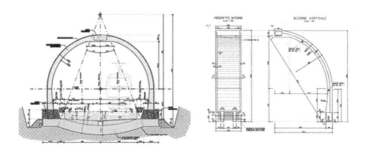

Figure 6. Genova Castello tunnel: narrow section and detail of the precast crown.

The difference with a bored tunnel is the opportunity of reaching the design depth by excavating from the top and reinstating the ground level by backfilling and compacting above the tunnel crown. The waterproofing is ensured through a PVC membrane applied directly on the reinforced concrete crown. The backfilling takes place directly on the waterproof and it acts also as foundation layer for the new overpass, Via San Pietro.

To temporarily support the excavation, it has been used a 168 mm diameter micro-pile retaining wall anchored with tie backs.

The precast solution results to be the most effective and practical alternative from a construction point of view, as this allow for a complete resignation from the installation of temporary ribs and formwork, but is also for the Work schedule as the Contractor has other 11 artificial tunnel along the line and the decision to precast part of the tunnel speeds up the activities and achieve the completion of the works at the time planned by the R.F.I. (Italian Railway Network).

5   PALERMO – PASSANTE: HARBOUR STATION

The project of the new urban rail connection under Palermo collides against the congested ground surface, given the high population density.

Both the alignment and the stations are fully underground. The tunnel, starting from the harbour, reaches the city - center below buildings and foundations, therefore it is necessary to provide a capillary monitoring system for the potential ground settlements that may occur, to statically preserve the structures above.

These works are realized by continuous Secant Piles walls and reinforced concrete slabs as top and base slabs: thus, the excavation will then be performed following the "cut & cover" method.

Since the excavation is performed below water table, ground improvement by means of jet grouting is one of the "key" factors to strengthen the purpose of reducing permeability of the ground and allow a largely dry and safe excavation.

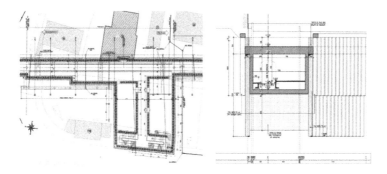

Figure 7.   Palermo Harbour Station: Plan and temporary cross section.

Due to the lack of lateral space and the need to support the excavation face from the load of the adjacent buildings, to achieve the required design level for the realization of the tunnel top slab, additional temporary secant pile walls are installed externally to the permanent ones. Both the permanent secant piles and the provisional ones, have a drilling diameter of 920 mm, while the final diameter of the piles will be 914 mm and they are placed at a 0.75 m spacing.

The length of the upper provisional piles is 10.0 m, while the length of the permanent piles is 15.0 m, such as the toe of the latter is 1.0 m deeper than the jet grouting block. The 6.5 m long grouted columns are carried out in between the secant piles and they have both a water tightness and structural function.

As shown in Figure 8b, differently to what commonly happens, even the primary piles are reinforced with a rectangular cage, in order to support the high loads.

As mentioned, the tunnel lays below the water table, and one of the risk of this type of structure is to create an undesirable "Dam effect", obstructing the correct drainage of water and leading to potential flooding. Therefore, a bypass system through a dewatering plant has been sized and implemented to prevent these phenomena in densely populated areas such as Palermo city center.

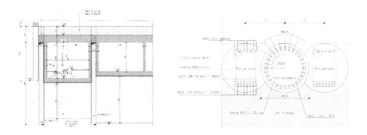

Figure 8.   Palermo Harbour Station: final cross section (a) and detail of secant piles (b).

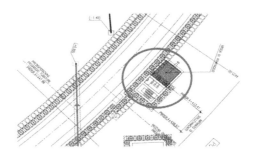

Figure 9.   Palermo Harbour Station: Plant of a water regimentation from uphill to downhill.

6 PALERMO – PASSANTE: POLYCLINIC STATION GA15

The aim of this paragraph is to present the structure for the Polyclinic Station artificial tunnel (GA15). The artificial tunnel is between the chainage 2+689.6 and 2+849.6 of the segment Central Palermo - Brancaccio (Carini).

Currently the railway alignment is an at-grade single rail track. The project involves the simultaneous widening and burial of the railway line, through the construction of the artificial tunnel, to be realized with the "cut & cover" technique; the main structure consists of a wall of secant piles; both the primary and secondary piles have a 90 cm diameter at a 80 cm spacing.

The top slab has a maximum thickness of 120 cm while the bottom slab is 80 cm thick.

Due to the presence of interfering buildings, as well as the mitigation of the existing railway line disruption, the construction sequence will be done on different stages. Two different microphases will be organized.

The microphase no. 1 will cover the demolition of interfering buildings and the execution of the first secant pile wall, sea-side; during this stage, only 2.00 m of the top slab will be cast, with the function of acting as a cantilever capping beam. Finally, it must be realized the temporary relocation of the existing railway.

The microphase no. 2 will cover the activation of the temporary line and the demolition of the existing one. Then, the construction of the second secant pile wall will be possible. During this stage the casting of top slab will be completed, through the realization of a joint. Once done, the excavation will be completed up to the design base slab level, as per a Top-Down approach.

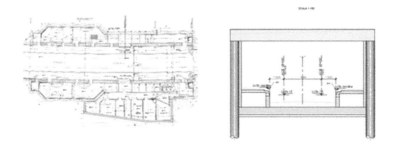

Figure 10.   Policlinic Station: Plan and typical section in line.

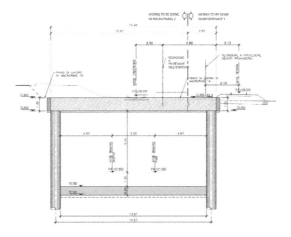

Figure 11.   Policlinic Station: Policlinic Station: detail of microphases.

7 PALERMO C. LE/ BRANCACCIO - CARINI: TOMMASO NATALE 2 TUNNEL GA32

This paragraph presents the artificial tunnel Tommaso Natale 2 (GA32) located between Cardillo and Isola delle Femmine on the widening of the line Palermo C.le/Brancaccio - Carini.

The tunnel is composed by 19 segments 30 m long and realized through "cut & cover", by means of ODEX micropiles as provisional work to support the excavation. The ODEX micro-piles have the advantage to be drilled and cased simultaneously.

The construction sequence comprises two stages not to interfere with the existing line, as shown in the previous paragraph. The difference consists of 2 bulkheads of ODEX micro-piles, which are only a temporary support to the excavation face and therefore the additional casting of an internal permanent lining is needed once the excavation is concluded. Another peculiarity is that the excavation of the tunnel itself needs to be divided in two stages upon installation of a temporary central bulkhead, to be demolished afterwards.

In the case of segment 13, due to the interference with the construction of a new sewer siphon along Via Serravalle, the central ODEX micro-piles are constrained by the base slab of the first stage through a special joint to let them collaborate even after the costruction of the siphon via concrete box jacking (Figure 13). Therefore, the design is influenced by the

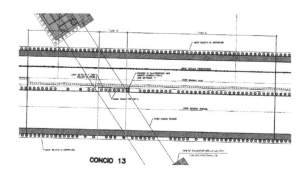

Figure 12.   Design Plan.

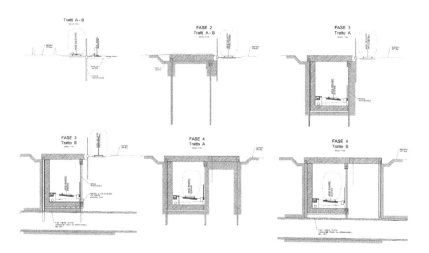

Figure 13.   Construction phases.

construction sequence to be implemented. Also, due to the presence of interfering utilities underground, it is prescribed to realize the micro-piles with a tolerance of 25 cm on their position.

8 MILAN-NAPLES HIGH SPEED RAILWAY LINE: FONTANELLATO TUNNEL GA13

The artificial tunnel "Fontanellato" GA13 on the Milano - Bologna High Speed railway line is located between the chainage 85+980.183 and 87+615.915, for a total length of 1636 m. Here the railway runs parallel to the A1 Milano - Bologna Highway and bypasses the small town of Fontanellato, hence a landscape reprofiling is needed to mitigate the visual impact.

The geotechnical characterization of the tunnel is highly complex: the superficial layer (about 25 m) is formed by silty clays interspersed with lenses of gravels and sands. The water table is approximately at the ground level.

The designed landscaping would have caused an increase of load over the foundation and a consequential issue with long-term settlements and displacements of the track. The first approach to consolidate the foundation was a 1 km long pile foundation, on a double row; since its high cost, a Value Engineering alternative was implemented and a solution with compensated foundation was developed.

The principle behind compensated foundation is the idea that the total load of the structure is balanced by the amount of excavated soil, to experience smaller settlements and provide a satisfying bearing capacity. Practically, the perfect compensation is hard to achieve, while the reached compensation ratio is in the range of 1.3 - 1.5.

The tunnel is divided into 37.5 m long segments, each one formed by a 14.0 × 8.5 m reinforced concrete box; construction and expansion joints are foreseen along the extension of the tunnel.

Th excavation is supported on both sides by temporary sheet piling, while the construction of the structure is independent, as per Bottom-Up approach.

Figure 14. Fontanellato Tunnel: plan view.

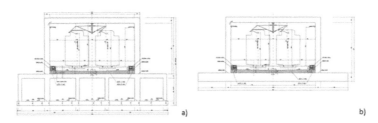

Figure 15. Comparison between adopted foundation.

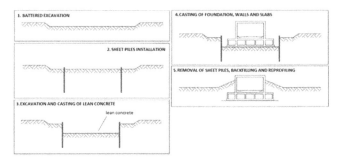

Figure 16. Construction sequence.

As shown in the Figure above, two different foundation are detected: the first one (Figure 15a) is a box caisson foundation, which is needed closer to the tunnel entrances as the compensation design depth is considerably higher than the Top of Rail; the second one (Figure 15b) is a traditional mat foundation.

The foreseen construction sequence, represented at Figure 16, is divided into the following phases:

1. Battered excavation up to sheet piling level of installation;
2. Installation of temporary sheet piles
3. Excavation up to the final bottom level;
4. Casting of permanent foundation and RC slabs and walls;
5. Removal of sheet piles and backfilling

9 CONCLUSIONS

Urban tunnels have a duty of paramount importance: to give the city back to the population, providing healthier space and availability for development and, at the same time to ensure an efficient transport system.

Five different artificial tunnels on the Italian railway system have been presented within this work. Each of them is included in medium-high urban context characterized by medium-high population density and proves to suit to the environment with challenging strategies and different design and construction approaches. It has been shown that the major concerns that usually affect both design and construction of an artificial tunnel were present on the analyzed project: interference with adjacent buildings and noise and vibration issues related to the works (Arcisate – Stabio Tunnel), availability of supplies and speeding up of activities (Genova -Ventimiglia Tunnel), presence of water table and water tightness (Palermo Harbour Station), avoidable disruption on the existing line (Palermo Polyclinic Station), influence of the construction staging on the structural design (Palermo Tommaso Natale Tunnel), ground geotechnical condition and settlements (Fontanellato Tunnel).

In most the case, it has been demonstrated that "cut & cover" is successfully used, but modified and adapted to the uniqueness of the site and of the project. It has been demonstrated that design of artificial tunnel cannot exclude lots of external conditioning and never can be adopted a "Copy & Paste" approach.

The progress of state of the art of technologies, materials, construction techniques and constructability in urban artificial tunneling can truly enhance design and engineering solution at any level of challenge and give back the "Genius Loci" to city dwellers.

REFERENCES

All the references presented in this article come from the design reports and drawings of the following projects:

Vv.Aa. 2016. Lavori di completamento del Nuovo Collegamento Arcisate – Stabio. GA01-GN01-GA02 Galleria Naturale di Induno, imbocco Lato Stabio.

Vv.Aa. 2014. Completamento Gallerie della linea ferroviaria nella tratta San Lorenzo al Mare – Andora. GA05 Galleria Castello Lato Genova.

Vv.Aa. 2018. Chiusura dell'anello ferroviario in sotterraneo nel tratto di linea tra le stazioni di Palermo Notarbartolo e Giachery e proseguimento fino a Politeama. GA12/FV02 Fermata Porto.

Vv.Aa. 2011. Nodo di Palermo Raddoppio Palermo c.le/Brancaccio – Carini. GA15 Galleria Artificiale Fermata Policlinico.

Vv.Aa. 2010. Nodo di Palermo Raddoppio Palermo c.le/Brancaccio – Carini. GA32 Galleria Artificiale Tommaso Natale 2.

Vv.Aa. 2007. Linea A.V. Milano – Napoli Tratta Milano – Bologna. GA13 Galleria Artificiale di Fontanellato.

Tunnels and Underground Cities: Engineering and Innovation meet Archaeology,
Architecture and Art, Volume 12: Urban
Tunnels - Part 2 – Peila, Viggiani & Celestino (Eds)
© 2019 Taylor & Francis Group, London, ISBN 978-0-367-46900-9

Design methodology of permanent linings of junctions at Crossrail project contract C121 (the Elizabeth line)

E.J. Sillerico, J. Suarez & R. Brierley
Mott MacDonald, London, UK

ABSTRACT: Crossrail Contract C121, awarded to Mott MacDonald, dealt with the design of the SCL (sprayed concrete lining) structures of five new underground stations in London. This paper describes the design methodology of permanent (secondary lining) structures such as junctions between platform tunnels and cross passages, shafts and cross passages and a combination of them. The paper provides a comprehensive explanation of the numerical modelling process for 3D complex geometry of junctions using FEM (finite element method) software, describing challenges like application of loads (ground and water pore pressures, surcharge, etc.) and how they are transferred from the primary to the secondary lining in the long term. A practical approach to assess the structural behaviour of different types of junctions and interpretation of results based on technical standards like RILEM and Eurocode 2.0 will be discussed as well as recommendations on bar reinforcement detailing.

1 INTRODUCTION

Mott MacDonald was appointed to undertake the design of Sprayed Concrete Lining (SCL) tunnels for the Crossrail project, a 43-km brand new running tunnel across London. Junctions are defined as the connections between 2 tunnels typically one of smaller diameter know as child opened into a bigger diameter tunnel known as parent. Tunnel junctions are one of the most complex SCL structures as a plane strain analysis used for singe tunnel sections is not enough to assess their structural behavior.

This paper presents a design methodology of tunnel junctions for both sprayed and cast concrete permanent linings. Based on ground properties, depth, geometry and boundary conditions; the aim of the methodology is to evaluate the structural behavior of the following types of junctions:

- Single junctions: Platform tunnels to cross passages.
- Single junctions: Access passage to shafts, for both top opening and bottom opening.
- Double junctions type 1: concourse tunnels to cross passages.
- Double junctions type 2: concourse tunnels to cross passages and headwall.
- Double junctions type 3: wraparound where platform tunnels are the child tunnel.
- Double junctions type 4 (Twin openings): Launch Chamber to ventilation ducts.

2 TUNNEL JUNCTIONS DESIGN METHODOLOGY

This design methodology comprises the following stages:

a) 3D FEM modelling process:
  - Geometry meshing: Challenges of the geometrical assembling (meshing size) of complex structures.

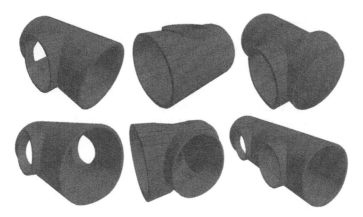

Figure 1. Different types of Junctions typically used in underground stations.

– Soil – structure interaction element: Design of the boundary conditions of the 3D model.
– Material properties.
– Loading conditions: loads and load combinations as per Eurocode 0 and Eurocode 1

b) Assessment of outputs of the 3D model.
    – Structural design as per RILEM and Eurocode 2.
    – Detailing of rebar as per Eurocode 2.

2.1 Description of junctions

Junctions are the connection between a main tunnel with a larger diameter known as "parent tunnel" and a second tunnel with smaller diameter known as "child tunnel". The main junctions are set up between platform tunnels and cross passages, and between concourse tunnels and cross passages. There are also other junctions such as concourse tunnels and shafts in which openings can be placed either at the crown or invert (such as for sumps and ventilation adits).

2.2 Finite Element Modelling (FEM) – STAAD Software

STAAD.Pro is a structural analysis and design program that is based on the Finite Element Method. It can efficiently generate finite element geometry for complex structures with openings such as junctions in tunnels. The modelling process and analysis carried out using STAAD.Pro provided the results of the analyses in terms of the predicted axial and shear forces, bending moment and deformation of the permanent lining under long-term conditions. To define the input of the methodology, the following parameters are to be defined:

2.2.1 Geometry
The geometric construction of 3D models for tunnel junctions is very time-consuming. For Finite Element (FE) models the Designer selects the appropriate mesh sizes to achieve the following:

– Accurate geometrical representation of the behavior of the selected junction.
– Finer mesh sizing at points of isolated loads to achieve accurate output.
– Coarser mesh at zones where construction is unlikely to change the stress conditions prior to the construction of the junction.
– The required level of numerical accuracy i.e. the fineness of the mesh.

2.2.2 Material

Concrete properties: Concrete Grade: C32/40 at 90 days, Density: 2500 kg/m^3, Poisson ratio: 0.2, Young's modulus: 17 GPa, Coefficient of thermal expansion: alpha: 10^{-5} K^{-1}

2.2.3 Boundary conditions

The interaction between soil and the structure is established through springs-beams attached to every node of the plates that form the 3D model. The radial beam-springs act in compression only as compressive stresses are transferred between the tunnel and the surrounding mass (as an equivalent axial stiffness between the concrete spring beams and the surrounding ground these beams).

2.2.4 Loading and load combinations

- Self-weight: Self-weight is applied on the whole structure except for the spring beams representing soil as these are not structural elements of the model.
- Earth pressure: The earth pressure acting on the secondary lining is determined in accordance with the results of numerical modelling using FLAC analysis for the long term (120 years). The radial contact pressure on the interface between primary and secondary lining consists of both 'active' and 'passive' load (bedding reaction from deformation of the tunnel lining).
- Pore Water pressures (PWP) ULS and SLS: The ULS (ultimate limit state) and SLS (service limit state) water pressure is depth-dependent and is taken from the water pressure profile coming as a result of the geotechnical study.
- Drying shrinkage: Shrinkage is determined according to Eurocode 2 where strains are a combination of drying and autogenous shrinkages. Shrinkage depends on concrete parameters, environmental conditions and lining thickness and it is simulated as a change in temperature.
- Temperature loads: In addition to the concrete temperature at the time of installation, the variable temperatures (Extrados (far face) and Intrados (near face)) depending on the season of the year are also considered.
- Load combinations: The load combinations are defined as per BS EN 1990:2002 (Eurocode 0) Basis of structural design for both ULS and SLS.

3 OUTPUTS & RESULTS

The outputs of the analysis in STAAD are based on the matrix displacement method which idealizes the structure into an assembly of discrete structural components (plates) as the loads are applied in as distributed loads on the element surfaces or as concentrated loads at the joints.

STAAD calculates member stresses specified at intermediate sections as well as at the start and end joints. These stresses include: *Axial stress (Si), Bending-y stress (Mi), Shear stresses* (SQi) and *Combined stress* (Top Combined SX & SY and Bottom combines SX & SY) which is the sum of axial-i and bending-i stresses as per Wood and Armed method.

Once the model is run, the post-process stage follows where certain or all load combinations can be selected as STAAD has calculated deformations, stresses and forces for each one of them.

3.1 Deformed shape

The first parameter presented in STAAD is the deformed shape which is the calculation of the displacements at every single node of the model and available for each individual load case and load combinations in ULS and SLS to comply with the requirements of deformation as per EC2 in SLS load combinations as shown in Figure 2.

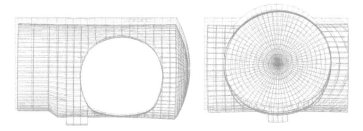

Figure 2. Typical deformed shape under a load combination case.

3.2 *Internal Forces: Axial Forces, bending Moments and shear forces*

STAAD presents the internal forces: axial forces and bending moments along two local axis: X and Y. When analyzing tunnels, it is recommended that the local axis x follows the "longitudinal" direction of the tunnel while local axis y follows the "hoop" direction of the tunnel.

STAAD presents the axial stresses, bending moments and shear stresses in both directions X (longitudinal) & Y (hoop). The axial and shear forces can be calculated by multiplying these values by the thickness of the element. Wood and Armed method is considered to determine the X and Y forces in plates.

4 STRUCTURAL DESIGN AND REINFORCED CONCRETE RC DETAILING

Steel quantity calculation is based on Eurocode 2. The detailing process is based on: Eurocode 2, RILEM TC 162-TDF, CIRIA C660, BS 8666:2005. The process comprises the assessment of the results of the numerical modelling:

– Moment Axial design capacity check to assess the need of rebar in the structure
– Stress contours obtained in the 3D model to assess zones under tensile stresses

An example is shown in Figure 3 below where areas requiring reinforcement are delineated based on the plotted contours in the 3D model:

Scheduling, dimensioning, bending and cutting of steel are designed as per BS8666 requirements.

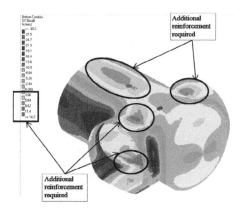

Figure 3. Stresses contour on near face.

5 SUMMARY OF OUTPUTS OF REPRESENTATIVE TYPES OF JUNCTIONS IN THE CROSSRAIL PROJECT

In this section the assessment of the structural behavior of representative tunnel junctions selected based on ground properties, levels and geometry of tunnels and boundary conditions are presented. The assessment is set up considering the following aspects:

a) Ground and water conditions: the parameters of the ground define how the ground loads are acting on the secondary lining.
b) Depth: the level of tunnel axis will determine the ground pressure that will be transferred to the secondary lining when this load is shared between both linings, primary and secondary, as is set up in the Numerical Modelling guideline.
c) Geometry:
   - Number of child tunnels: single or double junctions, e.g. double junctions typically got larger deflections and tensile stresses.
   - The ratio between the diameter of the smaller (child) tunnel and larger (parent) tunnel has a considerable influence on the design and typically varies between 0.4 and 0.8. Tunnel junctions with larger aspect ratios have been selected since tensile stresses are larger on the crown zone of the connection on near faces.
   - The location of the child tunnel opening with regards to the parent tunnel, i.e. this opening can be located at sides of the parent tunnel forming a typical junction or the openings can be located at crowns which is the case of shafts connection with cross passages.
   - Thicknesses of linings are important when junctions are cast in situ in different pours and, thus, the Early Age Thermal Cracking (EATC) ought to be considered in the design.

d) Boundary conditions: It is important to evaluate the potential impact of a nearby structural element such as headwalls, sumps and shafts on junctions. Headwalls, for example, have an important impact with regards to the reinforcement design when located very close (less than 2m) to the openings of child tunnels.

Table 1. Matrix of junction assessment.

Typology of junctions	Station	Type of ground	Tunnel axis level – depth	Pore water pressure (PWP)	Boundary conditions	Junction selected	main selection factor
Single junction	FAR*	London clay – LC3	106 mATD – 20 m	110 KPa crown to 50 kPa invert	Contraction joints all ends	STW1/ VA1	(worst ground conditions)
Shaft to tunnel	LIV*	London Clay – LC3	79 mATD – 35 m	200 KPa crown to 240 kPa invert	Contraction joint/ headwall	LS – CH7 to Shaft	(Worst condition in shear)
Double junction -Type 1	WHI*	London Clay – LC3	88 mATD – 24m	150 KPa crown to 120 kPa invert	Contraction joints and headwall	CP1a to EPW1	(bigger aspect ratio)
Double junction – headwall – Type 2	LIV* – NLL	London Clay – LC3	82 mATD – 32m	180 KPa crown to 220 kPa invert	Headwall in parent tunnel, in other ends, contraction joints	AP6a to AP10a/ AP10b	(loading conditions – PWP)
Wraparound (Double junction type 3)	LIV*	London Clay – LC3	79 mATD – 35m	190 KPa crown to 250 kPa invert	Headwall in child tunnel and contraction joints in other ends	CP5/CP6 to PTE/ PTEW	(thickness – EATC impact)

* FAR (Farringdon Station). LIV (Liverpool St Station). WHI (Whitechapel Station).

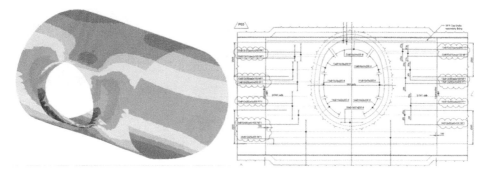

Figure 4.   Single Junctions: Platform tunnels to cross passages.

These factors determine the structural behavior of selected representative junctions in C121 SCL stations. This selection is summarized in the following table (matrix of junction assessment):

5.1   *Single junctions: Platform tunnels to cross passages. Farringdon Station STW1-VA1 junction.*

The model contours with tensile stresses higher than 3 N/mm2 are zones where the most critical tensile stresses are concentrated. These tensile stresses are bigger than the maximum tensile strength of the SFRC, thus additional reinforcement is needed on near faces. Furthermore, these contours define the limit length of reinforcement in terms of detailing to ensure the correct anchorage and lapping.
 Output: Near Face Stresses Hoop Direction

5.2   *Single junctions: Access passage to shafts. Liverpool Station. LS to CH7.*

Output: Shear stresses around the shaft connection
 The shear stresses are localized on the main tunnel where the shaft lining is punching and is governing the design against shear.

5.3   *Double junctions type 1: concourse tunnels to cross passages. Whitechapel Station. CP1 to EPW1*

The tensile stresses are concentrated at crown and invert connection between parent and child tunnel, so reinforcement must be provided on those areas. The bigger the aspect ratio the higher additional reinforcement is required.

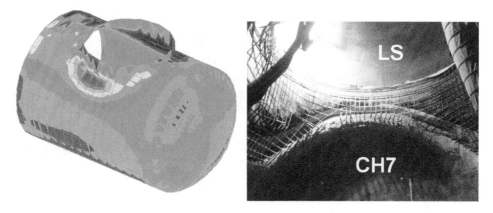

Figure 5.   Single junctions: Access passage to Shafts.

Figure 6.   Double junctions type 1: concourse tunnels to cross passages.

Output: Near Face Stresses Longitudinal Direction

5.4   *Double junctions type 2: concourse tunnels to cross passages and headwall. Whitechapel Station. CP1 to EPW1*

Output: Near face stresses longitudinal direction

This case has stresses concentrated on near face in longitudinal direction. Furthermore, the headwall is withstanding heavy compression on near face what means tensile stresses will be concentrated radially and L bars will be needed to be installed on far face.

5.5   *Double junctions type 3: wraparound where platform tunnels are the child tunnel. Liverpool Street Station. CP5/CP6 to PTE/PTW junction.*

These contours show tensile stresses around the headwall in far face what means reinforcement in hoop direction on far face will be required and heavy compressions on inverts on bottom faces what is a usual resultant from the pore water pressure loading. Tensile stresses are encountered on top faces of the invert.

Output: Far face stresses in hoop direction

5.6   *Double junctions type 4 (Twin openings): Launch Chamber to ventilation ducts. Liverpool Station. LCE to VD4 and VD5 junctions.*

Outputs: Stresses in hoop direction and far face stresses in longitudinal direction.

Figure 7.   Double junctions type 2: concourse tunnels to cross passages and headwall.

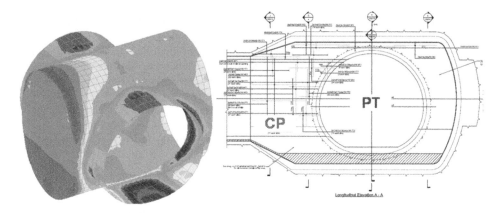

Figure 8.   Double junctions type 3: wraparound tunnels.

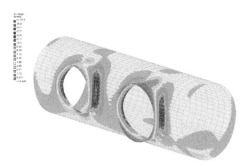

Figure 9.   Double junctions type 4: Twin openings.

The stresses in hoop show strong compressions between both child tunnels that have been checked against the maximum compression strength of the SFRC. This concentration of compression stresses is usual in junctions with two very close child tunnels. Furthermore, there is a concentration of tensile stresses on the far face in longitudinal direction between both child tunnels (above and below the tunnel axis). Longitudinal reinforcement is needed to be extended between both child tunnels, and between each child tunnel and joints located at the ends on the model.

6   CONCLUSIONS

This paper has presented a method of designing tunnel secondary linings of junctions with complex geometries. The following advantages of the method can be summarized as:

– The standardization of the design method allowed many different geometry schemes to be undertaken without repeating all modelling from the beginning. With a project as geometrically complex as Crossrail, with many different cross-passage, ventilation adits and access passage geometries, this allowed a great number of different junction types to be analyzed relatively quickly.
– The method allows the benefits of the two software used to be utilized to their full. In the case of FLAC, the design of the primary lining, where SLS states are not of interest, the soil-structure interaction is key. For the secondary lining, STAAD is able to add load cases

such as temperature, shrinkage, internal and accidental loads. The SLS load cases can be considered in more detail as they apply to the secondary lining.

– The method allows specific areas of junctions to be identified to ensure that heavy reinforcement, if needed at all, is limited to stress and bending moment concentrations. This minimizes working at height to fix reinforcement and is extremely important for sprayed linings as the reduction of reinforcement improves workmanship.

– The identification of areas with concentration of stresses is key to whether decide the real need for additional reinforcement or further analysis. Typically, further analysis is the preferred allowing stress concentrations considered to be unrealistic to be smoothed out over immediately adjacent parts of the structure, e.g. in tight corners around the immediate perimeter of the junction openings.

In using this method, patterns can be seen in the behavior of junctions:

– The near and far face bending patterns are typically very similar (although with differing magnitudes), as can be seen by the various displacement plots shown throughout this paper. An exception is egg-shaped parent tunnels, which can cause the moments immediately around the child tunnel perimeter to change tension face.

– It would likely be possible to define a standard spring stiffness for use in secondary lining models per project or per set of similar ground conditions and depths, without significant loss of accuracy.

– Double facing junctions, e.g. those from concourse to cross-passages presented here, cannot generally be assumed to behave as mirrored single junctions, especially as the aspect ratio child/parent grows.

REFERENCES

BS 8666:2005: Scheduling, dimensioning, bending and cutting of steel reinforcement for concrete — Specification.
CIRIA C660: Early-age thermal crack control in concrete.
Eurocode 2. BS EN 1992-1-1Design of concrete structures. Part 1.1: General rules.
Eurocode 2 BS EN 1992-1-2 Design of concrete structures. Part 1.2: Structural fire design.
RILEM TC 162-TDF: 'Test and design methods for steel fiber reinforced concrete'.

Tunnels and Underground Cities: Engineering and Innovation meet Archaeology,
Architecture and Art, Volume 12: Urban
Tunnels - Part 2 – Peila, Viggiani & Celestino (Eds)
© 2019 Taylor & Francis Group, London, ISBN 978-0-367-46900-9

Twin tunnels excavated in mixed face conditions

M.A.A.P. Silva & F.L. Gonçalves
Andrade Gutierrez Engenharia S.A., São Paulo, Brazil

A.A. Ferreira & H.C. Rocha
Cia. do Metropolitano de São Paulo Metrô-SP, São Paulo, Brazil.

ABSTRACT: This paper presents a final analysis of the excavation of two parallel tunnels for the Metro of São Paulo Line 5 Expansion, which were excavated by two 6.9m EPB TBM. This current paper focuses on the stretch where mixed face conditions in Precambrian materials were excavated, which had high influence on the TBMs performance, such as difficulties in keeping a stable face pressure, ground losses (high settlements), sinkholes, slow advances, clogging, and very frequent and long hyperbaric interventions. The purpose of this document is to present a technical explanation for the events occurred in the excavations in this stretch, and for the impact in terms of production.

1 INTRODUCTION

This paper presents an analysis of the excavation of two parallel tunnels for the Metro of São Paulo Line 5 Expansion, which were excavated by two 6.9m EPB TBMs. Table 1 shows the main characteristics of them.

Although these tunnels were already object of study by Silva et al. (2015, 2016) and by Comulada et al. (2016), this current paper focuses on the stretch where mixed face conditions in Precambrian materials were excavated, in order to evaluate the relationship between TBM operational factors and these materials, as well as its consequences. The authors above give more information about TBM performance in both soft sandy soils and stiff clays. Figure 1 shows a schematic tunnel alignment, being also highlighted the stretch where mixed face conditions were observed. shows the cutter head (CH) design and its main cutter tools used.

In this stretch, mixed face conditions could be the main reason for the impacts on the TBMs performance, such as difficulties in keeping a stable face pressure, ground losses (high settlements), sinkholes, slow advances, clogging, and very frequent and long hyperbaric interventions. The total length of this stretch was close to 450m, which were excavated in approximately 3 months (average of 4 rings per day).

2 GEOLOGICAL AND GEOTECHNICAL CONDITIONS

As described above, this stretch is strongly characterized by mixed face conditions in weathered gneiss and saprolite, and saprolite and residual soil. Some characteristics of these materials are presented below.

The residual soils are composed of sandy silt and clayey silt, with SPT values ranging from 2 to 30 blows. In these materials, water level ranged between 10 to 25 m above tunnel crown. Residual soils also contain some amount of cobbles and geological structures inherited from parent rock. It is highlighted that these materials can have high permeability to water (between 10^{-4} and 10^{-6} cm/s, according to Futai et al., 2012). Furthermore, in regard to air, it

Table 1. TBM characteristics.

Shield Characteristics	
Type	EPB
External diameter	6.89 m
Total CH opening	39%
Nominal Torque	3560 kNm
Exceptional Torque	4984 kNm
Breakout Torque	5340 kNm
Number of thrust cylinders	16x2
Total installed thrust force	60,801 kN
Maximum thrust force	50,000 kN
Cutters	Rippers, discs, scrapers (#41)
Minimum curve radius	250 m
Ring thickness	30 cm
Ring length	1500 mm
Ring segmentation	5+1

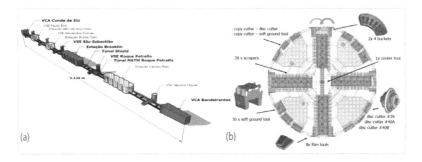

Figure 1. (a) Twin tunnels alignment and mixed face conditions stretch. (b) Cutter head design and main cutter tools used.

is also accepted that this permeability can be 10 to 100 times greater, which can make these materials very permeable and unstable when operating in transition mode (compressed air on top) and in hyperbaric conditions. Finally, Futai et al. (2012) have shown that these materials are very heterogeneous, which explains the difficulty in adjusting the foam parameters, as well as the high efforts necessary to have a good conditioned muck and to homogenize the material in the excavation chamber.

The saprolite term used here corresponds to the weathered rock impenetrable to the SPT sampler. Usually, saprolite has SPT values higher than 30 blows. The geomechanical behavior of this material is hard to be modeled, because it is highly heterogeneous and its behavior can be governed by the joints and discontinuities of the original rock. It is known from previous conventional excavations that saprolite does not have good response when excavated (very low geomechanical parameters when relieved) and it is associated to high ground losses.

As described by Oliveira et al. (2017), this saprolite (and some residual soils) has usually stiff to hard consistency and, thus, requires high amount of liquid addition (very liquid foam or water) in order to give it some plasticity. However, according to Hollman & Thewes (2012) clogging tendency methodology (Figure 2), adding liquids will take this material to the strong clogging tendency field and, thus, it will increase the probability of needing a hyperbaric intervention to clean the cutterhead.

The biotite gneisses are highly foliated and present steeply dipping NE-SW trending orientation. Uniaxial compressive tests carried out in weathered gneiss showed that strength can

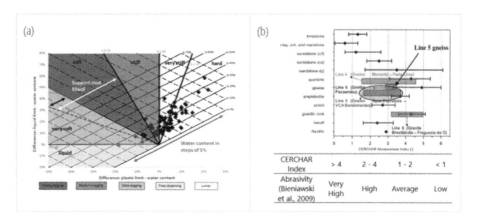

Figure 2. (a) Residual soil and saprolite clogging tendency (adapted from Oliveira et al., 2017). (b) Abrasivity Cerchar Index (Monteiro e Rocha, 2015b).

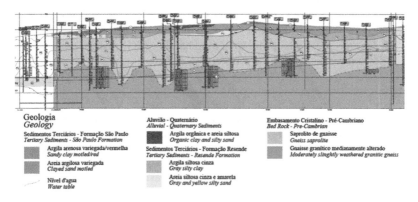

Figure 3. Geological profile (Monteiro & Rocha, 2015a).

easily vary from 5 to 60 MPa depending on the content of biotite bands. Figure 3 shows the geological profile from the stretch analyzed.

Recent studies presented by Monteiro & Rocha (2015) have revealed that the biotite gneisses also exhibit high CERCHAR abrasivity index (Figure 2b).

Because of the high heterogeneity and the presence of different materials in the excavation face in a transitional ground, the specific weight of the excavated material can constantly change. It usually fluctuated between 2.2 and 2.5 t/m³, which made the control of the excavation volume with the scales quite difficult.

3 THE IMPACT ON THE SHIELD PERFORMANCE

3.1 Conditioning

In the transitional zone, it often occurred three different materials in the excavation face – rock, saprolite and residual soil. Besides, these materials are very heterogeneous and, thus, very difficult to be conditioned into a homogeneous paste.

It must be remembered that water, air and foam agents (tensides and polymers) are added in the cutterhead front to condition the excavated material, in order to give to the mixture formed in the working chamber the workability required, without loosing its capabilities to counterbalance earth and water pressures. It must be also highlighted that air bubbles formed in these mixtures have an important function, once they are the ones to provide the workability mentioned. And thus, its structures must be stable enough to keep the air trapped.

The relationship between liquid (water plus tenside) and foam (water plus tenside plus air) injection is given by the foam expansion rate (FER). For cohesive materials, wet foam is required (FER < 6), while cohesionless materials require dryer and more viscous foam (FER > 8). Finally, it is worth to be remembered that the foam injected in the cutterhead prevents clogging and, because of that, it is very important to have a straight and continuous monitoring during the excavation, being constantly alert to any change in the geological conditions. In transitional ground, that need is even more important because of the everchanging conditions at the tunnel face.

In a specific stretch of the Line 5 Precambrian mixed face conditions, residual soil with cohesionless material occurred in the tunnel crown while there were weathered rock or a harder material in the bottom. The cohesionless residual soil disaggregated easily in the presence of water, so dryer foam was necessary to assure the stability of that layer. On the other hand, the harder material demanded a greater water volume to control the temperature during mining. That mixed face condition affected the balance between water and air injections in the excavation chamber.

Consequently, the cutterhead openings and the foam injection points were sometimes blocked, as well as high temperatures in the excavation chamber often occurred, which led to the use of "free" water injections in the working chamber (WI), with the aim to control temperature and give some plasticity to the muck.

Sometimes, extremely wet foam (FER < 4) was used in the bulkhead. However, the total flow injected through this line was not enough to give the plasticity needed for the muck and, thus, more "free" water injection in the chamber was constantly necessary.

Other attempts consisted in increasing the foam injection rate (FIR) to try to achieve higher workability, maintaining the dry foam to stabilize the cohesionless soil layer. In those cases, the air accumulation in the top of the excavation chamber packed the material in the bottom, leading again to the need of "free" water injection. However, the excess of "free" water injected in the chamber damaged the foam structure, releasing the air bubbles to the top of the working chamber.

Although the machines were not designed as hybrid machines, some attempts to inject bentonite though the foam lines during the advances were made. The bentonite showed some results in cooling the material and reducing the segregation. However, since the machine was not designed to work with high injection rates of bentonite, its available flow was not sufficient to allow a continuous advance. The injections were possible for only one or two rings.

However, the bentonite injection let at the foam lines is more susceptible to clogging. Thus, it was not possible to evaluate if the addition of bentonite would modify the chances of clogging.

In summary, rock and saprolite were not well conditioned. Those materials need water for cooling, while soft residual soils (SPT < 6 blows) need a dry and stable foam for a good muck conditioning and to keep the face stable. Regardless of the change in the foam parameters, with such different materials in the front, segregation always occurred in the chamber.

A higher cutterhead rotation speed could theoretically help to mix material in front and foam; however, in mixed face high rpm would increase the risk of cutting tools damage, as the impact of discs to the rock face at higher speeds increases, resulting in flat discs and thus creating the need for hyperbaric interventions. Besides, higher rpm also increases temperature in the chamber, which leads to stoppages, increase in the clogging possibility and also hyperbaric interventions.

Because of that, the machines could not work in EPB mode, but only in Transition Mode with compressed air on the top (Maidl et al., 2012), which enhanced the probability of an overexcavation, even maintaining a high face pressure (EP1 higher than 2.5 bars with the water pressure below 2.0 bars). Controlling the air bubble in the chamber was the key to avoid any possible face instability. Nevertheless, because of the high permeability in the transition zones, it was not rare having air escaping to the ground and face pressure fluctuations,

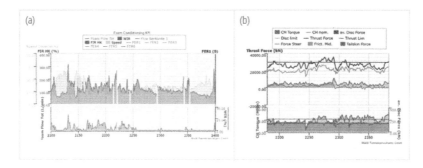

Figure 4. (a) Foam and water injection data in mixed face conditions. (b) Advancing TBM parameters.

leading to soil instabilities. In total, three sinkholes were associated with the operation in the mixed face stretch.

Since the geological conditions were always varying, foam parameters were also changing all the time, showing the difficulties in forming and keeping a homogeneous muck in the chamber. Figure 4(a) shows the relationship between average FIR, water injection rate (WIR) and velocity. It is clear the use of more foam and water in the stretch where lower velocities were achieved. In the same way, it is also clear that the faster the machine advanced, the lower FIR and WI were. The FIR values ranged between 100 and 130%, while the FER was between 5 and 8.

3.2 Digging

In Precambrian mixed face conditions, Shield advances are usually limited by the cutterhead rotation speed and the penetration rate, that limit the maximal forces applied in the discs in order to avoid damage. Ferber (2013) recommended the values shown in Table 2.

For the Line 5 Precambrian mixed face conditions, penetration rate was hardly higher than 10 mm/rev and rotation speed was around 2.0 rpm. Frequently, penetration values bellow 5 mm/rev were observed. With such low velocity, sedimentation in the chamber often occurred, making conditioning harder and, thus, having more difficult conditions during advance.

In critical stretches, penetration was limited by Torque or Thrust. In Figure 4(b), it can be seen an increase in discs forces, denoting a harder material in the front, which demands limiting the penetration in order to avoid disc damages. In this figure it also can be seen that the machine was advancing close to its torque limit.

In mixed face conditions, high cutting tool wear was observed, incurring in a high number of hyperbaric interventions to change them. The passage of the cutting tools through material of different resistance, promotes the impact of the tools, damaging them. That is one of the reasons to limit the cutterhead revolutions. Some damaged discs are shown in Figure 5(a). In this picture, it can be seem that the discs stopped rotating around their axis, leading to a high wear on them. The increase of steering forces was one indication that something was not working properly in the front and the discs were probably damaged.

Table 2. Recommended TBM advance parameters (Ferber, 2013).

Material	Netto-penetration (mm/min)	Netto-rotation speed (rpm)
Sand	30–40	1.2–2.0
Gravel and gravel sand	15–30	1.2–2.0
Compact clay	30–40	1.2–2.0
Stiff clay	40–50	1.2–2.5
Soft clay	40–50	2.0–3.0
- additionally boulders/stones	15–20	1.2–2.0

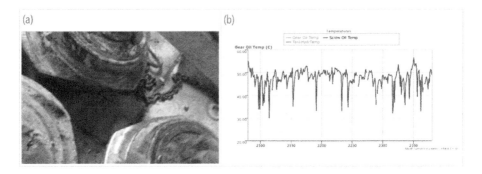

Figure 5. (a) Some damaged disc. (b) Temperatures observed in mixed face conditions.

A similar analysis can be done through Specific Energy (SE) evaluation, as shown in Equation 1. It can be understood as the necessary effort to advance the machine.

$$SE = \frac{E \cdot V + 2\pi \cdot R \cdot T}{A \cdot V},$$

(1)

where E = total thrust; V = velocity; R = rotation speed; T = torque; and A = cross section area.

As illustrated in Figure 6, in saprolite full face condition, SE ranged around 100 MJ/m³. In mixed face with rock and saprolite, values higher than 150 MJ/m³ occurred, denoting a very hard material in the front and high efforts necessary to advance. In soft residual soils, SE ranged between 20 and 70 MJ/m³. Thus, these high values can explain the high number of hyperbaric interventions in this stretch and also the high temperatures reached, as explained below in this paper.

Furthermore, it was clear that high SE values observed in Sao Paulo mixed face had close relationship with values registered in other geographical areas also characterized by tropical rock weathering, as presented for Singapore and Hong Kong mixed face conditions in Shirlaw (2015) and Shirlaw (2016). In Figure 6, it is clear the relationship between material "hardness" and velocity. The softer the material is (lower SE), the faster the machine advances.

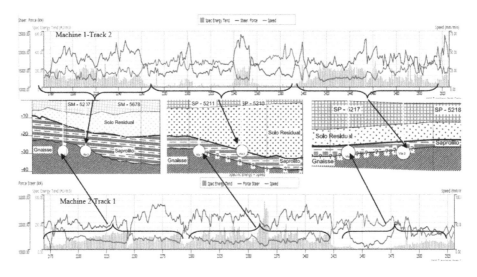

Figure 6. Differences in TBM's performances in term of SE, velocities and geology conditions.

On the other hand, the machine has more difficulty in advance the harder the material is (higher SE). These problems were more evident in mixed face with Saprolite and rock.

It is also important to report that the highest velocities were achieved after hyperbaric interventions, with new cutting tools and clean cutterhead. However, in some cases, velocities fell sharply after few rings, due to clogging or even cutting tools early damage.

Finally, once penetration was limited in order to decrease the chances of disc damages, heating was often observed. In critical stretches, the machine had to stop several times due to the high temperatures in the chamber, endangering the main bearing sealing and aggravating the muck sedimentation problem in the chamber. In some cases, temperatures in the chamber achieved values higher than 50° (Figure 5(b)), even when very wet foam and "free water" at the maximum flow rate were injected in the chamber. That high temperature caused vitrification of the material in the chamber, increasing the chances of clogging and resulting in the need of more hyperbaric interventions.

3.3 *Different performances and hyperbaric interventions*

In the mixed face stretch, the machines presented different performances, which can be related to a geological variance in the excavation conditions, as showed by Comulada et al. (2016) and Silva et al. (2016). Figure 6 also shows the specific energy difference between the machines and its relationship with the geological conditions. Variations in weathering and fracturing of the excavated material had important relevance in both TBM performances.

Other indicative of the performance difference is the number of hyperbaric interventions in each machine – 4 in Machine 1 and 14 in Machine 2. The total schedule impact was 56 days in Machine 2, besides two sinkholes and high ground losses (as shown in item 3.4).

Among the 14 hyperbaric interventions in Machine 2, only in three of them was not observed any ground instability. Four of them were aborted due to ground instabilities and 5 were possible only after the execution of ground treatment from the surface with chemical injections, as related ahead. Two of the interventions were due to cutterhead blocking. In Machine 1, there was no impact in the work schedule, but there was one sinkhole and also some high ground losses.

It is worth mentioning that the difficulties in carrying out the hyperbaric interventions can be associated with the saprolite heterogeneity. In some passages this material can be highly compact and with very low permeability, which results in a very thin filter cake layer. However, the existence of fissures and faults in some stretches makes it difficult to completely seal the front, causing compressed air leaks, which lead to pressure drops and destabilization of the front.

In the passages with residual soil in the crown, some interventions were aborted due to face instabilities. Despite the bentonite filter cake in the excavation front, long hyperbaric interventions led to difficulties in keeping hyperbaric pressures constants.

Soil treatments based on the use of colloidal silica (Figure 7 – on the left) were carried on both tracks to recompose the cavities opened by instabilities and, later, to allow shield advancing. This procedure had the purpose of reducing permeability and giving some apparent

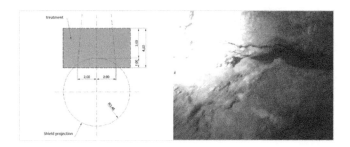

Figure 7.   Soil treatment with colloidal silica in front.

cohesion to the material in the front, notably the residual soil. Its success was directly associated with the total amount of product injected.

However, in the presence of ground discontinuities, the injected material flowed to these and sometimes did not seal the permeable residual soil, which can explain some instabilities verified during previously treated hyperbaric interventions. Figure 7 (on the right) reveals some injected silica colloidal in apparent discontinuity, verified during a hyperbaric intervention.

3.4 Settlement Analysis

Most of the ground losses in Precambrian materials were related to the operation mode (transition mode with applied compressed air), which might have led to some overexcavation. However, some settlements were also related to insufficient grouting and to the inherent material rheological properties, once settlements increase were observed even after a long distance after the TBM passage (Figure 8). Figure 9 shows values obtained from monitoring data.

Settlement analyses were made in order to evaluate remaining voids in the ground. Gaussian curve (Peck, 1969) and Yield Density Curve (Celestino & Ruiz, 1988) methodology, given by Equations (2) and (3) respectively, were used for this approach to estimate the ground loss percentage (Vs, in Table 3). Analyses were made in sections affected by face instability during hyperbaric interventions (M104-1 and M106-2), and with overexcavation observed during advance (M104-2; M105-both; and M106-1). Apparently, there were no operational problems in the excavation under the section M121.

$$\rho = \rho_{max} e^{\frac{-\lambda}{2} \frac{1}{i^2}} \tag{2}$$

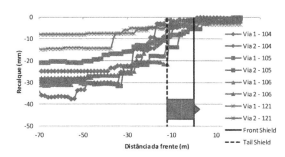

Figure 8.   Observed longitudinal observed settlements.

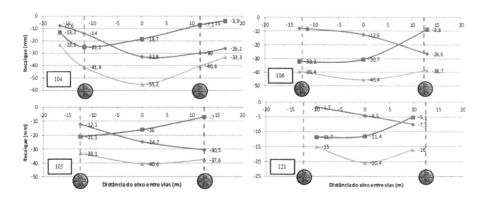

Figure 9.   Some observed transversal settlements.

Table 3. Gaussian and e approaches.

Gaussian Curve (Peck, 1969)

Section	ρ_{max} (mm)	i (m)	Vs (%)	R²
104-1	25.2	14.54	2.46	1.0
104-2	33.7	17.75	4.01	0.89
105-1	21.1	16.83	2.38	1.0
105-2	30.7	18.45	3.80	1.0
106-1	34.7	16.73	3.89	0.91
106-2	24.7	13.76	2.28	0.85
121-1	12.7	16.65	1.42	0.81
121-2	7.58	12.66	0.64	1.0

Yield Density Curve (Celestino and Ruiz, 1998)

Section	ρ_{max} (mm)	a (m)	b	Vs (%)	R²
104-1	25.1	16.5	2.98	2.69	1.0
104-2	30.2	22.9	6.0	3.88	0.97
105-1	21.1	19.2	2.66	2.77	1.0
105-2	30.5	21.4	2.71	4.43	1.0
106-1	32.2	21.1	5.8	3.82	1.0
106-2	26.6	12	1.30	6.22	0.99
121-1	12.0	18.5	5.5	1.26	0.98
121-2	7.58	14.5	2.65	0.75	1.0

$$\rho = \frac{\rho_{max}}{1 + \left(\frac{x}{a}\right)^b} \tag{3}$$

where ρ = settlement; x = front distance; i = inflection point; and a,b = calibration parameters.

Table 3 gives the obtained results. As can be seen, the ground losses calculated with the Yield Density Curve were higher. Further, it seems that this curve also gives a better settlement trough estimation, once the ground losses inferred from the data acquired in the machine [(inferred excavated volumes minus grouting volumes)/Theoretical excavated volume] in the critical stretch were ~7% in Track-1 and ~5% in Track-2. Thus, it seems that the Gaussian curve can underestimate ground losses estimations. Those high settlement values reported justified the need of soil treatment from the surface in the following sections with the same geological characteristic, as written before.

4 CONCLUSION

Analyzing soil conditioning and both TBM performances, it can be realized that the conditioning was the main challenge of excavating in Mixed Face Conditions. High efforts made in conditioning had little or no reflex in order to make TBM's advance easier. With some probability of success, maybe a hybrid machine could be more suitable for the excavation in such conditions. Some advances were tried with bentonite injection, but once the machines were not designed for that, a detailed analysis and its conclusions could not be made. Besides conditioning, high tool wear was observed, resulting in a high number of hyperbaric interventions. Furthermore, in terms of machine configuration, although a powerful machine would definitely help facing the problems experienced, it is not clear that it would avoid all of them under mixed face conditions. Difficulties in advancing are therefore related to the intrinsic characteristic of the excavated material, which can be considered as a unique material that cannot be treated as a soil, but not also as a rock.

Finally, it must be highlighted that differences were observed in both TBM performances, which can also be attributed to geological differences between each tunnel alignment, besides showing the difficulty in foreseeing TBM performance under these situations. Ground instabilities and high ground losses were consequences of boring in such difficult conditions.

ACKNOWLEDGMENTS

The authors would like to thank Metrô-SP and the Andrade Gutierrez – Camargo Correia Join Venture for the opportunity to publish this unique experience.

REFERENCES

Celestino, T.B. & Ruiz, A.P.T. 1998. Shape of settlement troughs due to tunnelling through different types of soft ground. *Feslbau* 16(2): 118–121.

Comulada et al. 2016. Experiences gained in heterogeneous ground conditions at the twin tube EPB shield tunnels in Sao Paulo Metro Line 5. In *Proceedings of the World Tunnel Congress*. United States: San Francisco.

Ferber, S. 2013. São Paulo – Metro Linha 5 Lote 3: Ground Conditioning and TBM Parameters. In *Herrenknecht Training Course, Herrenknecht AG*. Brazil: São Paulo.

Futai et al. 2012. Resistência ao Cisalhamento e Deformabilidade de Solos Residuais da Região Metropolitana de São Paulo. In Negro et al. (ed.), *Twin Cities: solos das regiões metropolitanas de São Paulo e Curitiba, São Paulo, ABMS*: 155–187.

Hollmann, F. & Thewes, M. 2012. Evaluation of the Tendency of Clogging and Separation of Fines on Shield Drives. *Geomech. & Tunnelling* 5: 574–580.

Maidl, B. et al. 2012. Mechanized shield tunnelling. In *2nd Edition. Berlin: Ernst & Sohn Verlag*.

Monteiro, M.M. & Rocha, H.C. 2015a. Personal Comunnication.

Monteiro, M. M. & Rocha, H. C. 2015b. A experiência do Metrô de São Paulo nos estudos de abrasividade de rochas: técnicas para previsão do consumo de ferramentas de corte em escavações subterrâneas. *Revista Engenharia* 625: 119–124.

Oliveira, D.G.G. et al. 2017. EPB Conditioning of Mixed Transitional Ground: Investigating Preliminary Aspects. In *WTC 2017*. Norway: Bergen.

Peck, R.B. 1969. Deep excavations and tunneling in soft ground. In *7th International Conference on Soil Mechanics and Foundation Engineering, Mexico City, State-of-the-Art volume*: 225–290.

Shirlaw, N. 2015. Pressurized TBM tunnelling in mixed face conditions resulting from tropical weathering of igneous rock. In *International Conference on Tunnel Boring Machines in Difficult Grounds (TBM DiGs), Singapore, 18–20 November 2015*.

Shirlaw, N. 2016. Mixed ground tunnelling in Hong Kong and Singapore. Presentation. In *Symposium on Underground Development and Technology, Singapore, November 2016*.

Silva et al. 2015. Twin tunnels excavated in sandy soils in a density urban area. In *Proceedings of the World Tunnel Congress*. Croatia: Dubrovnik.

Silva et al. 2016. Lições Aprendidas na Escavação de Túneis Paralelos da Linha 5 – Lilás do Metrô de São Paulo. In *Proceedings of the XVIII Congresso Brasileiro de Mecânica dos Solos e Engenharia Geotécnica, COBRAMSEG, 2016*. Brazil: Belo Horizonte.

Tunnels and Underground Cities: Engineering and Innovation meet Archaeology,
Architecture and Art, Volume 12: Urban
Tunnels - Part 2 – Peila, Viggiani & Celestino (Eds)
© 2020 Taylor & Francis Group, London, ISBN 978-0-367-46900-9

Effect of new structures and construction activities on old metro lines

E.V. Silva Espiña, L. Gil López & S. Sánchez Rodríguez
Geoconsult Ingenieros Consultores, Madrid, Spain

ABSTRACT: Since Madrid Metro network was inaugurated in 1919 the total length has been extended until the current 294 km. The construction activity on consolidated and old city centre in most cases brings with it the impact over the old metro infrastructure. New basements or high buildings designed near to the shallow masonry tunnels must be carefully studied to avoid damage, as well monitoring plans during construction must be followed. The number of interventions along the year can be high and this requires a proper management system.

1 INTRODUCTION

First Madrid Metro Line was put in service the 19th of October of 1919 and the network has been constantly growing and spreading from the very beginning. It has today a total length of 321,3 km. Two expansion plans of special importance were launched at the end of the past century: the "Madrid Metro Extension Plan 1995–1999", which consisted on the construction of 56 km of new lines and 32 stations, and the "Metro Extension of the Madrid Region, 2003–2007", which included new 59,08 km of Metro lines, 27,77 km of light rail; and 81 stations. Therefore, 30% of the existing network was developed during the two cited extension plans.

Since 2014, Geoconsult Ingenieros Consultores S.A. has provided technical support to Metro de Madrid under the contract "Geotechnical, structural and construction supervision for the analysis of the external impacts to the Metropolitan Railway Network", where interventions located on the infrastructure prior to the Extension Plan 2003–2007 are analyzed and supervised.

During this contract a total of 83 interventions or construction actions with potential impact have been studied. These cases can be divided in the following types:

- 30 cases of new buildings or constructions including demolition of the existing building and construction of new basements.
- 12 cases of existing buildings upgrade or rehabilitation with no demolition activities.
- 20 cases of mobile cranes surface operation above the Metro infrastructure.

Additionally, there were 18 cases that can be considered as 'special' and which consisted on pipe jacking works under the metro lines, Metro infrastructure repair interventions, monitoring plans, temporary mobile scenarios installation, etc. . .

In 48 of the cases, the interventions were exactly located above, whilst other 24 were located just adjacent; close to the infrastructure.

According to the distance to the infrastructure of the different cases, 40 of them were located less than 5 m to it, 24 were in the range of 5 to 10 m of distance and 10 were related with actions located 10 to 25 m of distance.

In 31 of the external interventions or actions analyzed, the design and fulfillment of monitoring plans was demanded, and in 10 on these cases the consultant participated on the supervision and continuous analysis of the recorded data of these plans.

Table 1. Madrid Metro network.

Line	Termini	Length (km)	Number of stations
Línea 1	Pinar de Chamartín – Valdecarros	23.86	33
Línea 2	Las Rosas – Cuatro Caminos	14.031	20
Línea 3	Villaverde Alto – Moncloa	16.464	18
Línea 4	Argüelles – Pinar de Chamartín	16	23
Línea 5	Alameda de Osuna – Casa de Campo	23.217	32
Línea 6	Circular	23.472	28
Línea 7	Hospital del Henares – Estadio Metropolitano – Pitis	32.919	31
Línea 8	Nuevos Ministerios – Aeropuerto T4	16.467	8
Línea 9	Paco de Lucía – Puerta de Arganda – Arganda del Rey	39.5	30
Línea 10	Hospital Infanta Sofía – Tres Olivos – Puerta del Sur	36.514	32
Línea 11	Plaza Elíptica – La Fortuna	8.5	7
Línea 12	MetroSur	40.96	28
Ramal	Ópera – Príncipe Pío	1.092	2
Metro Ligero 1	Pinar de Chamartín – Las Tablas	5.935	9
Metro Ligero 2	Colonia Jardín – Estación de Aravaca	8.68	13
Metro Ligero 3	Colonia Jardín – Puerta de Boadilla	13.699	16
Total		321.31	330

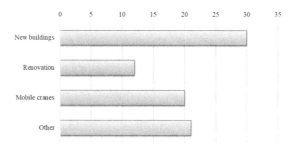

Figure 1. Resume of 2015–2018 cases of potential impact on Madrid Metro infrastructure.

2 CONSTRUCTION METHODS AND MATERIALS

Tunneling construction methods used in Madrid Metro were the most appropriate for each specific case according with geotechnical, alignment and functional conditions and with material and equipment available at each time. Therefore, different construction systems were used and the network was developed adopting cut and cover system as much as possible and boring mined tunnels with classic or traditional methods like Belgian (adapted as 'Madrid Traditional Method') and German tunneling methods; the latest for some of the station caverns. The huge extension of the last decades was based on the excavation using Earth Pressure Balanced Shields (E.P.B.). According to this, a wide range of construction materials has been used until date and tunnel linings or supports comprising masonry, mass concrete, reinforced concrete, segmental rings, etc...

It should be noted that the first E.P.B. machine used for boring Madrid Metro tunnels was launched during "Madrid Metro Extension Plan 1995–1999". From that time on, this construction system was systematically adopted where cut and cover system with embedded diaphragm o secant pile walls was not possible; limiting Belgian Method for short sections.

Metro Lines and sections located in Madrid Historic Centre are the oldest and were lined or supported by means of brick masonry mainly but also with the use of stone masonry.

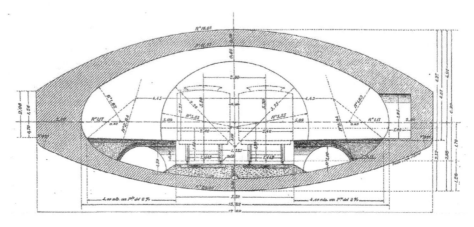

Figure 2.   Madrid Metro Line 2 station cross section (1925).

Figure 3.   Line 2. Original brick masonry lining recently reinforced with shotcrete.

Figure 4.   Line 4. Original brick masonry lining.

Figure 5.   Line 6. Reinforced concrete segmental lining used in the first section bored with EPB.

Figure 6.   Service tunnel with stone masonry side-walls and brick masonry at crown.

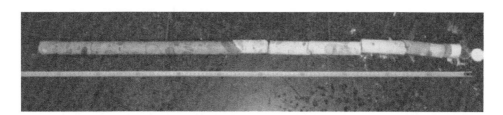

Figure 7.   Mass concrete lining sample (estimated 1920–1930).

Figure 8.   Brick masonry sample.

Examples of the use of these construction materials are Lines 1, 2, 3, 4 and 5. More recent techniques and materials were used in later network extensions located out of the historic center.

Besides, during the study of some of the cases of potential impact, lining samples were taken to analyze its state, thickness, etc... It was found that the state of conservation of the lining was quite good.

3   RELEVANT CASE STORIES

Following are described some relevant cases of potential impact to Madrid Metro studied during recent years where singular structural or geotechnical conditions appeared.

3.1   *Excavation below Line 2 tunnel slab level*

In this case a new building where the Top & Down method was adopted after demolition of the existing old one is studied. Singularity of this case is related to design of a embedded wall 100 m long parallel to existing Line 2 tunnel. Distance between both structures was less than 2 m in most of the affected area. Final building basement level depth was twice the tunnel track level. In this section, excavated during the twenties of the past century, the Metro tunnel lining is formed of brick masonry.

For the monitoring of movements in the interior of the tunnel an automated remote total station was installed collecting data during 24 hrs. Continuous interpretation of this

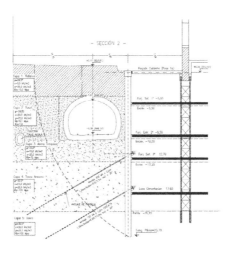

Figure 9.   Line 2 tunnel and new basement impact layout.

monitoring recorded data was carried out by geotechnical expert as well as other installed instruments were read and interpreted 2 or 3 times per week.

Additionally, site inspections were carried out depending on movements progress, as well as to check readings and the state of the infrastructure. During works supervision and winter 2017–2018 it was verified direct effect of thermal changes causing movements on new building slabs affecting to embedded walls and being finally transmitted to Line 2 tunnel lining. The mentioned displacements were not recovered after temperature raising and are still analyzed to study potential new movements in case of temperature changes.

3.2 *Grout leakage from micropiles near to Line 5*

Micropile deep foundation was designed for a new building located over Line 5 tunnel. The target of these micropiles was both the foundation of the building slab and getting a 'bridge effect' to protect the existing old tunnel from building new loads.

During micropile bond grouting phase, massive grout leakage was detected in one of the tunnel side walls although no measurable displacement was induced. Nevertheless, after leakage detection site inspection was carried out to verify monitoring readings and general state of the infrastructure. Grout leakage progressed through the different masonry elements what helped to avoid damage to lining. In case of having other type of lining, damages on it caused by pressure grouting, could have been more important.

3.3 *Interaction between existing and new lines or other tunnels*

During maintenance activities of an existing escalator located in a platform access, a cavity was detected below it. This cavity was located just over the new railway line bored below in

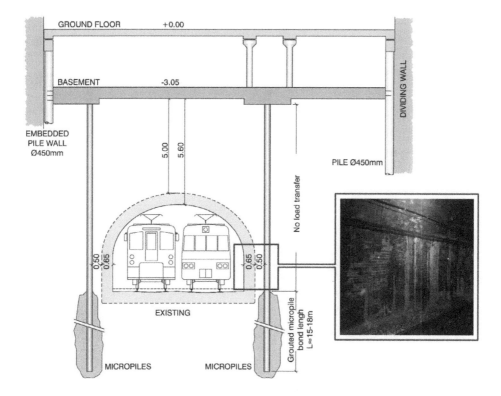

Figure 10.   Line 5 section affected by micropile grouting. Detail of sidewall.

recent years. This incident proves that even the most sophisticated construction methods can't avoid some geotechnical risk. In the case of E.P.B. tunneling, the control of face pressures and backfill grouting must be intense and rigorous.

It is interesting to remember that although Madrid soils are quite favorable for tunneling activities, frequently sand layers without cohesion, and sometimes water bearing, can be encountered. This material is quite prone to suffer over-excavation problems when working with E.P.B. machines.

Other factor to have in mind is possible leakage through lining joints or cracks, since water can drag surrounding ground and ease void appearance.

3.4 Damages on lining caused by drilling works

Lining damages caused by borehole drilling machinery have become one of the more dangerous risks affecting infrastructure. Thereby, during borehole drilling works without pertinent

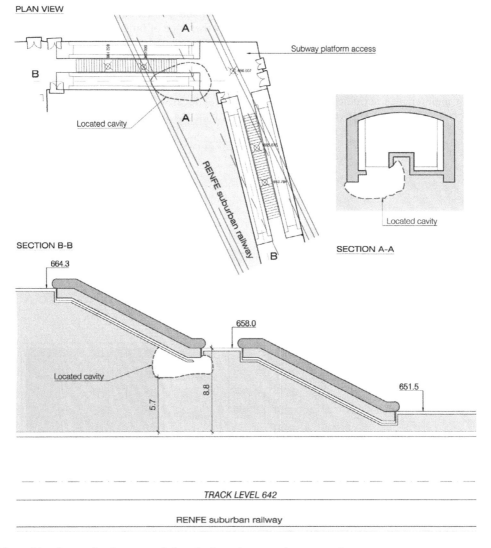

Figure 11.   Interaction between existing platform Access and new tunnel.

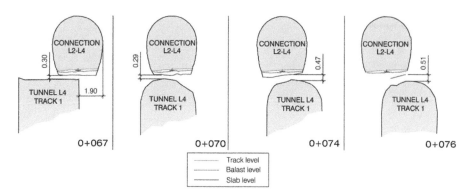

Figure 12.   Impact/interaction between existing lines.

studies and licenses some tunnel sections have been damaged while being in service. Geoconsult has been reported this frequent problem not only in the Madrid network.

Infrastructure owner can do nothing against these activities if 'official' procedures are not followed, therefore this type of damage or impact can be only limited with the help and awareness of technical staff in charge of borehole investigation works since there is a clear risk for both, the metro travelers and the operators.

According to the type of incident it can be better described as an operational problem rather than a geotechnical or structural one and its management to avoid or deal with is in charge of Metro personnel. Nevertheless, in case of incident, structural inspection is usually carried out.

3.5   Existing Metro lines interaction

During track maintenance works in one existing tunnel in operation it was found that base slab was tunnel lining crown of other tunnel in service located below.

Under this circumstance it was mandatory to check structural condition of both tunnels. For this purpose, methodology usually adopted for bridge load tests was followed, applying on existing tunnel crown expected loads of the above tunnel in operation and recording displacements. Structural condition of the tunnels was considered adequate since displacements induced under the worst scenario during load test were quite small.

4   IMPACT MANAGEMENT PROCEDURE DEVELOPED BY MADRID METRO

Every temporary activity or permanent intervention located less than 25 m from Metro infrastructure must be supervised following a specific working procedure.

This procedure starts with the license requested at Madrid City Council. In case this activity or intervention is located inside the so called 'public domain and railway easement zone', the Council shall inform the promoter about contacting Madrid Metro to get approval for the intervention or activity. For this, an impact report shall be requested.

Impact report has to deal with a double requirement. The first one is administrative and related with insurance coverage and all legal aspects. The second is technical and related with potential structural or geotechnical impact and the remedial actions if needed.

Following are shown scope and recommendations for the preparation of Impact Reports which are the result of accumulated experience on these studies. Nevertheless, it must be underlined that this can be considered as a reference to be adapted for each specific case.

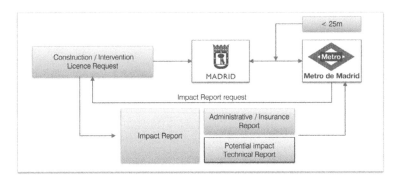

Figure 13.   Flow chart of the Intervention/construction license obtention and impact study process.

4.1   *Impact Report Scope*

The target of these reports is the study of the potential impact of the planned ground activities or construction works. Once the impact is determined it must be checked whether it is allowable by the infrastructure, measuring this according to standard safety factors and reference values of allowable displacement or deformations (for the latest, the reference values for monitoring of Metro de Madrid are usually considered, though this may change for each specific case).

For impact analysis, basic data related with geometry and location of the problem, work description and structural/geotechnical calculations must be included. Following are included the common impact types:

- Planned ground activities imposing temporary loads (mobile cranes, etc..)
- Planned activities (mostly building construction) imposing a new and permanent stress state on the environment and structure (rehabilitation or new building construction works)
- Unplanned activities (accidents, incidents, etc... not studied)

4.2   *Basic Data*

Basic data usually requested on every Impact Report is the following:

- Work description with drawings of the new structure or element causing impact.
- Work phases.
- Detailed location drawings with new structure and Madrid Metro infrastructure.
- Plan and profile view which must include critical or most unfavorable sections.
- Available geotechnical investigation (laboratory tests, boreholes, etc.).

In the case of mobile cranes imposing temporary loads, machinery description, work and maximum loads expected and entrance or exit routes of the equipment must be submitted for evaluation.

4.3   *Structural/geotechnical impact calculations*

The target of the calculations included in the report must be obtaining induced displacements and new stress state in order to verify that no damage, or only allowable impact, is induced. Thereby, depending on the sensitivity or risk of the intervention, complexity of calculation approach shall be different.

Accumulated experience has shown that analytic or linear-elastic models usually bring deformations more favorable than finally data read in monitoring whilst numerical models tend to provide conservative results in general. In the first case, additional checking is usually requested to verify structural behavior and allowable conditions.

Therefore, impact calculations shall be classified in two groups with particular requirements:

4.3.1 *Planned activities imposing temporary loads*

For this type of impacts a structural calculation is needed for each element affected due to change in load configuration. For this purpose, linear-elastic models are considered adequate.

In these cases, two situations can appear: The potentially impacted structure and materials are clearly known (remember an old infrastructure is analyzed) or, on the other hand, just limited data about materials, thickness, etc is available.

It is clear that in the first situation a safety factor can be obtained for impact evaluation, but in the second configuration the approach usually followed is the comparison of the induced loads caused by the actions included in the design codes with those expected to act during the temporary activity and not using in any case majority factors. Studying the problem by this approach allows to have a qualitative idea of the expected conditions compared with the current traffic loads.

4.3.2 *Planned activities imposing new permanent load configuration*

As cited above, for these types of impacts a numerical model for the study of ground-structure interaction for the new loads configuration is requested by Madrid Metro. Models shall be prepared for all geotechnical-geometrical sections needed and shall include all construction stages

To carry out the model, geotechnical ground properties shall be obtained from the specific geotechnical study. In any case, the obtained parameters must be check with the existing published data regarding Madrid specific soils.

Analysis shall study changes on stress state and load on Metro structures as well as displacements and deformations induced. These later shall be obtained for the different stages and used as reference for the monitoring plan to be carried out.

In case of having monitoring data from previous works, these values should be used for the proper calibration of the new models.

4.3.3 *Unplanned impacts*

These impact cases should be limited to those that did not follow the 'official' procedures. These impacts can have their specific conditions but theycan be usually classified according to the above described groups.

4.4 *Monitoring plan*

A monitoring plan must be implemented in case of actions causing a permanent change of the tensional state.

This monitoring plan should include:

- Core samples or manual trial pits in case there are uncertainties regarding the state, thickness, etc... of the existing lining or support.
- Surface topographic surveying of the action to be monitored.
- Monitoring instruments to be used.
- Monitoring sections proposed.

Figure 14.   Impact report basic content for the cases of temporary or permanent load change.

- Leveling surveying in rail track and type of existing track.
- For especially sensitive sections an initial state report of the potentially impacted area shall be prepared, including:
  - Detailed photographical report covering the entire impacted area
  - Especial attention shall be paid to existing damages prior to new actions. This must be included in specific pictures and drawings locating all of them.

If it is considered that there may be some type of specific impact it will be required to carry out a continuous monitoring of the structure.

For the cases of temporary actions imposing loads lower or more favorable than those considered by current design codes and no monitoring plan is considered mandatory, it is strongly recommended to carry out a photographical report of the Metro infrastructure potentially impacted in locations near to the planned temporary loads.

5 CONCLUSIONS

Throughout the work carried out for Metro Madrid, Geoconsult had the opportunity to study all the different impacts that can be caused over an existing underground infrastructure due to the development of the cities.

Thereby, construction or building rehabilitation, infrastructure upgrade works, surface actions with temporary loads higher than design loads, etc. are configurations that the existing structures may have to deal with but possibly these were not considered when designing or planning in the past.

On the other hand, the current design and construction techniques allow obtaining solutions adapted to each case, turning possible impacts to singular configurations without major transcendence for the existing infrastructure. In this way, the study of design approach of the more suitable solution in collaboration with the infrastructure owner allows to limit the degree of impact on it.

For this purpose, to verify that any potential impact has been analyzed and to limit as much as possible the damages on the metro infrastructure, a management system adequately implemented is strongly needed and recommended. The one currently followed in the Madrid Metro network has shown to be a key tool to reduce incidents but, of course, must be daily updated and improved.

REFERENCES

Mellis, M. Arnaiz, M., Trabada, J. & Díaz, J. M., 1995. Las infraestructuras de Metro y de Transporte en la Comunidad de Madrid. Comunidad de Madrid
Mellis, M. Arnaiz, M., Trabada, J. & Díaz, J. M., 2003. Metrosur. Comunidad de Madrid.
Gil L. Instrumentación, Control de ejecución y Método de Cálculo de una caverna. *INGEOPRES* Junio 1996.
Melis, M. 1997. Ampliación de la red del metro 1995–99. Estado de los trabajos en octubre de 1997. *Revista de Obras Publicas*, 144(3369): 9–14
Melis, M & Arnaiz, M. 2001 Ampliación de la Red del Metro de Madrid. *Revista de Obras Publicas*, 148 (3410): 19–49
Melis, M. 2003 Las infraestructuras del transporte de Madrid 1999–2003. Madrid Transport Infrastructure 1999–2003. *Revista de Obras Públicas*. 150 (3429): 9–19
Oteo, C. & Rodríguez, J. M. 1997. Subsidencia y auscultación en los túneles del metro de Madrid. *Revista de Obras Públicas*, 144 (3369): 49–68

Tunnels and Underground Cities: Engineering and Innovation meet Archaeology,
Architecture and Art, Volume 12: Urban
Tunnels - Part 2 – Peila, Viggiani & Celestino (Eds)
© 2019 Taylor & Francis Group, London, ISBN 978-0-367-46900-9

Soft ground tunneling below a mixed foundation building

D. Simic & B. Martínez-Bacas
Geotechnical Area Ferrovial-Agroman Engineering Services, Spain

ABSTRACT: A new railway infrastructure next to Barcelona city required the excavation of a 10,60 m diameter EPB TBM tunnel through the soft deltaic sediments close to the sea, consisting of a thick sequence of loose sands and soft silty clays below the water table. This paper addresses the main issues of the tunnel bore below a complex structure resting on heterogeneous direct and deep foundations, either short floating piles having their tip above the tunnel and deep column piles. Two different three-dimensional numerical models have been run to evaluate the movements of the building during the process of excavation of the tunnels, in order to capture both the specifics of the soil non-linearities and the structure deformability. The model results' have been compared to the settlement data obtained from the monitoring during excavation and the building behavior has been evaluated in term of the structural capacity and the foundation pile capacity during the tunnel advancement.

1 INTRODUCTION

1.1 *Description of the problem*

The crossing of the TBM machine below an existing building was a challenge regarding the limitations of the allowed movement of this complex building in operation, consisting of several structures built in different epochs with different foundation types. Figure 1 shows the parts of this facility: Entry Hall, Main building and Connection box. The overall dimensions of the structure are 88 m (width) x 49 m (length).

Part of the Main building and Hall rest on a Connection box, which is a caisson structure on a direct slab foundation. The rest of the building has deep foundations, with different concrete pile lengths between 6 m and 50 m.

Although the tunnel alignment does not interfere with any of the piles, it passes very close by affecting to the pile load distribution. Table 1 presents the data of the tunnel.

1.2 *Objective of the study*

In order to analyze the impact of the tunnel construction on the building some 3D models have been developed taking into account the stiffness of the superstructure to evaluate its behaviour when subject to the geotechnical stress-strain field induced by the excavation. In an initial stage, the movements obtained in the foundation of the structure building were exported to a specific structural model as additional load case. In these runs, preliminary soil parameters and face pressure were introduced.

However, once the TBM passed below the building and settlements were recorded, the TBM parameters could be established more accurately: pressure chamber, jack force, grout pressure to fill the gap between the shield and the soil. Also, the movements of the building measured with prims and inclinometers distributed along the foundation could be exploited to calibrate a second model.

In this way it was possible to develop a back-analysis and modify important factors that were not considered in the initial model. It seems that the restrain boundary conditions as well

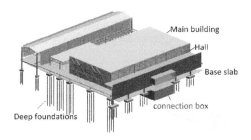

Figure 1. Sketch showing the heterogeneous foundation of the building.

Table 1. Tunnel data.

Tunnel shield	
Excavation diameter	10.60 m
Tail diameter	10.51 m
Volume loss	0.5%
Rings	
Number of segments per ring	7
Ring thickness	0.32 m
External diameter	10.24 m
Internal diameter	9.60 m
Ring width	1.60 m
Depth of centerline	19 m

as the stiffness of the overburden soil layers has influence in the settlement results and the load distribution along the piles as we can see in the next sections.

2 INPUT DATA FOR THE NUMERICAL MODELS

2.1 *Initial stage*

The first geotechnical model was developed with the program FLAC3D, in which it was simulated the building's shallow and deep foundations, soil layers and the tunnel construction phases (see Figure 2). The building structure was introduced considering beam and shell

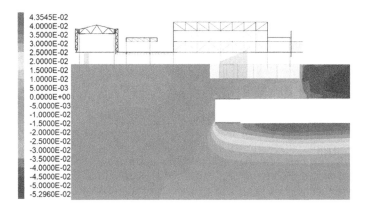

Figure 2. Vertical displacements (m) below the building.

elements. The dimensions of the T2 are 88 m (x-direction) x 49 m (y-direction). The dimensions of the model are 100 m in the longitudinal tunnel direction (y direction), 144 m in the cross section tunnel direction (x-direction) and the depth of the model is 75 m.

Figure 3 shows the longitudinal section of the tunnel (Table 1). It was considered that the TBM rests on the excavated annulus. On the other hand, it was assumed that the rings dip into the mortar, leaving a bottom gap of 10 cm.

Table 2 shows the values of the gaps considered into the model. These gaps were filled with grout during every sequential excavation advance phase.

The construction phases were the following ones:

Initial stage: the building with its pile foundations and soil layers are activated.

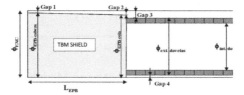

Figure 3. Longitudinal section of the TBM shield.

Table 2. Gaps of the model of the TBM.

Gaps	values
	cm
Gap1	3.0
Gap2	9.0
Gap3	26
Gap4	10.0

Table 3. Initial soil parameters Hardening soil model and Hardening soil model with small strain.

Parameters	UG0	UG1	UG2	UG3
Model	HS	HS	HS	HSsmall
Type	Drained	Drained	Drained	Undrained
γunsat (kN/m^3)	16.5	19.0	19.0	19.0
γsat (kN/m^3)	17.5	20.4	20.0	21.0
E_{50}^{ref}(MPa)	5	5	9	11
E_{oed}^{ref} MPa)	5	5	9	11
E_{ur}^{ref}(MPa)	25	25	27	33
ν_{ur}	0.33	0.4	0.35	0.4
$G_{0.7}$(MPa)	-	-	112.5	137.5
$\gamma_{0.7}$	-	-	2×10^{-4}	2×10^{-4}
m	0.5	0.5	0.5	0.5
$\phi'(°)$	22	24	32	-
c' (kPa)	5	10	5	-
s_u (kPa)				30+6z*
p^{ref} (kPa)	100	100	100	100
k_0^{NC}	0.62	0.59	0.7	-
k_0	0.62	0.59	0.7	1
OCR	1	1	1	1

*z is the depth from the top of the UG3 layer.

Table 4. Resistance of the piles.

Piles	Shaft resistance (kN/m)			End-bearing capacity (kN)		
	UG2	UG3	UG4	UG2	UG3	UG4
Hall-short-piles	160	62			293	
Main-long-piles	160	62	236		293	740
In situ-short-piles	204			798		

Advancement stage of 1 m for the TBM. The mortar that fills the gaps is activated at same time than the ring is activated. The stiffness of the mortar is varying in every stage. Considering an advancement of the TBM of 4 rings/hour, the Equation 1 presents the mortar set for the Young Modulus for the mortar (Thomas et al. 2001) is:

$$E_t = E_{28}(1 - e^{at})$$ (1)

where Et = Young Modulus for a time t; E_{28} is Young Modulus for 28 days and a is a constant of the age of the concrete. Taking into account the specific tests of the mortar results: E_{28}=1000 MPa and a=0.2.

In this model the mortar only fills the gap between the ring and the soil without applying any pressure to the injection of the mortar. This hypothesis is conservative, because normally the injection pressure compensates the settlement due to the passing of the TBM.

In this first model it was considered that the connection-box and the base slab surround it are not connected, it means that the connection-box has free movements in horizontal and vertical directions.

2.2 Initial soil parameters and TBM parameters

The soil layers according to geotechnical investigation carried out in the area it was the following one, from ground surface to the bottom:

UG0: Man made fill. Thickness 2 m
UG1: Continental clays and silts. Thickness 2 m
UG2: Sands. Thickness 10 m
UG3: Grey clays and silts. Thickness 34 m
UG4: Gravels and sands

The TBM goes through the layer UG3. The water level is 3 m depth. The chamber earth pressure applied was 200 kPa with an increment with depth of 15 kPa.

For checking the chosen soil parameters, two greenfield runs were done without/with injection of shield with mortar. The injection of the mortar was simulated reducing 50% the Gap1 and Gap2 (see Table 2). Without mortar between shield and soil the volume loss obtained was 1.24 % and the maximum settlement 42 mm. The greenfield simulation with mortar the volume loss was 0.64% and the maximum settlement 16 mm. These last results are in line with the results observed of the project of the Subway Line 9, that has been constructed in the same type of soils than the present project.

2.3 Pile parameters

For the pile behaviour 8 CPTU tests and 1 O-cell test were made available. From these tests the values of end-bearing capacity and shaft resistance were obtained for every type pile of the model.

The building has deep foundation with different concrete pile length: below some parts, circular piles 0.65 m of diameter (in situ-short-piles) and 6 m length, Main building piles with square section of 0.4 m side and length of 50 m (Main-long-piles), Hall building piles with square section of 0.4 m side and 8 m length (Hall-short-piles).

3 MODEL CALIBRATION AFTER TUNNEL EXCAVATION

3.1 *New input data*

Once the TBM passed below the building it was decided to re-run the model incorporating the following data:

Settlements measured by the prims and the inclinometers.
The chamber earth pressure applied during the construction.
The grout pressure applied during construction.

These data help us to recalibrate the model analyzing the piles behaviour as well as comparing the building settlement with the preliminary model.

Figure 4 presents the monitoring settlements measured by the prisms placed on the hall over connection box, during the passing of the TBM, being the range between +6 mm and -6 mm (negative value is settlement). The first model shows a settlement of -13 mm much higher than the measured in-situ data.

Figure 5 and Figure 6 present scatter values of the chamber earth pressure and grout pressure applied respectively. The earth pressure was measured with 7 pressure cells placed into the TBM face. The grout injection pressure was measured with 6 injection lines. For the recalibrated model, it was selected the average values of the ring 1037, that is approximated in the middle of the building. Hence, the input face pressure was 240 kPa with an increment with depth of 15 kPa/m. The input grout pressure was 300 kPa with and increment with depth of 10 kPa/m.

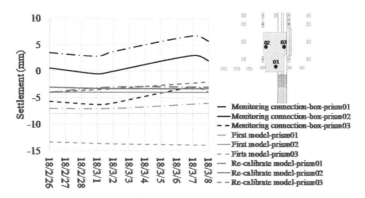

Figure 4. Longitudinal section of the TBM shield

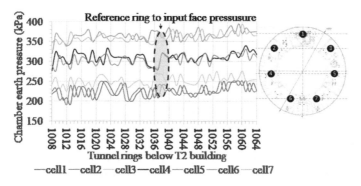

Figure 5. Chamber earth pressure measure during TBM passed below the building

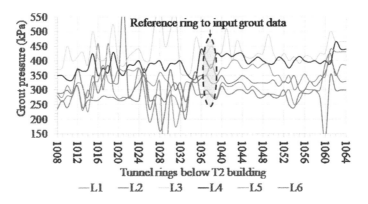

Figure 6. Grout pressure measured during the pass of the TBM below the building

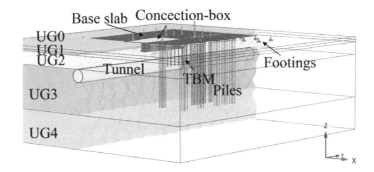

Figure 7. Recalibrated model

Figure 7 shows the geometry of this model in Plaxis 3D, introducing the new data. From this model it can be concluded that, on one hand, the boundary conditions of the connection box have very important effect on the results. In this respect, the surrounding base slab supplied certain fixing in horizontal plane as well as in vertical directions due to the pile foundations. On the other hand, the stiffness of the soil layer UG2 was increased to avoid excessive settlements observed in the first model. Figure 4 shows how the settlement of the connection box for the recalibrated model is reduced to -5 mm. These values are closer to the measured values.

The dimensions of the model are 220 m in the longitudinal tunnel direction (y direction), 200 m in the cross-section tunnel direction (x-direction) and the depth of the model is 75 m. The number of zones 433000. The thickness and the type of soil layers are identical to the first model (see section 2.2).

The tunnel was simulated considered the gaps between the soil and shield, the grout pressure applied and the face pressure.

This model also incorporated the long and short piles of the foundation (Table 4), the individual footings, the connection-box and the base slab (see Figure 1). Furthermore, it was applied the vertical load of every column of the building, as it can see in Figure 7.

3.2 Soil parameters for the recalibrated model

According to the results of the monitoring settlements of the connection box, the triaxial results of the working area and the previous experience of the Subway Line 9 in Barcelona, it was decided the simulated the layers UG0, UG1 and UG4 as Mohr Coulomb and the layers more significant regarding the movements of the building as Hardening Soil with small strain. Table 5 and Table 6 shows the soil parameters calibrated.

Table 5. Parameters Hardening soil with small strain soil model.

Parameters	UG2	UG3
Condition	Drained	Undrained
γunsat (kN/m^3)	16.70	15
γsat (kN/m^3)	20.50	19.2
E_{50}^{ref} (MPa)	15	3.3
E_{oed}^{ref} MPa)	15	3.3
E_{ur}^{ref} (MPa)	45	9.9
$G_{0.7}$ (MPa)	56	33
$\gamma_{0.7}$	0.2e-3	0.1e-3
ν_{ur}	0.2	0.2
m	0.5	1
$\phi'(°)$	30	22
c' (kPa)	1	1
p^{ref} (kPa)	100	100
k_0^{NC}	0.47	0.53
OCR	1	1

Table 6. Parameters Mohr Coulomb soil model.

Parameters	UG0	UG1	UG4
Condition	Drained	Undrained	Drained
γunsat (kN/m^3)		15.3	19.0
γsat (kN/m^3)	20.4	19.5	21.0
E'(MPa)	5	5	160
ν	0.4	0.35	0.33
$\phi'(°)$	21	29	38
c' (kPa)	9	2	1
k_0	0.64	0.49	1

4 DISCUSSION OF THE RESULTS

4.1 Ground settlements

Figure 8 and Figure 9 compare the soil displacements, showing a much better agreement of the new runs with the monitoring. In these the connection box and base slab are structurally connected as well and soil parameters of UG2 and UG3 layers are introduced.

Figure 8 compares the transversal settlements of the monitoring prisms with the first model and the recalibrated model, along two sections. The settlements were measured at z=0:

Section1 goes through the connection box and base slab.
Section2 goes through the base slab only.

The first model shows settlement up to -23 mm, however the re-calibrated model reaches -5 mm in line with the prism data (±5 mm). Figure 9 compares the monitoring inclinometer measurements with the model data, showing good correlation as well.

4.2 Pile behaviour during tunnel excavation

4.2.1 Evolution of the pile settlement as the TBM advances

An important issue to be analyzed with the 3D models was the impact of the tunnel excavation to the pile's axial capacity. In this paper two cases are studied: a short pile of the hall building (with its tip above the tunnel crown) and a long pile close to the tunnel that belongs to the Main building (bearing on a deep stratum below the tunnel), as shown in Figure 10.

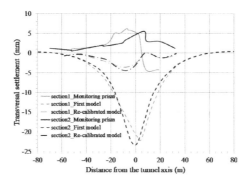

Figure 8. Comparison of ground settlements from monitoring

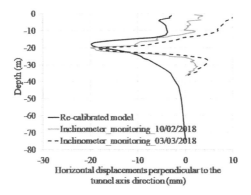

Figure 9. Comparison of inclinometer measures and recalibrate model

Figure 11 shows the tunneling induced settlement for the selected piles. The lateral distance of the piles regarding the tunnel axis is (in tunnel diameters), for the short pile 0.6Dt and for the long pile 0.7Dt. This figure shows the Pile settlement/pile diameter(Dp=451mm) versus distance tunnel face/tunnel diameter(Dt). The x-axis 0 corresponds to the location of the pile. Initially, it is observed that when distance of the tunnel face is ±5Dt away from the pile location, the increment of the settlement is negligible. This finding is in line with Dias and Bezuijen (2015). Furthermore, as Lee and Ng (2006) and Willianson (2014) find, the pile settlements increase substantially from -1Dt to +1Dt. The settlement for the short pile reaches 5% of the pile diameter and, for the long pile, 1%

4.2.2 Evolution of the pile forces as the TBM advances

Figure 12 and Figure 13 show the progressive changes in the axial force of selected piles during the tunnel advance, for both the short pile and long piles. The pile position is represented at x-axis, negative x-axis indicated that the tunnel face has not reached the pile yet, whilst positive values indicate the tunnel face is beyond the pile.

Due to the ground displacements induced by the excavation, the soil settlement relative to the pile generates a negative skin friction on the pile shaft, which increases the pile axial load. Figure 12 shows the progressive increase of the shaft load, negative skin friction and total axial load for the short pile belong to the Hall building. From -1Dt to +1Dt a significant increment is detected, and beyond +1Dt the values are nearly constant. In this case due to the pressure of the grout through the tail of the TBM, the soil surrounding the pile toe moves upwards and giving a slight increment of the toe load and a reduction of the negative skin friction. This effect is shown in Figure 14a, where the distribution of axial forces along the pile are represented for various stages of tunnel excavation. The negative skin friction mobilized

6236

on the bottom section of the pile during the passing of the TBM (0Dt and +1Dt) induces tension forces in that section.

For the long pile, Figure 13 shows that from -2Dt (-19.6 m) to +1Dt (9.6m) an important increment of the side load, negative skin friction and total axial load. Beyond +1Dt the values are nearly constant.

In contrast to the short pile behaviour, Figure 14b shows that the negative skin friction for the long pile is mobilized on the upper part of the pile, above the tunnel crown.

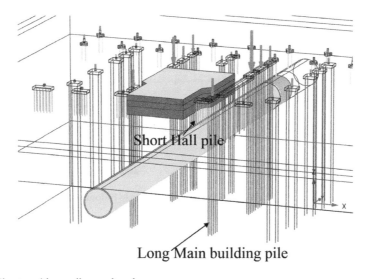

Figure 10. Short and long piles analyzed

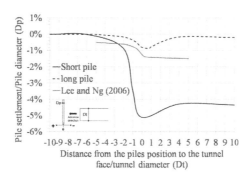

Figure 11. Pile settlement during the advance of the TBM excavation

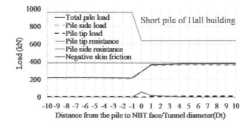

Figure 12. Pile loads due to tunnels excavation for short pile close to tunnel axis (0.6Dt)

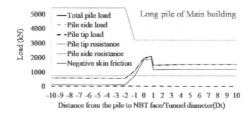

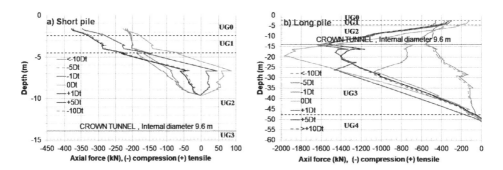

Figure 13.   Pile loads due to tunnels excavation for long pile close to tunnel axis (0.7Dt)

Figure 14.   Evolution of the axial load along the pile: a) short pile b) long pile

5   CONCLUSIONS

The tunnel excavation below a building with mixed foundations (slab and piles) is a complex interaction problem that requires the use of numerical models capable of simulating the structural elements of the building and the pile behaviour, including the development of the shaft friction and the end bearing and the non-linear soil behaviour. From such models, the following conclusions can be obtained:

It is important to reflect the real conditions of the structural elements, as they have an important impact in the building forces and settlements.

The forces of the piles vary along the tunnel advance, depending on their length and relative position in respect of the tunnel.

The zone of the major influence of the TBM on the pile forces occurs when the tunnel face is located between -1Dt from the pile location and +1Dt beyond the pile location.

For both long and short piles, a negative skin friction is mobilized, and the total axial load of the pile increases with respect to its initial value.

REFERENCES

Lee, G. T. K. and Ng, C. W. W. (2006). Three-dimensional numerical simulation of tunnelling effects on an existing pile. In *Proceedings of the 5th International Symposium on the Geotechnical Aspects of Underground Construction in Soft Ground* (Bakker, Bezuijen, Broere and Kwast (eds)). Amsterdam, pp. 139–144

Williamson (2014) Tunnelling effects on bored piles in clay. *PhD thesis*, University of Cambridge, U.K.

Dias, T. G. S. and Bezuijen, A. (2015) Data Analysis of Pile Tunnel Interaction. *Journal of Geotechnical and Geoenvironmental Engineering*, Vol. 141 (12), pp. 1–15.

Tunnels and Underground Cities: Engineering and Innovation meet Archaeology,
Architecture and Art, Volume 12: Urban
Tunnels - Part 2 – Peila, Viggiani & Celestino (Eds)
© 2019 Taylor & Francis Group, London, ISBN 978-0-367-46900-9

Small shield operation for pre-support method at underground space enlargement

M. Sugimoto, T.A. Pham & J. Chen
Nagaoka University of Technology, Nagaoka, Japan

T.N. Huynh
Vietnam National University Hochiminh City, University of Technology, HCMC, Vietnam

L.G. Le
Can Tho University, Can Tho, Vietnam

A. Miki & K. Kayukawa
Ando-Hazama Co. Ltd., Tokyo, Japan

ABSTRACT: To connect two existing shield tunnels by non-open cut method, the pre-support method at underground space enlargement using several small diameter shield tunnels was proposed. To examine its possibility, the simulation on the articulated shield behavior was carried out by the kinematic shield model, using the estimating shield operational parameters. As a result, it was found that 1) the calculated shield steering conditions, such as, the copy cutter length and range, and the articulation angle and direction, are reasonable from the viewpoint of theory and site experience; and 2) when the setting shield operation data including the shield steering conditions are in use, the calculated shield behavior has a good agreement with the planned one.

1   INTRODUCTION

Recently, a pre-support method at underground space enlargement, as shown in Figure 1, was proposed, to connect two existing shield tunnels by non-open cut method. The pre-support method was planned to be constructed using several small diameter shield tunnels with a sharp curve in three-dimensional space, which launch from the main tunnel and ramp tunnel, as shown in Figure 2 (Miki et al. 2016). But the small diameter tunnels with 3 dimensional sharp curve has never been constructed before, since it is difficult to control the shield following the planned alignment. To realize this method, it is necessary to examine the shield operation method preliminarily by the simulation of shield behavior using the shield operation data.

Shield behavior has been studied by both statistical and theoretical methods. To predict and control the shield behavior based on the equilibrium conditions of force and moment acting on the shield, the latter method is used. To clarify the shield behavior and the behavior of the surrounding ground, numerical methods, such as Finite element method (FEM) and Discrete element method, have been adopted (e.g., Finno & Clough 1985, Komiya et al. 1999, Kasper & Meschke 2004, Melis & Medina 2005, Alsahly et al. 2013). These methods cannot represent the overcut effect, which is considered to be the predominant factor affecting the shield behavior from the viewpoint of practice (e.g., Clough et al. 1983, Rowe et al. 1983, Hansmire & Cording 1985, Fujita 1989, Lee et al. 1999, Festa 2013). Furthermore, numerical methods taking into account the gap between excavated surface and tunnel lining have been proposed (Lee et al. 1992, Rowe & Lee 1992, Takeda et al. 1998, Date et al. 1999). However, these numerical methods cannot simulate the immediate ground movement during excavation,

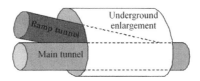

Figure 1. Pre-support method for underground space enlargement.

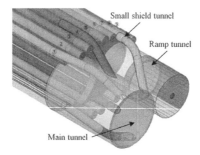

Figure 2. Small diameter shields for pre-support method.

since the shield movement is required as known conditions in these numerical methods and it depends on the past shield movement and the past excavated area.

The authors have proposed the kinematic shield model, taking into account shield tunnel engineering practices, namely, the excavated area, the tail clearance, the rotation direction of the cutter disc, sliding of the shield, ground loosening at the shield crown, and the dynamic equilibrium condition (Sugimoto 1995, Sugimoto & Sramoon, 2002, Sramoon et al. 2002, Sugimoto et al. 2007). This model is a function of shield operations, shield behavior and ground properties.

On the other hand, shield direction is controlled by jack, copy cutter, and articulation mechanism in practice. But since these three functions have a high co-linearity to shield behavior, it is difficult to obtain a unique solution at once. To solve this problem, the authors have developed a shield steering method (Huynh 2015).

In this research, to examine the possibility of the small diameter shield tunnelling for the pre-support method at underground space enlargement, the simulation on shield behavior was carried out, using the estimating shield operational parameters under the setting analysis conditions. This paper briefly introduces the calculation method on shield steering parameters and the kinematic shield model, shows the calculated results, and finally examines the possibility of the pre-support method at underground space enlargement quantitatively, comparing the calculated shield behavior with the planned tunnel alignment.

2 METHODOLOGY

To examine a shield operation method preliminarily, the following method have been developed (Huynh 2015):

1. The copy cutter length and range and the articulation angle and direction are determined uniquely, based on geometric conditions under some constraint conditions (Chen 2008). This method can deal with horizontal and vertical curved alignments at the same position, which is called 3D compound alignment.
2. After determining the steering parameters on the copy cutter and the articulation of the shield, the initial jack force is calculated by a sequential analysis using the kinematic shield model.

3. Those values are modified so that the calculated shield behavior has a good agreement with the planned one.

In this section, the method for articulated shield is described.

2.1 Shield steering

From the viewpoint of geometric conditions, shield is classified into three types according to the maximum length among $L_{M1} - L_{CSE}$ (Type1), L_{CSE} (Type2), and $L_{M2} - L_{CSE}$ (Type3), where L_{M1}=length of front body; L_{M2}=length of rear body; and L_{CSE}=length from crease center to segment front end.

The followings are the calculation conditions, based on geometric conditions: (Sugimoto et al. 2018).

1. The center of the segment end, P_{CSE}, follows the planned tunnel alignment.
2. The axis direction of the rear body is the tangential direction of the planned tunnel alignment at P_{CSE}.
3. The articulation angle is determined so that the copy cutter length at the concave side of the curve is minimized for type 1, and the articulation angle is determined so that the copy cutter length at the convex side of the curve is minimized for types 2 and 3.
4. The copy cutter length and range are determined so that the shield body does not push the ground, that is, the shield exists inside the excavated space.

The flowchart to determine the articulation condition and the copy cutter condition on a 3D compound alignment is shown in Figure 3, using the above calculation conditions.

2.2 Shield behavior simulation

2.2.1 Kinematic shield model

The kinematic shield model is composed of five forces: force due to self-weight of machine f_1, force on the shield tail f_2, force due to jack thrust f_3, force on the cutter disc f_4, and force on the shield periphery f_5, as shown in Figure 4 (Chen et al. 2008). In case of the articulated

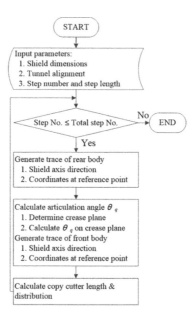

Figure 3.   Flowchart of numerical procedure.

shield, f_1 and f_5 act on both sections of the shield, f_3 comes from shield jacks and articulation jacks, and f_2 and f_4 act on the rear section and the front section, respectively.

The force on the shield periphery f_5 is due to the earth pressure acting on the shield skin plate and the dynamic friction force on the shield skin plate. Since the earth pressure is relied on the ground deformation, to estimate the earth pressure on shield periphery, the following were assumed:

1. The shield is regarded as a rigid body.
2. The ground reaction curve, which shows the relationship of the distance from the original excavated surface normal to the shield skin plate, U_n, and the coefficient of earth pressure, K, in vertical and horizontal directions, can be represented by the following functions, as shown in Figure 5 (Sramoon & Sugimoto 1999).

$$K_i(U_n) = \begin{cases} (K_{io} - K_{i\min}) \tanh\left[\frac{a_i U_n}{K_{io} - K_{i\min}}\right] + K_{io} (U_n \leq 0) \\ (K_{io} - K_{i\max}) \tanh\left[\frac{a_i U_n}{K_{io} - K_{i\max}}\right] + K_{io} \left(U_n 0\right) \end{cases} \quad (i = v \text{ or } h) \quad (1)$$

where a = the gradient of the functions $K(U_n)$ at $U_n = 0$, which is defined as the coefficient of subgrade reaction k divided by the overburden earth pressure σ_{vo}; the subscripts v, h = the vertical and the horizontal directions, respectively; and the subscripts o, min, max = the initial, lower limit, and upper limit of the coefficient of earth pressure, respectively. Here it is noted that K at $U_n = 0$ is the coefficient of earth pressure at rest, K_0.
3. The coefficient of earth pressure in the normal direction at any position, K_n, can be interpolated by using K_v and K_h as

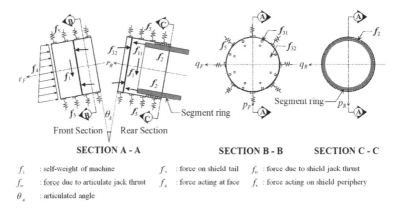

f_1 : self-weight of machine $\quad$ f_3 : force on shield tail $\quad$ f_D : force due to shield jack thrust
f_D : force due to articulate jack thrust $\quad$ f_a : force acting at face $\quad$ f_5 : force acting on shield periphery
θ_A : articulated angle

Figure 4. Model of loads acting on articulated shield.

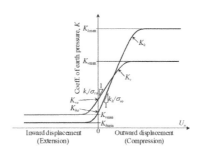

Figure 5. Ground reaction curve.

$$K_n(U_n, \theta) = K_v(U_n)\cos^2\theta + K_h(U_n)\sin^2\theta \qquad (2)$$

where θ = the angle measured from the bottom to the point in the transverse cross section.

4. The earth pressure normal to the shield periphery, σ_n, can be estimated from

$$\sigma_n = K_n(U_n, \theta)\sigma_{vo} \qquad (3)$$

5. The displacement of the excavated surface normal to the shield skin plate, d_n, is defined as the arrows in Figure 6.

It should be noted that the position of the original excavated surface and the shield skin plate can be determined by the measured shield position and rotation or by the calculated ones using the input data on shield operation. Therefore, U_n can be calculated geometrically by taking account of the trace of the cutter face and the position and rotation of the shield at present. Considering the displacement of the excavated surface is similar to a contact problem in FEM.

2.2.2 Simulation algorithms

Here, the following coordinate systems were used to model each force, as illustrated in Figure 7. The global coordinate system C^T is selected so that the x-axis is vertically downward and the y and z-axes are on a horizontal plane. The machine coordinate system C^M is selected so that the p-axis is vertically downward without the rotation of shield and the r-axis is in the direction of the machine axis. Here, the origin of C^M was selected at the center of the articulation on the shield axis.

The shield behavior is represented by the shield movement and the shield postures (yawing angle ϕ_y, pitching angle ϕ_p, and rolling angle ϕ_r), as illustrated in Figure 7. Since the change of ϕ_r is limited in practice, the factor of shearing resistance due to the cutter torque α_{SG} was adopted as the parameter instead of ϕ_r. The shield behavior during excavation can be obtained by solving the following equilibrium conditions:

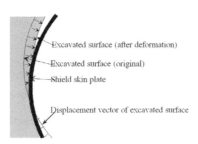

Figure 6. Displacement of excavated surface.

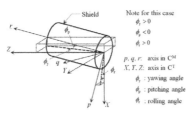

Figure 7. Definitions of a coordinate system.

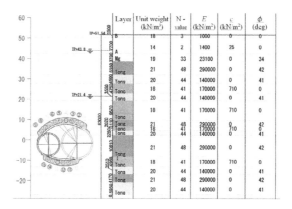

Figure 8. Geological profile and soil properties.

Table 1. Ground properties.

Ground layer	Tong	Tonc	Tons
Unit weight (kN/m^3)	21	18	20
Cohesion (kN/m^2)	0	710	0
Internal friction angle (deg)	42	0	41
K_{ha} (= K_{hmin})	0	0	0
K_{ho}	0.331	1	0.344
K_{hp} (= K_{hmax})	5	5	5
K_{va} (= K_{vmin})	0	0	0
K_{vo}	1	1	1
K_{vp} (= Kv_{max})	5	5	5
k_h (MN/m^3)	60.0	51.2	55.0
k_v (MN/m^3)	60.0	51.2	55.0
Coeff. of friction	0.1	0.1	0.1

$$\left[\begin{array}{c} \sum_{i=1}^{5}\left(\boldsymbol{F}_{Fi}^{M} + \boldsymbol{F}_{Ri}^{M}\right) \\ \sum_{i=1}^{5}\left(\boldsymbol{M}_{Fi}^{M} + \boldsymbol{M}_{Ri}^{M}\right) \end{array} \right] = 0 \qquad (4)$$

where $\boldsymbol{F}$ and $\boldsymbol{M}$ are the force and moment vectors, respectively; the subscripts F and R denote the front and rear sections of the shield, respectively; the superscript M shows the machine coordinate system; and the subscripts 1 to 5 represent the component of force f_1 to f_5, respectively.

2.2.3 Index of shield behavior

Since a shield performs in three-dimensional space, the shield behavior has six degrees of freedom, so the traces of shield in horizontal and vertical planes, ϕ_y, and ϕ_p represent the shield behavior adequately.

3 APPLICATION

3.1 Site description

To construct the pre-support at underground space enlargement, 15 small diameter shield tunnels, which launch from the main tunnel and the ramp one, were planned. The overburden depth is approximately 53.2 m, and the groundwater level is GL-30.14 m, as shown in Figure 8. Here, the tunnel passes in Toneri gravel layer (Tong), Toneri clay layer (Tonc), and Toneri sandy layer (Tons), of which the properties are shown in Table 1. The ground reaction curves K-U_n relationships for Tong layer, which is obtained by substituting the corresponding values in Table 1 into Eq. (1), is shown in Figure 9 as an example. This means that the tunnel is constructed in the stiff soil. In this paper, No. 2 shield tunnel is taken as an example. No.2 shield tunnel launches from the main tunnel to the vertical upward direction with the elevation angle of 30 degrees, and passes a vertical curve with a radius of 10 m from 17.8 m distance to 23.0 m distance. The analysis length is approximately 38.8 m. The dimensions of the shield and the tunnel are shown in Table 2, that is, the shield is an articulated shield with 2.14 m in outer radius and 4.575 m in total length. Since L_{CSE} = 1.260 m; $L_{M1} - L_{CSE}$ = 0.975 m; and $L_{M2} - L_{CSE}$ = 1.080 m, the shield is categorized to type 2.

3.2 Shield operation

The input data used for simulation were estimated as shown in Figure 10. The tunneling operations are jack thrust F_{3r}, horizontal jack moment M_{3p} (+: right turn), vertical jack moment M_{3q} (+: downward), cutter face (CF) rotation direction (1: counterclockwise direction, viewed from the shield tail), copy cutter length CCL, area of applied copy cutter CC range, and articulation angle in vertical direction θ_{CV} (+: upward), which are employed to control the shield position and the shield rotation during excavation. The shield rotation is defined as yawing angle ϕ_y (+:right turn) and pitching angle ϕ_p (+: downward). The excavation conditions are shield velocity v_s, slurry pressure σ_m, and slurry density γ_m in the chamber, which is usually controlled to stabilize the tunnel face.

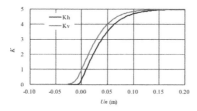

Figure 9.   Ground reaction curve (Tong).

Table 2.   Dimensions of shield and tunnel.

Item	Component	Value
Shield	Outer radius (m)	1.070
	Total length (m)	4.575
	Length of front body (m)	2.235
	Length of rear body (m)	2.340
	Length between crease center & segment front end (m)	1.260
	Self-weight (kN)	232
Shield jack	Number of jacks	8
	Radius of jack (m)	0.810
Segment	Outer radius (m)	1.00
	Width (m)	0.75, 0.30
Tunnel	Horizontal curve radius (m)	10.00
	Slope (ascend)	0.577
Ground	Ground water level (m)	GL-30.14
	Overburden depth (m)	53.20

F_{3r} is applied to the shield to drive the shield forward against earth pressure at the face and friction on the shield skin plate, as it advances. M_{3p} is applied to negotiate the horizontal moment due to the shear resistance on the cutter disk and the normal earth pressure around the skin plate. Since the planned tunnel alignments are vertical downward curves, M_{3q} is mainly applied against the vertical moment due to the normal earth pressure around the skin plate, the earth pressure on the cutter disc and the self-weight of the shield. Here, F_{3r} of approximately 4 ~ 5 MN was obtained by the sequential analysis using the kinematic shield model without jack force (F_{3r}, M_{3p}, and M_{3q} are zero). On the other hand, M_{3p} and M_{3q} were assumed to be zero as the initial values, since the sequential analysis, which does not consider the equilibrium conditions, provides eccentric values of M_{3p} and M_{3q}, especially at a sharp curve.

The sign of cutter torque defines the CF rotation direction, which generates the shear resistance on the cutter disc, and its rotation direction causes the shield rolling around its axis. Therefore, the rotation direction of the cutter disc is alternately controlled to maintain the use of facilities inside the shield. CCL and CC range were set, based on the required overcut as shown in "2.1 shield steering". The copy cutter is used to increase the excavated area around the cutter disc. θ_{CV} was calculated following "2.1 shield steering" and corresponds to the vertical curve radius of the tunnel alignment. The use of articulation of shield is to fit the skin plate to the area excavated by the cutter disk and copy cutter. The copy cutter and the shield articulation can reduce the ground reaction force acting on the skin plate and makes a shield easily translate or rotate.

ϕ_y is close to zero because the shield is on an almost vertical plane. ϕ_p shows that the rotation of the front body to follow the planned vertical tunnel alignment, that is, the ascent straight line and downward curve.

Shield velocity v_s of 0.025 m/min and muck density γ_m of 13.5 kN/m^3 were set, based on the experience. To stabilize the face, chamber pressure σ_m approximately 230 kPa is applied based on the lateral earth pressure at the tunnel face.

It is noted that the shield steering data on shield jack and copy cutter in Figure 10 are the modified values so that the calculated shield behavior has a good agreement with the planned one, as described at the first paragraph in "2 METHODOLOGY".

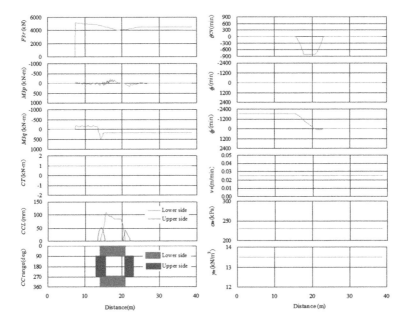

Figure 10.   Shield operation data.

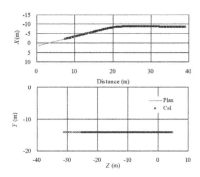

Figure 11. Calculated and planned shield traces.

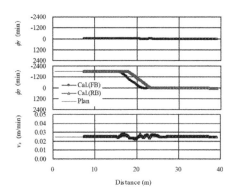

Figure 12. Calculated and planned shield behavior.

3.3 *Shield behavior*

The shield behavior was simulated from 7.4 m distance to 38.8 m distance. Figure 11 shows the planned alignment and the calculated traces of the shield on vertical and horizontal planes. The calculated and planned time-dependent parameters ϕ_y, ϕ_p, and v_s are shown in Figure 12. From Figure 11, the following were found: 1) the shield trace on vertical and horizontal planes have a good agreement with the planned one; and 2) the deviation in horizontal and vertical plane is less than 0.1 m. From Figure 12, the following were found: 1) the ϕ_y and ϕ_p of the front body have a good agreement with the planned one; and 2) the v_s is close to the planned one, but the v_s fluctuates within 5 mm/min at the vertical curve.

4 CONCLUSIONS

To examine the possibility of the small diameter shield tunnelling for the pre-support method at underground space enlargement, the simulation on the articulated shield behavior was carried out by the kinematic shield model, using the estimating shield operational parameters as shown in Figure 10. As a result, the followings can be concluded:

1. The calculated steering conditions, that is, the copy cutter length and range, and the articulation angle and direction, are reasonable from the viewpoint of theory and site experience.
2. When the setting shield operation data including the shield steering parameters are in use, the calculated shield behavior has a good agreement with the planned tunnel alignment.

REFERENCES

Alsahly, A., Stascheit, J., & Meschke, G. 2013. Computational framework for 3D adaptive simulation of excavation and advancement processes in mechanized tunnelling. *Computational Methods in Tunnelling, EURO:TUN 2013*: 85–96. Bochum: Ruhr University.

Chen, J. 2008. Study on calculation method of overcutting space and articulation angle during curved excavation by articulated shield, PhD thesis, Nagaoka Univ. of Technology, Niigata, Japan.

Chen, J., Matsumoto, A., & Sugimoto, M. 2008. Simulation of articulated shield behavior at sharp curve by kinematic shield model. *Geotechnical Aspects of Underground Construction in Soft Ground*: 761–767. Leiden, The Netherlands: Balkema.

Clough, G. W., Sweeney, B. P., & Finno, R. J. 1983. Measured soil response to EPB shield tunneling. *J. of Geotech. Eng.* 109(2): 131–149.

Date, K., Igarashi, H., Sasakura T., & Yakeyama K. 1999. A study of posture change prediction of multi-shield tunneling machine. *J. Japan Soc. Civil Eng.*, 630/VI-44: 39–53 (in Japanese).

Festa, D., Broere, W., & Bosch, J. W. 2013. On the effects of the TBM-shield body articulation on tunnelling in soft soil. *Computational Methods in Tunnelling, EURO:TUN 2013*: 109–118. Bochum: Ruhr University.

Finno, R. J., & Clough, G. W. 1985. Evaluation of soil response to EPB shield tunneling. *J. Geotech. Eng.*, 111(22): 155–173.

Fujita, K. 1989. Underground construction, tunnel, underground transportation: Special lecture B. *Proc. 12th Int. Conf. on Soil Mech. Found. Eng.*, 5: 2159–2176. Rio de Janeiro.

Hansmire, W. H., & Cording, E. J. 1985. Soil tunnel test section: Case history summary. *J. Geotech. Eng.*, 111(11): 1301–1320.

Huynh, T. N., Chen, J., & Sugimoto, M. 2015. Analysis on shield operational parameters to steer articulated shield. *15th Asian Regional Conference on Soil Mechanics and Geotechnical Engineering (15ARC)*. Fukuoka, Japan, ISSMGE.

Kasper, T., & Meschke, G. 2004. A 3D finite element simulation model for TBM tunneling in soft ground. *International Journal for Numerical and Analytical Methods in Geomechanics* 28(14): 1441–1460.

Komiya, K., Soga, K., Akagi, H., Hagiwara, T., & Bolton, M. D. 1999. Finite element modelling of excavation and advancement processes of a shield tunneling machine. *Soils & Foundation* 39(3): 37–52.

Lee, K. M., Rowe, R. K., & Lo, K. Y. 1992. Subsidence owing to tunneling: I. Estimating the gap parameter. *Canadian Geotech. J.* 29: 941–954.

Lee, K. M., Ji, H. W., Shen, C. K., Liu, J. H., & Bai, T. H. 1999. Ground response to the construction of Shanghai metro tunnel-line 2. *Soils & Foundation* 39(3): 113–134.

Melis, M. J., & Medina, L. E. 2005. Discrete numerical model for analysis of earth pressure balance tunnel excavation. *Journal of Geotechnical and Geoenvironmental Engineering* 131(10): 1234–1242.

Miki, A., Nagura, H., Kayukawa, K., Shimbara, K., Ehara, A., & Sugimoto, M. 2016. Development of underground enlargement method using multiple shield tunnels at great depth and high water presser. *Proc. of Tunnel Engineering* 26: 1–12 (in Japanese).

Rowe, R. K., Lo, K. Y., & Kack, G. J. 1983. A method of estimating surface settlement above shallow tunnels constructed in soft ground. *Canadian Geotech. J.* 20: 11–22.

Rowe, R. K., & Lee, K. M. 1992. Subsidence owing to tunneling: II Evaluation of prediction technique. *Canadian Geotech. J.* 29: 941–954.

Sramoon, A., & Sugimoto, M. 1999. Development of a ground reaction curve for shield tunnelling. *Proc., Int. Symp. on Geotechnical Aspects of Underground Construction in Soft Ground*: 437–442. Rotterdam, The Netherlands, Balkema.

Sramoon, A., Sugimoto, M., & Kayukawa, K. 2002. Theoretical model of shield behavior during excavation II: Application. *Journal of Geotechnical and Geoenvironmental Engineering*, 128(2): 156–165.

Sugimoto, M. 1995. Modeling of acting load on shield. *Underground Construction in Soft Ground*: 273–278. Rotterdam, The Netherlands, Balkema.

Sugimoto, M., & Sramoon, A. 2002. Theoretical model of shield behavior during excavation I: Theory. *Journal of Geotechnical and Geoenvironmental Engineering* 128(2): 138–155.

Sugimoto, M., Sramoon, A., Konishi, S., & Sato, Y. 2007. Simulation of shield tunnelling behavior along a curved alignment in a multilayered ground. *Journal of Geotechnical and Geoenvironmental Engineering* 133(6): 684–694.

Sugimoto, M., Tanaka, H., Huynh, T. N., Salisa, C., Le G. L., & Chen, J. 2018. Study on Shield Operation Method in Soft Ground by Shield Simulation. *Geotechnical Engineering Journal of the SEAGS & AGSSEA* 49(2): 182–191.

Takeda, H., Kusabuka, M., Yoshida, T., Tanaka, H., & Kurokawa, N. 1998. Finite element analysis of general contact problems application for excavation of shield tunnel. *J. Japan Soc. Civil Eng.* 603/III-44: 1–10 (in Japanese).

Tunnels and Underground Cities: Engineering and Innovation meet Archaeology,
Architecture and Art, Volume 12: Urban
Tunnels - Part 2 – Peila, Viggiani & Celestino (Eds)
© 2020 Taylor & Francis Group, London, ISBN 978-0-367-46900-9

Evaluation and numerical interpretation of measured pipe umbrella deformations

A. Syomik
ETH, Zurich, Switzerland (currently: Maidl Tunnelconsultants GmbH & Co. KG, Munich, Germany)

R. Rex
ETH, Zurich, Switzerland (currently: gbm, Limburg, Germany)

A. Zimmermann
Rothpletz, Lienhard + Cie AG, Olten, Switzerland

G. Anagnostou
ETH, Zurich, Switzerland

ABSTRACT: This paper analyses the load-bearing behaviour of pipe umbrellas by means of 3D numerical calculations considering the interaction among pipes, ground, temporary support and face reinforcement or improvement. With the purpose of validating a numerical model, data from an in-place chain-inclinometer in a Swiss tunnel is evaluated and interpreted numerically. The limitations of the measurements and potential sources of error in the evaluation of the monitoring data are discussed. The numerical simulation results map the observed pipe behaviour reasonably well. Face reinforcement or improvement is found to considerably affect the ground deformations ahead of the tunnel face and thus also the loading of the pipes. Furthermore, the numerical analyses indicate that the degree of structural utilization of the pipes is rather low in the present tunnel case, meaning that safety against failure is high and confirming the applied simplified design models to be conservative.

1 INTRODUCTION

Pipe umbrellas aim to mitigate the risk of instability of the unsupported span and thus to increase safety for workers and equipment in the excavation area during tunnel construction. A pipe umbrella consists of a set of sub-horizontal longitudinal steel pipes, systematically introduced into the ground prior to excavation along the periphery of the tunnel face and subsequently injected with cement grout. The installation of the pipes ahead of the face allows for support of the tunnel crown and walls for the following excavation rounds, until the temporary support is applied and develops its load-bearing capacity (Carrieri *et al.* 1991).

First applications of pipe umbrellas for tunnel excavation in difficult ground conditions can be traced back to Italy in the 80's (Bruce *et al.* 1987) and were thereafter widely spread abroad. Over the years several studies on the topic have been conducted by numerous authors using analytical (Anagnostou 1999, Volkmann and Schubert 2010), semi-empirical (Eckl 2012) and numerical approaches (Volkmann *et al.* 2006, Eclaircy-Caudron *et al.* 2006, Eckl 2012), centrifuge (Kamata and Mashimo 2003), small-scale tests (Shin *et al.* 2008) as well as *in situ* measurements (Volkmann 2004). However, the structural design of pipe umbrellas is yet commonly based on simplified design methods and assumptions, whose reliability is still uncertain. Due to the lack of existing codes, regulations, guidelines and recommendations as well as objective design criteria (Peila 2013, Oke 2016) an assessment of the adequacy of common design methods is important. Thus, the goal of the present paper is to improve understanding of the main mechanisms and influencing parameters governing the load-bearing behaviour of pipe umbrellas and to provide valuable insights for their structural design. For this purpose, the load-bearing behaviour is investigated by means of field measurements and 3D stress analyses.

2 PROJECT OVERVIEW

Tunnel Burg is a 500 m long road tunnel of the Küssnacht-bypass project in central Switzerland. The tunnel was constructed conventionally at a depth of 5 - 15 m underneath buildings, in a slope that strikes parallel to the tunnel axis (Figure 1). The tunnel was excavated full-face with the following measures: inclined and stepped tunnel face; short round lengths; quick closure of the temporary lining (steel fibre and mesh reinforced shotcrete and lattice girders) at the invert after each excavation step; pre-support by pipe umbrellas (Table 1); face-reinforcement by 20 m long fiberglass bolts installed every 12 m of advance; advance drainage of the ground by 18 m long boreholes (Zimmermann and Schneider, 2017).

Figure 1.   Plan view of tunnel Burg and location of investigated section of pipe umbrella stage no. 8.

Table 1.   Parameters of the pipe umbrellas.

Number of pipes per stage	39
Pipe length	15 m
Pipe spacing (at location of installation)	0.35 m
Installation angle	5°
Pipe profile and grade of steel	ROR 139.7/8, S355

Table 2.   Parameters of the "lake deposits" (based upon laboratory and *in situ* tests).

	Estimated based upon laboratory and *in situ* tests	Values considered in the numerical computations	
		Range	Best-fit set
Unit weight γ	22 kN/m^3	22 kN/m^3	22 kN/m^3
Stiffness modulus			
1st loading $E_{s,1}$	20–70 MPa		
unloading-reloading $E_{s,2}$	> 200 MPa		
Young's modulus E'		20–100 MPa	30 MPa
Poisson's ratio v'		0.3	0.3
Friction angle φ'	20°	20°	20°
Cohesion c'	50–180 kPa	20–500 kPa	40 kPa
Dilatancy angle ψ'		0°	0°
Undrained shear strength s_u	50–220 kPa		
Hydraulic conductivity k	1e^{-8} m/s		

The present study focuses on pipe umbrella no. 8 (marked red in Figure 1). The ground in this area consists of glacially preloaded "lake deposits" (stiff, overconsolidated clay, CM after USCS; Table 2, 2nd column). The groundwater table is located above the tunnel crown. Figure 2a shows the sequence of the excavation and support works in the area of pipe umbrella no. 8. After the application of the auxiliary measures and the installation of the monitoring devices, a tunnel section of total 12 m length was excavated in nine stages (Table 3).

Pipe umbrella design was based on a simplified model, considering a simply supported beam with uniform loading. Beam length was chosen based on construction process; the loading was estimated based on silo theory with reduced soil cohesion values.

Table 3. Construction process in the area of pipe umbrella no. 8.

27.02-01.03.2017	Installation of pipe umbrella stage no. 8
02.03-04.03.2017	Installation of face bolts
05.03.2017	No works
06.03.2017	Drilling works for advance drainage and installation of RH-extensometer
07.03/08.03.2017	Full face excavation and support installation over 1.2 m round length
09.03.2017	Excavation and support installation over 2.4 m in top heading followed by 1.2 m in bench/invert
10.03-14.03.2017	Excavation and support installation over 1.2 m in top heading followed by 1.2 m in bench/invert (shifted) Sunday, 12.03.2017: no works
15.03.2017	Excavation and support installation over 2.4 m in top heading
16.03.2017	Excavation and support installation over 3.6 m in bench/invert

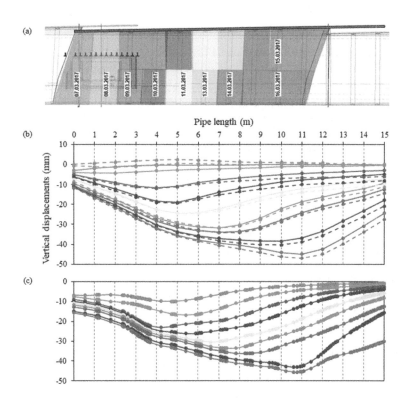

Figure 2. (a) Construction sequence in the area of pipe umbrella no. 8; (b) measured and, (c), numerically computed bending curves of the crown pipe after each excavation step.

3 EVALUATION OF INCLINOMETER DATA

The displacements of the pipe umbrella were monitored in real-time by an in-place inclinometer chain with 1 m gauge length, installed in the crown pipe prior to excavation. The sensors provide an individual tilt output for each gauge interval (specifically, the average angle change over the gauge interval; commonly presented as mm offset/meter). For the integration of the recorded inclinometer data to pipe displacement profiles, the displacements of a reference point located on the pipe have to be monitored with sufficient resolution by independent measurements. In the present case, the absolute displacements of the starting point of the crown pipe were not continuously monitored during tunnel construction and, therefore, plausible assumptions for the reference point and the corresponding displacements have to be made.

Figure 2b presents the evaluated displacement profiles of the crown pipe after each construction step (the colours correspond to those of Figure 2a) based on two different assumptions concerning the reference point. The solid lines consider as reference point a point that is located on the shotcrete lining directly below the pipe begin; the displacements of this point were measured geodetically (not continuously and not automatically recorded). The dashed lines consider the displacements of the pipes in the two previous pipe umbrellas (no. 6 & 7), taking the pipe deepest ahead of the tunnel face as a fixed point. Both evaluation methods present inaccuracies. Insufficient data resolution, detachment of the pipe from the tunnel lining (occasionally observed *in situ*, especially during the first excavation rounds) and the deformations of the ground ahead of the tunnel face introduce errors. The effect of taking the pipe deepest as a fixed point is clearly visible for the first two rounds (Figure 2b, dashed lines); the evaluated displacements show a slight uplift of the pipe over several meters, followed by a large downshift far ahead of the tunnel face from excavation step 2 to step 3 (from 08.03.2017 to 09.03.2017). However, the possibility of other influencing factors (such as certain consolidation of the ground or the drilling of additional boreholes for face bolts or advance drainage) cannot be excluded.

4 NUMERICAL INTERPRETATION

4.1 *Numerical model and key assumptions*

The load-bearing behaviour of the pipe umbrella no. 8 was investigated numerically by means of 3D stress analyses, performed by the finite element code PLAXIS 3D, version 2017.0. Figure 3 shows the computational domain with the finite element mesh, which consists of approximately 260'000 nodes and 180'000 soil elements.

The ground is discretized by 10-node tetrahedral elements and considered as a linearly elastic, perfectly plastic material obeying the Mohr-Coulomb yielding criterion with a non-associated plasticity flow rule. Table 2, 3rd column, shows the studied parameter range. The initial *in situ* stress field was taken as lithostatic, considering several values of the coefficient of earth pressure.

Due to the low advance rates of 1–2 m/d and the permeability of 10^{-8} m/s, drained conditions are assumed for the present analysis (*cf.* Anagnostou and Kovári 1996). The temporary shotcrete lining (Figuer 4) is discretized by 6-noded shell elements, assuming a linear elastic, isotropic material behaviour with properties according to Table 4. For the sake of simplicity, the characteristic "saw-tooth" longitudinal profile of the lining is neglected in the numerical model.

The stiffness development of the shotcrete lining during the construction process (hardening; application in two layers) was taken into account by assuming that the mechanical properties depend on the distance behind the tunnel face, considering an "idealized" elastic modulus for the shotcrete.

To simulate the three-dimensional load-bearing behaviour of the pipe umbrella and to capture the interaction between the pipes and the surrounding ground, each pipe is modelled individually by discrete, so-called "embedded beam" elements, implemented in PLAXIS with properties according to Table 4. Grouting of the pipes may increase ground strength and stiffness locally in their vicinity, but this was not considered in the numerical analyses, because

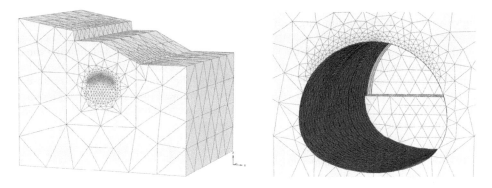

Figure 3. Computational domain with finite element mesh used for the numerical calculations.

Figure 4. Numerical model of the tunnel lining and pipe umbrella.

Table 4. Tunnel support parameters.

Shotcrete lining (shell elements)		
Young's modulus	$E_{green}/E_{hardened}$	7.5/15 GPa
Lining thickness	t_I/t_{II}	20/35 cm
Poisson's ratio	v	0.2
Umbrella pipes (embedded beam elements)		
Axial stiffness	EA	695'096 kN
Bending stiffness	EI	1'512.6 kNm^2

in situ tests of single umbrella pipes showed that the grout infiltrates very little into the ground (Zimmermann and Schneider 2017). Pipe-grout composite behaviour (which would result in a higher stiffness of the pipe umbrella) was not considered either, because there is no bond between grout and pipe intrados (smooth pipe walls; grout shrinkage) and the grout stiffness during early loading is anyway low. In the present case, grouting of the pipes aims mainly to prevent buckling of the steel profiles.

Considering the big quantity of face bolts (approximately 50 per face reinforcement stage), discrete modelling of each bolt would result in excessive calculation time. Therefore, the focus was set on modelling the main effect of face bolts, which is providing axial support to the tunnel face. The bolts are simulated by prescribing a uniform pressure at the tunnel face. Considering the average effective bolt length and the bond strength at the grout - ground interface (approximately 100 - 150 kPa for the present clayey ground), a support pressure of about 50 - 100 kPa was considered as statically equivalent. For the sake of simplicity, the face support pressure is taken constant over all excavation steps.

In the present numerical simulation, a construction of 36 m of tunnel length in total, consisting of three 12 m-excavation stages of pipe umbrella was considered. To account for the three-dimensional stress relief and redistribution in the ground due to tunnelling, the excavation and support sequence are simulated step by step, *i.e.* by successively deactivating the clusters representing the ground ahead of the face (core) to be excavated over the round length and simultaneously activating the shell elements behind the unsupported span, representing the temporary shotcrete lining. The core is therefore discretized into slices, 1.2 m thick in the longitudinal direction. The "embedded beam" elements representing the pipe umbrella are activated prior to the simulation of each construction stage.

4.2 *Back-calculation of pipe displacements*

First, in order to validate the numerical model, a back-analysis of the evaluated displacement profiles of the crown pipe (Figure 2b) is performed, considering the actual construction process within pipe umbrella no. 8 (Table 3). Young's modulus and cohesion of the ground, face support pressure as well as coefficient of horizontal *in situ* stress were systematically varied within the plausible range, whereby several iterations were performed in order to find a parameter set that maps well the monitored behaviour.

The numerical calculations reproduce reasonably well the measured displacements in spite of the modelling simplifications with regard to constitutive behaviour of the ground, face support and the temporary shotcrete lining. Figure 2c shows the calculated bending curves of the crown pipe for the "best-fit" parameter set (face support pressure 50 kPa; coefficient of earth pressure 0.7; other parameters according to the last column of Table 2). The best-fit coefficient of earth pressure (0.7) is lower than expected for an overconsolidated deposit, but can be explained on account of the morphology of the ground surface (reduced lateral confinement due to post-glacial erosion).

The order of magnitude and the main characteristics of the development of the bending curve with the excavation progress are captured well by the numerical analyses, particularly for the intermediate and last excavation steps.

The rather big differences in the first two steps can be traced back to the evaluation uncertainties of the monitored data (see Section 3) and to the simplified modelling of the face reinforcement. The assumption of a constant support pressure over all excavation steps slightly overestimates the displacements (especially ahead of the tunnel face) in most of the simulated excavation steps; in reality, the effectiveness of the fiberglass bolts decreases with the successive excavation rounds due to the decrease in the anchorage length (*cf.* Anagnostou and Serafeimidis 2007). Therefore, the accuracy of numerical results could be probably further improved by assuming a gradually decreasing face support pressure. The adequacy of the simplified numerical modelling of face bolts by an equivalent pressure on the tunnel face (*cf.* Peila 1994) was indicated by Dias and Kastner (2005). Nevertheless, this assumption may as well be a cause for certain deviations.

A further reason for the differences between the computed and the observed displacements is the simplified modelling of the ground as a constant stiffness material. Disregarding the stress-dependency of the stiffness and the difference between loading/unloading stiffness seems to have an effect especially on the displacements caused by the two final excavation steps, *i.e.* the excavation of the top heading, followed by that of the bench and invert over a length of 3.6 m; excavation of the top heading causes a heave of the ground elements, which contributes to the non-uniform shape of the displacement curves ahead of the face shown in Figure 2c.

4.3 *Parametric study*

Starting from the best-fit parameters determined above ("reference case") a parametric study was conducted in order to identify the key influencing factors and quantify their effect on pipe deformations and loading. Within the parametric study, the geomechanical parameters of the ground, the initial stress state, the mechanical properties of the tunnel lining and the magnitude of face support pressure were varied.

Ground stiffness, ground strength and face reinforcement (represented in the numerical analyses by the Young's modulus E', the cohesion c' and the face support pressure p, respectively) were found to be key influencing factors for fixed round length. Figure 5 shows the effect of face support pressure and ground cohesion on the pipe bending curve. The influence of these parameters is expectedly nearly equivalent (*cf.* Anagnostou and Kovári 1993). An increase in face support pressure or in soil cohesion reduces the stress relief ahead of the tunnel face, leading to smaller ground deformations (Peila 1994, Dias and Kastner 2005) and pipe displacements and to higher loading of the pipe umbrella (Figure 6) and the tunnel lining. The apparently paradox effect of cohesion is known from the literature (Cantieni and Anagnostou 2011). Furthermore, the numerical results show a nonlinear relationship between ground stiffness (represented by Young's modulus E') and pipe stresses and displacements, which agrees with the results obtained by Eckl (2012). However, Young's modulus does not affect the pipe displacement profile qualitatively.

4.4 *Evaluation of the degree of structural utilization of the pipes*

Next, the numerically calculated loading level of the pipes (and based thereupon also the degree of their structural utilization) will be compared with results obtained by evaluating the monitoring data. Considering that pipe stresses change with the progress of excavation, the bending moment is evaluated in three representative locations: at the edge of the temporary lining (M_1), at the field over the unsupported span (M_2) and ahead of the tunnel face (M_3; Figure 6). The degree of structural utilization is hereby defined as $U = \sigma_{Ek}/f_{yk}$, where σ_{Ek} and f_{yk} denote the elastically computed stress and the yield strength of the steel, respectively.

The utilization degree is rather low (max. 50%; Figure 6). The highest values result for a high ground cohesion c' of 500 kPa (elastic ground behaviour) or a high face support pressure p of 300 kPa (approximately equal to the *in situ* horizontal stress); the peaks in the utilization degree over date lines occur for constructions steps with a big round length at the top heading (2.4 m on 9.3.2017 and 15.3.2017). Besides the face pressure and ground cohesion, the round length has expectedly a considerable effect on the loading of the pipes. The utilization degree is max. 30% for the best-fit parameters (cohesion of 40 kPa and face support pressure of 50 kPa), both based on the measured deformations as well as on the computations. Consequently, there is a high safety margin against pipe failure, which means that the simplified model of a simply supported beam under uniform silo loading, which was used for the pipe umbrella design in the present case, is conservative.

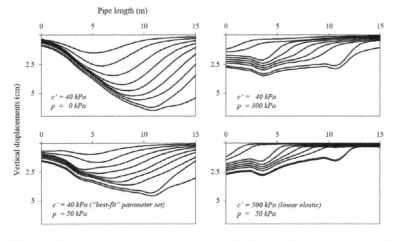

Figure 5.   Influence of face support pressure and ground cohesion on pipe displacement profiles.

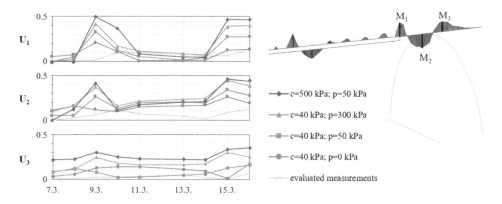

Figure 6. Evolution of the degree of structural utilization of the pipes at three characteristic locations.

5 CONCLUSIONS

In-place chain-inclinometers represent a useful tool for geotechnical monitoring of tunnels excavated under the protection of pipe umbrellas, providing real-time information on the development of ground deformations behind and ahead of the tunnel face, allowing to evaluate the structural utilization of umbrella pipes during tunnel construction, thus contributing to construction safety, especially when tunnelling below sensitive infrastructure on the ground surface. However, the assumption of a fixed point at pipe deepest, which is commonly made (in the absence of geodetic measurements) in order to determine the bending curve of the pipe from the inclinometer data, may introduce errors and represents a source of uncertainty in the interpretation of the monitoring data. An accurate assessment of the load-bearing behaviour of pipe umbrellas requires, therefore, the evaluation of the primary data, *i.e.* the changes in inclination. To adequately allocate the measured pipe deformations to particular construction stages, all secondary effects and conditions during tunnel construction (such as consolidation due to drainage or installation of face reinforcement) have to be carefully considered in the evaluation.

Spatial numerical simulations with adequate modelling of the ground, the temporary tunnel lining, the auxiliary support measures as well as of the construction stages are able to reproduce the observed behaviour. The magnitude and the development of pipe displacements for the documented excavation process and support sequence of tunnel Burg are captured reasonably well. This indicates that 3D numerical analyses are suitable for investigating the actual load bearing behaviour of pipe umbrellas.

The latter is found to be influenced by almost all input parameters of the 3D-model: (i) the excavation and support sequence; (ii) the mechanical properties of the temporary tunnel support and their time-dependency; (iii) the ground parameters; and, (iv), other auxiliary measures such as face reinforcement or ground improvement.

Face support and soil cohesion are important because they govern the ground deformations ahead of the tunnel face and thus the development of pipe deformations and stresses.

Furthermore, the present investigation shows, that the simplified models, which were used designing this particular pipe umbrella, are conservative. Specifically, the evaluation of the field measurements and the numerically estimated stresses in the pipes show that the degree of structural utilization of the pipes in the present case study is in the range of 10–30% only, which indicates a high safety against pipe failure. Simplified structural models are not suitable for analysing the actual load-bearing behaviour of pipe umbrellas but can be used for a conservative design.

ACKNOWLEDGEMENTS

The authors express their acknowledgements to the client of tunnel Burg the civil engineering office of canton Schwyz (CH) and to the design office Rothpletz, Lienhard + Cie AG for providing all necessary data and project documentation as well as for the assistance during editing.

REFERENCES

Anagnostou, G. 1999. Standsicherheit im Ortsbrustbereich beim Vortrieb von oberflächennahen Tunneln. In Städtischer Tunnelbau. Bautechnik und funktionale Ausschreibung. Berichte des Internationalen Tunnelbau-Symposiums vom 18. März 1999 in Zürich (pp. 85–95). ETH Zürich, IGT.

Anagnostou, G. & Kovári, K. 1993. Significant parameters in elastoplastic analysis of underground openings. In Journal of geotechnical engineering, 119(3),pp.401–419.

Anagnostou, G. & Kovári, K. 1996: Face stability conditions with Earth Pressure Balanced shields. In Tunnelling and Underground Space Technology, 11 (2), 165–173.

Anagnostou, G. & Serafeimidis, K. 2007: The dimensioning of tunnel face reinforcement. In: Bartak, Hrdina, Romancov & Zlamal (eds) "Underground Space - the 4th Dimension of Metropolises", 291–296. Taylor & Francis Group, London.

Bruce, D. A., Boley, D. L., & Gallavresi, F. 1987. New developments in ground reinforcement and treatment for tunnelling. In RETC Proceedings, Volume 2, Chapter 51, pp. 811–835.

Cantieni, L. & Anagnostou, G. 2011. On a paradox of elasto-plastic tunnel analysis. In Rock mechanics and rock engineering, 44(2),pp.129–147.

Carrieri, G., Grasso, P., Mahtab, A. & Pelizza, S. 1991. Ten years of experience in the use of umbrella-arch for tunnelling. In Proc. SIG Conf. On Soil and Rock Improvement, Milano, Vol. 1, pp. 99–111.

Dias, D. & Kastner, R. 2005. Modélisation numérique de l'apport du renforcement par boulonnage du front de taille des tunnels. In Canadian geotechnical journal, 42(6),pp.1656–1674.

Eclaircy-Caudron, S., Dias, D., Chantron, L. & Kastner, R. 2006, March. Numerical modeling of a reinforcement process by umbrella arch. In International Conference on Numerical Simulation of Construction Processes in Geotechnical Engineering for Urban Environment (NSC06), Bochum (Germany).

Eckl, M. 2012. Tragverhalten von Rohrschirmdecken beim Tunnelbau im Lockergestein. Doctoral dissertation, Technical University Munich TUM, Centre for Geotechnics, Chair and Testing Office for Foundation Engineering, Soil Mechanics, Rock Mechanics and Tunnel Construction.

Janin, J.P. 2012. Tunnels en milieu urbain: Prévisions des tassements avec prise en compte des présoutènements (renforcement du front de taille et voûte-parapluie). Doctoral dissertation, Institut national des sciences appliquées de Lyon.

Kamata, H. & Mashimo, H. 2003. Centrifuge model test of tunnel face reinforcement by bolting. In Tunnelling and Underground Space Technology, 18(2–3), pp.205–212.

Oke, J. 2016. Determination of nomenclature, mechanistic behaviour, and numerical modelling optimization of umbrella arch systems. Doctoral dissertation, Queen's University Department of Geological Sciences & Geological Engineering, Kingston, Ontario, Canada.

Peila, D. 1994. A theoretical study of reinforcement influence on the stability of a tunnel face. In Geotechnical & Geological Engineering, 12(3),pp.145–168.

Peila, D. 2013. Forepoling design (Lecture). Ground improvement pre support & reinforcement short course. Geneva: International Tunnelling and Underground Space Association (WTC 2013).

Shin, J.H., Choi, Y.K., Kwon, O.Y. & Lee, S.D. 2008. Model testing for pipe-reinforced tunnel heading in a granular soil. In Tunnelling and Underground Space Technology, 23(3),pp. 241–250.

Volkmann, G. 2004. A contribution to the effect and behavior of pipe roof supports. In Proceedings of the ISRM Regional Symp. EUROCK (pp. 161–166).

Volkmann, G., Button, E. & Schubert, W. 2006, January. A contribution to the design of tunnels supported by a pipe roof. In Golden Rocks 2006, The 41st US Symposium on Rock Mechanics (USRMS). American Rock Mechanics Association.

Volkmann, G.M. & Schubert, W. 2010. A load and load transfer model for pipe umbrella support. In Rock mechanics in civil and environmental engineering, pp.379–382.

Zimmermann, A. & Schneider, A. 2017. Südumfahrung Küssnacht, Tunnel Burg: Bauhilfsmassnahmen für innerstädtischen Tunnelbau. In Kolloquium Bauhilfsmassnahmen im Untertagbau, ETH Zürich, Professur für Untertagbau, 30.11.2017.

Tunnels and Underground Cities: Engineering and Innovation meet Archaeology, Architecture and Art, Volume 12: Urban Tunnels - Part 2 – Peila, Viggiani & Celestino (Eds)
© 2019 Taylor & Francis Group, London, ISBN 978-0-367-46900-9

Special design considerations for underpinning systems of existing structures due to tunnelling

Y.C. Tan, W.S. Teh & C.Y. Gue
G&P Geotechnics Sdn. Bhd., Kuala Lumpur, Malaysia

ABSTRACT: Protective work such as underpinning and ground improvement are routinely used to reduce tunnelling induced impacts, ensuring safety and serviceability of vulnerable existing building structures. This paper presents two case studies where the construction of the new Klang Valley Mass Rapid Transit- Line 2 in Kuala Lumpur, Malaysia, crosses very closely below the existing structures. The first case involves the underpinning of a 5-storey building located above the proposed tunnel alignment with some of its piles located within the tunnel horizon. Pile removal was designed to be carried out while the building was occupied. In the second case study, the same tunnel crosses beneath an existing reinforced concrete retention pond of a pumping station where partial pile removal was required to allow unobstructed access for the TBM. Despite the improved ground support from jet grouting, additional analyses were required as the underpinning work changes the response of the reinforced concrete structure.

1 INTRODUCTION

Construction of the Sungai Buloh-Serdang-Putrajaya (SSP) Klang Valley Mass Rapid Transit Line 2 (herein referred to as KVMRT2) in Kuala Lumpur, Malaysia began in the second half of 2016 and is in progress at the time of writing this paper. Scheduled to be fully operational by year 2022, it will serve a total of 37 stations over an alignment of 52.2km. Within this alignment, 13.5km of the alignment will be underground, connecting 11 underground stations.

Following the success of the world's first variable density tunnel boring machine (TBM) in KVMRT Line 1 (Bäppler et al., 2001), the new Line 2 utilises the same TBMs for its twin tunnel drives along the underground alignment. The 6.684m external diameter tunnel is made up of seven 275mm thick G50 precast segmental concrete linings.

While the TBMs are capable of addressing a variety of soils and rock of various conditions, tunneling through reinforced concrete piles is not advisable as the ductile rebars are likely to bend instead of being crushed and cored through, potentially jamming the TBM cutterhead.

One of the simplest ways to work around this issue is to design a tunnel alignment that does not clash with foundations of high-rise buildings or structures with pile foundations. However, in an urban environment, due to restrictions of land use or land acquisitions on top of any other technical or financial requirements, there will be situations where it is unavoidable. In these cases where buildings will remain in service throughout, underpinning and pile removal of existing structures are necessary for TBM passage.

This paper presents two case studies for the underpinning and pile removal of existing structures. It aims to highlight some of the less obvious but equally critical considerations for each specific site requirements.

2 CASE STUDY 1- UNDERPINNING DESIGN OF 5-STOREY EDUCATION QUARTER

2.1 *Site background*

The education quarter is a 5-storey reinforced concrete building with a ground floor car park and 4-storey residential quarters for a government primary school, Sekolah Kebangsaan Jalan Raja Muda. The KVMRT2 line consists of twin tunnels (Northbound and Southbound tunnels), that are parallel to each other at this location. The Southbound tunnel cuts under part of the existing education quarter building, indicating potential obstructions from the existing foundations to the tunnel construction works as shown in Figure 1.

Due to unavailability of foundation information of the building, investigation work, including trial pits, boreholes and parallel seismic tests were carried out. The trial pits revealed that the building is supported by 300mm square reinforced concrete (RC) piles. Based on parallel seismic tests conducted on two piles at different columns, the estimated pile lengths were found to be 24.5m and 32m respectively. The depth of tunnel boring machine (TBM) cutterhead is approximately 9m to 16m below ground level (Crown and invert of TBM extrados). As such, all the piles within the TBM extrados need to be removed. Since the building itself it currently occupied and will remain as such throughout tunnel construction, foundation underpinning work is required to protect the building, and ensure safety and long-term serviceability of the building.

The pile removal zone considers a TBM cutter head diameter of 6.684m, including uncertainties of 100mm TBM driving tolerance and pile verticality of 1 in 75 from true vertical position in the direction of the tunnel. All piles within the aforementioned zone will be removed and no underpinning pile can be carried out in this region. Figure 2 shows the photo of the affected columns. A total number of 4 piles (one for each column) have been identified for removal.

2.2 *Geological conditions*

The building is underlain by Kuala Lumpur Limestone formation with overburden soil typically consisting of mainly silty/gravelly SAND or sandy SILT of low SPT-N values (generally less than SPT-N 20). Figure 3 shows a typical geologic section. The subsurface investigation results show that the bedrock is generally encountered at 34m to 45m below ground level which is approximately 18m to 29m below invert level of KVMRT2 tunnel. Based on the nearest borehole information, the groundwater level is about 4.0m below ground level.

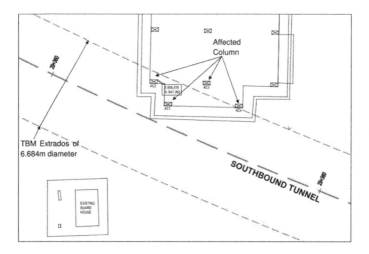

Figure 1.   Plan view of affected columns in relation to the southbound tunnel.

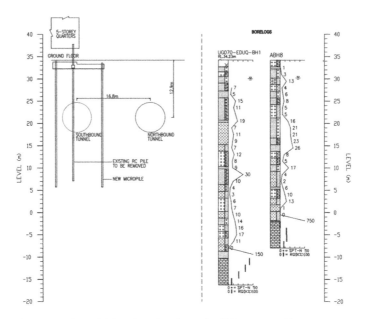

Figure 2. Typical cross section with borelog showing subsoil SPT-N profile.

Figure 3. Photo of affected columns.

2.3 Loading assumptions

Due to the absence of loading information for the existing building, the column loadings were estimated based on simple tributary area method. The loads derived by the authors were based on an assumed surcharge load of 15kPa per storey and 10kPa for roof loading. These estimated loads were independently assessed by two other consultants and as a means of assessing the final adopted values, the maximum estimated loading obtained from three parties on each column was conservatively chosen to represent the column loading for analysis.

2.4 Foundation underpinning scheme and design considerations

The concept of the proposed foundation underpinning scheme consists of a transfer slab which distributes the column loads to new micropiles beyond the no-pile zone. This meant that the transfer slab had to be designed for a wide effective span of 9.3m, with a thickness of 1.0 to 1.5m. The micropiles are 300mm diameter with design rock socket length of 2.5m. Due to karstic characteristics of the limestone formation, the bedrock level of the site is uneven, causing the installed pile lengths to vary from 28.5m to 50.1m.

To minimize the impact of the loading of the piles to the KVMRT2 tunnel, the micropiles are de-bonded up to 2m above tunnel invert level using bituminous membrane sheet (applied with grease) attached to the permanent casing. In addition, the permanent casing also serves as protection to the micropiles from the pile removal (i.e. coring works)as well as tunneling works.

Overlapped grouted columns were adopted as a temporary earth retaining system to facilitate the excavation and construction of transfer slab. It deserves to be highlighted that pile removal wouldonly be carried out after the load transfer structure is put in place. The method of pile removal will be further discussed in the subsequent section.

It is important to consider the reaction of working piles and transfer slab at different stages of construction to ensure adequacy of the underpinning system. This includes but is not limited to the following checks:

i. Impact of coring of existing pile (pile removal method) on newly constructed micropiles and adjacent existing RC piles.
ii. Expected pile head movement and transfer slab deflections post load application; permanent condition.
iii. Building impact assessment due to tunneling (Impact of tunneling on newly installed micropiles)

The new micropiles are expected to settle after mobilization of building loads to the pile and again during tunneling due to volume loss. As such, pre-loading on the newly installed micropiles is necessary to control the movements and differential settlements of the building.

These micropiles will be preloaded to the combined self-weight of the transfer slab, column load and weight of backfill above the transfer slab. This is achieved through the use of reaction frames where they are anchored to the transfer slab, providing the required reaction to transfer loads to the micropiles via hydraulic jack (see Figure 4). Upon reaching the targeted preload, the gap between the micropile and transfer slab will be filled with non-shrink grout to complete the load transfer process.

2.5 Methods of pile removal

Two feasible options to remove existing piles were considered. Option 1 was to remove the pile manually by creating an access via hand-dug caisson shaft with horizontal mined adit, while Option 2 was to core the piles in inclined direction from ground surface. In order to avoid prolonging the planned tunneling schedule, the tunnel intervention method to remove the pile from TBM cutterhead during tunneling works was not taken into consideration.

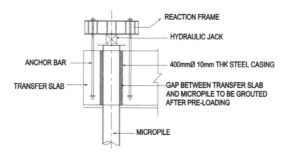

Figure 4. Pre-loading jack configuration.

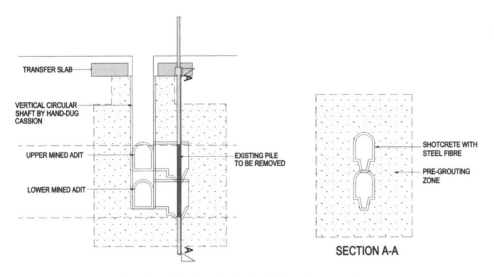

Figure 5.   Caisson vertical shaft and horizontal mined adit for pile removal.

2.5.1 *Option 1: Manual cutting of pile via caisson vertical shaft and horizontal mined adit*

Given that the length of the pile removal is approximately the diameter of TBM cutter head, a single adit would be excessively large and not cost-effective. Therefore, two levels of horizontal horse-shoe shaped mined adit with localized deepened excavation; below the pile location, has been proposed as shown in Figure 5.

In view of high groundwater table above tunnel horizon, pre-construction ground treatment (i.e. jet grout block) is required to ensure stability and dry condition inside the shaft and adits during excavation. Once grouted block has gained strength, the vertical shaft will be excavated first, followed by mining of horizontal adit towards the pile before pile cutting and removal. The horizontal mining would start from the lowest adit which is subsequently backfilled upon completion of the interim pile removal before repeating the process for the upper adit.

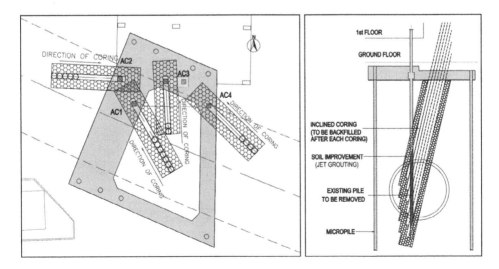

Figure 6.   Pile removal from ground surface using coring rig (Left: Layout plan; Right: Typical section).

6262

Based on KVMRT Line 1 experience, this option had been successfully implemented by local contractors as described in Khoo et al. (2015). The working space required for this option is small and therefore feasible for a site with headroom constraint. However, the drawback of Option 1 is the slow excavation progress. Considering the target was to remove all obstacle piles prior to tunnel arrival within tight a schedule, this option was not adopted.

2.5.2 *Option 2: Inclined coring of piles*

As the affected piles are located at the edge of the building, there is sufficient open space along-side the building to make use of a drilling rig for pile removal. The core diameter is 350mm (slightly larger than the existing RC square pile size of 300mm) and angle of coring ranges from 13° to 16° from vertical (see Figure 6).

In order to ensure stability of drilled hole during coring work, ground improvement by means of jet grouting block was carried out for an extent of 600mm surrounding the edge of the inclined drilled hole. Once the grouted block has gained strength, coring is to be carried out at the planned direction and distance. The void left by the coring will be backfilled with cement? grout.

The next coring can only begin once the minimum grout strength of 1MPa is achieved to ensure stability of adjacent drilled hole. Minor adjustments of coring direction on site may be required subject to coring results. The acceptance of the pile removal will be subject to verification of the extracted material.

It should be noted at that unlike Option 1, the pile removal from Option 2 may not be as complete or thorough, even within the tunnel horizon. However, this is considered manageable for the tunnel operation as long as the majority of the pile material (especially the pile reinforcement), can be removed; preventing jamming of TBM's system. This option has been adopted as it offers a much shorter pile removal duration as compared to Option 1.

3 CASE STUDY 2- DATO' KERAMAT PUMPING STATION

3.1 *Site background*

Constructed in 2001, the Dato' Keramat pumping station is a manmade reinforced concrete retention pond located at Kampung Dato' Keramat, next to Klang River. As part of a wider flood mitigation scheme initiated by the Department of Drainage and Irrigation Malaysia (JPS), the 1400 m² wide two-tiered retention pond was designed to retain an operation volume of approximately 3700 m³ of water, diverted from the adjacent Klang river.

The retention pond itself is supported by 200 mm square RC piles in a grid pattern with 2 m centre-to-centre spacing. An elevated pump house control room is situated on the south-west corner of the retention pond. No as-built information was made available. Therefore, it was expected for the actual foundation installed on-site to differ slightly from the construction drawings. Based on the construction drawings, pile lengths of 18 m are expected.

Both north and southbound tunnels of KVMRT2 will cross the retention pond along its width in a stacked alignment as shown in Figures 7 and 8. The axis depth for the shallower southbound tunnel is approximately 16.5m below ground level(mbgl) where it has a gentle gradient which was higher on the north side of the retention pond while the deeper northbound tunnel is relatively level with an axis depth of approximately 28 mbgl. Since the piles are within the southbound tunnel's horizon, pile removal is required for affected areas (i.e. along the tunnel alignment with a width of 3m to each side of the tunnel extrados). The adopted pile removal method consists of loosening the soil surrounding the pile with a steel casing (internal diameter of 330mm) before lifting it out from its position.

3.2 *Subsoil condition*

The pump house is situated on limestone formation where the underlying ground is of alluvial soil. This consists of silty/gravelly SAND and CLAY. Prediction of rock head levels in

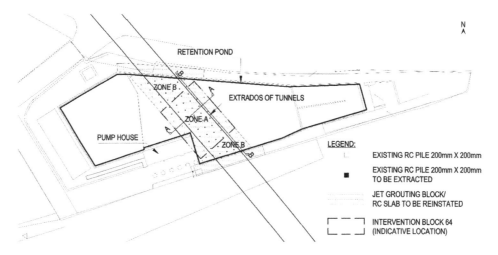

Figure 7. Layout of Dato' Keramat Pumping Station.

limestone formation is generally very difficult due to its karstic features, but the role it plays in this case study is less important given that the rock head level recorded by nearby boreholes are much deeper than the areas of concern. Groundwater level was relatively high with an adopted design level of 2 mbgl. The interpreted cross section A-A of the analysis is shown in Figure 8 below.

3.3 Design considerations and proposed solution

Ground improvement via jet grouting was proposed to support the slab post pile removal. In the initial stages, it was only intended for the jet grout block (JGB) to extend 3m from the southbound tunnel extrados. Subsequently, it was decided to utilize the scheme to serve as an intervention block for both tunnels; facilitating the maintenance of the TBM. Thus, the JGB at the central portion (Zone A) under the retention pond was extended further to encompass the deeper northbound tunnels (see Figure 8). Zone B as indicated in Figure 7 represents the area outside of the intervention block where JGB extends 3m below the invert of the southbound tunnel. Piles outside of these

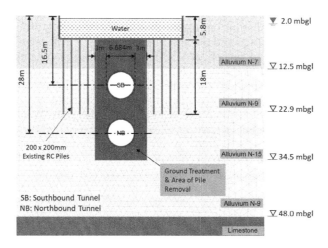

Figure 8. Section A-A of Dato' Keramat Pumping Station.

zones remain in place. Note that the construction drawing records that each pile has a working load of 300kN.

Tunnelling within the improved ground is beneficial in terms of the stress reductions on the tunnel linings at a given volume. This is because the JGB is designed to be self-supporting, allowing forces to arch over the circular tunnels. JGB is expected to achieve an unconfined strength of 1 MPa in 28 days with stiffness of 150MPa as well as a permeability within the region of 1×10^{-7} m/s. Along with the increased stiffness and strength of the JGB, magnitudes of soil movements toward the tunnels would similarly be less, reducing surface settlements.

While the above summarizes the pros of having ground improvement, the improved stiffness introduces an additional problem. The retention pond will resume operation upon reinstatement of the base slab, after pile removal. The base slab is therefore expected to take full water loading. Even though the settlements directly above JGB will be very small due to the its high stiffness, the sides which are seated on the existing loose/soft ground will deform much more under the same area load, creating differential settlement. This is exacerbated with the high stiffness of the piles immediately adjacent to JGB. Even though the existing slab within zones A and B will be demolished, the adequacy of its initially design steel reinforcements would need to be reassessed to cater for this hogging moment. A two-dimensional finite element analysis (FEA) was carried out using a commercial geotechnical FEA software, Plaxis 2D in order to evaluate the potential slab and pile settlement that may occur. A similar analysis was carried out to ascertain the slab settlement at Zone B. Structural forces for the base slab were not adopted directly from the FEA given the unique shape of the retention point and the skewed angle that the tunnel passes through it.

Soil spring stiffness was then computed by dividing the allowable bearing capacity of JGB with the settlements obtained from Plaxis. Similarly, the spring stiffness of the pile is taken as the working load divided by pile settlement. This was then modelled in a dedicated structural FEA software, SAFE 12. This allowed for a more realistic approach whereby local stiffening effects (e.g. corner of slab and wall of retention pond) can be accounted for.

A full model of the retention pond was not necessary as the area of concern lies is relatively small. The authors note that the response of the slab towards the extreme sides of the retention pond were less representative but the model was sufficiently wide to capture the slab response over Zones A and B.

In line with expectations, the edges of JGB experienced large hogging moments (refer to Figure 9). Across zone A, the results showed negligible sagging moment while in zones B, due to a lower stiffness from the shorter length of JGB, the result indicated a sagging moment at midspan and hogging moments towards the sides; similar to zone A.

The maximum hogging moment of 117.7 kNm/m was recorded in near the south border of zones A and B. This has exceeded the allowable moment capacity of the existing slab based on the existing reinforcement design at 95kNm/m. At these locations, additional reinforcement was required during reinstatement.

The adopted bearing capacity of JGB was taken as 300 kPa. This was based on previous local experience on similar ground conditions. Nonetheless, it is important to highlight that the actual bearing capacity could vary on site, owing soil variability as well as uncertainties in workmanship, this is in-line with the load cases considered for piled raft foundation (Tan et al., 2004).

To cater for this, a sensitivity study was carried out by varying the soil stiffness. This was essentially achieved by increasing or decreasing bearing capacity by a prescribed amount varying from 100 kPa to 400 kPa. The respective maximum bending moment is summarized in the table below.

Due to the high spring stiffness of the piles at the sides of JGB, a high bearing capacity or stiffness of JGB would result in a smaller contrast in stiffness. This would eventually yield smaller differential displacements and thus hogging moments. It can be observed from Table 1 that the results follow this trend.

Fluctuations for maximum sagging moments on the other hand were relatively small at less than 7% across the range of bearing capacities. This is due to the fact that the maximum

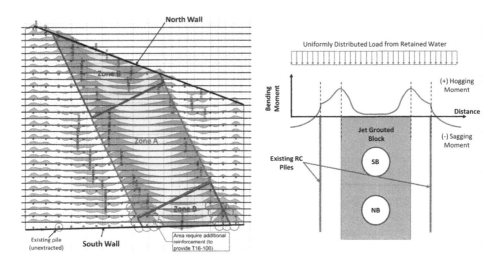

Figure 9. Results of bending moment analyses for the RC slab (left) with sketch of bending moment diagram illustrating the effects of the stiffen ground response with jet grouted block (right).

Table 1. Summary of maximum bending moments from sensitivity analyses.

JGB Bearing Capacity (kPa)	Max. Hogging Moment (kNm/m)	Max. Sagging Moment (kNm/m)
100	147.9	-146.9
150	107.9	-141.6
200	98.4	-141.7
300	86.5	-146.4
400	82.4	-151.9

sagging moment generally occurred at midspan of JGB which was less affected by the varying relative stiffness of the piles located at a distance.

Given the affected locations were relatively small, the final slab reinforcement design was based on the worst-case scenario with a bearing capacity of 100kPa. The authors believe that it was justified to take a conservative approach to the design to account for the uncertainties at a relatively small construction cost.

4 CONCLUSIONS

This paper has presented two case studies where the existing structures were required to be underpinned for pile removal to allow passage for the TBM. In the first case study, various considerations were given to the underpinning and pile removal techniques as the 5-storey residential building remained occupied throughout the construction. The second case study discussed the double-edged effect of ground improvement as an underpinning solution where the improved ground stiffness imposes additional hogging moments on the RC slabs of the pumping station.

It is hoped that this paper will help in highlighting the importance of careful deliberation, particularly for less obvious secondary effects during the designing process of protective works due to tunnelling. It is important to note that the aforementioned protective works are still on-going at the time of writing this paper, therefore, the possibility of minor field revisions remain.

ACKNOWLEDGEMENTS

The authors would like to convey the deepest gratitude to our colleagues, Ir. Low Chee Leong, Ir. Noor Azlina bt. Azhari, Mr. Ooi Qi Wei, Mr. Soh Yang Wee and team members of KVMRT2 of G&P Geotechnics Sdn. Bhd. for their out-of-the-box thinking, value-added engineering and technical support for the abovementioned projects.

REFERENCES

Bäppler, K., Battistoni, F. and Burger, W. 2018. Variable density TBM- combining two soft ground TBM technologies. Georesources Journal, 21–27.

Khoo, C.M., Ooi, T.A., Ng, S.W.F., and Feng, Q.Y. 2015. An innovative approach of mined adit design and construction for removal of existing piles. International Conference and Exhibition on Tunnelling & Underground Space (ICETUS 2015), 3–5 March 2015, Kuala Lumpur, Malaysia, 268–271.

Tan, Y.C., Chow, C.M. and Gue, S.S. 2004. A design approach for piled raft with short friction piles for low rise building on very soft clay. Proceedings of the 15th Southeast Asia Geotechnical Conference (SEAGC), 22–26 November 2004, Bangkok, Thailand.

Tunnels and Underground Cities: Engineering and Innovation meet Archaeology,
Architecture and Art, Volume 12: Urban
Tunnels - Part 2 – Peila, Viggiani & Celestino (Eds)
© 2019 Taylor & Francis Group, London, ISBN 978-0-367-46900-9

Arcueil-Cachan, Grand Paris Express Metro Station

S. Telhawi & L. de Saint-Palais
Setec tpi, Paris, France

G. Chapron
Setec terrasol, Paris, France

J.-P. Vaysse & B. Mougel
Ar.Thème, Paris, France

J. Royer
Société du Grand-Paris, Paris, France

ABSTRACT: Arcueil-Cachan metro station (ARC) is one of 16 stations of the line 15 South of the Grand Paris Express (GPE). It is in connection with the existing train station of the same name (line B of the Parisian Regional Express Network). The west section of the line 15, called "section 3", entirely underground, extends over 13 km, from the Ile de Monsieur access shaft (included) at the Villejuif Louis Aragon metro station (included). The studies were entrusted to the engineering group led by Setec and composed of Ingerop and six architectural agencies, including Ar.Thème, architect of ARC station.

1 GENERAL CONTEXT

The ARC metro station is located in the town of Cachan in the Val-de-Marne department (94), near the Carnot Avenue (Department Road n°157). To promote intermodality, the station was implanted near and under the existing rail network (Line B). ARC station is partly close to existing buildings (building of 5 floors, individual houses) which had to be maintained. The station environment is also composed of several major networks in direct interface with it, mainly: sanitation, water, telecom, gas, etc. The sensitivity of the buildings and the line B structures to ground settlements and vibrations created by the station works was studied to minimize the impact on this existing infrastructure. Recommendations from the RATP, transport infrastructure manager, and various networks owners were integrated by specific verifications made to insure that works doesn't affect their stability during and after construction.

2 DESIGN CONSTRAINTS

2.1 *Architectural design*

The ARC metro station is at the heart of a district undergoing transformation with a major real estate development and a redefinition of public spaces at the articulation of new urban polarities. The superstructure building of the ARC station is designed as an autonomous entity of the related operation overlooking it, at the foot of a wide plaza.

Figure 1.   Outside view of the passenger building of the station and the real estate project above.

The general morphology of the station spaces is linked to the search for coherence, legibility and fluidity in the articulation of the different volumes housing the different sequences of the passenger's itinerary:

– An urban sequence, with the metro station and the associated operation, inscribed in its environment both as a public building, an urban landmark, as transport equipment for local service (to and from the existing surrounding districts, to and from the new districts under development, as an interconnection space for intermodality: bus, soft modes, Line B, etc.

– An "access" sequence, crosswise to the station, guiding the passenger from the plaza to and inside the station. The main access reception room is located at the entrance to the building on level 0. It is the link between the various passenger flows, between line 15 and the city and with the Line B passenger areas. These access thresholds are largely glazed to accentuate the fluidity between exterior and interior public spaces.

– A "longitudinal" sequence of the routes to and from the metro platforms marked on the ground floor by the two large urban windows on each side of the station to mark the main axis of the internal routes. This sequence is centered on a vertical circulation shaft illuminated by a large central glass roof, which opens deep towards the platforms.

– A correspondence sequence with the passage under the Line B's bridge and the interconnection by the two large lateral volumes to the Line B tracks (urban signals along the itinerary).

The proposed material for the station's facings (terracotta) is omnipresent in the station spaces. Depending on the situation, these facings will be treated in different ways, grooved, with more marked bosses creating more or less strong shadows and more or less expressed textures.

Figure 2.   Different views of the architectural design of the station.

The outer skin of the building "turns" inside out to create the different functional subsets that inhabit it. This continuity facilitates the perception that the public space enters the spaces of the metro to the deepest part of the station.

2.2 Functional design

The station is built to respond to various functional constraints:

- New metro rolling stock, flow matrix by 2030's horizon and reinforced intermodality (Line B, bus, taxis, etc.).
- Mixed mechanization lifts and escalators, fire safety and public safety exits.
- Various functional premises for staff station: operation, services, shops, security, etc.
- Technical rooms: electricity (red and orange networks in figure 3 below), air-conditioning (green and brown networks), electromechanical, maintenance areas and water treatment (blue networks). The station houses one of the line's electrical room and one of the ventilation plants of the tunnel.

2.3 Real estate project

The future station is part of an urban area intended to mutate by creating buildings projects above and in immediate vicinity of the ARC station as shown in figure 4. The design takes into account the future construction of a 7-floors office building located above the passenger building.

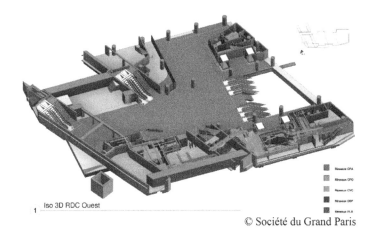

© Société du Grand Paris

Figure 3. Isometric view of the level 0.

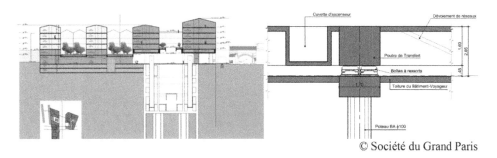

© Société du Grand Paris

Figure 4. Structural principle of the connection building-station.

- A structural measures are taken in the metro station to permit the realization of this building based on feasibility studies done by the station designer:
- Building load transfer on the station structure (beams of 1.60 m thick as shown in the figure 4 on the right). In order to reduce vibrations and structure-borne noise effectively and sustainably, this transfer is designed by spring elements (thickness of 0.45 m in the figure 4 on the right). The real estate promoter will realize a transfer floor above the upper slab of the station passenger building to bring the office building loads back to the supports provided in the station structure (columns of 1.00 m diameter).
- Construction of various parts of this tertiary building: technical rooms, hall, fire escape.
- Connection hoppers.

This future real estate project is planned to be completed after the completion of ARC station using supports above the passenger building while the station is in use.

At this stage, the building's promoter is not chosen yet but the bearing capacity for the future building is already defined and his foundations already realized by the station builders. The future promoter should respect the prescriptions given by the station designer: bearing capacity and stiffness of each used support, the expected settlement for each part of foundation, the position of safety exits and the architectural constraints for the building's facades.

2.4 Geotechnical context

The two main geotechnical risks for ARC station are:

- Uncertainty about the presence of old underground quarries. Injection works were necessary to realize the diaphragm walls and to safely execute the excavations into the station box. In the eastern part of the station, a huge quantity of mud was lost during the realization of the diaphragm wall due to the lack of knowledge of quarries characteristics (about 7000 m^3 in total and 450 m^3 per panel in maximum).
- The deformability and the swelling of the overconsolidated Plastic Clay.

The plastic Clay strongly constrains the design of the retaining solutions. It creates very important earth pressures on the diaphragm walls developed very early during earthworks due to the unloading of the soil. This phenomenon is difficult to effectively countered by conventional support elements. It induces relatively large displacements whose limitation is necessary because of the sensitive structures around the station (especially Line B).

In the design phases, a temporary shaft of 29.00 m depth and 4.00 m diameter was realized in the center of the future station to study the behavior of the soil layers. A gallery was also realized in the plastic clay. Collected parameters were helpful to design the retaining system. The shaft and the gallery were filled and will be demolished during the construction of the station.

2.5 Realization constraints

© Société du Grand Paris

Figure 5. View of the eastern shaft.

Figure 6.   View of diaphragm wall works under RER line B's bridge.

The construction site has a very small area (approximately 5000 m^2). The pedestrian and automobile circulation were maintained to insure the continuity towards Carnot avenue.

The eastern part of the station is a starting point for the tunnel boring machine (TBM) designed to build a double-trach tunnel with an internal diameter of 8.7 m. This TBM is planned to realize the tunnel section between ARC station and Villejuif Louis Aragon metro station. A temporary diaphragm wall was necessary to separate this shaft of 42.00 m length from the rest of the station.

On the West side, under the Line B railway, the station is created below a 40.00 m long bridge that has been shifted with temporary supports to realize the diaphragm walls under reduced height (6.50 m). When these foundations are realized, the bridge will be set on its final supports. The earthworks of the station can thus begin without disrupting rail traffic.

The TBM that will realize the section between Robespierre access shaft and the ARC station is planned to be disassembled in the center part of the station using openings implanted in the floors.

The context of realization was also complicated due to the buried structures present on site: reinforced concrete piling of the demolished buildings, networks structures, etc.

2.6   Planning elements

The station is on the critical path of the section's works. The planning includes many preliminary works of demolition and networks displacements. These works began on 2015 and finished in 2016.

The line B's bridge was constructed in 2017 in parallel to the realization of diaphragm walls of the eastern shaft. The eastern TBM departure in planned in the beginning of 2019.

By the end of 2018, the totality of diaphragm wall of the station will finish. The construction of internal floors will start in 2019 to provide them gradually to the contractors in charge of the networks and the equipment into the station.

3   DESCRIPTION OF THE STRUCTURE

The graphic studies of the station were carried out by a 3D software. All the structural elements have been modeled to allow the detailed design of station's networks and architectural elements.

The BIM software was helpful to resolve clashes between the different specialties by offering a collaborative working tool to structure engineers, MEP engineers, architects, designers and contractors.

3.1   General characteristics

The underground part of the station that gives access to the subway platforms and houses the infrastructure rooms, is a rectangular structure of 109 m long and about 24 m wide. The width was reduced to 16 m under the line B bridge to optimize the central span.

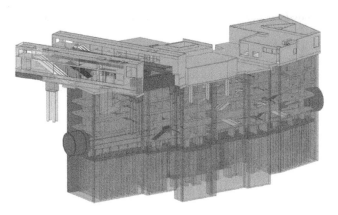

Figure 7.   3D Modeling of the station.

The station is developed on 4 floors, including platforms at ~ 25 m under the ground level. The subway platforms are asymmetrical (4.30 m width for La Defense direction and 5.15 m for Noisy-Champs direction).

3.2   Diaphragm walls

The presence of plastic clays at the bottom of the excavation defines the geotechnical context. During the design phases, various solutions were studied, using numerical modellings, to conceive diaphragm walls able to give acceptable bending moments and reduce the settlements around the structure: different thick (1.20/1.50 m), using pre-loading temporary props, different steel props configurations and reinforced concrete T-beam for the diaphragm walls (external and internal).

Finally, the deformable and inflatable plastic clay layer leads to a heavy retaining solution with diaphragm walls of 1.2 m thick with transverse reinforcements of 0.80 thick to limit deformation.

These reinforcements will be realized from the ground level but only the part below the lower slab is made of concrete to block the diaphragm wall before excavation. The rest of reinforcements has been filled with treated cement.

The interaction between the diaphragm walls of the ARC station and the drilled piles of the line B's bridge is one of the most complicated subjects in design. An instrumentation is put on the railways in addition to a reinforced technical coordination with the designers of that bridge. The settlements are limited all around the diaphragm wall according to the French railway standards for the line B (threshold of vigilance: 5 mm, alert threshold: 8 mm, intervention threshold: 13 mm and slowing threshold: 22 mm). The admissible thresholds (in vertical, in horizontal and the slope) for the existing buildings were determined by the sensitivity of these buildings for damage classes (architectural, structural and functional damages).

The length of the diaphragm wall is about 50 m. Some panels are deeper due to the vertical loads created by the supports of the 7-floors building or the line B's structure.

3.3   Internal structures

The slab covering the station ensures the stability at the head of the diaphragm walls. In reinforced concrete, it is of constant thickness of 1.00 m. A free height of at least 1.50 m is left above allowing the passage of the networks.

The internal floors of the station are supported by reinforced concrete columns. In the station, these columns are supported by drilled and cast-in-place foundations. The differential settlement was estimated because the real estate project above the station has different types of foundation.

The floors are composed of 1.00 m and 1.50 m thick reinforced concrete beams and thinner slabs resting on the beams. The floors ensure permanent stability between the diaphragm walls. Longitudinal beams connected to diaphragm walls transfer the earth pressures on the principal beams which allow a self-balancing of the structure. The floors are completed by reinforcements around the openings for lifts, escalators and technical networks of the station.

The 1.00 m thick lower slab is made with an underground in order to counter the swelling of clays. This part of structure is accessible for future inspection with access hatch.

The ducts outside the station (air intake, discharge, ventilation and decompression of the station and tunnel) are concrete structures ~ 40-80 cm thick. The structure of connection with the line B platforms, located 8.00 m higher than the ground floor level of the station, are also concrete structure. The two structures are connected by expansion joints.

Over the entire station, the structure was developed architecturally. Careful consideration has been given to the architectural design of the concrete's texture of the interior columns, a highly visible elements of the design.

All structures are fire-stable 2 hours. Structures and partition walls in relation to other constructions (existing or futures) are 2 hours firebreak.

4 CONCLUSIONS

The design of a metro station in a dense urban area is a real challenge. The station must fit into its environment and respond to a long list of constraints: administrative, technical, architectural and functional.

Engineers, architects and contractors should take into account the existing structures and networks around the works zones and realize a sustainable equipment able to welcome the passengers of the future metro line.

Tunnels and Underground Cities: Engineering and Innovation meet Archaeology,
Architecture and Art, Volume 12: Urban
Tunnels - Part 2 – Peila, Viggiani & Celestino (Eds)
© 2019 Taylor & Francis Group, London, ISBN 978-0-367-46900-9

Design concept of a large TBM tunnel having extensive diameter with pioneering three deck orientation under extreme water pressure conditions

H.T. Tunçay & K. Elmalı
Yüksel Proje Uluslararası A.Ş., Ankara, Turkey

ABSTRACT: 3 Deck Great Istanbul Tunnel Project aims to connect Europe and Asia under the Bosporus via Metro and Highway systems in a single tunnel. The constructability plays a vital role. In the Bosporus crossing section of the project there are many engineering challenges such as; extreme hydrostatic pressure, variable soil conditions, extensive diameter of the tunnel, limitation of the tunnel construction time, design of transition structures and TBM launching and receiving structures including logistic area requirements under dense urban patterns, setting out a common alignment for two different means of transport systems having discrete origins and destinations and a limited time frame for the overall design. This paper will present its readers the encountered problems and developed solutions during the design process.

1 INTRODUCTION

The 3-Deck Great Istanbul Tunnel Project is a combined highway and metro system connecting two continents under Istanbul Strait (the Bosporus) in Istanbul, Turkey. The project consists of 16.5 km highway and 31 km metro line with 14 stations. The Bosporus crossing section of the project is 4.3 km long and through a single TBM tunnel with 16.8 m outside diameter. The tunnel contains 2×2 lane highway and double track metro.

Marmaray commuter rail system immersed tube tunnel under the Bosporus was holding the record of the deepest tunnel in İstanbul below sea level with a depth of 60 m until Eurasia Highway system TBM tunnel constructed at 106 m below sea level. The 3-Deck Great Istanbul Tunnel Project Bosporus crossing tunnel designed at 130 m below the sea level is expected to break the record not only in Istanbul but also many in the world.

The metro section of the project will be located between İncirli (European side) and Söğütlüçeşme (Asian side). In addition, metro line will provide connection to the Istanbul New Airport at the European Side and Sabiha Gökçen Airport at the Asian Side. The highway section of the project will be located between Hasdal (European side) and Çamlık (Asian side). With the realization of the project, travel time between the European Side and the Asian Side will be 14 minutes by using the highway and 6 minutes by using the metro system.

This paper will present its readers the encountered problems and developed solutions for the Bosporus crossing section of the project during the design process to achieve a robust design under challenging conditions.

2 AIM OF THE PROJECT

The Ministry of Transport and Infrastructure General Directorate of Infrastructure Investments (UAB-AYGM) conducted a study to improve transportation capacity between two continents in Istanbul and developed a mega transport project with combined metro and

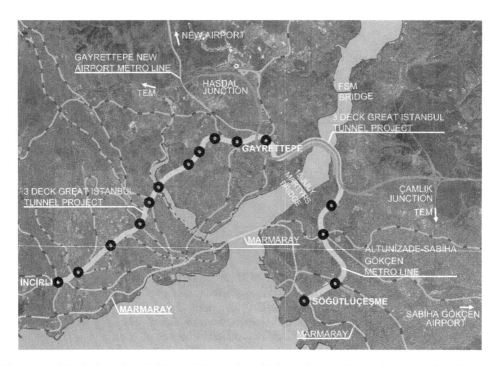

Figure 1.   3-Deck Great Istanbul Tunnel Project both highway (orange line) and metro (yellow line) sections including combined Bosporus crossing (green line) (author).

highway systems. This study was announced to the public in February 2015 (Republic of Turkey Ministry of Transport and Infrastructure 2015). In this study, existing and future transportation demand between the two continents was assessed. According to the transportation survey, demand for the crossing, which was 1.3 million in 2015, was estimated to reach 3.8 million in 2023. Total capacity provided by the existing bridges and tunnels will be

Table 1.   Population of Istanbul and daily transportation demand for intercontinental crossing (author).

	2000	2014	2023
Population*	10	14	17
Daily crossing demand*	0.8	1.3	3.8

* millions.

Table 2.   Existing daily capacity for intercontinental crossing (author).

	2014	2023
15th July Martyrs Bridge	355,000	355,000
Fatih Sultan Mehmet Bridge	475,000	475,000
Yavuz Sultan Selim Bridge	-	475,000
Eurasia Tunnel	-	235,000
Marmaray	125,000	900,000
Metrobus	160,000	200,000
Maritime	235,000	250,000
Total	1,350,000	2,890,000

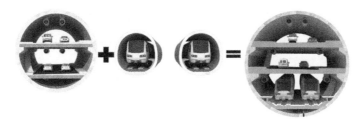

Figure 2.   Unique design of TBM tunnel for combined transport systems (author).

insufficient in 2020. As a result, it was realized that a new high-capacity transportation system is required.

A new high capacity transportation system was developed by the Ministry of Transport and Infrastructure consisting of 16 km long 2x2 lane highway system with a daily capacity of 120,000 vehicles and 31 km long double track metro system having 15 stations with a daily capacity of 1,5 million passengers.

Bosporus crossing section of this project was anticipated as a three-deck TBM tunnel to combine both transport systems in a single structure, which is unique for many aspects in the world.

3   DESIGN DEVELOPMENT PHASE, CHALLENGES AND SOLUTIONS

In the Bosporus crossing section of the project there are many engineering challenges such as; extreme hydrostatic pressure, variable soil conditions, extensive tunnel diameter, limited tunnel construction time, design of transition structures and TBM launching and receiving structures including logistic area requirements under dense urban patterns, setting out a common alignment for two different means of transport systems having discrete origins and destinations and a limited time frame for the overall design.

3.1   *Geophysical survey and soil investigations*

Due to limited design time frame of ten months, all design activities were scheduled as fast track. Site survey and soil investigations on the Bosporus were the critical activities of the schedule. Obtaining working permits, mobilization, performing survey and investigations, and evaluation of results was occupying a major part of the time frame. Therefore, these activities should be performed once on the most suitable alignment corridor.

Based on former studies of three existing tunnel projects under the Bosporus (Marmaray Commuter Railway Project, Eurasia Highway Tunnel Project and Melen Water Tunnel Project), there are mainly two types of soil encountered as Holosen sequence consisting of sand, silt and clay deposits on top underlined by Trakya formation consisting of sandstone, siltstone, mudstone and shale alternation with rare conglomerates. TBM tunnel need to be excavated in the Trakya formation because of its large diameter. Therefore, optimum alignment should pass through higher rock levels to keep the depth as low as possible.

Geophysical study to investigate the seafloor morphology of the Strait of Istanbul carried out by Hacettepe University (Gökaşan et al., 2006) contributed significantly to estimate the location of higher rock levels. For the verification of rock levels, geophysical site survey was conducted in the area determined based on that study. 1.5 km2 bathymetrical measurements and 54.22 km of high-resolution geophysical measurements were made and the results were compared. It was realized that rock levels along concept design corridor are deeper than 250 m below sea level which is also consistent with the previous geophysical study. As a result, significant rerouting for the corridor became necessary. Two new corridor alternatives were developed where the maximum rock levels are around 130 m.

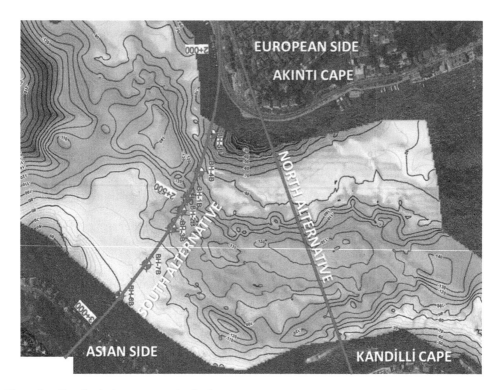

Figure 3. Geophysical survey map (author).

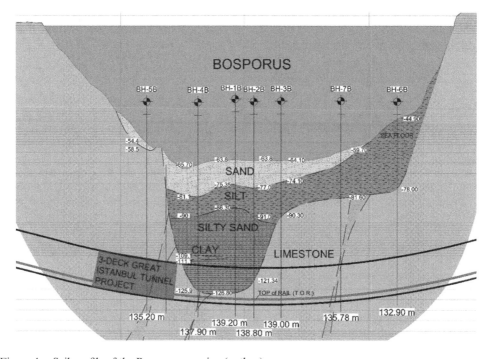

Figure 4. Soil profile of the Bosporus crossing (author).

Rock levels are observed at 135-140 m below the sea level at the north alternative and 120-130 m below the sea level in the south alternative. Borehole locations for soil investigations were identified on the south alternative, which is advantageous due to its shorter length and lowest depth for rock level.

The RV Fugro Scout, research and survey vessel, started soil survey works in Bosporus on 28 July 2017 to investigate the soil characteristics and rock levels. In the scope of this investigation, seven borehole drillings around 80 m below the sea floor were conducted. As a result, the deepest rock level on the alignment was determined as 126.80 m below the sea level and the correlated and idealized geological profile was revealed.

3.2 *TBM tunnel design and optimization of cross section*

Under consultancy of Tunnelconsult Engineering Ltd., manufacturers of large TBM machines all over the world were identified as the potential manufacturers. In this context, Herrenknecht from Germany, Hitachi Zosen and JIMT from Japan were consulted. Immediately after the soil investigation results were obtained the data was shared with these potential TBM manufacturers to allow them to investigate the possibility of manufacturing the largest TBM machine in the world and constructability of the tunnel under these soil conditions and hydrostatic pressure of 13 bars. Results of their investigations were discussed in a series of meetings and their comments and confirmations were obtained to take into the consideration in the design. Based on manufacturers' recommendations and geological investigation results, tunnel alignment was located mainly in Trakya formation except for 300 m long section at the sag point in impervious sandy clay formation.

At the same time, structural analysis was conducted by Tunnelconsult Engineering Ltd. to determine segment thickness and orientation. According to that study, segment thickness was determined as 65 cm and segment orientation determined as 10+1 segment including key.

In the concept design, metro tracks were in the middle deck and highway lanes were at the top and the bottom decks of the tunnel. With this orientation, since the provision of vertical structural elements was not possible, required middle deck thickness increased due to heavy metro system loads. Vehicle clearance requirements with this orientation and increased deck thickness resulted in a tunnel diameter of 19 m which may exceed the constructability limits of the existing technology.

A new orientation was developed to decrease the tunnel diameter by locating the metro system at the bottom deck. With this new orientation, vertical structural elements were provided to decrease the span length and the thicknesses of intermediate decks. As a result, the outer diameter of the tunnel was decreased to 16.8 m.

In this project, a pilot tunnel was introduced. It was anticipated to run in parallel to 3 Deck Tunnel. This concept was proposed to identify the risks such as; unexpected soil conditions,

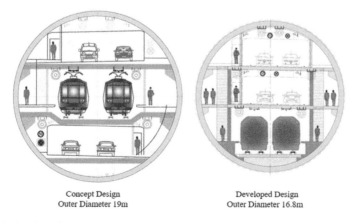

Concept Design
Outer Diameter 19m

Developed Design
Outer Diameter 16.8m

Figure 5.   Optimization of tunnel cross section (author).

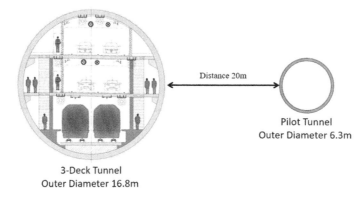

Figure 6.   Pilot Tunnel (author).

inactive faults crossings etc. that might be encountered during the TBM drive of the three-deck tunnel. Manufacturing period of the required conventional TBM and construction period of a pilot tunnel would be much shorter than the three-deck tunnel due to its small dimensions. By this way, it would be possible to construct the pilot tunnel onsite prior to the commencement of three-deck tunnel construction. The pilot tunnel would contribute significantly to identification of exact soil conditions to be faced on the alignment, determination of soil improvement measures and implementation of soil treatments from pilot tunnel prior to the main TBM operations. In the future, this pilot tunnel would be used for relocation of overhead energy transmission lines, which cause visual pollution on the Bosporus.

3.3   *TBM launching-receiving structures and transition structures.*

Among the many challenges of the project, limitation of tunnel construction time of the main TBM, design of transition and TBM launching-receiving structures including logistic area requirements under dense urban patterns were three important ones to overcome in the design.

In the concept design, the main TBM tunnel length was 6.5 km. Tunnel length was decreased to 4.3 km by positioning launching-receiving structures close to the Bosporus as much as possible. TBM advance rate was estimated as 4 m per day. Reduction of 2.2 km in length resulted in a decrease in construction time by 1.5 years and considerable savings in the total cost of the tunnel consequently.

Due to extensive space requirement which was not possible to construct by tunnelling method, a cut & cover structure was required to provide the transition between three-deck tunnel section and tunnels of metro system and two-deck tunnel of highway system. Transition length required in the concept design was around 300 m. With the new orientation of the cross-section, transition length was decreased to 80 m, which resulted in less land requirements. The depth of the transition structures was maintained at around 50 m, which allowed sufficient soil cover on top of the main tunnel.

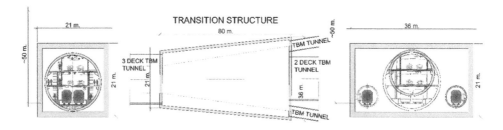

Figure 7.   Transition and TBM Launching and Receiving Structure (author).

Transition and TBM Receiving Structure
European Side

Transition and TBM Launching Structure
Asian Side

Figure 8. Layout of Transition and TBM Launching and Receiving Structures (author).

It was decided that transition structure could also be used for launching-receiving operations of TBM during construction stage. Having the advantage of a single structure for launching-receiving operations of TBM and transition with reduced length, it became possible to locate these structures within the municipal areas along the route without any need for land acquisition.

3.4 *Alignment design*

Great effort was spent to set out a common alignment by using the most critical geometrical design requirements for two different means of transport systems having discrete origins and destinations in compliance with structural, geotechnical and tunnelling requirements such as sea crossing corridor at high rock levels, suitable locations for transition structures and

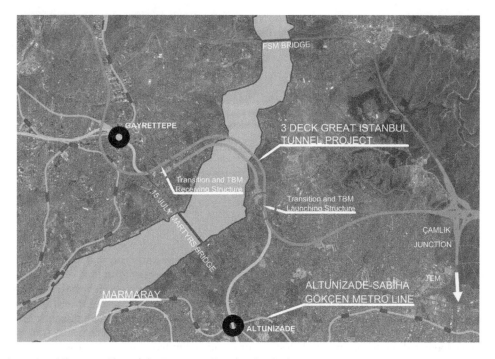

Figure 9. Alignment Plan of the Bosporus Crossing (author).

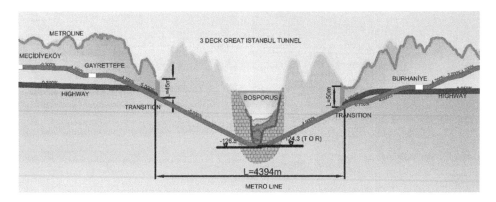

Figure 10. Alignment Profile of the Bosporus Crossing (author).

Figure 11. artistic rendering of the cross section of the tunnel (author).

nearby metro stations to be integrated to existing metro lines on each side at desired elevations.

4 CONCLUSION

3 Deck Great Istanbul Tunnel Project aims to connect Europe and Asia under the Bosporus via Metro and Highway in a single tunnel. In the Bosporus crossing section of the project there are many engineering challenges such as; extreme hydrostatic pressure, variable soil conditions, extensive tunnel diameter, limited tunnel construction time, design of transition structures and TBM launching-receiving structures including logistic area requirements under dense urban patterns, setting out a common alignment for two different means of transport systems having discrete origins and destinations and a limited time frame for the overall design. Due to limited design time frame, all design activities were scheduled as fast track.

TBM tunnel diameter was minimized by reorientation of metro and highway decks and inner structural elements. Potential TBM manufacturers investigated constructability of the largest TBM tunnel under extreme hydrostatic pressure. To eliminate the effects of hydrostatic pressure around 13 bars to be faced during construction, tunnel alignment was positioned in low permeable rocks and impervious soil layers. Tunnel length was minimized by locating transition and TBM launching-receiving structures close to the Bosporus as much as possible. TBM Launching-receiving area and transition length requirements for transport systems were optimized. A single structure was designed to reduce construction area within municipal properties to avoid land acquisition.

As a result, the 3-Deck Great Istanbul Tunnel Project stands out as an exceptional state-of-the-art engineering design project with its unique characteristics.

REFERENCES

Gökaşan, E. et al. 2006, Factors controlling the sea floor morphology of the Strait of İstanbul: Evidences of an erosional event after last glacial maximum, *Journal of the Earth Sciences Application and Research Centre of Hacettepe University*, 3(27): 143–161.

Republic of Turkey Ministry of Transport and Infrastructure 2015, *İstanbul'un yeni mega projesi, 3 Katlı Büyük İstanbul Tüneli.*

Tunnels and Underground Cities: Engineering and Innovation meet Archaeology,
Architecture and Art, Volume 12: Urban
Tunnels - Part 2 – Peila, Viggiani & Celestino (Eds)
© 2019 Taylor & Francis Group, London, ISBN 978-0-367-46900-9

Introduction of an extra-large undersea shield tunnel in composite ground: Maliuzhou traffic tunnel in Zhuhai, China

S.Y. Wang & H.M. Wu
Shanghai Tunnel Engineering Co., Ltd., Shanghai, China

Y. Bai & C. Zhou
Tongji University, Shanghai, China

ABSTRACT: Large-diameter (with diameter above 12m) shield tunneling underwater usually encounters more difficulties compared with small-section tunnels. Maliuzhou Traffic Tunnel in Zhuhai is the third major link between Hengqin Island and the mainland, which will put an end to the isolation problem in typhoon seasons. This linkage is able to promote tourism in Zhuhai and the long-term development of the city and can also help saving human lives in extreme climates. Meanwhile, the construction of the Maliuzhou Traffic Tunnel has conquered the challenges in designing and building an extra-large shield tunnel in composite ground and undersea circumstances. This paper introduces the construction strategies and main technical innovations in this project. The designing methods and constructing approaches used in this project will provide theoretical support and technical guidance to similar projects.

1 INTRODUCTION

Shield method has become a competent tunnel construction method, especially for large-section underwater tunnel. After numerous engineering practices plus technical advances, shield tunneling is now spoken highly of by tunnel industry, for its high safety, high mechanization level and high adaptability to difficult ground. Large-diameter(usually with diameter above 12m) shield tunneling underwater usually encounters more difficulties compared with small-section tunnels(with diameter below 10m), such as face stability control, cutter wear mitigation in long-distance shield tunneling, water-proofing and shield seal (Min et al., 2015, Zhu et al., 2018). It is expected that more engineering practice will contribute to the improvement of shield technology. This paper introduces the construction of a super large diameter shield tunnel, the Maliuzhou Traffic Tunnel in Zhuhai, China. The tunnel design, shield construction difficulties as well as main technical properties are detailed.

2 PROJECT OVERVIEW OF MALIUZHOU TRAFFIC TUNNEL

2.1 *Basic Tunnel Design*

Maliuzhou Traffic Tunnel, with a diameter of 14.93m, is a twin-tube, single-floor, six-lane dual-carriageway tunnel. A Herrenknecht slurry-balanced shield is used in the construction. The shield launches at the working shaft on the south coast. The boring machine makes a U-turn in the working shaft on the north coast to continue to excavate the east line tunnel after finishing the west line. The plan layout and longitudinal profile of Maliuzhou Traffic Tunnel is shown in Figure 1. The tunnel has an outer diameter of 14.5m, and an inner diameter of 13.3 m. The segment ring has a width of 2.0m, which is made of concrete with the strength in Grade C55 and the impermeability in Grade P12. The segments are generally wedge shaped and assembled

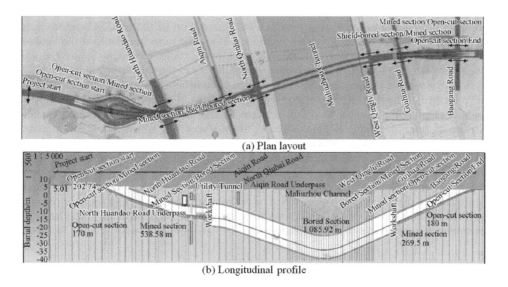

(a) Plan layout

(b) Longitudinal profile

Figure 1.   Plan layout and longitudinal profile of Maliuzhou Traffic Tunnel.

with staggered joints. The west and east tube have the lengths of 1,082m and 1,081m respectively; meanwhile, the lateral clearance between the two tubes is 6.5 to 15m. The tunnel has the minimum alignment radius of 1,170m, and the burial depth of 8.5 to 15.5m.

2.2 *Geological Conditions*

The geological condition of the project is composite with soft soil on the top and hard rock at the bottom, which is a typical ground in the south part of China. The TBM tunnel crosses the Maliuzhou Channel through the composite ground consisted of ②2 clay, ②3 medium coarse sand with clay, ②4 silty clay, ④medium coarse sand, and ⑤gravelly clayed soil. Additionally, in some regions, the ground contains shallow-buried high strength rock including ⑥fully, strongly and moderately weathered granite. Meanwhile, numerous obstacles are observed at the tunneling working face, such as plastic drainage panels, rip-rap stones and spherical weathered rock (boulder). The maximum uniaxial compressive strength of the bedrock is up to 80.4 MPa, and the average value is 43. 8 MPa. The uniaxial compressive strengths of rip-rap stones are generally larger than 60 MPa, with the maximum value being 120 MPa. The Maliuzhou Channel is about 600 m wide, and the water level difference is about 1. 5 m due to sea tides and can even reach 3 m in rainstorm seasons.

3   ENGINEERING CHALLENGES AND CONSTRUCTION STRATEGIES

3.1 *Engineering Challenges*

As the first extra-large undersea shield tunnel in composite ground in China, Maliuzhou Traffic Tunnel has the following challenges:

(1) The composite ground has soft soil on the top and hard rock at the bottom. In some sections of the tunnel alignment, excavating faces are located in composite ground with soft clay, medium coarse sand, gravelly clay and granite. The cutting tools must be able to cut both soft and hard stratum. Moreover, the controlling parameters of the tunneling must ensure a smooth driving of the shield and avoid the consequences of the shield getting jammed or clay becoming clogged on the cutter head.

(2) There are numerous obstacles along the tunnel alignment. Along the axis, there used to be a municipal road. The vacuum pre-loading drainage consolidation method and rip-rap

compaction method are used to strengthen the foundation of the road. Therefore, lots of plastic drainage panels and riprap stones are left along the tunneling route at the land area. Meanwhile, there are also lots of spherical weathered rocks (boulders) along the whole alignment, which has brought unprecedented challenges to the shield tunneling.

(3) New slurry mixture design is applied in undersea circumstances. Since the Maliuzhou Channel is located at the Pearl River estuary, with the contents of chloride, the level of magnesium ion and calcium ion is higher than that in normal water. These contents will cause the deterioration of the slurry and decrease the quality of the slurry film, which endangers the stability of the excavating face. Thus, it is critical to design a new slurry mixture.

(4) As the first extra-large shield tunnel in South China reaching 14m, this presents difficulties in areas such as shield launching and receiving, tunneling under shallow overburden, crossing under embankment and controlling the stability of the composite ground, which all contribute to the increased risks of shield tunneling. Besides, the strong rocks in the region aggravates the wear of cutter heads and cutting tools, yet it is extremely risky to replace cutting tools under large pressured underwater.

3.2 Construction Strategies

In order to deal with the engineering challenges, the construction team have developed several methods such as using targeted design of the shield, applying excavation pretreatment, and improving the ground adaptability of cutting tools.

3.2.1 Targeted Shield Design

The cutter head of the shield has a diameter of 14.93m and an opening ratio of 57%. The cutter is composed of 127 disc cutters, 182 scrapers, 43 pre-cutting cutters, 16 peripheral scrapers, 8 diameter maintaining cutters, 1 copy cutter, 5 tool wear detection devices and one device for detecting the abrasion of the cutter head steel structure. Meanwhile, the shield is equipped with a Seismic Scattering Profile (SSP) advanced detection device to collect and evaluate the data of the front ground conditions continuously during the shield tunneling. Furthermore, it is able to detect obstacles with a diameter larger than 800mm in the scope of 40m ahead of the cutter head. The layout of the cutting tools is shown in Figure 2 and the sketch of the SSP advanced detection is shown in Figure 3.

Figure 2. Layout of cutting tools.

Figure 3. Sketch SSP advanced detection.

High performance jaw crushers (Figure 4) crush the stones entering into the slurry chamber and prevent them from accumulating at the mud discharge pipe inlet, thus ensuring smooth transportation of the muck. The crushing efficiency of the crusher is in concert with the driving speed of the shield. To prevent large gravels from clogging the mud discharging pipes, there is a gravel trapper (Figure 4) in front of the muck pump to collect large gravels entered from the inlet.

3.2.2 *Excavation Pretreatment Combination*
When a TBM drives through composite ground, the cutterhead and cutting tools are abraded rapidly due to stiff bedrock and boulders. When this occurs, frequent interventions are needed to replace the cutting tools. If the shield gets jammed, the cutter head is damaged. If the intervention condition is not met, the construction efficiency is greatly affected. Therefore, the pre-excavation reinforcement measures are taken to avoid these risks and to ensure smooth tunneling.

3.2.3 *Elaborate Geological Survey*
Geophysical methods that frequently used for land physical detection, such as shear wave reflection method and cross hole electrical resistivity tomography, are not applicable in the Maliuzhou Channel areal. In order to determine the distribution of bedrocks and boulders in the undersea section, the construction team uses the SSP method to survey the geological conditions above water (as shown in Figure 5). The geological profiles and 3D wave velocity database of the surveyed area are developed by using the SSP method. It is discovered that there are 25 boulders/bedrocks distributed along the TBM driving route, which are classified according to their sizes and strengths. There are nine boulders/bedrock with a diameter larger than 3m and hardness over Grade 2 (i. e. in which wave velocity exceeds 2 500m/s). Based on the survey results, the basic distribution of boulders and bedrocks within the undersea section are determined, which provides accurate position information for the subsequent drilling-and-blasting processes.

(a) (b)

Figure 4.   (a) Jaw crusher; (b) Gravel trapper.

(a) (b)

Figure 5.   (a) Wave detection cables; (b) Seismic data collection.

3.2.4 Underwater Detonation

Based on the geophysical results, the bedrocks intruded in the tunneling face under the channel are all dynamited into pieces. Blasting holes are drilled using a drilling machine that is fixed on the work vessel. A steel casing pipe is hammered into the stone in advance, followed by a PVC pipe that is pressed in the steel casing pipe to support the hole after it is excavated. Milli-second Nonel detonators are used in the blasting holes, and instantaneous electric detonators are used to trigger the explosion. The standard diameter of the emulsion explosive (Figure 6) is 60mm. The blasting hole spacing is 0.8 to 1.2m. In order to ensure the quality of blasting, the depth of the blasting hole below the tunnel bottom is no less than 1.5m. The underwater blasting is shown in Figure 6.

3.2.5 Blasting Hole Sealing

Once a complete passage is formed in the ground through the blasting holes, slurry gushing tends to happen during the shield tunneling. Therefore, the blasting holes and the passages formed afterwards must be sealed. For the blasting holes drilled by machines, small explosive packs are installed in the ground above the tunnel. The amount of explosive is controlled to be just enough to break the PVC pipes, so that the PVC pipes would not form a passage connecting the riverbed and the tunneling face. The passage formed by the pulling out of the steel casing pipes are sealed by pouring expansive soil balls and stones when pulling out the pipes. The blasting hole passages formed after blasting in the under-channel sections are sealed by reinforcing the riverbed with high pressure rotary grouting. The rotary spraying rigs are carried by modified vessels. The blasting hole passage sealing under the Channel is shown in Figure 7.

3.2.6 Improvement of ground adaptability of shield cutting tools

In order to deal with the silt ground and plastic drainage panels existed extensively in the land section, before the shield launching, 46 of 127 disc cutters are replaced by shell bits, which are cooperated with the pre-cutting cutters to excavate. In order to ensure the efficiency of the excavation in the rock sections under the Channel, during the time that the shield is under

(a)   (b)

Figure 6.   (a) Emulsion explosive and detonators fixing;(b) Underwater blasting.

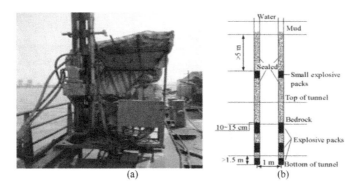

(a)   (b)

Figure 7.   (a) Blast hole sealing; (b) Blast hole passage sealing under channel.

(a) Arm #7 Disc cutter #83   (b) Arm #11 Disc cutter #83   (c) Arm #15 Disc cutter #83

Figure 8.   Under-channel inspection of the condition of cutter tools in the excavation chamber.

examination in the south embankment reinforced area, 23 of the 46 shell bits in the central area of the cutter head are replaced by disc cutters, while the rest of the shell bits remain unchanged. The pretreatment measures ensure smooth boring in the west line rock area, and avoids unpredictable accidents such as shield jamming and riverbed collapsing. When the shield is crossing the first rock section, the driving speed is controlled at about 10mm/min; after the third rock section, the driving speed increases to about 20mm/min. In that case, the torque and thrust are still relatively stable. The length of most of the muck transported is within 10cm, and a few reaches 20cm.

After completing the west line, the examination shows that 40 of the 104 disc cutters were unevenly worn. The damage is distributed in 3m range of the cutter head periphery, and the largest uneven wear amount is about 40mm. The cutting rings of 7 disc cutters are dropped off, which are distributed at the junction of the front surface and the surface of the cutter head. Besides, all scrapers are severely worn. According to the driving parameters and the abrasion indices of the cutting tools in the west line, the cutting tool configurations are improved before the shield launching in the east line,after the following adjustments are made: (1) All cutting tools in 3m range of the cutter head periphery are switched to 18-inch heavy disc tools; (2) Scrapers are switched to better ones with inlaid alloy; (3) 33 shell bits are added on the front surface of the shield at 10mm lower than the 17-inch disc cutters, with the inlaid alloy remaining the same as shell bits in the central area, to assist the disc cutters.

The driving speed of the modified shield is controlled to be 20mm/min to cross the first rock section of the east line, and the muck transportation is efficient. In the second rock section of the east line, rocks that intruded in the tunneling face are as high as 5.8m and are more intact and of higher strength. In order to check the wear condition of the cutting tools, and to ensure that the shield can cross the second rock section of the east line smoothly, an intervention chamber is constructed in the improved ground between the 210th ring and the 219th ring. The intervention, including the inspection in the excavation chamber and the changing of cutting tools, lasts 16 days. During that period, 14 disc cutters, 10 sets of peripheral scrapers, and 4 front scrapers are changed. With the driving speed of 6mm/min, the torque after intervention is significantly reduced. The intervention under the Channel ensures that the shield passes the second section smoothly and is able to move forward continuously until breakthrough.

4   CONCLUSIONS

To meet the challenge of having "soft soil on top of hard rock", a result of this project's unique situation of being in an undersea environment and composite ground, a number of technical innovations are developed and applied. A shield design in composite ground and excavation pretreatment are utilized to minimize possible tunneling problems. A series of control techniques are also developed for the composite ground during the shield tunneling. These techniques ensure the success of the tunnel excavation and promote the technical development of extra-large shield tunnels.

REFERENCES

Min, F. and W. Zhu, et al. (2015). *"Opening the Excavation Chamber of the Large-Diameter Size Slurry Shield: A Case Study in Nanjing Yangtze River Tunnel in China."* Tunneling and Underground Space Technology 46: 18–27.

Zhu, Y., Li, L.WuH. (2018). *Extra-Large Undersea Shield Tunnel in Composite Ground: Maliuzhou Traffic Tunnel in Zhuhai.* Tunnel Construction, 38(3): 494–500.

Tunnels and Underground Cities: Engineering and Innovation meet Archaeology, Architecture and Art, Volume 12: Urban Tunnels - Part 2 – Peila, Viggiani & Celestino (Eds)
© *2019 Taylor & Francis Group, London, ISBN 978-0-367-46900-9*

Research on the construction control for the metro tunnel under passing buildings in Jinan

X.Y. Xie, Y.L. Zhang, B. Zhou & L. Zeng
Department of Geotechnical Engineering, Tongji University, Shanghai China

ABSTRACT: The tunnel of Wang-Pei interval underpasses 6 residential buildings in a concentrate manner. The construction time of buildings is mostly from the 70s to the 80s of the last century. It's very difficult to control subsidence and protect building. This paper uses PLAXIS 3D to simulate the construction. The calculation results and the monitoring data are compared. The results show that: (1) the building is slightly raised firstly; (2) then rapidly settles when tunnel passes; (3) slightly lifted before settlement stable; (4) the building is slightly inclined towards the central axis of the tunnel; (6) the maximum settlement is 14.1mm, which is consistent with the initial monitoring data. And this project adopts monitoring program of automatic and manual monitoring, and developed a real-time detection technology for grouting behind segment by using radar, timely grouting again to reduce the ground loss. Eventually, "micro settlement" was basically achieved and the building was well protected.

1 INTRODUCTION

With the rapid development of urban construction in Jinan, the problem of urban traffic congestion is becoming increasingly serious, and the establishment of urban high-speed track transportation network has become the preferred way. Shield tunnels have become the better choice of subway construction because of their safety, high efficiency, environmental protection and other advantages. However, due to the complex urban environment and its own construction methods, subway tunnels often have to go through municipal pipelines or existing buildings. It is very easy to cause surface uplift or subsidence, lightly causing uneven settlement or tilt of buildings, and seriously causing building cracking, even endanger the safety of people's lives. Therefore, strict construction measures are needed to reduce settlement and ensure the risk can be controlled. (Bai et al., 2014, Zhu et al., 2014)

There are many related researches on the settlement control when tunnels underpass buildings. The research methods can be roughly divided into empirical formula method (Peck, 1969), theoretical analysis method (Lu et al., 2007), numerical simulation method (Si, 2013) and model test method. Xie X.Y. (Xie et al., 2016) introduced the settlement controlling of the Nanning Metro Line 1 crossing the railway station, and used the three-dimensional finite element software MIDAS/GTS to establish a numerical model to analyze the values of the equivalent layer parameters. Finally, the automatic monitoring system was proposed. The system combines the instant messaging platform to realize the automatic processing and release of monitoring information, which provides a good solution for the shield construction management system. Zhang et al. (Zhang et al., 2016) used the FLAC3D software to analyze the construction parameters of the S-line large-diameter mud-water shield and the settlement of the buildings with the FLAC3D, and compared the numerical calculation results with the on-site monitoring. Then they proposed corresponding settlement control measures. Wang Jian (Wang et al., 2016) discussed some risks of shield tunnel construction in Jinan rail transit network. Through numerical simulation of rail transit

R1 line shield construction, the influence of different shield tunneling construction parameters on the ground surface was analyzed. And he proposed the settlement risk control system of the shield crossing the bridge to reduce the environment impact of tunnel construction.

The Wang-pei interval of Jinan Rail Transit R3 line mainly passes through the water-rich silty clay stratum. The soil and the hydrogeological conditions of the project have high quality. However, the buildings, just like Suning Appliance building and dormitories of fertilizer factory, have a long history, which the weathering of the wall is serious. Thus the ground subsidence control and building protection are difficult. In this paper, the large-scale three-dimensional finite element software PLAXIS 3D is used for numerical simulation to analyze the surface deformation law of tunnels under passing buildings, and combines with the real-time grouting detection system and automatic settlement monitoring system to control the construction in real time and adjust the construction parameters in time. With strict control of risks, the goal of "zero settlement" has basically reached.

2 ENERGING BACKGROUND

The section between Wangsheren Station and Peijiaying Station on R3 line of Jinan urban rail network has two single tunnels with parallel arrangement, and the EPB shield has been used. On the plane, the tunnel starts from Wangsheren Station and moves eastward along the North Industrial Road, then turns northward through the curve section with a radius of 700m and a radius of 1500m, and then advances to Peijiaying Station through villages and farmlands. The interval tunnel through the moat and the Longji River, partially under passes the Agricultural Bank, Suning Application Building, and Jinan Fertilizer Factory Dormitories, as shown in the Figure 1.

The outer diameter of the straight section of the tunnel is 6.4 m. The lining width is 1.2 m, which thickness is 0.3 m. The distance between the two tunnels is about 14 m ~ 16.6 m. The covering soil thickness at the top of the structure is about 10.5 m ~ 21.9 m, and the groundwater level is about - 6.3 M. The thickness of the filling layer varies greatly from 0.5m to 6.5m due to the uneven distribution of the strata along the way, and the underlying strata are 9-1 silty clay, 10-1 silty clay and 14-1 silty clay, respectively. The underlying bedrock of the site is diorite and limestone, with different buried depth, rock integrity, weathering degree and strength. The mechanical parameters of each soil layer are shown in Table 1. Tunnel mainly passes through silty clay layer. According to field pumping test, the comprehensive permeability coefficient of silty clay layer is 4.3 *10^{-3} ~ 5.49 *10^{-3} cm/s, and the permeability is strong. The longitudinal profile of Wang-pei line is shown in Figure 2.

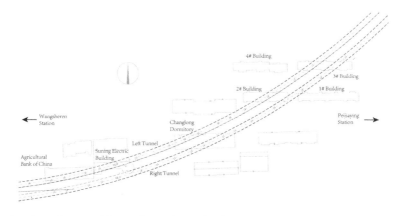

Figure 1. location of Wangpei interval tunnel crossing buildings.

Table 1. Typical soil properties.

name	ρ	μ	W	E_{s1-2}	c	Φ	k
	g • cm^{-3}	-	%	MPa^{-1}	kPa	°	*10^{-6}cm/s
①₁Plain fill	1.94	-	21.7	5.82	18.13	14.14	-
①₂Miscellaneous fill	1.95	-	-	-	-	-	-
⑨₁Silt clay	1.98	0.3	24.4	5.25	30.6	12	2.35
⑩₁Silt clay	1.96	0.35	25.6	5.58	31.7	11.7	8.07
⑭₁Silt clay	1.93	0.25	25.7	6.19	33.9	12.5	5.91
⑭₄Gravel	2.07	0.2	-	20	-	-	-
⑲₁Weathering diorite	1.84	0.25	34.2	25	15	25	-

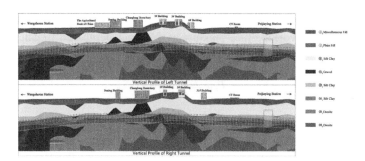

Figure 2. The longitudinal profile of Wangpei metro line.

3 NUMERICAL SIMULATION

The tunnel will first cross the Agricultural Bank of China (ABC) and Suning Appliances Building (SAB), and then pass through the dormitories of fertilizer factory. The area the tunnels pass through is large, with complex ground environment. And it's difficult to make numerical simulation of the whole area. So this chapter will use PLAXIS 3D, finite element software, to simulate the underpass area of ABC and SAB, to study the general settlement law and the method to control settlement, which can provide reference for construction.

PLAXIS 3D is a three-dimensional finite element analysis software for small deformation and stability problems in geotechnical engineering. It can deal with the structure deformation and construction. It has many advanced analysis tool and processing function. Its core analysis theory is reliable and has been verified and approved by international academic and engineering circles for a long time. The latest version of the software, 2017.1, can well simulate the classical TBM construction method. The tunnel editor module can directly set up excavation, grouting, lining and other construction stages. And it also can simulate the face pressure, grouting pressure, ground loss, etc.

3.1 Numerical model

According to the construction plan and monitoring data, the main factors considered in the simulation are as follows:

1. Because the maximum of monitoring value of surface subsidence is less than 12 mm, it's a small deformation problem. The elastic model with Mohr-Column failure criterion is assumed for soil. In order to simplify the model, the homogeneous and horizontal distribution of soil layer are assumed. The plain fill and the miscellaneous fill are combined to, and use the weighted average value of the parameters. The model considers the effect of

groundwater. The saturation gravity is adopted for the soil below the groundwater level. The physical and mechanical parameters of the soil are shown in Table 1.

2. The model section size is 150m × 50m, and the longitudinal length is 200m. The model boundary conditions are set by default: the bottom surface is all constrained; the top surface is free; the four sides are applied with normal constraints.

3. The tunnel radius is 3.05m and the lining thickness is 0.3m. In order to simplify the model and improve the calculation efficiency of the simulation, the lining ring width is assumed to be 3m. The curvature radius of curve segment is 710m. And the model simulates construction process at the distance of 150m. The distance between two tunnels is about 15m, and the right line tunnel advances the left in 100m. The shell of shield machine and lining are simulated by linear elastic shell element. The detailed parameters are shown in Table 2.

4. In order to improve the calculation efficiency, it is assumed that the building and the foundation are coordinate deformation. The building is simplified to a uniform load, and the foundation is assumed for solid unit.

5. the reference value of working face pressure at reference depth can be calculated by equation (1) and the result is 142.6 kN/m². [9]

$$P_{ref} = \sum K_i \gamma h_i \tag{1}$$

$$K_0 = 1 - \sin\varphi \tag{2}$$

Where h is the depth of overlapped soil; γ is the unit weight of soil, kN/m³; φ is the fraction angle.

6. The principle of shield excavation is shown in Figure 3 [8]. The shield shell is conical, which length is 9m and maximum diameter is 6.66m. The ground loss can be simulated by surface contraction with 0.033% increasing along the axis, and the reference point is the shield tail with the maximum contraction value. The radial surface load that deviates from the center of the ring is applied to simulate the effect of grouting.

Table 2. Mechanical properties of materials.

name	γ	μ	E	c	Φ
	kN • m⁻³	MPa⁻¹	kPa	°	
Fill	19.4	0.3	60	18.13	14.14
⑨₁Silt clay	19.8	0.3	52.5	30.6	12
⑩₁Silt clay	19.6	0.35	55.8	31.7	11.7
⑭₁Silt clay	19.3	0.25	61.9	33.9	12.5
⑲₁Weathering diorite	18.4	0.25	250	15	25
TBM shell	247	0.2	2×10⁵	-	-
Lining	27	0.2	3.1×10⁴	-	-
Foundation	24	0.25	2×10⁴	-	-

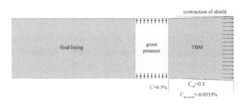

Figure 3. Excavation principle of shield machine.

6294

3.2 *Numerical results*

According to the above calculation model, the law of building settlement is simulated and analyzed, and the influence of different grouting pressure on settlement is compared.

1) deformation results Figure 4 respectively shows the vertical settlement after two tunnels excavation finished. After the right line tunnel passed, the largest settlement is point A, which is 10.4mm. The settlement trough satisfies the law of 'peck' settlement curve. The settlement decreases from the tunnel axis to both sides, and there is a slight uplift at point D. Thus there is some risk of inclination for SAB. After the left line tunnel excavation finished, the settlement increased, but the inclination decreased. The he max settlement value of building is about 14.2 mm.

Figure 5 shows the comparison between the settlement simulation results and the monitoring data of two buildings. JGC-9, 10 and 11 are the manual monitoring curves, and others are the simulation results of settlement. The simulation results show that the settlement is mainly divided into two stages, corresponding to the left and right tunnel crossing successively, which is consistent with the overall trend of the monitoring results, but the absolute value still has errors. Due to the right line tunnel partly crossing building, with point C and D uplifting, the building has a risk of inclination. 2) influence of grouting pressure

Synchronous grouting system of TBM is an important measure to control ground deformation and building subsidence. Figure 6 is the surface subsidence curve of cross section (section where AC point is) within different grouting pressures. It shows that the higher grouting pressure, the smaller vertical subsidence; if the grouting amount is too large, it may cause surface uplift, and even cause building cracking. Settlement value is large in the middle and decreases to both sides, which conforms to peck formula. However, influenced by the building, the difference of ground load and stiffness is obvious. And in lateral, the settlement of the building is obviously larger than that of the other side without building, which leads to the

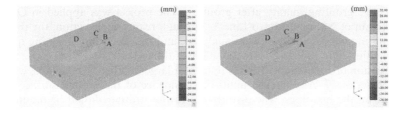

Figure 4.   Vertical displacement contour during tunneling.

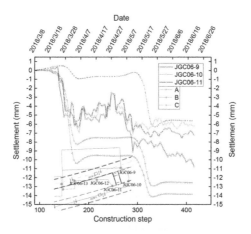

Figure 5.   Surface settlement comparison of simulation results and monitoring data.

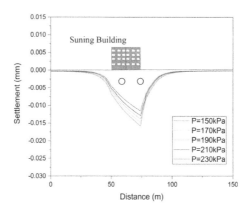

Figure 6. Settlement comparison of AC cross section within different grouting pressure.

risk of inclination. Therefore, the grouting quantity should be increased when the right tunnel under passes, and it can't be increased when the left tunnel passes.

4 REAL-TIME GROUTING DETECTION SYSTEM FOR TBM

Synchronous grouting system of shield tunnel is an important part of shield machine. It is an important measure to control ground deformation and settlement synchronously. However, the traditional method can reduce settlement by setting proper grouting pressure and quantity. The grouting quality is difficult to control and can't be detected real-time.

Zeng L., Xie X. Y. (Zeng, L. & Xie, X.Y., in prep). combined ground penetrating radar and TBM to detective grouting quality after grouting. The project was applied in Qingxiushan-Boyi Road section of Nanning Metro Line 3. Through real-time detection, the thickness and density of grouting behind the wall were analyzed, and the secondary grouting quantity and pressure were adjusted in time to ensure the safety of the project.

Wang-pei section of Jinan metro R3 has to cross a large number of buildings, which are old and weathered seriously. Any uneven uplift or subsidence of the ground surface will have a serious impact on the buildings. The disturbance to the stratum should be minimized in the construction process. Therefore, the radar detection system of grouting behind the lining can be applied in this section to detect the quality of grouting behind the wall in real-time and reach the 'zero settlement'. In this project, the grouting radar equipment is optimized and upgraded. The mechanism design of the second generation equipment is more reasonable, the scanning range is larger, and the algorithm analysis result is more accurate.

4.1 Framework of real-time grouting detection system

The depth of synchronous grouting is only about 1 m behind the lining. The electrical properties of concrete, grouting liquid and soil can be measured by radar. There are obvious differences in the electrical properties of the three media, which includes speed, attenuation rate and relative dielectric constant. At the same time, when the electromagnetic wave passes through the interface of different media, it will produce obvious reflection wave. Therefore, the thickness and compactness of the grouting behind the tunnel lining can be analyzed by using radar to emit and receive reflected electromagnetic wave. And the problem can be found in time, which provides a reference for controlling the grouting quantity.

The synchronous grouting real-time detection system is divided into hardware system and software analysis system. The system framework is shown in Figure 7. Detection mechanism and servo extension are installed on the shield machine truss, where is 4 rings behind the assembly position. The equipment includes track, synchronous belt, transmission mechanism,

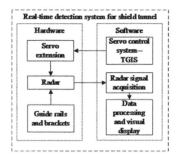

Figure 7. Framework of real-time grouting detection system for tunnel.

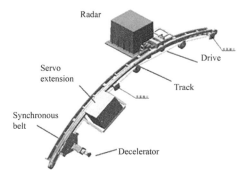

Figure 8. Simulation graph of detection equipment.

detection antenna, servo extension and drive and reducer. The supporting mechanism is composed of several assembled guideways and brackets. The equipment effect diagram is shown in Figure 8.

The software system includes the control software of the mechanism, the radar signal acquisition and processing module, and the visualization software of the grouting test results. The software is installed in the operating room of the shield machine, which the shield driver convenient to use. And, the software interface is very concise and clear. The parameters are already set at the time of debugging without any additional settings during using. Operators can scan the grouting by radar in the excavation break. If the grouting quality behind the lining is poor or there are cavities, they can guide the secondary grouting in time to avoid risk.

4.2 Engineering Application

1) Tunneling and grouting parameters

According to the tunneling parameters of 100 meters after the start tunneling and the test section, combined with the stratum conditions of the crossing section, the construction parameters of the crossing section are set as shown in Table 3 and 4. Cement mortar is used for synchronous grouting. The grouting quantity is 5.5–6.0m^3 per ring. The grouting quantity is adjusted in real time according to the ground monitoring situation to determine whether

Table 3. Proposed values of the tunnel excavation parameters.

Soil pressure (bar)	Thrust (T)	Advance speed (mm/min)	cutting wheel revolving speed (r/mm)	Grouting (m^3/ring)	Excavation (m^3/ring)
1.25	1200	20 ~ 40	1.0 ± 0.1	5.5 ~ 6.0	≤54

Table 4.   Mixture ratios of synchronous grouting.

Volume (m³)	Water (kg)	Cement (kg)	Sand (kg)	Fly ash (kg)	Bentonite (kg)
1	360	150	850	400	100

Figure 9.   The installation of Ground Penetrating Radar Equipment.

secondary grouting is needed. Secondary grouting is an assistant method to reduce the surface subsidence. It can make up for the shortage of synchronous grouting by twice (or more) grouting to the lining around 10 rings after the shield assembled. And it is necessary to inject cement-water glass double slurry to control land subsidence on the basis of synchronous grouting. Secondary grouting slurry ratio is water: sodium silicate = 3:1 (weight ratio); water: cement = 1:1 (weight ratio); cement slurry: sodium silicate slurry = 1:1 (volume ratio). Grouting pressure is generally 0.3 to 0.4MPa.

2) System application

As shown in Figure 9, equipment is installed on the shield machine of right line tunnel. The detection radar equipment moves along with the shield machine, which remotely controlled in the operating room between assembling segments and finishes the scanning.

Table 5.   Grading evaluation results of each sampling point.

Grouting thickness (m)	≤0.092	>0.092& ≤1.65	>0.092
Evaluation of point	Under-grouting	Normal	Super-grouting

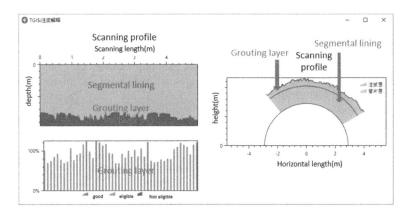

Figure 10.   The 1150st ring grouting interpretation image.

According to the construction scheme, the grouting volume per ring is about 5.5–6.0 m³, so the average equivalent thickness of grouting layer is about 0.14m. It is assumed that 120% and 66% of the average thickness are the control indexes of grouting quality at each sampling point, as shown in Table 5. After getting about 1566 sampling points in each ring, software calculates the equivalent thickness of each sampling point, and counts the number of super-grouting and under-grouting sampling points. When the number of under-grouting sampling points exceeds 250, it is considered that under-grouting in this ring and needs secondary grouting in time; when the number of super-grouting sampling points exceeds 100, it is considered that the ring is super-grouting; otherwise, grouting is normal.

Taking the 1150th scanning results in April 17, 2018 as an example, the visual image is shown in Figure 10. According to statistics, there is zero under grouting point and 67 super grouting points, so we can think that the grouting is normal.

5   AUTOMATIC SETTLEMENT MONITORING SYSTEM

Building SAB and ABC were built new, and the structure was healthy. Conventional manual monitoring could be carried out. However, many old buildings such as the dormitories of chemical fertilizer plant, have a great risk in tunneling, and there are a lot of sundries around them. It is difficult to distribute artificial monitoring points and the frequency is low. Therefore, it is necessary to adopt automatic settlement monitoring.

According to the requirements of construction design and specifications, each building has 4 settlement monitoring points, 2 direction inclination monitoring points and some crack detection points. The monitoring layout of building 2# is shown in Figure 11, for example.

In April-August 2018, continuous automatic monitoring was carried out in the underpass construction. The settlement monitoring data of the No. 2 dormitory of the chemical fertilizer plant can be seen in Figure 12. The instrument is affected by the environmental temperature, and the data fluctuate regularly. The data have great changes when the two tunnels are crossed successively.

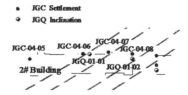

Figure 11.   Layout of monitoring points of building 2#.

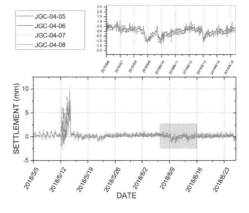

Figure 12.   Monitoring results of settlement for 2# building.

From May 11th to May 15th, the right tunnel crosses the 2# building. Due to the excessive grouting pressure and the work face pressure, the building is heavily uplifted up to 10mm. Then the grout gradually diffuses and the building is gradually restored. From June 6th to June 10th, the left tunnel goes through the same building. Because of reducing the grouting amount, the building has obvious settlement, the maximum settlement is -1.9 mm. The radar detection result of the grouting behind the wall indicates that the secondary grouting is needed and the settlement is gradually restored. After the completion of the double tunnel, the settlement of the building gradually becomes stable, and the total settlement is about -0.5mm.

6 CONCLUSION

Based on the numerical simulation of the buildings under passed by the shield tunnels of Jinan metro R3 line, this paper analyzes the law of settlement and deformation of the buildings, and puts forward the settlement control technology which combines the radar detection system with the automatic monitoring system.

1. Three-dimensional shield tunneling simulation by PLAXIS 3D shows that the settlement trough satisfies the peck settlement rule. And the building inclines to the center of the tunnel when the right tunnel crosses, so the grouting quantity should be increased appropriately. Then the inclination of the building will gradually recover when the second tunnel crosses. It is not advisable to increase grouting to avoid aggravating the inclination of buildings.
2. The advanced type of radar detection system for grouting behind the lining is adopted in the tunnel crossing section, which can detect the grouting quality in the tunneling break. The visual interface is easy to operate and the result of analysis is intuitive.
3. In order to control the risk and ensure the safety of the building, a monitoring scheme combining remote automatic deformation monitoring and manual monitoring is adopted to monitor the settlement and tilt of the building in real time, and feedback the deformation information through the mobile platform in real time. However, the monitoring instrument is obviously affected by temperature, which may make error and reduces the credibility of the data.

REFERENCE

Bai, Y., Yang, Z. H. & Jiang, Z. W. 2014. Key protection techniques adopted and analysis of influence on adjacent buildings due to the Bund Tunnel construction. *Tunnelling and Underground Space Technology*, 41, 24–34.

Lu, H., Zhao, Z. M., Fang, P. & Jiang, X. L. 2007. Analytical method of image theory used to calculate shield tunneling induced soil displacements and stresses. *ROCK AND SOIL MECHANICS*, 28, 45–50.

Peck, R. B. 1969. Deep excavations and tunnelling in soft ground. *Proc.int.conf.on Smfe*, 225–290.

Si, X. Y. Study on the influence of shield tunnel working face balance pressure on soil deformation. National Highway Engineering Geology Science and technology information network 2013 technical exchange meeting, 2013.

Wang, G. F., Wang J., Lu, L. H. & Wang, W. L. 2016. Risk Analysis and Control Research of the Subway Shield Originating Construction. *Construction Technology*, 45, 91–95.

Xie, X. Y., Wang, Q., Liu, H., Li, J. & Qi, Y. 2016. Settlement control study of shield tunnelling crossing railway station in round gravel strata. *Chinese Journal of Rock Mechanics and Engineering*, 3960–3970.

Zhang, Y. Z., Wang, S. G. & Min, F. L. 2016. Analysis and Control of Ground Settlement Caused by Slurry Shield Tunneling Crossing Buildings in the Weisanlu Yangtze River Tunnel. *Journal of Disaster Prevent and Mitigation Eng*, 959–964.

Zhu, H. H., Ding, W. Q., Qiao, Y. F. & Xie, D. W. 2014. Micro-disturbed construction control technology system for shield driven tunnels and its application. *CHINESE JOURNAL OF GEOTECHNICAL ENGINEERING*, 36, 1983–1993.

Tunnels and Underground Cities: Engineering and Innovation meet Archaeology, Architecture and Art, Volume 12: Urban Tunnels - Part 2 – Peila, Viggiani & Celestino (Eds)
© 2020 Taylor & Francis Group, London, ISBN 978-0-367-46900-9

Undergrounding the railway line and mending an extremely anthropized territory maintaining the railway in operation

F. Zambonelli & G. Li Puma
Italferr S.p.A, Rome, Italy

ABSTRACT: This article describes the solution chosen for the construction of the "Nodo di Palermo" artificial railway tunnel, whose layout was to follow an existing railway line crossing a highly urbanized area (interfering therefore with important road axes, regulated by level crossings) and was to ensure continuity of the railway operation. This led to the simultaneous burial of the railway line with the doubling operation. The narrowness of the available spaces, not sufficient for a deviation of the track compatible with the final arrangement of the underground work, led to an artificial tunnel solution. The "cut & cover" method was used. Artificial tunnel cover slabs were divided into two portions, essentially half in the longitudinal sense of the tunnel and realized in two phases, ensuring the possibility, through temporary deviations, to keep the railway line in operation. The article shows the status quo ante, the design choices and main executive phases.

1 INTRODUCTION

The contract "Nodo di Palermo" consists in the realization of three sections (A, B, C) of railway (Figure 1). This paper focuses on "C" section. The project scheduled two types of railway lines, one of which on the surface, and the other underground with both single-barrel tunnel and double-barrel one; the entire project also scheduled the suppression of twenty-two level crossings, the construction of one new station and nine new train stops.

Open sky works (trenches, artificial tunnel, stations/train stops, entrance well) to be executed in high density urban context, as Palermo, interferes with countless restrictions and conditionings: confined spaces, utilities, accesses to buildings, commercial activities. Moreover, the choice has been taken of keeping railway in operation during the renewal. The narrowness of the available spaces, not sufficient for a deviation of the track compatible with the final arrangement of the underground work, led to the adoption of an artificial tunnel solution performed through several phases and effectively allowed the existing railway line to be maintained in operation during all phases. The intervention was carried out following the so-called "cut & cover" method, which allows the rehabilitation of the territory before the complete construction of the tunnel (after the realization of the lateral bulkheads and the roof slab). Furthermore, in order not to disrupt the existing railway line and allow for temporary line deviations, the casting of the roof slab was divided into two slots, each one half of the transversal section of the tunnel.

In the paper, the secant pile technology, the cut & cover method and the odex pile technology and their application to this intervention are described in section 2 and 3 4. In section 5 the traffic during works is discussed. Main conclusions are summed up in section 6.

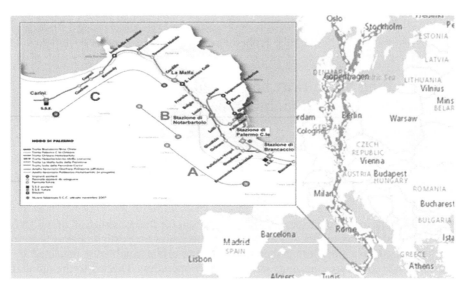

Figure 1. Scandinavian – Mediterranean corridor (TEN T) and "NODO DI PALERMO" focus.

2 USE OF CASE SECANT PILES TECHNOLOGY (CSP)

Case Secant Piles (CSP) (Bringiotti, 2006; Bringiotti, 2010) was adopted to support the excavation of the tunnel.

The adoption of this technic, rather than the usage of diaphragm walls as retaining system with cut and cover method, revealed to be strategically decisive and effective, even over the expectation.

Indeed, although its intrinsic limits (limited resistance to bending moment against rectangular sections, depth of 33–35 meters hardly achievable), this technique presented numerous advantages, namely a better and well organized layout of the construction site, with absence of tanks and fluids, clean areas, reduced vibrations; absence of misalignment, good verticality within tolerances; eventually the quality of surface after excavation allowed for the usage of walls without any casting internal final lining. A picture of the "NV 22" artificial tunnel during excavation is shown in Figure 2.

Figure 2. NV 22 artificial tunnel.

3 USE OF HALF SLAB IN HIGH DENSITY URBAN CONTEXT

In this paragraph the planned worksite operational schemes are described. A sequence of phases was foreseen during the construction works, studied to maintain the rail traffic in a usual/normal condition.

After the construction of large diameter piles (called phase 1 and not described in detail in this paper) a CSP bulkhead and half of the roof slab were built, on top of which was realized a deviation and relocated the existing railway, so avoiding line disruption (Phase 2, Figure 3).

Successively the other CSP bulkhead was realized and the roof slab was completed through a second stage casting as shown in Figure 4. The last phase included the excavation inside the walls, and the realization of the foundation slab as shown in Figure 5.

Finally, the new set-out of the line on a double track was realized and completed. Once tunnel construction work was completed, rail traffic was relocated in the new line inside the tunnel and the renovation outside was carried out.

4 USE OF ODEX PILES TECHNOLOGY

Some segments of the new line, which as already mentioned is crossing the city of Palermo, featured particularly narrow available spaces. For these segments, the use of ODEX pile technology (Tanzini, 2002) as retaining and supporting system was preferred to the CSP pile wall as provisional containment constructions and support of tunnel roof. In the final phase it was realized the foundation slab and the concrete walls.

To realize the second half roof slab it was necessary to install steel bar inside the first half slab and to joint, with dedicated sleeves, the reinforcement of the second half slab on the first

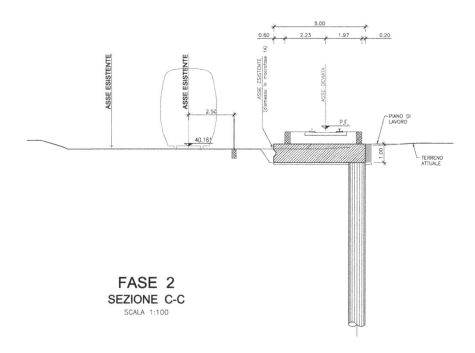

FASE 2
SEZIONE C-C
SCALA 1:100

Figure 3.   construction of "half tunnel" "Tommaso Natale 1"and deviation of existing railway on the roof slab.

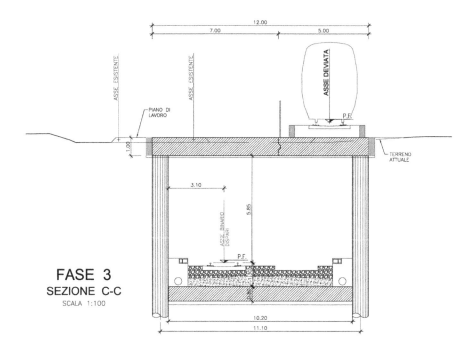

Figure 4.   construction of tunnel "Tommaso Natale 1" second half.

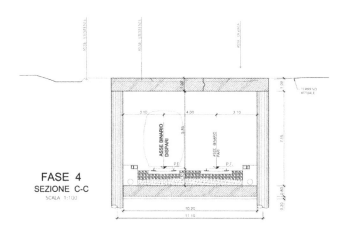

Figure 5.   deviation of railway inside the tunnel "Tommaso Natale 1".

lapping rebars (Figure 6). Figure 7 and Figure 8 show same phases as Figures 4 and 5 referred to ODEX piles technology.

4.1   *ODEX technology tools and characteristics*

"ODEX" technology can be used for the realization of bulkhead's small piles in case of loose soils, with the presence of big stone element alternate to minor granulometry. This technology implies the combined use of a hummer to drill down and a steel temporary casing.

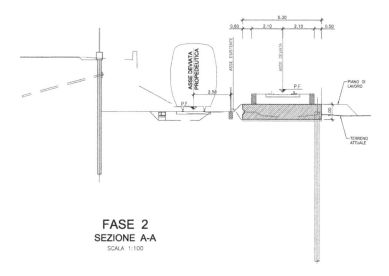

FASE 2
SEZIONE A-A
SCALA 1:100

Figure 6. construction of "half tunnel" "Tommaso Natale 1" and deviation of existing railway on the roof slab with ODEX technology.

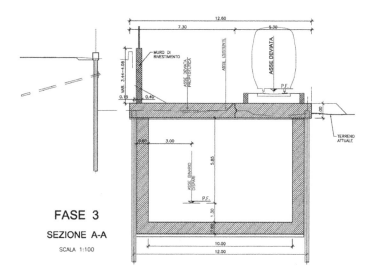

FASE 3
SEZIONE A-A
SCALA 1:100

Figure 7. construction of tunnel "Tommaso Natale 1" second half with ODEX piles.

At the beginning, a little reamer is closed and then, when system begins drilling, the same swings out to enlarge the hole and to allow the advancement of the steel casing. Figure 9 shows ODEX technology excavation tools scheme and principle and an example of ODEX piles bulkhead.

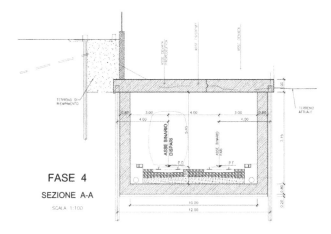

Figure 8. deviation of railway inside the tunnel "Tommaso Natale 1" constructed with ODEX piles.

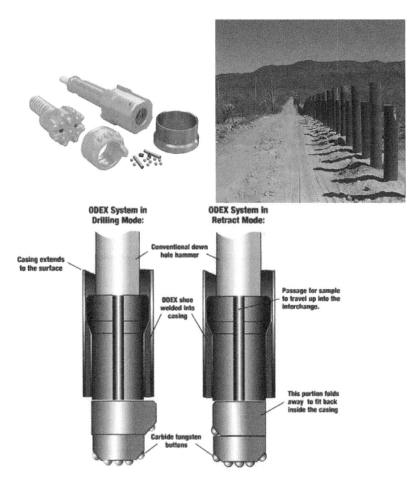

Figure 9. ODEX technology excavation tools scheme and principle and an example of ODEX piles bulkhead.

5 RAILWAY TRAFFIC IN HALF TUNNEL WITH CONSTRUCTION WORKS IN PROGRESS

In a particular situation, it was necessary to deviate railway traffic from the existing subway after the realization of a first CSP bulkhead and half roof slab supported by the other side by an odex piles bulkhead. The other half tunnel was realized with railway traffic deviated on the first half tunnel. Figure 10 to Figure 12 illustrate the three phases adopted for this case.

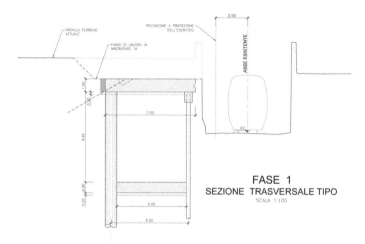

Figure 10. Phase 1: construction of first half tunnel.

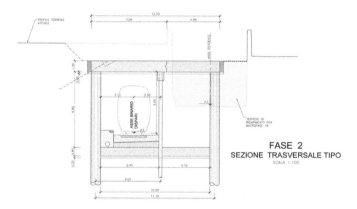

Figure 11. Phase 2: railway traffic deviated in the first half tunnel and construction of the second half tunnel.

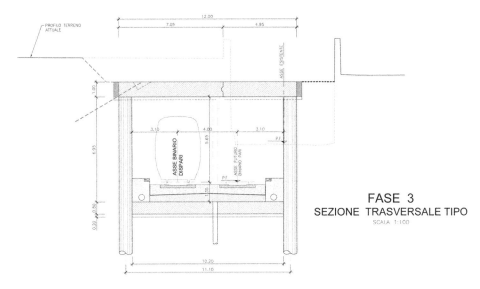

Figure 12.   Phase 3: normal condition in operation.

6   CONCLUSION

The construction of the Nodo di Palermo section of the railway, relying to CSP and ODEX technology, required the use of several civil engineering methods and technics to cope with specifications and constraints of this railway section (urban context, keeping railway in operation,).

This paper illustrates rationales and implementation of techniques used for artificial tunnels (as particularities of "cut and cover method") which allowed to work in total safety, within timetable, in very restricted spaces, sometimes with railway traffic in operation.

REFERENCES

Bringiotti, M. 2006 *Recenti Cantieri innovativi in Italia*, Parma, Quarry & Construction, GeoFluid, Edizioni PEI.
Bringiotti, M. 2010 Nuove metodologie di esecuzione dei pali trivellati, In *Proceedings of Nuove regole per una vera qualificazione, AIF, XVIII Mostra Internazionale GeoFluid, 3–6 October 2018, Piacenza, Italy*.
Tanzini, M. 2002 *Perforazioni a scopo geotecnico e tecniche di consolidamento*, Flaccovio Editore.

Tunnels and Underground Cities: Engineering and Innovation meet Archaeology,
Architecture and Art, Volume 12: Urban
Tunnels - Part 2 – Peila, Viggiani & Celestino (Eds)
© 2019 Taylor & Francis Group, London, ISBN 978-0-367-46900-9

A simulation study on the interaction between backup structure and surrounding soil in a large section pipe curtain-box jacking project

X. Zhang, L.S. Chen, Z. Zhang, Y.H. Liu & J. Niu
STEC Shanghai Urban Construction Municipal Engineering (Group) Co. LTD, Shanghai, China

Y. Bai
BY Civil Engineering Consulting Co., LTD, Shanghai, China

ABSTRACT: Construction of underpass crossing existing heavy traffic lanes in dense urban area is challenging. The pipe curtain-box jacking method (PCBJ) is an innovative method for underpass construction and has been applied in several projects, because it causes a low ground disturbance. In large section PCBJ projects, the interaction between backup structure and surrounding soil is related to the stability of the whole soil-structure system, which determines the success or failure of the project. This article presents a case study of an ongoing project of a large and shallow PCBJ underpass (19.6m wide, 6.4m high and 6.3m buried depth) that crosses the Mid-ring Road in Shanghai China. Three-dimensional numerical modeling is performed, replicating the actual condition of the project to the greatest extent. This article also analyzes the deformation of the soil within the area affected by the jacking force, and the stress state of the backup structure.

1 INTRODUCTION

Pipe curtain-box jacking method (PCBJ) is often used for underpass construction beneath existing roads. This PCBJ method features a low ground disturbance compared with open trenching method and is especially suitable for underpasses projects crossing existing railway and highways. In PCBJ method, a series of steel pipes are firstly jacked into the ground, and are connected with each other by special locking-joints, forming a spatial frame structure. This frame structure supports the ground above and stop the ground water intrusion. Afterwards, concrete boxes are casted in the launching shaft and pushed forward into the ground, while soil being excavated inside the box structure. The jacking system provides a high jacking force for box movement, and jacks are usually equipped against the shaft retaining walls.

Tianlin Underpass, an ongoing project which crosses the Mid-ring Road in Shanghai, is on the culvert jacking stage. Tianlin Underpass connects Xuhui District and Minhang District and crosses beneath Mid-ring Road in Shanghai. This underpass is expected to relieve traffic congestion in other crossing roads near the construction site. The project starts from Gumei Road in the west and Guiping Road in the east, including a buried section of 290m and an open section of 272m (see Figure 1). The buried section is designed to be a rectangular tunnel with 3 motorized lanes and 2 non-motorized lanes. The 86m long crossing section is being constructed with the PCBJ method. The scale of the tunnel cross section is 19.6m wide and 6.4m high, with an average overburden of about 6.3m (from ground surface to the top of the pipe curtain roof, see Figure 2).

To begin with, working shafts on the west and east sides of Mid-ring Road need to be constructed. The size of the west launching shaft is designed to be 27.2m long and 27.2m wide, and the size of the east receiving shaft is designed to be 13.2m long and 27.2m wide. Both of

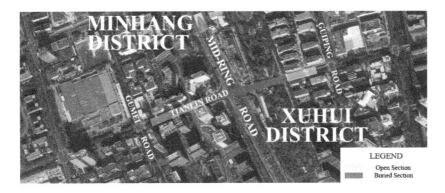

Figure 1. Bird view of the Project.

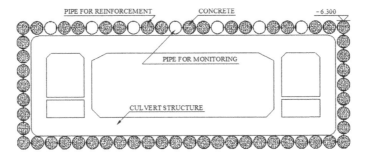

Figure 2. Cross section of the Underpass.

the launching shaft and the receiving shaft is retained by 1.6m thick diaphram walls on each side (Wang, et al, 2017, see Figure.3).

After the completion of working shafts, pipe curtain was constructed using mini pipe jacking machines. Section by section, the culvert structure was casted in launching shaft before jacking. Now, the box culvert is being jacked with the aid of a full cross-section EPB boring machine, and under protection of the pipe curtain. The underpass will be completed after final reception of the culvert boring machine and the replacement of friction reducing slurry.

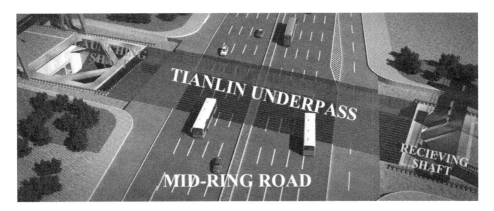

Figure 3. Bird view of the Underpass.

Table 1. Physical and mechanical parameter of site soils.

Layer Number	Soil Name	Thickness of layer (m)	Water ratio (%)	Moist unit weight (kN/m³)	c (kPa)	φ (°)
②	silty clay	1.34	32.2	18.5	19	19
③	muddy silty clay	6.39	40.6	17.6	12	18
④	muddy clay	6.78	50.0	16.8	11	11.5
⑤$_{1-1}$	silty clay	6.91	35.9	17.9	14	19.0
⑤$_{1-2}$	silty clay	6.57	35.1	18.0	15	19.0
⑤$_2$	sandy soil	7.88	28.9	18.5	5	31.5
⑥	silty clay	2.67	23.9	19.5	39	19.5
⑦$_{1-1}$	sandy soil	9.52	26.1	19.3	4	34.0
⑦$_{1-a}$	silty clay	4.30	30.0	18.9	21	19.5
⑦$_{1-2}$	silt	19.51	26.9	18.7	3	35.0

Table 2. Surrounding buildings and simulated ground overloads.

Building number	Simulated overloads (kPa)	Building number	Simulated overloads (kPa)
1	60	5	390
2	90	6	135
3 (6/7 storey)	90/105	7	285
4	180	8	210

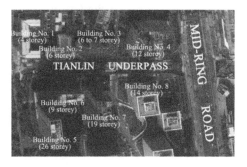

Figure 4. Site surrounding buildings.

1.1 Engineering geology

Located on typical soft clay in Shanghai, the underpass is faced with high ground water level and strict ground settlement requirement. According to hydrological survey and geological survey, the elevation of site ground surface is 4.560m, the ground water level is 1.0m below the ground surface, and the engineering geology condition is shown in Table 1.

1.2 Surrounding condition and parameters for simulation

Tianlin Underpass is a project in dense urban area. Many existing buildings, structures and underground pipelines may have great impacts on construction procedure of the Underpass. Representatively, eight nearby buildings have to be investigated (see Table 2and Figure 4). For equivalent simulation, each building is replaced by ground overloads. Since all of these buildings are made of reinforced concrete, the ground overloads are calculated according to building heights, namely, 15kPa for each storey.

2 CALCULATION OF CULVERT JACKING FORCE

2.1 *Experience from completed project*

The jacking resistance consisted of two major parts, i.e., the face resistance and side friction, as Equation (1) shows.

$$F_j = F_f + F_s \tag{1}$$

In Equation (1), F_j = total jacking resistance; F_f = face resistance; and F_s = side friction.

Before Tianlin Underpass, engineers from Shanghai have gained experience from another PCBJ project not far from the Tianlin Road. Beihong Underpass, a tunnel constructed in 2004, is only 3.6km from Tianlin underpass. The major difference between these two underpasses are from aspect of cross section scale, pipe interlocks and boring machine. To reduce jacking resistance, anti-friction slurry was grouted to the gap between the culvert and the pipe curtain, as Figure 5 shows.

Beihong Underpass was a 126m long, 34.2m wide and 7.85m high tunnel. In pipe curtain, steel pipes were connected with each other by external interlocks. The culvert was jacked into

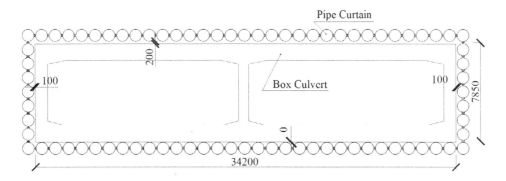

Figure 5.   Cross section scale and pipe curtain-box culvert gap in Beihong Underpass (unit: mm).

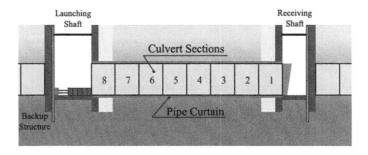

Figure 6.   Overview of culvert jacking in Beihong Underpass.

Table 3.   Culvert jacking resistance record.

Culvert Section Number	1	2	3	4	5	6	7	8
Starting resistance (kN)	70000	128000	150000	163000	177000	206000	237000	251000
Ending resistance (kN)	60000	85000	128000	130000	132000	144000	163000	192000

Table 4. Surrounding buildings and simulated ground overloads.

Position	Jacking resistance (kN)	Position	Jacking resistance (kN)
Top side	4983.37	Bottom side	182268.85
Wall side	2147.78	Total	189400.00

position with the aid of a grid boring machine, and consists of 8 sections with different length, as Figure 6 shows. Engineers recorded the jacking resistance while jacking the box culvert, as Table 3 shows.

It's easy to know from Table 3 that the maximum jacking resistance happened when engineers tried to jack Section 8. The maximum resistance $F_j = 251,000$kN.

In order to have a better understanding of culvert jacking resistance, engineers buried earth pressure sensors to monitor the tunnel face resistance. The result showed the face resistance $F_j = 61,600$kN when the total resistance F_j reached its peak, which means the total side friction $F_j = 189400$kN.

2.2 Component analysis of culvert jacking resistance

It would be easy to understand that the total side friction consisted of 3 parts: top side friction, wall side friction and bottom side friction, as Equation (2) shows.

$$F_s = F_{st} + F_{sw} + F_{sb} \qquad (2)$$

In Equation (2), F_{st} = top side friction; F_{sw} = wall side friction; and F_{sb} = bottom side friction.

The bottom side friction was different from the top side friction and the wall side friction due to its special contact with the pipe curtain. The culvert bottom contacted with both the pipe curtain and the anti-friction slurry, while the other sides only contacted with anti-friction slurry. According to model test (Xiao, et al., 2005) and calculation based on buried depth, detailed jacking resistance is shown in Table 4.

The result in Beihong Underpass culvert jacking shows that the major jacking resistance is from the bottom side of the box culvert, with friction from both the pipe curtain and the anti-friction slurry. The contact area of the concrete-slurry is 1.45 times than that of concrete-steel (Zhang, 2018). The result would be convincing when used in Tianlin Underpass.

2.3 Final confirmation of culvert jacking resistance

According to experience gained in Beihong Underpass, the final culvert jacking resistance can be confirmed. When jacking the culvert in Tianlin Underpass, the face resistance can be calculated by passive earth pressure for conservative estimation. Meanwhile, the side resistance can be calculated according to the result mentioned in section 2.3. The result of side friction estimation (Zhang, 2018) is shown in Table 5.

Conclusively, the estimated total resistance of culvert jacking in Tianlin Underpass is 128075kN.

Table 5. Culvert jacking resistance estimation in Tianlin Underpass.

Component	Value (kN)	Component	Value (kN)
Face resistance	42769.24	Wall side friction	1160.63
Top side friction	2390.18	Bottom side friction	81754.95

3 THREE DIMENSIONAL SIMULATION ON THE INTERACTION BETWEEN BACK UP STRUCTURE AND SURROUNDING SOIL

In order to make a thorough analysis of back up structure in Tianlin Underpass, a large-scale three dimensional model is built using MIDAS GTS NX software. The model also takes surrounding buildings into consideration. To begin with, Cartesian space coordinate system is chosen and the ordinate origin is defined at the start of the ground section.

In transverse direction, the model scaled from -140m in the south and 150m in the north. In longitudinal direction, the model scaled from -50m in the west and 285m in the east. Vertically, the model scaled from -57.68m to 4.56m, referring to yellow sea height datum of China. This modeling range takes all of the surrounding buildings mentioned in part 1.2 into consideration. The underpass structure starts at 0m in the west and ended in 285m in the east, ranging from ground section to the middle of launching shaft.

3.1 *Modeling object and culvert jacking simulation*

The detailed 3D solid model includes the underpass structure, the reinforced area behind the backup wall, and the friction layer between underpass structure and surrounding soil. Arranged longitudinally along the underpass structure, the friction layer is defined as solid model with low shear stiffness, for the purpose of simulating the relative sliding along the interfaces. The underpass structure, reinforced area and friction layer are shown in Figure 7 to Figure 11.

To be consistent with the existing building overloads, boundaries are sketched on the surface of soil entity, and overloads are defined according to building heights, as figure 13 shows. The culvert is jacked on the working platform in launching shaft, and the jacking reaction force is imposed on the bottom face of underpass structure, as Figure 13 and 14 shows. The

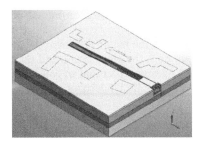

Figure 7. Bird view of numerical model.

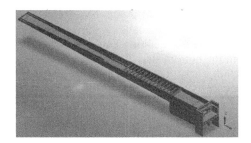

Figure 8. Tianlin Underpass structure.

Figure 9. Friction layer modeling.

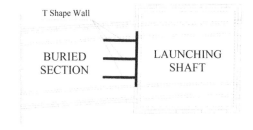

Figure 10. T shape wall on blueprint.

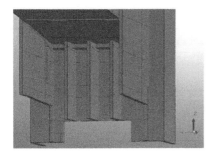

Figure 11. T shape wall modeling.

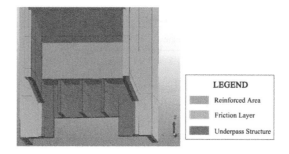

Figure 12. Position of different model components.

culvert jacking force is simulated as uniformly distributed pressure calculated by dividing culvert jacking resistance by jacking reaction face area.

3.2 Soil definition and boundary condition

Tianlin Underpass is an underpass with strict ground settlement requirement, in other words, plastic zones and large deformation should be avoided. Therefore, the whole model is computed using isotropic elasticity constitutive relation. To optimize simulation quality and computational cost, soils with similar physical and mechanical property are merged, and parameters are defined as weighted average, as Table 6 shows.

Except east boundary of the underpass structure, X displacements are fixed in perpendicular boundary planes of X axis. In transverse direction, Y displacements are fix in perpendicular boundary planes of Y axis. In vertical direction, Z displacement is fixed in bottom boundary of the entire model.

3.3 Model meshing

To achieve better meshing robustness and efficiency, hexahedral centered hybrid element is introduced in 3D model meshing. The model meshing is shown in Figure 15 and Figure 17, and meshing for underpass structure is shown in Figure 16.

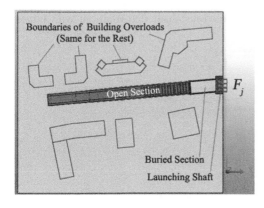

Figure 13. Bird view of simulated model.

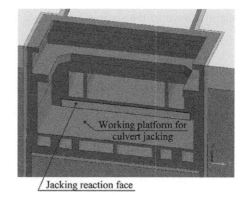

Figure 14. Culvert jacking simulation.

Table 6. Material parameters in simulation.

Material number/name	Elasticity modulus (kPa)	Poisson's ratio	Lateral pressure coefficient
Structural concrete	31, 500, 000	0.30	——
Reinforced area	120, 000	0.32	0.30
①, ②	21, 311	0.36	0.42
③, ④, ⑤$_{1\text{-}1}$	15, 124	0.38	0.52
⑤$_2$, ⑥, ⑦$_{1\text{-}1}$	47, 766	0.33	0.39
⑦$_{1\text{-a}}$	26, 900	0.35	0.45

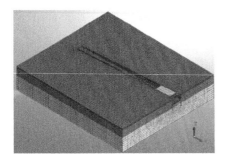

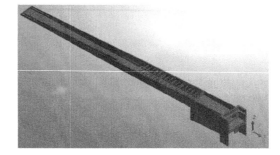

Figure 15. Bird view of model meshing.   Figure 16. Underpass structure meshing.

3.4 *Analysis and post-processing*

The analysis includes into two steps, the gravity activation step and culvert jacking step. In gravity activation step, gravity is activated and displacement triggered by gravity has been reset to zero. In culvert jacking step, the jacking reaction force is activated so that the response of backup structure can be analyzed.

The stress state of underpass structure in gravity activation step is shown in Figure 18. The displacement response of backup structure in culvert jacking step is shown in Figure 19 to Figure 21.

It can be seen from Figure 18 that the junction part of buried section and working shaft will have a comparatively high principal stress, especially on the top of the diaphram walls, and the inner side of retaining beam in the launching shaft.

It can be seen from Figure 19 to Figure 21 that under the load of jacking reaction force, the largest displacement occurs on the jacking reaction face, and the maximum displacement is no bigger than 2mm. In launching shaft, the west wall bulges out, and the south and north wall slightly move to the shaft. The T shape wall has an obvious shear deformation, with larger

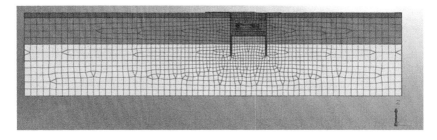

Figure 17. East elevation of model meshing.

Figure 18.　Principal stress of underpass structure in gravity activation step.

Figure 19.　Displacement response in underpass structure in culvert jacking step.

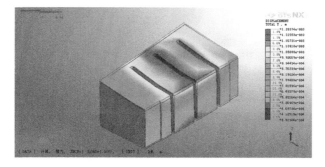

Figure 20.　Displacement response in T shape wall in culvert jacking step.

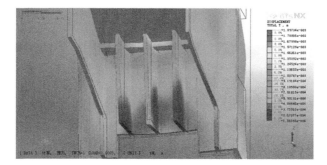

Figure 21.　Displacement response in reinforced area in culvert jacking step.

displacement on its top and smaller displacement on its bottom. Except deformation in a thin layer on the top, rigid motion is the main displacement in reinforced area.

4 CONCLUSION

The backup structure of culvert jacking is special in Tianlin Underpass. Instead of conventional backup structure, the buried section is used to bear the jacking reaction force. To have a better understanding of the jacking reaction, a large-scale three dimensional model is built for analysis.

The result shows that under the jacking reaction force, the deformation of backup structure is under control. The maximum displacement of the whole structure is about 1.9mm, which is acceptable under the strictly required ground settlement.

In backup structure, the specially designed T shape wall will bear most of the jacking reaction force and have an obvious shear deformation. The reinforced area behind the T shape wall will bear comparatively smaller jacking reaction force due to its rigid motion. The simulated result and backup structure design can be a reference for projects with big horizontal force in working shafts.

REFERENCES

Wang, H., Zhu, J.H. & Zeng, Y.J. 2017. Construction organization plan of Tianlin Underpass Project: 1–94.
Xiao, S.G., Xia C.C., Li X.Y., et al. 2005. Experimental study on coefficient of friction between a box culvert and mixture composed of thixotropic slurry and clay or steel pipe during culvert being pushed by pipe-roof. *Chinese Journal of Rock Mechanics and Engineering* 24(15): 2746–2750.
Zhang X. 2018. *Study on the interaction between backup structure and surrounding soil in a large section pipe curtain-box jacking project.* Tongji University Master's Thesis, Shanghai China: 1–100.

Tunnels and Underground Cities: Engineering and Innovation meet Archaeology,
Architecture and Art, Volume 12: Urban
Tunnels - Part 2 – Peila, Viggiani & Celestino (Eds)
© 2019 Taylor & Francis Group, London, ISBN 978-0-367-46900-9

Responses of an existing metro station to adjacent excavations in soft clay: Case Study

D.M. Zhang & X.H. Bu

Key Laboratory of Geotechnical and Underground Engineering of Education, Tongji University, Shanghai, China

Department of Geotechnical Engineering, College of Civil Engineering, Tongji University, Shanghai, China

ABSTRACT: It's inevitable to excavate foundation pits in the proximity of the existing metro station with the rapid development of subway in congested urban area. In this paper, the response of an operating metro station to the adjacent excavation was investigated. Zoned excavation method and an improved construction procedure of conducting excavation and basement construction alternately were adopted. The vertical displacements of rail tracks, entrance structure and convergence of shield tunnel were extensively analyzed. The results show that the excavation leads to the heave of station, while the construction of basement imposed reloading effect on the environment, which could balance excavation unloading effect in a certain degree. Therefore, the vertical displacements of metro station were within the acceptance limit. The entrance experienced tremendous settlements with a maximum value of 21.75mm, which is different from the behavior of metro station. The shield tunnel moved horizontally to 5.30mm because of lateral stress relief.

1 INTRODUCTION

As an efficient way to ease the pressure brought by traffic jam, subway has been constructed in urban congested area. For the safety of the metro structure and normal operation of the train, the deformation of the metro structure should be controlled within a certain range. However, for the lack of sufficient land, it's inevitable to excavate foundation pits in the vicinity of the operating metro line which may lead to detrimental effects on the metro structure. Especially when the foundation pit is large in scale, the deformation of station induced by excavation need to pay more attention.

Considerable studies have been conducted to investigate the responses of the shield tunnel to adjacent excavations (Chang et al. 2001; Huang et al. 2013, 2014). Only limited studies have researched the performance of stations during nearby excavation (Liao et al. 2016; Jia et al. 2016). Liu et al. (2016) reported a case history in which the metro station located within excavation zone moved upwards significantly with a maximum value of 9.6 mm due to stress relief resulting from the deep excavation. Li et al. (2017) examined the deformation of a newly built metro line with the excavation process. Deep excavations led to enormous upwards movements of the rail tracks and hence noticeable differential deformations were happened at the connection between the metro station and shield tunnels.

Excavation of foundation pits has significant impacts on the deformation of the station and the studies of protective measures to reduce effects of excavation were of importance. Tan et al. (2015) investigated the responses of the metro station and nearby twin shield tunnels in service induced by an oversized excavation, in which the zoned excavation method was adopted. The metro station settled uniformly along the transverse direction and inclined slightly along the longitudinal direction, which indicated the station possessed a good structural integrity. Xing et al. (2016) analyzed the effects of dewatering and excavation on the

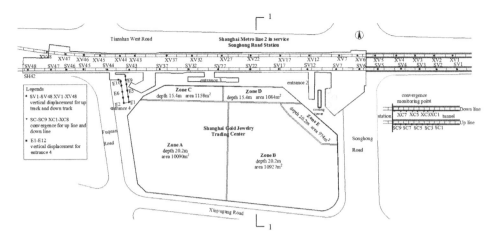

Figure 1. Plan view of project site with instrumentation layout.

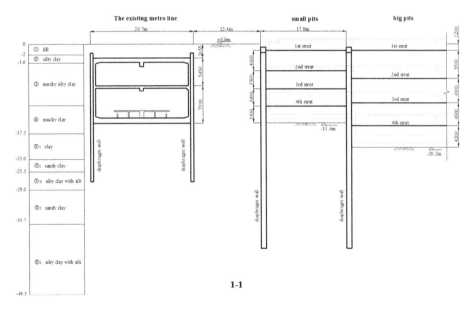

Figure 2. Typical cross sections of the foundation pits and the existing metro line and geometry.

subway station and a protection scheme was proposed. The protective measures, including composite anchor piles, cement mixing piles and a reinforced concrete beam, effectively controlled the deformation of subway station. Zhang et al. (2017) introduced a new construction method with bipartition walls (BPW), which aimed to decrease the adverse impacts on the surrounding environment. Numerical simulation and field observation proved the beneficial effect of the proposed construction method.

This paper presented a case history in which an improved construction procedure was adopted to reduce the adverse effect on the station caused by adjacent excavations and basement construction. The responses of metro station were examined attentively through an extensive filed instrumentation program. This research results should provide some useful references for similar projects in the future.

2 PROJECT DESCRIPTION

2.1 *Project overview*

The project is located in Shanghai for a gold jewelry trade center. In the project, a basement was in the construction plan and foundation pits with depths of 15.4m and 20.2m needs to be excavated correspondingly. As shown in Figure 1, the foundation pits were approximately rectangular in shape with excavation geometry of 236m×140m. The excavation of large-scale foundation pit was bound to have a significant influence on the surrounding environment. Meanwhile, an operating metro station was located in the north side of the project and within the influence range. The minimum distance between the main structure of the station and foundation pits was approximately 13m. In addition, the auxiliary structure of the station was closer to the foundation pit, including the station entrance, the wind well and so on, of which the entrance 4 was 7.4m from the excavation boundary. Considering that the large scale of foundation pit and distance between the pits and metro station, it's necessary to pay more attention on the adverse impact induced by excavation. In order to reduce the risk and guarantee the operational safety of the metro line, the foundation pit was divided into 5 separated zones by partition walls.

Figure 2 presents the typical cross sections of the new pits and the adjacent Metro station along the direction 1-1. The station is a two-story basement structure with a roof-slab depth of 2.66m and a base-slab depth of 15.50m, most of which was above the pits bottom. The width of the station was19.5m and supported by 0.6m-thick diaphragm walls on both sides. The retaining structure and the partition wall of the foundation pit were also 1m-thick diaphragm walls installed to 40m deep underground.

2.2 *Zoned excavation and construction procedure*

As mentioned above, the method of zoned excavation was adopted during construction. The foundation pits were divided into 5 zones of A, B, C, D and E, respectively. The excavation zones far from the metro station are larger than that near the station. The excavation zone A, B and E are 20.2m deep and braced by four layers of concrete struts. The zone C and D with small area was excavated to 15.4m and braced by one concrete strut and three layer of steel struts. The width of zone C and D is approximately 16.0m. All zones were excavated independently using the bottom-up method.

In addition to zoned excavation method, the construction procedure has been improved in this project. The excavation of foundation pits and construction of basement of different zones were conducted simultaneously. Figure 3 presents the detailed construction procedure of the project. It was started with excavation of zone A at 2015/10/30. After the completion of excavation and concreting of the bottom slab, the struts were dismantled and the underground structure was constructed from the bottom. Until the fourth-story slab in zone A was constructed, excavation in zone B was started. As soon as zone B was excavated and the bottom slab was casted, the zone D in the small pits was started to be excavated and the basement construction of the Zone B was carried out at the same time. Repeat this procedure and excavate zone E and zone C sequentially. Construction of basements in all zones was completed up to 2017/2/17. The basement can balance unloading effect in a certain degree and avoid the large deformation of the soil caused by all zone excavation. Thus, the impact on the station can be reduced. In order to better analyze the deformation of stations in different stages, the whole construction procedure is divided into **5 stages**.

2.3 *Geological condition and soil reinforcement*

Prior to the construction, a series of geological survey was carried out to obtain the soil parameters of the project sites including drilling, standard penetration, static penetration, cross plate shear test. According to the survey results, the strata within the 80m depth range were all Quaternary loose sediments as a consequence of the impact of sedimentation of the Yangtze River. Figure 4 presents the typical soil profile and different soil parameters of layers below ground surface. The shallow fill layer was at the top, followed by the 1.6m thick silty. Below layer ②, three thick layer including mucky silty clay layer, mucky clay layer and clay layer was ranged from -3.6m to -23.0m which has the characteristic of high water content, large void ratio, low strength, high sensitivity and poor stability and was a typical very soft clay. Under the very soft clay layer, sandy clay and silty clay with silt layer distributed alternatively. Located at the termination depth of filed survey was mostly sand layer.

It can be seen from the distribution of soil layer that the soft clay with poor soil quality was mainly distributed in the scope of foundation pits excavation. Therefore, it was necessary to reinforce the soil before the excavation was carried out for the sake of decreasing deformation of the metro line. Fig. 5 illustrates the soil reinforcement measures used in the project. In the small excavation zones, triaxial cement mixing piles were driven from -5.5m to -12.4m in the form of striped shape, from -12.4m to -21.4m using all-round reinforcement. In the big excavation zone, only the soil within the range of 10m from the diaphragm walls was strengthened. In addition to soil reinforcement in pits, when constructing diaphragm walls in the soft soil, the two sides of the soil would take place the displacement to the wall, thereby affecting the quality of the wall. Trench face was reinforced before the construction of diaphragm wall by triaxial cement mixing piles.

2.4 Instrumentation

In order to capture the displacements and deformation of subway stations and shield tunnels caused by excavation to avoid the large deformation that will damage the operation of

zone	construction activities	2015			2016												2017				
		10	11	12	1	2	3	4	5	6	7	8	9	10	11	12	1	2	3	4	5
A	excavation and bottom slab concreting																				
	basement construction																				
B	excavation and bottom slab concreting																				
	basement construction																				
C	excavation and bottom slab concreting	stage 1						stage 2				stage 3			stage 4		stage 5				
	basement construction																				
D	excavation and bottom slab concreting																				
	basement construction																				
E	excavation and bottom slab concreting																				
	basement construction																				
	later monitoring																				

Figure 3. Construction procedure for the whole process.

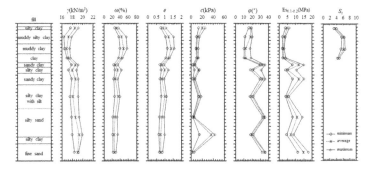

Figure 4. Soil profile and geotechnical parameters at the project site

Note: γ=unit weight; ω=water content; e=void ratio; c=cohesion obtained from direct shear test; φ=friction angle obtained from direct shear test; $Es_{0.1-0.2}$=compression modulus; S_t=soil sensitivity.

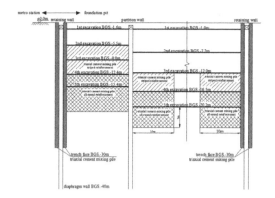

Figure 5. Soil reinforcement along the direction 1-1.

subway system, long-term instrumentation programs were conducted during the whole pro-cess of pro-jects including the excavation and construction of underground structure.

Figure 1 shows the three monitoring projects installed at stations and shield tunnels. 48 pairs settlements monitoring points marked as SV/XV1-48 was layout along the up and down tracks through level instruments to get vertical displacements. For the existing tracks located within the excavation zone, the interval of points (SX/XV7-43) is 6m. For other points the interval of settlement markers is about 24m due to less influence of excavation. At the same time, the convergence of the tunnel is measured by total station, and 17 points with an interval of approximately 6m were installed on the up line and down line, respectively. 12 settlement points are set at entrance 4 to record the settlement of entrance with the excavation of founda-tion pit for study the displacements of attachment structure.

3 FIELD OBSERVATION

3.1 *Vertical displacements of rail tracks*

Figure 6 shows the vertical displacements of the up track and down track along the longitu-dinal direction. The metro station was in the heaving state from the excavation of zone A to the completion of construction, which is consistent with the results founded by Liu et al. (2016) and Li et al. (2017). At the same time, it's obvious that differential displacements appeared in the station along the longitudinal direction. The construction procedure is pre-sented in Figure 2 and the whole construction procedure is divided into 5 stages. In order to analyze the field observation data, the constructive activities in every stage is listed in Table 1.

Zone A was the first to be excavated and the corresponding monitoring points SX/XX25-SX/XX46 experienced larger heaves than other points and the maximum value of displacements was 3.72mm, consequently the station tilting at the completion of stage 1. During stage 2, zone B was excavated and underground structure was continued to be constructed. A noticeable

Table 1. Construction activities in every stage.

Stage	Starting time*	Ending time*	Construction activities
Stage 1	2015/10/30	2016/01/28	excavation of zone A
Stage 2	2016/01/29	2016/07/21	excavation of zone B basement construction of zone A
Stage 3	2016/07/22	2016/10/06	excavation of zone D, E basement construction of zone A, B
Stage 4	2016/10/07	2017/02/17	excavation of zone C basement construction of zone B, C, D, E
Stage 5	2016/02/18	2017/05/01	later monitoring

* the date format is yyyy/mm/dd.

6323

heave increments with a maximum value of 2.88mm was occured in the section of the station near zone B while the other section of the station undergone settlements, which indicated that the excavation changed the stress stage of surrounding environment to cause the heave of station, while the construction of basements had the significant impact to the metro station as well to induce the settlement of the station. What's more, the inclination along the longitudinal direction was still existing however the inclined direction varied due to the different vertical displacements between the sections of the metro station. Meanwhile, the settlements of up track were more obvious than that of down track, which leaded to the tilt along the transverse direction. The combined action of tilt along the transverse and longitudinal direction would greatly increase the risk of vehicle operation. During stage 3 and stage 4, the excavation of small pits and the basement construction were carried out at the same time. Although small pits were closer to the metro station, there was no significant change in the vertical displacement of all monitoring points because the small size of excavation. It can be inferred that the small size excavation and the reloading effect of basement construction were the main reasons that prevented the further development of station deformation. The excavation resulted in the heave of metro station and basement structural weight could balance the unloading effect for the station. At the stage of later monitoring, some points moved downwards, especially the points of the up tracks. In addition, the vertical displacement of monitoring points marked by red boxes was smaller than that of both sides, which is evident in the up track. The monitoring points was located near attachment structure, such as entrance 4, and the different displacement was due to the effect of attachment structure, the vertical displacement of which will be illustrated in the following section. The tunnel heaved during stage 1 and settled during other stages and there has been differential settlements between the station and nearby excavation.

It can be concluded that the excavation caused the station to heave. The unloading effect induced by excavation lead to the soil under the foundation bottom. And as a result, the diaphragm walls of foundation pits and metro station uplifted due to the friction between the wall and soil. Therefore, the metro station heaved. What's more, benefiting from the improved construction procedure, the vertical displacements of metro station were within allowance limits while the inclination along the longitudinal and transverse direction should not be ignored.

As mentioned in the above, the monitoring points was not moving upwards during the whole construction process, which suggested that the excavation of the foundation pit and the basement construction have different influence on the surrounding metro structure. In order to illustrate the response of the metro station to the adjacent construction in detail. The vertical displacements development with time of point SX/XX 16, 30, 38, 8 and 44 was showed in Figure 7, representing the left, middle and right section of metro station and the connections with attachment structure, respectively.

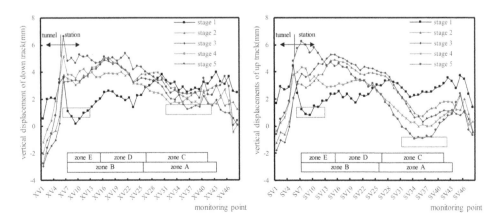

Figure 6.   Vertical displacements of the up and down rail tracks along the longitudinal direction.

It can be seen from Figure 7, the varied trend of all points with the construction process were approximately similar. All the points moved upwards quickly, especially the points 38 and 44. The maximum value of heave was 5.27mm, which was the effect of excavation of zone A. As soon as the casting of base slab in zone A, all points reached peak and turned to settlement, which demonstrated the importance of timely pouring base slab. During stage 2, only the construction of basement was implemented in the primary period and all points was in the state of moving downwards. It can be concluded that excavation contributed to the heave of metro station and the basement construction resulted in the settlement of metro station. From the mechanical perspective, the excavation and structural construction applied the unloading effect and loading effect on the soil layer under the station, which controlled the displacement of metro station. At the later period of stage 2 and stage 3, zone B and D was excavated and points was starting to move upwards again. From 2016/3/22 to 2016/8/19, the point SX16 experienced a heave of 5.97mm and increased to the maximum value of 6.63mm. After the excavation of Zone D, all the monitoring points stayed stable. During the entire process, the vertical displacement fluctuated as a result of multiple construction processes simultaneously. What's more, it is evidence that the variation magnitude of displacements of up track was large than that of down track. This was because the influence of excavation and structural construction on the excavation was in relation to the distance.

3.2 Vertical displacements of entrance 4

Figure 8 shows the vertical displacement development of the entrance 4 on behalf of the attachment structures. The vertical displacements of points at both sides of entrance 4 was of similar value. Therefore, only points at one side of entrance were selected and illustrated in detail.

Different from the station, the entrance undergone slight upward movement at the beginning and then settled until the completion of field observation. At last, the maximum settlement of all points reached 21.75mm, which exceeded the local allowable limit of 20 mm. Compared with the station, the different displacement curves of entrance 4 were due to the different position between metro station and entrance. Entrance was closer to the retaining wall and the depth of entrance was smaller than that of station. Hence, the entrance was surrounded by the soil in the range of settlements. During the whole excavation process the entrance moved down gradually as the settlement of soil. Meanwhile, it can be seen that the point far from the station undergone more settlements than that close to the station. On the one hand, the farther away from the station, the closer to the ground and the station and the greater settlements of surrounding soil. On the other hand, the station was moving upwards during excavation, which could exert additional force on the connection between the entrance

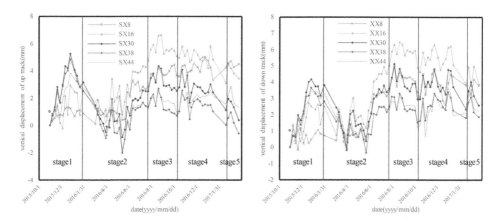

Figure 7. Vertical displacements development of up track and down track.

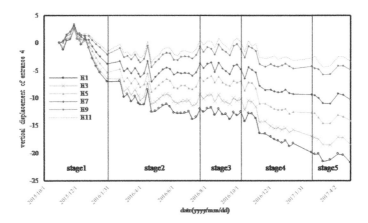

Figure 8.   Vertical displacements development of entrance 4.

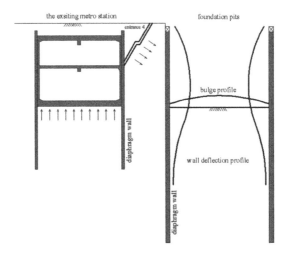

Figure 9.   Schematic illustration for the displacements of the station and entrance.

and the station. This could explain the phenomenon aforementioned that the vertical displacements of rail track closer to the connection were obviously less than that of other areas.

It's obvious that the settlement happened in stage 1, 2 and 4, corresponding to excavation of Zone A, B and C, accounted for the total settlements. Taking the monitoring point E1 as an example, the settlements during these three stages were 6.44mm, 5.55mm, 7.36mm, respectively. While the total settlement value was 21.75mm, which indicated the oversized excavation with little distance had more significant effects on the settlement of the attachment structure.

As shown in Figure 9, the deformation of metro station was influenced by the soil under the foundation bottom and the entrance was affected by the soil at the side of foundation pits. Hence, the metro station heaved while the entrance undergone the settlements during the excavation.

3.3   *Convergence of shield tunnel*

The typical convergence development of shield tunnel nearby the metro station is shown in Figure 10, in which the positive magnitudes mean increment of the tunnel diameter and the

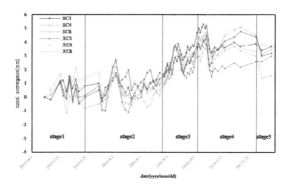

Figure 10.   Convergence development of up track and down track.

negative magnitudes indicate reduction of tunnel diameter, respectively. According to the distance from the metro station, three groups of monitoring points are selected.

Both the variation tendency and the value of convergence were similar for all the monitoring points. Due to the lateral unloading induced by excavation, the soil on the sides of shield tunnel moved towards the foundation pits and lateral earth pressure experienced the transition between the static state and active state. As a result, the tunnel elongated in the horizontal direction. During the first two stages, the excavation-induced convergence is not significant. It can be concluded that the impact of the excavation of zone A and zone B was relatively small, which were with tremendous area but were far from the tunnel. During stage 3 and stage 4, there was a greater increase in the convergence of shield tunnel due to the excavation closer to the tunnel, which also demonstrated that the lateral unloading effect of the excavation was more related with the distance. The maximum convergence of the tunnel is 5.30mm.

4   CONCLUSION

Based on the field observation, the responses of metro line and the attachment structure to adjacent deep excavation with an improved construction procedure are investigated. The vertical displacements of the rail tracks, entrance and the convergence of shield tunnel are illustrated in association with the construction procedure. According to the analysis of field data, the following conclusions can be drawn:

- The metro station heaved with a maximum value of 6.63mm while the entrance on behalf of attachment structure settled significantly with the settlements reaching 21.75mm due to adjacent construction. The distinction was in relation to the different position to the excavation of the metro station and entrance. The excavation exerted unloading effect leading to the heave of station. However, the soil on the sides of retaining wall moved down due to the stress relief which influenced the displacement of entrance.
- Benefiting from the improved construction procedure adopted in this case, the vertical displacements of metro station were within the acceptance limit. The construction of underground structure imposed reloading effect on the surrounding environment and resulted in the settlement of metro station.
- Due to sequential excavation of different excavation zone, the inclination along the longitudinal direction was observed. Meanwhile, the differential displacements were happened along the transverse direction.
- For the shield tunnel, the tunnel squat was observed as a result of lateral stress relief induced by excavation. The excavation closer to the tunnel lining had obvious impacts on the tunnel de-formation and the measured maximum convergence reached 5.30mm.

ACKNOWLEDGEMENT

This study is financially supported by National Natural Science Foundation of China (Grants No. 41772295 and No. 51478344).

REFERENCES

Chang, C.T., Suna, C.W., Duannb, S.W., & Hwang, R.N. 2001. Response of a taipei rapid transit system (TRTS) tunnel to adjacent excavation. *Tunnell. Underground Space Tech.* 16(3): 151–158.

Huang, X., Huang, H.W., & Zhang, D.M. 2014. Centrifuge modelling of deep excavation over existing tunnels. *Geotech. Eng.* 167(1): 3–18.

Huang, X., Schweiger, H.F., & Huang, H.W. 2013. Influence of deep excavations on nearby existing tunnels. *Int. J. Geomech.* 13(2): 170–180.

Hu, Z.F., Yue, Z.Q., Zhou, J., & Tham, L.G. 2003. Design and construction of a deep excavation in soft soils adjacent to the Shanghai Metro tunnels. *Can. Geotech. J.* 40(5): 933–948.

Li, M.G., Wang, J.H., & Chen, J.J., & Zhang, Z.J. 2017. Responses of a Newly Built Metro Line Connected to Deep Excavations in Soft Clay. *J. Perform. Constr. Facil.* 31(6).

Liao, S.M., Wei, S.F., & Shen, S.L. 2016. Structural Responses of Existing Metro Stations to Adjacent Deep Excavations in Suzhou, China. *J. Perform. Constr. Facil.* 30(4).

Liu, G.B., Huang, P., Shi, J.W., & Ng, C.W.W. 2016. Performance of a Deep Excavation and Its Effect on Adjacent Tunnels in Shanghai Soft Clay. *J. Perform. Constr. Facil.* 30(6).

Tan, Y. Li, X., Kang, Z.J., Liu, J.X., & Zhu, Y.B. 2015. Zoned excavation of an oversized pit close to an existing metro line in stiff clay: case study. *J. Perform. Constr. Facil.* 29(6).

Jia, F.Z., Wang, L.F., Lu, W.Q., & Yang, K.F. 2016. Influence of foundation pit excavation on adjacent metro station and tunnel. *Rock and Soil Mechanics* 37(S2): 673-678+714.

Xing, H.F., Xiong, F., Wu, J.M. 2016. Effects of Pit Excavation on an Existing Subway Station and Preventive Measures. *J. Perform. Constr. Facil.* 30(6).

Zhang, Z.J., Li, M.G., Chen, J.J., Wang, J.H., & Zeng, F.Y. 2017. Innovative Construction Method for Oversized Excavations with Bipartition Walls. *J. Constr. Eng. Manage.* 143(8).

Tunnels and Underground Cities: Engineering and Innovation meet Archaeology,
Architecture and Art, Volume 12: Urban
Tunnels - Part 2 – Peila, Viggiani & Celestino (Eds)
© 2019 Taylor & Francis Group, London, ISBN 978-0-367-46900-9

Numerical investigation of vertically loaded pile on adjacent tunnel considering soil spatial variability

J.Z. Zhang, H.W. Huang & D.M. Zhang
Department of Geotechnical Engineering, College of Civil Engineering, Tongji University, Shanghai, China

ABSTRACT: An adjacent loading pile in congested urban areas will inevitably cause an adverse effect on existing tunnel. Meanwhile, it is commonly accepted that soil exhibits spatial variability. A comprehensive study is performed on the effect of loading pile on the response of adjacent tunnel embedded in spatial variability soil. Herein, the random finite difference method (RFDM) is used. Monte Carlo simulations (MCS) approach is combined with finite difference analysis. Elastic modulus of soil is considered as a random variable. The soil is modeled with horizontally stratified anisotropic random fields using K-L expansion technique. An assessment method of tunnel deformation affected by adjacent loaded pile is proposed. The horizontal and vertical convergence and distortion degrees (α and β) can effectively evaluate the tunnel deformation in this situation. The results show that there is a significant amplification effect with the loading increase when considering soil spatial variability.

1 INTRODUCTION

In congested urban environment, existing tunnels are bound to be affected by other engineering activities. Under soft ground condition, structures are supported by piles, as illustrated in Figure 1. As shown in Figure 1, the environment around the existing tunnels has many uncertainties, such as spatial variability of soil and loading pile. A loading pile in crowded urban areas will inevitably cause an adverse effect on adjacent tunnel.

In the past, many studies main focused on the interaction of tunnel and pile (Beladjal and Mertens, 2009, Franza et al., 2017, Lueprasert et al., 2015). In recent years, there has been an increasing attention on the effect of pile loading on adjacent tunnel. Lee (2012) analyzed the response of a single pile and pile groups to tunnel using the three-dimensional numerical model. Lee and Bassett (2007) proposed the influence zones for 2D pile–soil-tunnelling interaction based on model test and numerical analysis. Lueprasert et al. (2015) established the tunnel influence zone subject to pile loading using the three dimensional finite element analysis. Lueprasert et al. (2017) analyzed the tunnel deformation due to adjacent loaded pile and pile-soil-tunnel interaction. Although many efforts have been made, but they are all purely deterministic.

Homogeneous soil will ignore the spatial variability of soil parameters, so that the results are only the state of the mean. The average level of results may miss the true failure mechanisms and ignore the weakest part of soils in the sense of randomness of soil properties. Therefore, it is necessary to consider the soil spatial variability on probabilistic analysis. Huang et al. (2017) analyzed the influence of spatial variability of soil Young's modulus on tunnel convergence in soft soils. Haldar and Babu (2008) analyzed the effect of soil spatial variability on the response of laterally loaded pile in undrained clay. Nevertheless, few previous researches have been devoted to the effect of loaded pile on adjacent tunnel considering soil spatial variability.

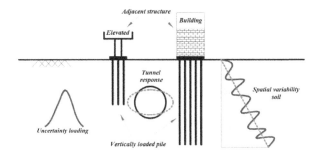

Figure 1. The uncertainty environment around the existing tunnels.

This paper is organized as follows. First, the FDM for modeling tunnel and adjacent loaded pile are presented. Second, the framework of RFDM program and assessment deformation method used in this study is introduced. Third, several cases are implemented to demonstrate how the tunnel responses under the load pile considering the spatial variability of soil properties. Last, some conclusions are obtained.

2 FINITE DIFFERENCE ANALYSIS

2.1 *Finite difference mesh and boundary conditions*

In this study, finite difference analyses are performed using the FLAC3D software. There are five possible ways to model a loaded pile, as shown in Figure 2. Ting-Kai et al. (2012) showed that the result of five models are similar. Meanwhile, model (V) involves the least computational efforts. Therefore, it will be adopted in this study. The Sp means the distance of two piles. In this study, four meter is adopted. The numerical model considered in the present work is shown in Figure 3. In this paper, A shield tunnel embedded into a spatially varied soil with its outer diameter D=6.2 m, internal diameter 5.5m, thickness 0.35m and depth H=17m is considered. As shown in Figure 3, the soil domain is set as 8.06 D in width and 6.04 D in depth to avoid the boundary effects and minimize effects on the analysis results. The pile with a 1m diameter, and its length is 20m, which is the height of tunnel center. The clearance between pile and tunnel of 0.5m is used in this study (Lueprasert et al., 2017, Lee and Bassett, 2007, Jongpradista et al., 2013). The plane-strain condition is assumed for this finite difference method analysis. The drained condition is assumed for the soils in this paper. There are 4332 soil zones and 8808 grid points.

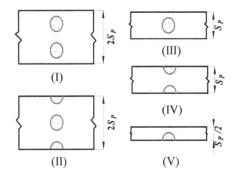

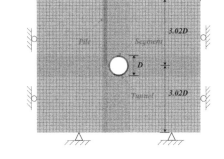

Figure 2. Five possible ways to construct the plan strain model.

Figure 3. Geometry of the finite difference model of shield tunnel.

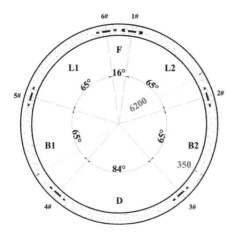

Figure 4. A typical shield-driven tunnel in Shanghai.

Figure 5. The interface of simulating tunnel in the model.

A typical shield-driven tunnel in Shanghai is shown in Figure 4. It is consists of 6 prefabricated segments, which contains one top block (F), one bottom block (D), two standard blocks (B1 and B2) and two adjacent blocks (L1 and L2). Two segments are connected by bolts. Detailed simulation of all joint components is obviously too time consuming for numerical convergence. In order to gain a balance between the computational time and simulation accuracy, the interface elements in FLAC3D are employed to simulate the normal and tangential behavior of joints. The segments and interface used in simulation tunnel is shown in Figure 5.

2.2 Constitutive model and model parameters

The soil in this paper is modeled as an elastic-plastic medium, following the Mohr-Coulomb yield criterion, which is most widely used in numerical simulations, in particular with the random field analysis(Huang et al., 2017). The concrete lining and pile were assumed to be a linear elastic material model. The pile and the soil are connected through interface elements. Details of input parameters used in the numerical analyses are summarized in Table 1 and Table 2.

Table 1.   Soil parameters adopted in finite difference modeling.

Parameter	Notation	Value	Unit
Elastic modulus	E_s	25	MPa
Cohesion	c	17	kPa
Friction angle	φ	15	°
Density	ρ_s	1800	kg/m^3
Poisson's ratio	v_s	0.33	—

Table 2.   Material properties of the pile and tunnel lining.

	Elastic modulus(GPa)	Poisson's ratio	Density(kg/m^3)
Tunnel lining	34.5	0.2	2500
Pile	31.5	0.18	—

2.3 Numerical modeling procedure

The simulation process for the investigation of the responses of the tunnel lining due to a loaded pile was divided into three stages in this study. The first stage is the initial stress balance. The initial stress condition is established using the gravity. The second stage is simulating the tunneling process and pile construction. The soil in the tunnel is given a null model and the interface is activated. The soil at the pile location is given the material properties of the pile, meanwhile the interface of the pile and soil is activated. The deformation of tunnel in this stage is cleared. The last stage is simulating the pile loading. In this study, five levels of loading is adopted. The every level loading is 600kN, total loading is 3000kN.

3 METHODOLOGY

3.1 Modeling of spatial variability

There is a lot of uncertainty in the construction of tunnel engineering. Traditional design method usually adopts single safety factor to consider many uncertainty factors. It is fails to consider the effect of spatial variability on engineering safety risk. Scale of fluctuation is an important concept of geotechnical parameters in the random field modeling. It can well reflect the spatial variability of the soil. In this study, the correlation matrix is built with the anisotropic exponential autocorrelation function:

$$\rho(\Delta x, \Delta y) = \exp\left[-2\left(\frac{(\Delta x)^2}{\delta_x^2} + \frac{(\Delta y)^2}{\delta_y^2}\right)\right] \tag{1}$$

where Δx and Δy are horizontal and vertical distances between the two points, respectively, δ_h and δ_v are correlation distances in horizontal and vertical direction, respectively, and $\rho(\Delta x, \Delta y)$ is the correlation coefficient between two points; The correlation distance quantifies the distance within which the soil properties exhibits relatively strong correlation. A smaller correlation distance indicates a stronger spatial variability. The Karhunen-Loeve expansion technique is used to discretize the random field in this study.

In this study, only the elastic modulus E_s is considered to be a spatially random property. Random fields of soil E_s are generated and mapped into finite difference analysis. As soil properties tend to be more similar in the horizontal direction than in the vertical direction, δ_h is usually larger than the δ_v (Phoon et al., 2016). In order to analysis the effect of soil spatial variability, the cases are set as shown in the Table 3.

3.2 Framework to realize RFDM program

The framework to realize random field difference method program in this study is shown in Figure 6. The procedures of this program for predicting the effect of loaded pile on adjacent tunnel considering soil spatial variability can be summarized in the following steps:

Table 3. Scales of fluctuation in anisotropic random fields.

| Case | COV | Scale of fluctuation | | Anisotropic ratio | | |
		δ_h /m	δ_v /m	$\delta h/\delta v$	$\delta h/D$	$\delta v/D$
Case1	0.3	30	0.5	60	4.84	0.08
Case2	0.3	30	1.5	20	4.84	0.24
Case3	0.3	30	4	7.5	4.84	0.65
Case4	0.3	30	30	1	4.84	4.84

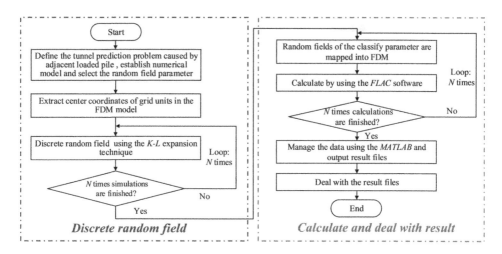

Figure 6. Flowchart illustrating the proposed framework for optimization of the random field analysis.

Step 1 is to define the tunnel prediction problem caused by adjacent pile and establish the numerical model. Meanwhile, the random field parameter should be selected in this step.

Step 2 is to discrete the random field parameter using the K-L expansion technique. In order to realize the goal, the center coordinates of grid units in the FDM model should be extracted. In this step, N times simulations (random field parameter) should be finished.

Step 3 is to map the random field parameter into the FDM model. Then, using the FLAC software to calculate the model and get the result. In this step, N times calculations should be finished.

Step 4 is to manage the result using the MATLAB code and output the result files.

3.3   Determination about number of MCS runs

The number of MCS has a great effect on the estimated result when using the MCS method. A small probability event may not happen if the number is too small. On the contrary, the computational efficiency would be lower if number is too large. Hence, we should find a suitable number to meet our demands and improve the efficiency of calculation.

Figure 7 depicts both the variations of calculated mean and COV of convergence ΔD_h against the running number of MCS simulations. It is observed from Figure 7 that both the variation of mean and COV of ΔD_h are not sensitivity with the MCS number when N is beyond 300. Hence, converged solutions of these two statistics can be roughly achieved at

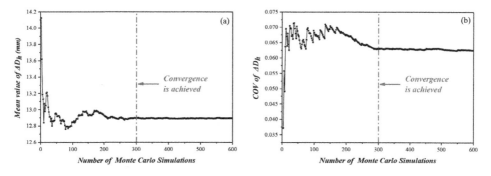

Figure 7. The mean value and COV of horizontal convergence with various simulations (a) Mean value; (b) COV.

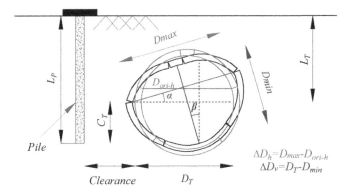

Figure 8.   The schematic of the assessment method for tunnel deformation proposed in this study.

300 MCS runs. Thus, 300 MCS runs are employed for the subsequent finite difference analyses in this paper.

3.4 Assessment method of tunnel deformation due to adjacent loaded pile

In order to more properly reflect the impact of the load pile on the adjacent tunnel, an assessment method of tunnel deformation is proposed. In this problem, the tunnel not only undergoes conventional deformation, but also due to the asymmetric stress condition, the tunnel will rotate and have an unsymmetrical distortion degrees. In engineering practice, due to the presence of tunnel joints, the maximum horizontal deformation often occurs at this location. Convergence indicates the difference between the distance between the maximum deformation point after the deformation of the tunnel and the two points before the deformation. The distortion degree indicates the angle between the line connecting the maximum deformation point of deformed tunnel and the horizontal and vertical directions. The distortion degree may be a significant problem which should be focused especially in spatially varied soil. In this sense, the horizontal and vertical convergence can be calculated as shown in Figure 8. As shown in Figure 8, α and β means the horizontal and vertical distortion degree.

4   RESULTS AND DISCUSSION

4.1 The tunnel deformation of deterministic analysis

The typical deformation of tunnel is shown in Figure 9. It is also the result of deterministic analysis. It should be noted that the deformed shapes with a magnification ratio of 1:100 are illustrated in the figure. As shown in the Figure 9, the tunnel has an unsymmetrical settlement under the loading pile. The settlement of the tunnel close to the pile is significantly larger than the other side. Therefore, the tunnel has a distortion degree.

Generally, a pile under loading causes the in situ stress to change and thus soil displacement, which inevitably exerts influence on nearby existing structures(Lueprasert et al., 2017). Therefore, the contour of horizontal and vertical displacement are illustrated in Figure 10 in order to obtain the movement of soil nearby the pile and tunnel. From the Figure 10 (a), we can know the main affected area of the loaded pile is at the pile tip. In the top of pile, the soil will move to the pile position. On the contrary, the pile will have a compacting effect on the bottom of pile. From the Figure 10 (b), it can be seen that the soil around loaded pile has a large settlement. This is the reason why the tunnel has an unsymmetrical settlement. The tunnel near the pile side has a large displacement under the driving of the soil.

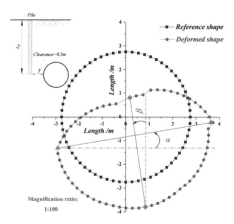

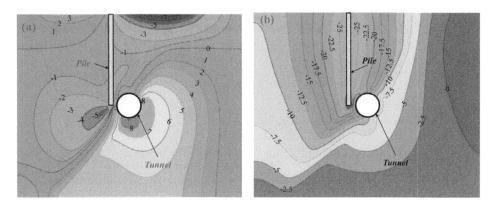

Figure 9. The deformation of tunnel effected by adjacent loaded pile.

Figure 10. The contour of horizontal and vertical displacement with loaded pile (a) horizontal displacement; (b) vertical displacement.

4.2 Influence of pile loading on the tunnel convergence

In order to get the effect of pile loading on the adjacent tunnel convergence, the lognormal fitting lines of ΔD_h with pile loading different are shown in Fig .11. It can be seen from the figure that as the loading increases, the mean value of tunnel convergence also gradually increases. It is interesting that as the loading increases, the distribution type of tunnel convergence also changes.

When loading is small, the distribution is relatively more concentrated, and when loading is large, the distribution is relatively more dispersed. In other word, the variation of tunnel convergence is gradually increases with the increase of loading.

It is also can be clearly seen from the Figure 12 that there is a significant amplification effect with the loading increase. Although the soil spatial variability is same, the standard deviation (STD) of tunnel convergence is different. The amplification value is about 1.61~4.28 when loading increases from 600kN to 3000kN.

4.3 Influence of pile loading on the distortion degrees of adjacent tunnel

In this section, the effect of pile loading on the distortion degrees of adjacent tunnel will be considered. Figure 13 shows the lognormal fitting line of horizontal distortion degrees α value

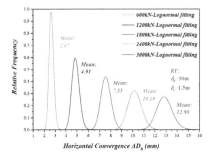

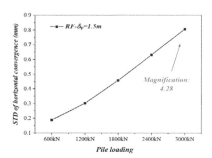

Figure 11. The lognormal fitting line of ΔD_h with pile loading different.

Figure 12. The influence of pile loading on the variability of tunnel convergence.

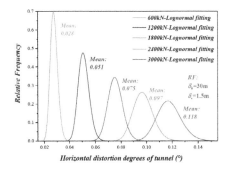

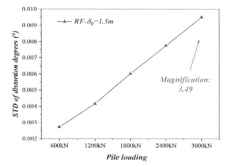

Figure 13. The lognormal fitting line of α with pile loading different.

Figure 14. The influence of pile loading on the variation of distortion degrees.

with pile loading different. It is similar with the effect on tunnel convergence. With the increase of loading, the mean value of α value increasing and the distribution is relatively more dispersed.

Figure 14 shows the influence of pile loading on the variation of distortion degrees of tunnel. The amplification value is about 1.53~3.49 when loading increases from 600kN to 3000kN. The amplification value is relatively smaller than tunnel convergence.

4.4 Influence of vertical SOF on the deformation of adjacent tunnel

Figure 15 shows the histogram of horizontal convergence of tunnel in different vertical SOF. The horizontal SOF is same in the four cases and the loading is all 3000kN. As shown in Figure 14(a), when the δ_v=0.5m, the mean value is 12.88mm, the range of distribution is about 11.5–15mm. As shown in Figure 14(b), when the δ_v=1.5m, the mean value is 12.90mm, the range of distribution is about 10.5–16mm. As shown in Figure 14 (c), when the δ_v=4m, the mean value is 12.96mm, the range of distribution is about 9.5–17mm. As shown in Figure 14(d), when the δ_v=30m, the mean value is 13.21mm, the range of distribution is about 9–19mm. With the δ_v increases, the mean value of ΔD_h also increases. Meanwhile, the range of distribution is larger, which means the variability is more significant. If the spatial variability is neglected, the most dangerous situation may be ignored. The spatial variability of soil has a relatively small impact on the mean value and a relatively greater impact on the distribution.

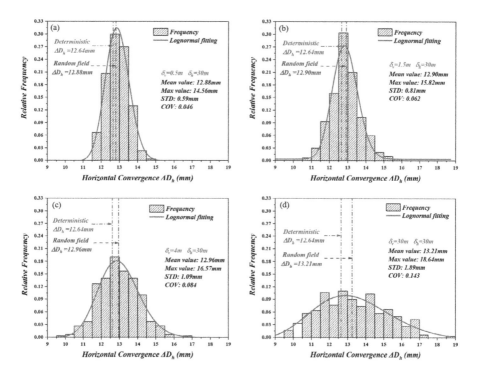

Figure 15. Histogram of horizontal convergence of tunnel in different vertical SOF (a) δ_v=0.5m.

5 CONCLUSION

This paper presents a numerical analysis of the vertically loaded pile effect on adjacent tunnel considering the spatial variability of the soil. The spatial variability of the Young's modulus is modeled with random fields. Based on the results of numerical simulations, the following conclusions can be summarized as follows:

(1) The tunnel deforms as a rotated ellipse shape when the pile tip is same with the tunnel center. The pile tip will have a compacting effect on the adjacent tunnel. Due to an unsymmetrical settlement of tunnel sides, the tunnel will have a distortion degrees.

(2) An assessment method of tunnel deformation affected by adjacent loaded pile is proposed. The ΔD_h, ΔD_v and distortion degrees (α and β) can effectively evaluate the tunnel deformation due to an adjacent pile under loading on one side. Meanwhile, the larger joint dislocation above the opposite side of the pile may be pay more attention in the engineering practice.

(3) There is a significant amplification effect with the loading increase. In the case of the same soil spatial variability, the amplification value is about 1.53~4.28 when loading increases in this study.

(4) The vertical SOF of soil has a relatively small impact on the mean value and a relatively greater impact on the distribution pattern. If the spatial variability is neglected, the most dangerous situation may be ignored.

ACKNOWLEDGEMENT

This study is substantially supported by the Natural Science Foundation Committee Program (No. 51608380, 51538009), by Shanghai Rising-Star Program (17QC1400300) and by

Shanghai Science and Technology Committee Project (17DZ1204205). Hereby, the authors are grateful to these programs.

REFERENCES

Beladjal, L. & Mertens, J. 2009. Investigation of Tunnel-Soil-Pile Interaction in Cohesive Soils. *Journal of Geotechnical & Geoenvironmental Engineering*, 135, 973–979.

Franza, A., Marshall, A. M., Haji, T., Abdelatif, A. O., Carbonari, S. & Morici, M. 2017. A simplified elastic analysis of tunnel-piled structure interaction. *Tunnelling & Underground Space Technology Incorporating Trenchless Technology Research*, 61, 104–121.

Haldar, S. & Babu, G. L. S. 2008. Effect of soil spatial variability on the response of laterally loaded pile in undrained clay. *Computers & Geotechnics*, 35, 537–547.

Huang, H. W., Xiao, L., Zhang, D. M. & Zhang, J. 2017. Influence of spatial variability of soil Young's modulus on tunnel convergence in soft soils. *Engineering Geology*, 228, 357–370.

Jongpradista, P., Sawatparnich, A., Suwansawat, S., Youwai, S., Kongkitkul, W. & Sunitsakul, J. 2013. Development of tunneling influence zones for adjacent pile foundations by numerical analyses. *Tunnelling & Underground Space Technology Incorporating Trenchless Technology Research*, 34, 96–109.

Lee, C. J. 2012. Three-dimensional numerical analyses of the response of a single pile and pile groups to tunnelling in weak weathered rock. *Tunnelling & Underground Space Technology*, 32, 132–142.

Lee, Y. J. & Bassett, R. H. 2007. Influence zones for 2D pile–soil-tunnelling interaction based on model test and numerical analysis. *Tunnelling & Underground Space Technology*, 22, 325–342.

Lueprasert, P., Jongpradist, P., Charoenpak, K., Chaipanna, P. & Suwansawat, S. 2015. Three dimensional finite element analysis for preliminary establishment of tunnel influence zone subject to pile loading. *Maejo International Journal of Science & Technology*, 2015, 209–223.

Lueprasert, P., Jongpradist, P., Jongpradist, P. & Suwansawat, S. 2017. Numerical investigation of tunnel deformation due to adjacent loaded pile and pile-soil-tunnel interaction. *Tunnelling & Underground Space Technology*, 70, 166–181.

Phoon, K. K., Prakoso, W. A., Wang, Y. & Ching, J. 2016. *Chapter 3 Uncertainty representation of geotechnical design parameters.*

Ting-Kai, N., Hai-Yang, X. U. & Liu, H. S. 2012. Several issues in three-dimensional numerical analysis of slopes reinforced with anti-slide piles. *Rock & Soil Mechanics*, 33, 2521–2526+2535.

Tunnels and Underground Cities: Engineering and Innovation meet Archaeology,
Architecture and Art, Volume 12: Urban
Tunnels - Part 2 – Peila, Viggiani & Celestino (Eds)
© 2019 Taylor & Francis Group, London, ISBN 978-0-367-46900-9

Author Index

Printed and bound by CPI Group (UK) Ltd, Croydon, CR0 4YY

18/10/2024

01776250-0008